Studies in Systems, Decision and Control

Volume 295

The series "Studies in Systems, Decision and Control" (SSDC) covers both new developments and advances, as well as the state of the art, in the various areas of broadly perceived systems, decision making and control–quickly, up to date and with a high quality. The intent is to cover the theory, applications, and perspectives on the state of the art and future developments relevant to systems, decision making, control, complex processes and related areas, as embedded in the fields of engineering, computer science, physics, economics, social and life sciences, as well as the paradigms and methodologies behind them. The series contains monographs, textbooks, lecture notes and edited volumes in systems, decision making and control spanning the areas of Cyber-Physical Systems, Autonomous Systems, Sensor Networks, Control Systems, Energy Systems, Automotive Systems, Biological Systems, Vehicular Networking and Connected Vehicles, Aerospace Systems, Automation, Manufacturing, Smart Grids, Nonlinear Systems, Power Systems, Robotics, Social Systems, Economic Systems and other. Of particular value to both the contributors and the readership are the short publication timeframe and the world-wide distribution and exposure which enable both a wide and rapid dissemination of research output.

** Indexing: The books of this series are submitted to ISI, SCOPUS, DBLP, Ulrichs, MathSciNet, Current Mathematical Publications, Mathematical Reviews, Zentralblatt Math: MetaPress and Springerlink.

More information about this series at http://www.springer.com/series/13304

Mostafa Al-Emran · Khaled Shaalan ·
Aboul Ella Hassanien
Editors

Recent Advances in Intelligent Systems and Smart Applications

Springer

Editors
Mostafa Al-Emran
Department of Information Technology
Al Buraimi University College
Al Buraimi, Oman

Khaled Shaalan
Faculty of Engineering and IT
The British University in Dubai
Dubai, United Arab Emirates

Aboul Ella Hassanien
Faculty of Computers and Artificial Intelligence
Cairo University
Giza, Egypt

ISSN 2198-4182 ISSN 2198-4190 (electronic)
Studies in Systems, Decision and Control
ISBN 978-3-030-47410-2 ISBN 978-3-030-47411-9 (eBook)
https://doi.org/10.1007/978-3-030-47411-9

This Springer imprint is published by the registered company Springer Nature Switzerland AG
The registered company address is: Gewerbestrasse 11, 6330 Cham, Switzerland

Preface

The field of intelligent systems and smart applications has extremely evolved with new trends during the last decade. Several practical and theoretical findings are growing enormously due to the increasing number of successful applications and new theories derived from numerous diverse issues. The applications of these trends have been applied to various domains, including education, travel and tourism, health care, among others. This book is dedicated to the intelligent systems and smart applications area in several ways. First, it aims to provide and highlight the current research trends in intelligent systems and smart applications. Second, it attempts to concentrate on the recent design, developments, and modifications of intelligent systems and smart applications. Third, it aims to provide a holistic view of the factors affecting the adoption, acceptance, or continued use of intelligent systems and smart applications. It is important to understand these issues in order to determine future needs and research directions. Fourth, this edited book aims to bring scientists, researchers, practitioners, and students from academia and industry to present the recent and ongoing research activities about the recent advances, techniques, and smart applications and to allow the exchange of new ideas and application experiences.

This book is intended to present the state of the art in research on intelligent systems, smart applications, and other related areas. The edited book was able to attract 60 submissions from different countries across the world. From the 60 submissions, we accepted 35 submissions, which represents an acceptance rate of 58.3%. The accepted papers were categorized under five different themes, including information systems and software engineering, knowledge management, technology in education, emerging technologies, and social networks. Each submission is reviewed by at least two reviewers, who are considered specialized in the related submitted paper. The evaluation criteria include several issues, such as correctness, originality, technical strength, significance, quality of presentation, and interest and relevance to the book scope. The chapters of this book provide a collection of high-quality research works that address broad challenges in both theoretical and application aspects of intelligent systems and smart applications. The chapters of this book are published in *Studies in Systems, Decision and Control Series* by Springer, which has a high SJR impact.

We acknowledge all those who contributed to the staging of this edited book. We would also like to express our gratitude to the reviewers for their valuable feedback and suggestions. Without them, it would not be possible for us to maintain the high quality and the success of the *Recent Advances in Intelligent Systems and Smart Applications* edited book. Therefore, on the next page, we list the reviewers along with their affiliations as a recognition of their efforts.

Al Buraimi, Oman — Mostafa Al-Emran
Dubai, United Arab Emirates — Khaled Shaalan
Giza, Egypt — Aboul Ella Hassanien
April 2020

List of Reviewers

Andrina Granić, University of Split, Croatia
Garry Wei-Han Tan, Taylor's University, Malaysia
Gonçalo Marques, Universidade da Beira Interior, Portugal
Hussam S. Alhadawi, Ton Duc Thang University, Vietnam
Ibrahim Arpaci, Tokat Gaziosmanpasa University, Turkey
Lee Voon Hsien, Universiti Tunku Abdul Rahman, Malaysia
Leong Lai Ying, Universiti Tunku Abdul Rahman, Malaysia
Luigi Benedicenti, University of New Brunswick, Canada
Mohammed Al-Sharafi, Universiti Malaysia Pahang, Malaysia
Mohammed N. Al-Kabi, Al Buraimi University College, Oman
Mohanaad Shakir, Al Buraimi University College, Oman
Mona Mohamed Zaki, The University of Manchester, UK
Noor Al-Qaysi, Universiti Pendidikan Sultan Idris, Malaysia
Reham Marzouk, Alexandria University, Egypt
Tariq Rahim Soomro, Institute of Business Management, Pakistan
Vitaliy Mezhuyev, FH JOANNEUM University of Applied Sciences, Austria

Contents

Information Systems and Software Engineering

A Systematic Review of Metamodelling in Software Engineering

Murni Fatehah, Vitaliy Mezhuyev, and Mostafa Al-Emran

Abstract Metamodelling has become a crucial technique to handle the complexity issues in the software development industry. This paper critically reviews and systematically classifies the recent metamodelling approaches to show their current status, limitations, and future trends. This systematic review retrieved and analyzed a total of 1157 research studies published on the topic of metamodelling. The retrieved studies were then critically examined to meet the inclusion and exclusion criteria, in which 69 studies were finally nominated for further critical analysis. The results showed that the main application domains of metamodelling are the cyber-physical and safety-critical systems development. Moreover, the majority of used approaches include metamodels formalization, adding spatial and time semantics, and considering nonfunctional properties. Further, the main trends of metamodelling development include the support of complex systems, behavior modeling, and multilevel modeling. The results of this systematic review would provide insights for scholars and software engineering practitioners looking into the state-of-the-art of metamodelling and assist them in improving their approaches.

Keywords Metamodelling · Software engineering · Systematic review

M. Fatehah
Faculty of Computing, Universiti Malaysia Pahang, Gambang, Malaysia
e-mail: murnifatehah@gmail.com

V. Mezhuyev
Institute of Industrial Management, FH JOANNEUM University of Applied Sciences, Werk-VI-Straße 46, 8605 Kapfenberg, Austria
e-mail: vitaliy.mezhuyev@fh-joanneum.at

M. Al-Emran (✉)
Department of Information Technology, Al Buraimi University College, Al Buraimi, Oman
e-mail: mustafa.n.alemran@gmail.com

M. Al-Emran et al. (eds.), *Recent Advances in Intelligent Systems and Smart Applications*, Studies in Systems, Decision and Control 295,
https://doi.org/10.1007/978-3-030-47411-9_1

1 Introduction

Extensive applications of software in daily life help users to save time, money, and energy. However, the development of modern tools requires more and more efforts from the side of software specialists and companies. Safety, security, interoperability are just a few issues that arise during the software development process. To address the complexity of modern software, the resources needed for its development should be multiplied.

Model-Driven Engineering (MDE) is an assuring practice in addressing software complexity with the application of Domain-Specific Modelling Languages (DSMLs), transformation engines, and code generators [1]. The MDE approach is used to increase productivity and simplify the software design by incorporating domain knowledge into the software development process. The domain model is a model that integrates both the properties and behavior of a specific domain. MDE allows making a domain knowledge explicit and formal.

In MDE, a model has become a crucial representation of an idea and rules of a software system. However, a model development itself needs a model design at higher levels of abstraction, which are called metamodels [2]. Metamodels help to define the constraints, rules, and the relationship between concepts of a domain model. There are several definitions of the metamodelling in scientific literature. One of the first definitions is that a metamodelling can be interpreted as a modeling process, which has a higher level of conceptualization and logic than a standard modeling process [3]. Metamodelling can be also defined as a modeling framework that consists of the metamodel, model and an instance of the model, where a metamodel defines the syntax of the modeling language [4]. Some articles define the metamodel as a standardized DSML for software modeling [5, 6].

The application of metamodelling approach to domain-specific software engineering has become a common practice nowadays. Metamodeling has become a crucial technique in the process of software development [7]. Metamodelling is used to define DSMLs for software systems modeling, where the result of the metamodelling (a DSML) is used to carry out the domain-specific properties and behavior [8]. Metamodelling defines rules and methods for creating metamodels as a part of organized and systematic metamodels development process. Many studies are devoted to the development of new and application of existing metamodelling approaches. However, to the best of our knowledge, there is no recent study devoted to the critical analyses of the state-of-the-art of the metamodelling. To solve this problem, a systematic literature review (SLR) was conducted to summarize knowledge on the current applications and limitations of the metamodelling in software engineering. By deriving evidence from the analyzed articles, this research is believed to assist software engineers in the development of new and existing metamodeling techniques.

2 Methodology

Prior research suggested that software engineering researchers should adopt the evidence-based practice in their study, which is known as Evidence-Based Software Engineering (EBSE) [9]. A systematic literature review (SLR) is the recommended approach to perform an EBSE practice. This study follows the SLR guidelines proposed by Kitchenham et al. [9], and other systematic reviews conducted in the past [10–13].

2.1 Research Questions and Motivation

Over the years, different metamodels, metamodelling approaches, and techniques have been proposed to find the best way of software design. Unlike the existing studies of metamodelling in software engineering (SE), the proposed research questions of the present study are:

RQ1. What are the purposes and application domains of current metamodelling research?
RQ2. What are the trends of the metamodelling development in software engineering?
RQ3. What is the quality of publications that reflect the metamodelling research in SE domain?
RQ4. What are the limitations of the current studies and the prospects for future research?

The motivation for formulating RQ1 is to analyze the research purpose of the selected articles along with the application domain of the recent metamodelling approaches. RQ2 discusses the current trends of the metamodelling to define a general tendency for its development. To identify the quality of the current metamodelling research (RQ3), a quality assessment questionnaire was used. Finally, RQ4 discusses the limitations of the current metamodelling approaches in the SE domain, along with future improvements.

2.2 Search Process

An extensive search was conducted in the digital libraries of scientific literature to answer the formulated research questions. Journal articles, workshop articles, conference papers, and books were included in the search process. Firstly, five digital libraries were selected as platforms for the search process. These libraries are the ScienceDirect, IEEE Xplore, Clarivate Analytics (formerly known as ISI Web of Knowledge), ACM Digital Library, and SCOPUS. Next, the keywords for the search

Table 1 Keywords definition

A	B	C
Metamodelling	Meta-modelling	Software engineering
Metaprocess modelling	Meta-process modelling	
Metadata modelling	Meta-data modelling	

process were defined. As the use of general keywords results in a large number of research papers, specific keywords were formulated with the aim of finding the most relevant studies. The search string was defined as "(A OR B) AND C", where the keywords are shown in Table 1.

The search process also takes into account a different spelling of the word "modelling" in British and American versions of English (i.e., the search query was repeated for "modelling" and "modeling"). The motivation to use the more general keyword "Software engineering" comparatively to the "Model Driven Engineering" is to cover other possible applications of the metamodelling (e.g., validation, standards compliance, and risk management). At the same time, MDE is considered as a part of SE, so the search string does not need to include both terms.

2.3 *Inclusion and Exclusion Criteria*

A total of 1157 papers were found as a result of the search process since specific keywords were applied. Hence, the inclusion criteria are crucial to find the papers relevant to this research. The importance of the keyword "software engineering" can be illustrated in the example that metamodelling now has intensive applications in different domains such as spatial correlation modeling (also known as Kriging metamodel) [14]. The collected papers have undergone eight phases of selection through the use of inclusion and exclusion criteria, as outlined in Table 2. The review process and the number of studies identified at each stage were undertaken according to the preferred reporting items for systematic reviews and meta-analysis (PRISMA), as depicted in Fig. 1.

2.4 *Data Collection*

The relevant information was extracted from selected papers to facilitate the data analysis process. The data were categorized into three different types. The explanation of extracted data and their category is shown in Table 3.

Table 2 Inclusion and exclusion criteria

Phases	Inclusion/exclusion criteria
1	Search engine results after executing a search query
2	Date of publication since 2012
3	Full-text access
4	Published in the English language
5	Published in software engineering venues
6	Not duplicated
7	Not summaries of workshop, panel, or tutorials
8	Title, abstract, or full text contains a search string with the predefined keywords in Table 1

2.5 Search Strategy

Although the keywords used were specific, a number of irrelevant papers to the research topic were found. Hence, the exclusion criteria were applied to identify the relevant papers. Figure 2 and Table 4 show the distribution of papers retrieved by the application of search keywords and their distribution among the digital libraries after applying the exclusion criteria. It can be seen that the IEEE dominates the initial papers retrieved from the digital libraries with a percentage of 46% of the total collected papers. On the contrary, ACM Digital Library is the highest contributor to the final amount of papers, representing 31% of the total remaining papers. This result shows the effectiveness of applying the inclusion and exclusion criteria. A validity threat of misinterpretation of the primary studies was mitigated by the first two authors of this study. The candidate studies also went through multiple reviews. The inclusion of each article into the final list of relevant primary studies was checked during the concluding agreement meeting among the authors.

3 Results and Discussion

The answers to the formulated research questions are discussed in this section.

> **RQ1**. What are the purposes and application domains of current metamodelling research?

Although there were 1157 papers found published on the metamodelling since 2012, only 69 papers were found relevant to this study after applying the exclusion criteria. Tables 5, 6, 7, 8, 9 and 10 describe the research purpose of the selected papers. The analysis of the studies shows that the scholars have focused on several common issues, including formalization, safety and security aspects of the systems, multilevel modeling, behavior modeling, and processes improvement.

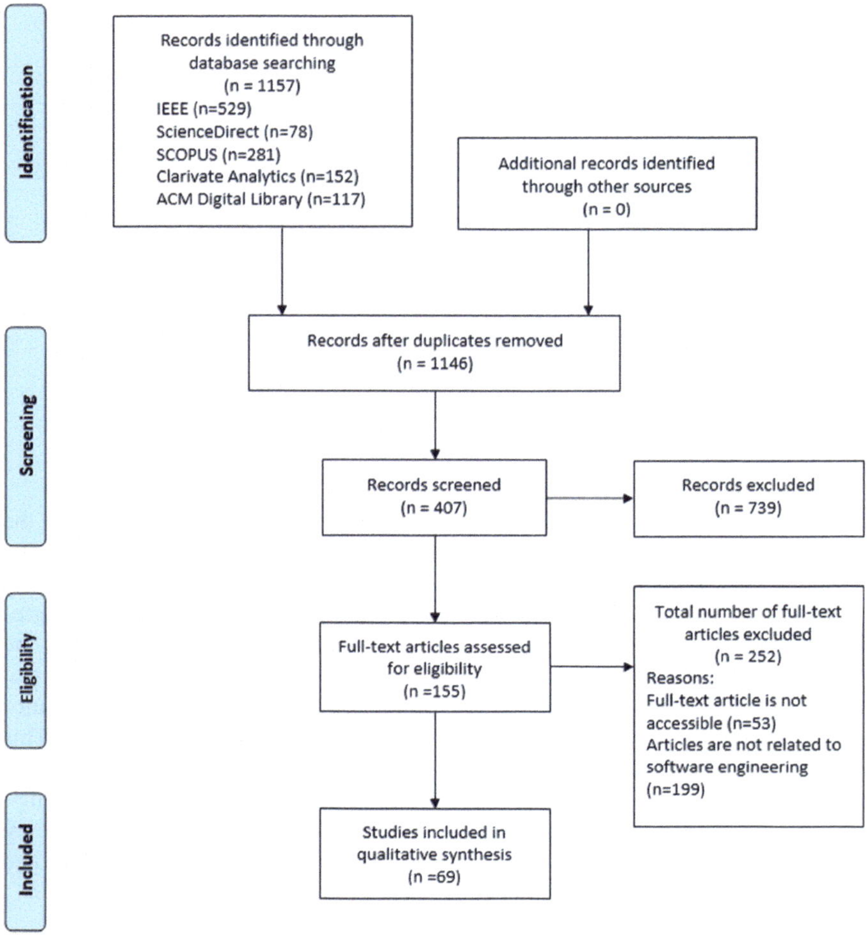

Fig. 1 PRISMA flowchart

Table 3 Extracted information

Category	Data
Basic information	• Title • Author(s)
Publication information	• Source • Year
Research information	• Application domain • Research purpose • Used metamodels and tools • Industrial application • Limitation of research • Future work

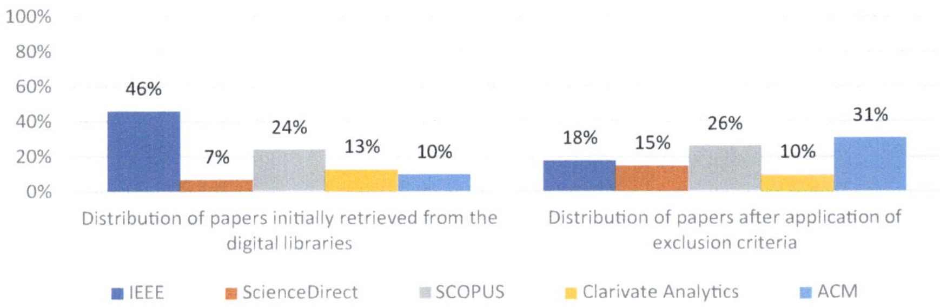

Fig. 2 Distribution of the papers retrieved from digital libraries

Table 4 Distribution of initially retrieved and selected papers across digital libraries

Digital library	Number of papers initially retrieved	Number of papers after applying the exclusion criteria
IEEE Xplore	529	12
ScienceDirect	78	10
SCOPUS	281	18
Clarivate analytics	152	8
ACM digital library	117	21
Total	1157	69

The advantages of adopting metamodelling in the software development lifecycle have been taken notice in software engineering. This claim is proved by the multiple metamodelling studies in various SE domains. Figure 3 illustrates the main application domains of metamodelling in several specific categories. This figure provides evidence regarding the industry willingness to adapt metamodelling in software design, analyses, and validation. Cyber-physical systems development and the Internet of Things (IoT), together with modeling safety and security aspects of the systems, are recognized as the most important application domains of metamodelling. Figure 4 demonstrates the types of systems involved in metamodelling applications.

RQ2. What are the trends of the metamodelling development in software engineering?

The current trends of metamodelling development can be divided into three parts, including the analysis of metamodeling improvement directions, studying the use of metamodels, and tools. Based on Tables 5, 6, 7, 8, 9 and 10, the development of metamodelling approaches has various ways. However, the analysis shows that the most frequent approaches are the formalization of metamodelling approaches, conceptual modeling, and multilevel modeling. Researchers also concentrate on the

Table 5 Research purpose description for the studies published since 2017

References	Research purpose	Approach/domain
[15]	To develop a framework that supports the behavior and structural complexity of combat system simulation	Simulation and software validation
[16]	To enhance the semantics of UML templates in OCL through the development of an aspectual template	Formalization, constraints, and software development
[17]	To simplify the development of domain-specific modeling tools	Code generation and software development
[18]	To develop the UML profile for modeling cognitive behavior	Development, cognitive behavior, and UML
[19]	To develop a flexible tool-supported modeling approach that augments a sketching environment with lightweight metamodelling capabilities	Development and sketching environment
[20]	To transform the Open Platform Communications Unified Architecture (OPC UA) to the UML model	Transformation and development of UML profile
[4]	To facilitate the simulation application in software development	Simulation and software development
[5]	To transform the machine learning element to domain modeling	Transformation and Cyber-Physical Systems (CPS)
[6]	To present the constraints validation model benchmark for a large graph model	Constraints, validation, and safety-critical systems
[21]	To create a metamodel for modeling a smart environment through functional and data perspective	New metamodel, CPS, Internet of Things, and smart environments
[22]	To increase the modeler productivity by task automation	Automation, workflows, and software development
[23]	To adopt a dual deep modeling approach as conceptual data modeling	Formalization, multilevel modeling, and conceptual modeling
[24]	To provide a uniform formalization on sub-model and sub metamodel through constraint properties	Formalization, constraints, and software development
[25]	To propose a theory for multi-level conceptual modeling	Multi-level modeling and conceptual modeling

methods for metamodels reuse, adaption, and integration. Since the present study analyzed both the basic research and applied research, Fig. 5 also presents the improvement of the syntax and semantics and the metamodel development inside an existing meta-metamodel (as a development of the new UML profiles).

Figure 6 summarizes the most used metamodelling tools since 2012. This figure shows that Eclipse Modelling Framework (EMF) is the most used tool. Table 11

Table 6 Research purpose description for the studies of 2016

References	Research purpose	Approach/domain
[26]	To investigate the need for strict metamodelling practice	Formalization and safety-critical systems
[27]	To develop and maintain modeling language and generator with test-driven development	Code generation and software validation
[28]	To adopt Software Product Lines (SPL) at the metamodel level	Metamodel reuse and software development
[29]	To define graphical syntax by designing metamodel to support graphical modeling language	New metamodel and software development
[30]	To specify the safety compliance needs for a critical system	Standards compliance and safety-critical systems
[31]	To discover basic elements in the modeling of software development	Multilevel modeling and software development
[32]	To define an abstract framework for multilevel modeling	Multilevel modeling and conceptual modeling
[33]	To support the security pattern specification and validation	Formalization, validation, and safety-critical systems
[34]	To develop the metamodel that merges evidence, goal, and risk into software development	Risk management and safety-critical systems
[35]	To create a modeling language that supports abstract syntax and dynamic semantic for the cyber-physical system	Formalization, CPS, and intelligent systems
[36]	To create an interface for safety evaluation and fault simulation in an automotive system	Evaluation, validation, and safety-critical systems
[37]	To represent metadata in the buildings using the proposed schema	Metadata, information systems, CPS, and intelligent houses
[38]	To create a metamodel development method with different mathematical semantics	Formalization (spatial and timing semantics) and CPS
[39]	To introduce behavior modeling for automatic discrete event system specification model	Formalization and behaviour modeling
[40]	To investigate the capabilities of simulation metamodelling in handling decision support tool	Simulation and decision support
[41]	To develop a modeling language for data description	New metamodel, metadata, and information systems
[42]	To provide information on the simulation model characteristic and its limitations	Simulation and intelligent systems

Table 7 Research purpose description for the studies of 2015

References	Research purpose	Approach/domain
[43]	To understand the requirements and needs of modeling stakeholders	Requirement engineering and conceptual modeling
[44]	To prove the efficiency of simulation metamodelling in healthcare analysis	Simulation and CPS (Healthcare)
[45]	To support runtime synchronization between the model and program state	Synchronization and software development
[46]	To evaluate the software modeling language elements	Metamodel evaluation and conceptual modeling
[47]	To define the execution semantic in the concurrent model	Execution semantics and software development
[48]	To discuss the situation when multilevel modeling is beneficial	Multilevel metamodelling and software development

Table 8 Research purpose description for the studies of 2014

References	Research purpose	Approach/domain
[49]	To develop a framework for modeling security patterns specification and validation	Validation and security domain
[50]	To use the ontological approach to address a semantic gap in security patterns	Semantics and security domain
[51]	To define the archetypes-based framework for evolutionary, dependable, and interoperable healthcare information systems	Metadata and CPS (Healthcare)
[52]	To develop SysML-based automating building energy system for modeling and analysis	Automation and CPS (energy system)
[53]	To improve the change impact analysis in software requirements	Analyses and requirement engineering
[54]	To formalize coordination among metamodels	Formalization and CPS (healthcare)
[55]	To define a seamless chain for software structural, functional, and execution modeling	Transformation and CPS (embedded systems)
[56]	To address the conflict in information system design using a conceptual modeling approach	Multilevel metamodelling and software design

Table 9 Research purpose description for the studies of 2013

References	Research purpose	Approach/domain
[57]	To support the theory of relation and reasoning in different approaches to requirement modeling	Formalization and requirement engineering
[58]	To define the rules to create a risk management system model	Risk management and safety-critical system
[59]	To improve the component-based approach for specifying DSML's concrete syntax	Syntax, semantics, and software development
[60]	To integrate safety analysis in the system engineering process	Safety analyses, system engineering, and safety-critical systems
[61]	To evaluate the existing software architecture viewpoint language	Evaluation and software design
[62]	To propose a tool that supports the software process reuse and automates their execution	Process reuse and automation and software processes
[63]	To introduce a model-driven process-centered software engineering environment	Software processes and software development
[64]	To identify the right time and scenario to use multilevel modeling	Multilevel modeling and conceptual modeling
[65]	To develop a metamodel that supports constraint in a database management system	Constraints and information systems
[66]	To define a template for the rationale of architecture design decision	Decision support and software requirements and design
[67]	To develop a technology that supports rapid domain-specific language development	Automation and software development
[68]	To support the metamodel development based on the graphical representation	Metamodel generation and sketching environment
[69]	To define a programming and modeling language that treats requirement and architecture concepts explicitly	Formalization and requirement engineering
[70]	To generate test cases and metamodelling correctness automatically	Code generation, formalization, and software validation
[71]	To develop a metamodel that generalizes DSML and MDE	Models execution and software development

Table 10 Research purpose description for the studies of 2012

References	Research purpose	Approach/domain
[72]	To provide a behavior model reasoning at different levels of abstraction	Formalization, behavioral modeling, and CPS (Healthcare)
[73]	To provide a framework for metadata editing and development	Metadata and geographic information systems
[74]	To provide an approach that supports model and constraint evaluation	Automation, model adaptation, and software design
[75]	To propose a metamodelling in the adaptation of Service-Oriented Architecture (SOA) roles in the traditional organization structure	Model adaptation and web technologies (SOA)
[76]	To combine the agent-oriented software development and MDE paradigms	Integration and artificial intelligence
[77]	To develop a metamodel that supports concurrent task tree modeling and execution	Code execution and concurrent processes
[78]	To improve the search engine ability with metamodelling	Improvement and web-technologies
[79]	To investigate the need for a multilevel business process	Multilevel modeling and business processes
[80]	To address the challenge requirements specification for a graphical modeling language	Syntax and requirement engineering

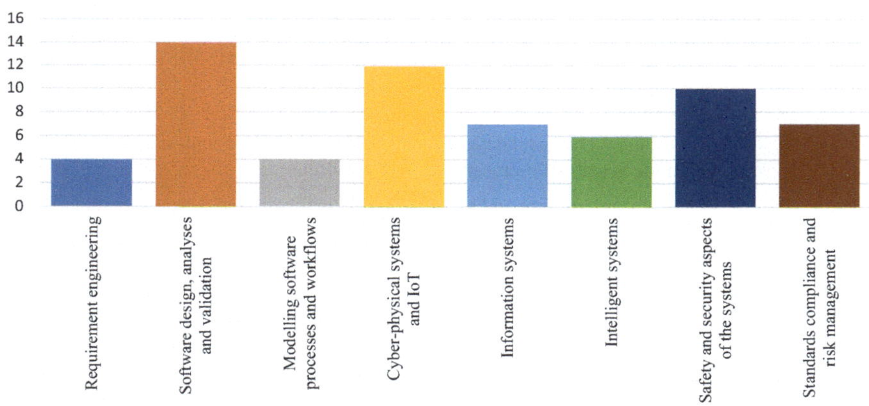

Fig. 3 The main application domains of metamodelling

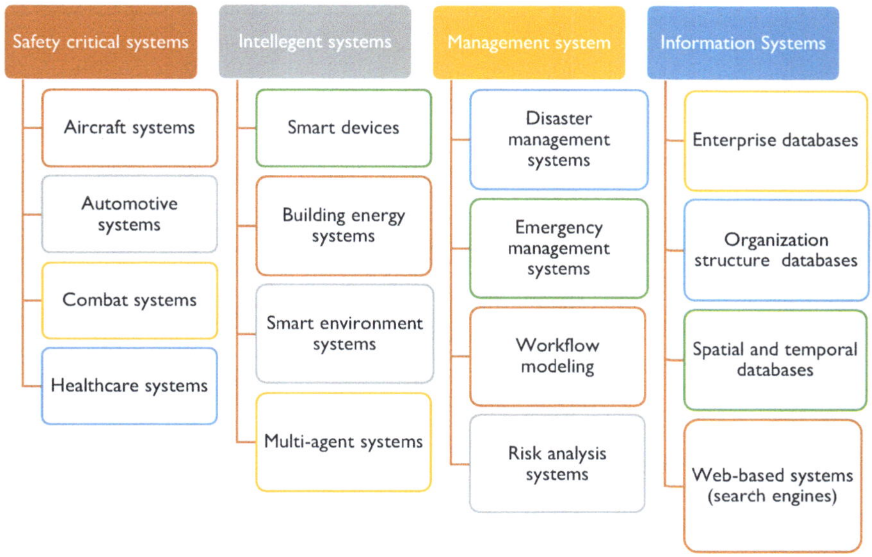

Fig. 4 Types of systems used in metamodeling applications

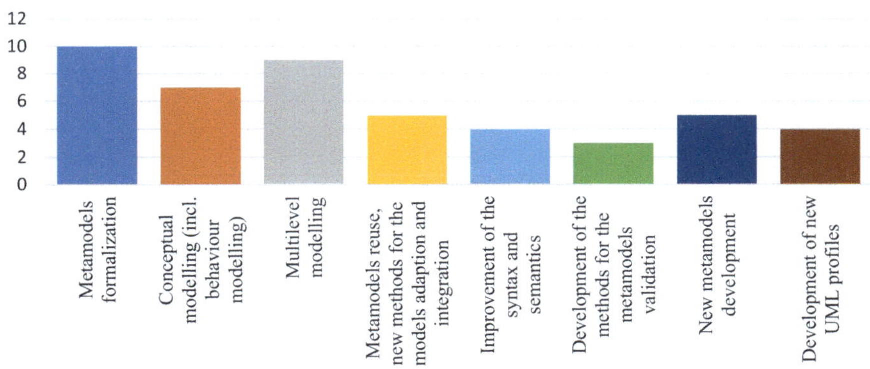

Fig. 5 The main trends of metamodeling development

shows the advantages and disadvantages of EMF. The researchers noticed that the main advantage is the various software environments and hardware platforms support. The EMF includes a constraint definition language, which can be used to define the model metrics and automatically detect deviations from the model design heuristics. Moreover, the Eclipse framework offers a variety of tools, which are portable for different platforms.

Other than modeling tools, this review study also studies the most used meta-metamodels and metamodels. Based on Fig. 7, ECORE is the most used meta-metamodel. ECORE offers many advantages, such as compatibility to support the

Fig. 6 Metamodelling tools

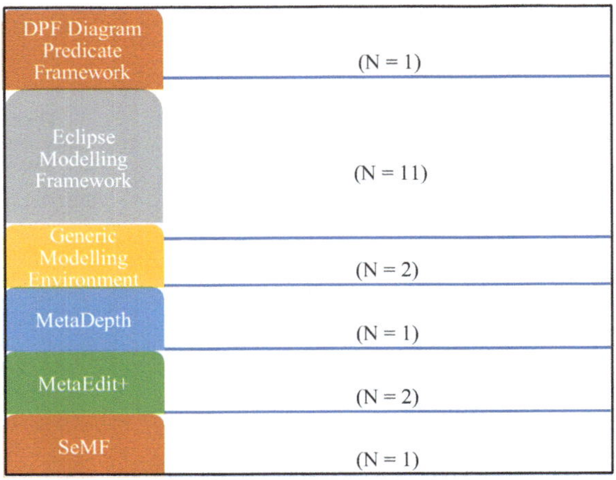

Table 11 Advantages and disadvantages of eclipse modelling framework

Advantages	Disadvantages
• Provides various platforms for a language definition • Ease of use in any software environment • Easy to define a graphical model • Easy to develop a graphical editor	• Does not allow to model a system behavior • Does not support a unique numeric identification • Fixed number of meta-levels

Fig. 7 Most referenced metamodels

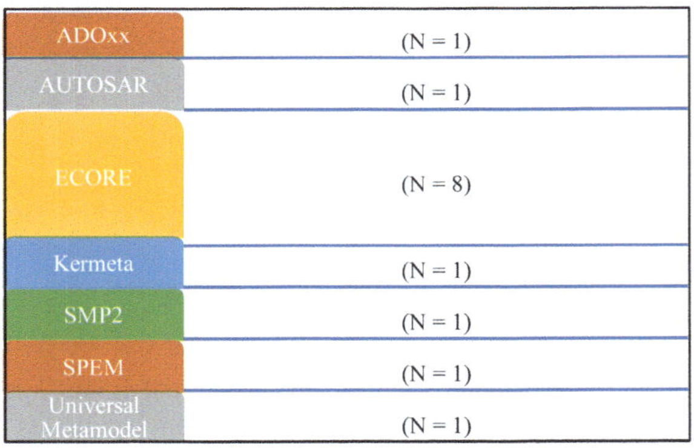

various platforms. The advantages and the motivation of ECORE are described in Table 12, along with the other metamodel descriptions and researchers' motivations. While all considered metamodels are used to define concrete and abstract syntax of the models, Table 12 is dedicated to show their specific features.

Table 12 Existing metamodels descriptions and motivations

Core metamodel	Descriptions	Motivations
ADOxx	• Used to specify a platform with multiple deployment options	• Support for an independent declarative language
AUTOSAR	• Specify architectures for automotive systems	• Standardize architecture and methodology of an automotive system
ECORE	• Describes three layers of model abstraction • Can describe the safety and security concepts of a particular system • Needs expanding by constraint language • Defines metamodel and the rules for the automation of tool generation • Can support software methodology Prometheus language • The definition of concrete textual syntax is supported	• To support model development for component-based systems • To follow the EMF compatibility metamodel • To support metamodel collaboration and various tools • To describe the modeling element of Prometheus methodology • To place the proposed methodology in EMF • To map structural features of a model to abstract syntax
Kermeta	• Metamodel that supports specific design pattern interpretation	• Can support model interpreter, compiler, and checker
Simulation Model Portability standards 2 (SMP2)	• Support automatic code generation from the platform-independent model	• This metamodel support model composability
Software and Systems Process Engineering Metamodel (SPEM)	• Initiate a model-driven process	• Support for a specific aspect of the model-driven development concept
Universal Metamodel	• Use the diagram element for a specialized class	• Metamodel that supports model-management systems

RQ3. What is the quality of publications that reflect the metamodelling research in SE domain?

Each selected study was evaluated according to journal ranking, which divides papers into four quartiles (i.e., Q1, Q2, Q3, and Q4). Q1/Q2 covers the journals of the highest impact factor, while Q4 covers the lowest ones. Conference papers were also considered. Figure 8 shows an increasing number of the impact of journal articles published between 2012 and 2017. These results acknowledge the increasing amount of metamodelling approaches in the scientific literature. The selected papers were also undergone six quality evaluation questions, as presented in Table 13. Three possible answers (i.e., “Yes”, “Partially”, and “No”) were used to evaluate the selected articles against the six quality evaluation questions. Figure 9 shows the quality assessment results of the selected papers.

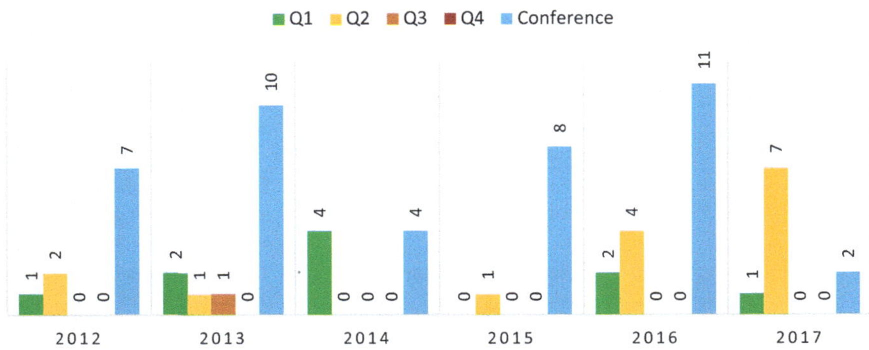

Fig. 8 Articles ranking per year

Out of the total number of selected articles (N = 69), all the articles mentioned the study purpose and application domain of the metamodelling approach either with a wide explanation or by a simple statement. 13% of the selected papers can be classified as basic research that extends the state-of-the-art in metamodeling, while 62% of them apply metamodeling to some domain. More than 60% of the selected articles explained in detail the methodology of research, while 30% of them did not elaborate on it in a structural way. Additionally, 27 articles mentioned and explained the use of a proven meta-metamodel, while only 21 articles mentioned the meta-metamodel. The remaining 21 articles just mentioned the use of an existing meta-metamodel without a clear explanation. Further, 37% of the considered studies clearly showed the applicability of the metamodelling approach in the software development industry, while 19% of them just mentioned the possible industrial applications. Moreover, about 25% of the selected studies indicated the future work and limitations of their studies, which could help to determine the trends of the metamodelling approach. 28% of the selected studies just mentioned either the possible future work or limitations of their studies, while more than 40% did not mention the future work and limitations.

RQ4. What are the limitations of the current studies and the prospects for future research?

The answers to this question provide an overview of the limitations and future work of the reviewed studies. The overview is divided into two parts, including the analysis of limitations and the proposed future improvements. First, the limitations of the reviewed studies were analyzed in Table 14. Second, based on these limitations, future works were also analyzed. A brief overview of future proposals is described in Table 15. Tables 14 and 15 present the limitations and future proposals for the scholars who only reported the limitations and future works in their studies.

There are several common issues highlighted in the selected research papers. Table 14 allows to deduct the limitations of metamodelling. While its application is recognized as an effective, many authors noticed that manual coding is still required.

Table 13 Quality assessment questionnaire

QA1 Did the study consider the purpose and application domain of the considered metamodelling approach? • Yes: it explains the purpose of metamodelling approach and its application domain • Partially: it explains the purpose of metamodelling approach, but not the application domain • No: it does not mention the purpose and application domain
QA2 Did the study extend the state-of-the-art in metamodeling? • Yes: it extends the state-of-the-art in metamodeling • Partially: it shows some new results in the metamodelling domain • No: it applies an existing metamodeling approach to some domain
QA3 Did the study consider and give details on the methodology used? • Yes: it clearly explains the methodology used • Partially: it lists a few methods but not in a structural way • No: it did not discuss the methodology used
QA4 Did the study use a proven meta-metamodel? • Yes: it names the meta-metamodel and how it is used in the research • Partially: it names the meta-metamodel but did not describe how the meta-metamodel is used • No: it mentions the meta-metamodel name without explanation
QA5 Did the study consider an industrial use of the metamodelling approach? • Yes: it discusses the use of metamodelling approach in the industry • Partially: it mentions the possible use of metamodelling in the industry • No: it does not mention the industrial use of metamodelling
QA6 Did the study consider the future work and discuss limitation? • Yes: it explains the possible future elaboration of metamodelling approach and its limitations • Partially: it mentions either future works or limitations • No: it did not mention the limitations nor the possibility of future work

To support the design of cyber-physical systems (e.g., intelligent houses), further metamodel formalization and expansion by spatial and time semantics are required.

Based on Table 15, the researchers highlighted several common directions for the improvement of metamodelling. The important directions include metamodel formalization, support of constraint language, and allowing formal specification and automatic reasoning. To solve these arising issues, the authors of this study proposed to improve the conceptual foundation of metamodelling approach. Metamodel

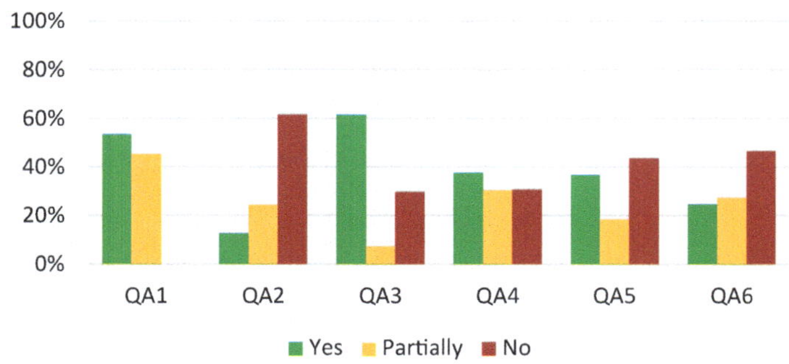

Fig. 9 Quality assessment results

Table 14 The limitations of reviewed studies

Limitations description	References
The proposed metamodel did not support spatial information	[19]
The application of metamodelling is effective, but manual coding is still needed	[15]
The implementation of automation components depends on the metamodel process used to define the software process line	[62]
In-depth knowledge of UML is needed to apply metamodelling	[78]
The proposed metamodel can only support a limited system analysis	[52]
The metamodel is limited to the predefined requirement functionality	[53]
Limited possibility to support the changing learning algorithm	[5]
Limited usability of the proposed model editor	[76]
Cannot support large domain-specific systems	[26]
Open platform communications unified architecture is not aligned with UML	[20]
The proposed metamodelling strategy cannot be validated for large data	[43]

adaptations and supporting wider domains are the topics that have seen the light to be explored. The problems associated with metamodel adaptation could be solved by further elaboration of the metamodelling architecture, especially elaborating the multilevel metamodelling.

Another area for further research is the behavior metamodels, and this could be handled by supporting process-centered software engineering. In that, the integration of the behavior model and structural model is considered as a possible direction for further research. The employment of metamodelling for safety-critical and cyber-physical systems is a special and highly demanded domain of research. This could be further expanded through the application of metamodels for the certification of safety-critical systems. The limitation of metamodelling approach in supporting large data is one of the common problems faced by many researchers. Although the proposed metamodels are effective, their functionality is still limited to small data.

Table 15 Summary of future work in the reviewed studies

Proposals for future work	References
Adapt multi-level modeling with potencies to standard modeling, data representation, and tool	[23]
Improve the metamodelling to be able to support various domains	[6, 15, 54]
Apply the behavior metamodel to other modeling formalisms	[39]
Broadening the process-centered metamodel to other model-driven development approaches	[63]
Improve the metamodelling to support constraint language	[4]
Evaluate the effectiveness of the proposed framework through complex case studies for safety evaluation purposes	[30, 36, 50, 53]
Provide evaluation for the visual and textual properties of the proposed metamodel	[73]
Expand the approach to the common model repository	[78]
Improve the metamodel conceptual foundation	[43]
Improve the model binding flexibility	[62, 64]
Improve the non-functional requirements	[26, 61, 76]
Incorporate the theory into a unified foundation ontology	[25]
Extend the metamodel by scheduling model	[44]
Simply the certification of a critical system	[34]
Develop a formal specification of the security pattern	[33]
Support an approach for the full generation of domain-specific graphical modeling tools	[17]
Apply the approach in the goal-oriented requirement model	[57]
Support the design of the proposed behavior model with a structural model	[72]
Apply the approach to a more complex scenario for smart environments modeling	[21]
Utilize the model for simulation performance and integrate safety concerns into systems engineering processes	[60]
Improve the results of automated reasoning using metamodelling	[70]
Use the metamodel to define a concrete abstract syntax for graphical modeling languages	[29]
Develop suitable graphical editor and metamodel description in a component-based approach for specifying DSML's concrete syntax	[59]
Use the aspectual template to determine the model type for enhancing the semantics of UML templates in OCL	[16]

In comparison with the approaches used in software quality assessment, the methods for the metamodels evaluation are weekly considered in the literature. This problem stems from the inefficient evaluation of employing the approaches to support complex systems, which can be recognized as an important direction for future work. The researchers intended to improve the non-functional requirements of metamodelling approach such as flexibility, interoperability, and reusability features for their future attempts. These proposals would guide the possible directions of metamodelling improvement and its alternative applications in the SE discipline.

4 Limitations of the Study

Although the present review study is believed to yield significant results to the existing literature of metamodelling, it also posits some limitations that need to be stated. These limitations can be classified into four parts, including search process, inclusion and exclusion criteria, analysis of trends and approaches, and quality assessment.

In terms of the search process, only six search engines were selected for articles collection. For this reason, the collected articles were limited to these digital libraries only. Future studies may include more data sources to increase the probability of retrieving more articles. Concerning the inclusion and exclusion criteria, it was decided that only articles published from 2012 would be included. This condition allowed us to focus on the recent research published on the topic, and support the answer to the question of "What are the trends of the metamodelling development in software engineering?". Other than that, summaries of workshops, panels, and tutorials were excluded from this research. This is because this type of studies does not discuss the research problems on metamodelling in software engineering. Thus, their inclusion cannot help to answer the research questions. Further, this review study mixes the papers that define metamodels for a given purpose with the papers that extended metamodeling with new capabilities. Although this limitation results in retrieving a large amount of papers, it allows us to learn the current landscape of metamodeling from different points of view.

In terms of trends and approaches analysis, the classification of the concepts in the employed approach might lead to some overlapping. For instance, Fig. 3 represents the main application domains of metamodelling, which correspond to the stages of SE (such as requirement engineering and software design) and systems types (such as cyber-physical systems and information systems). Moreover, this study did not distinguish between meta-metamodels and metamodels in analyzing the trends. With regard to quality assessment, the analysis of the journal articles was based on index quartiles (Q1, Q2, Q3, and Q4). In fact, we have focused on the reputation of the venue, rather than on the quality of the published paper. This study also did not consider the quality of conferences, as all the conferences were combined into one general conference category. In addition, the quality assessment of the research papers was limited to only six quality questions. Thus, the quality assessment limits the scope but did not change the conclusion of this review study.

5 Conclusion

Metamodelling has become a common practice in software systems engineering. Together with the proven meta-metamodels, metamodels, and tools, there are new approaches to the metamodelling appeared. The aim of this systematic literature review was to analyze the state-of-the-art of metamodelling in software engineering. In line with this objective, this paper studied the research articles published on the metamodelling starting from 2012.

The study showed that metamodelling was adopted in various SE domains. However, the most current applications of metamodelling are cyber-physical and safety-critical systems development. Additionally, metamodelling has been applied in novel domains, including standards compliance, risk management, and decision-making processes. Scholars in the domain also concentrated on the methods for metamodels reuse, adaption, and integration. Due to the fact that this study analyzed both the basic research and applied research, new metamodels development and improvement of the syntax and semantics of existing metamodels are recognized as important topics. Several problems in the reviewed research papers allowed to find the trends in metamodelling approach development. The results of this review study suggested that perspective research plans need to support complex system modeling, formalization, behavior modeling, multilevel modeling, adding spatial and time semantics, and considering nonfunctional features such as flexibility and interoperability. We believe that the outcomes of this review study would provide constructive insights into the current applications, methods, trends, limitations, and future directions of metamodelling.

References

1. Schmidt, D.C.: Model-driven engineering. IEEE Comput. **39**, 25–31 (2006)
2. Ernst, J.: What is metamodeling, and what is it good for? (2002)
3. Van Gigch, J.P.: System Design Modeling and Metamodeling. Springer Science & Business Media, Berlin (2013)
4. Durak, U., Pawletta, T., Oguztuzun, H., Zeigler, B.P.: System entity structure and model base framework in model based engineering of simulations for technical systems. In: Proceedings of the Symposium on Model-driven Approaches for Simulation Engineering (2017)
5. Hartmann, T., Moawad, A., Fouquet, F., Le Traon, Y.: The next evolution of MDE: a seamless integration of machine learning into domain modeling. In: Proceedings—ACM/IEEE 20th International Conference on Model Driven Engineering Languages and Systems, MODELS 2017 (2017)
6. Szárnyas, G., Izsó, B., Ráth, I., Varró, D.: The train benchmark: cross-technology performance evaluation of continuous model queries. Softw. Syst. Model. **17**, 1365–1393 (2018)
7. Mezhuyev, V., Al-Emran, M., Fatehah, M., Hong, N.C.: Factors affecting the metamodelling acceptance: a case study from software development companies in Malaysia. IEEE Access **6**, 49476–49485 (2018)
8. Shukla, S.: Metamodeling: what is it good for? IEEE Des. Test Comput. **26**, 96 (2009)

9. Kitchenham, B., Pearl Brereton, O., Budgen, D., Turner, M., Bailey, J., Linkman, S.: Systematic literature reviews in software engineering—a systematic literature review. Inf. Softw. Technol. **51**, 7–15 (2009)
10. Al-Saedi, K., Al-Emran, M., Abusham, E., El Rahman, S.A.: Mobile payment adoption: a systematic review of the UTAUT model. In: International Conference on Fourth Industrial Revolution (2019)
11. Saa, A.A., Al-Emran, M., Shaalan, K.: Factors affecting students' performance in higher education: a systematic review of predictive data mining techniques. Technol. Knowl. Learn. **24**, 567–598 (2019)
12. Al-Qaysi, N., Mohamad-Nordin, N., Al-Emran, M.: A systematic review of social media acceptance from the perspective of educational and information systems theories and models. J. Educ. Comput. Res. **57**(8), 2085–2109 (2020)
13. Al-Emran, M., Mezhuyev, V., Kamaludin, A., Shaalan, K.: The impact of knowledge management processes on information systems: a systematic review. Int. J. Inf. Manage. **43**, 173–187 (2018)
14. Kleijnen, J.P.C.: Regression and Kriging metamodels with their experimental designs in simulation: a review. Eur. J. Oper. Res. **256**, 1–16 (2017)
15. Li, X.B., Yang, F., Lei, Y.L., Wang, W.P., Zhu, Y.F.: A model framework-based domain-specific composable modeling method for combat system effectiveness simulation. Softw. Syst. Model. **16**, 1201–1222 (2017)
16. Vanwormhoudt, G., Caron, O., Carré, B.: Aspectual templates in UML. Softw. Syst. Model. **16**, 469–497 (2017)
17. Naujokat, S., Lybecait, M., Kopetzki, D., Steffen, B.: CINCO: a simplicity-driven approach to full generation of domain-specific graphical modeling tools. Int. J. Softw. Tools Technol. Transf. **20**, 327–354 (2018)
18. Zhu, Z., Lei, Y., Zhu, Y., Sarjoughian, H.: Cognitive behaviors modeling using UML profile: design and experience. IEEE Access **5**, 21694–21708 (2017)
19. Wüest, D., Seyff, N., Glinz, M.: FlexiSketch: a lightweight sketching and metamodeling approach for end-users. Softw. Syst. Model. **18**, 1513–1541 (2019)
20. Lee, B., Kim, D.K., Yang, H., Oh, S.: Model transformation between OPC UA and UML. Comput. Stand. Interfaces **50**, 236–250 (2017)
21. Cicirelli, F., Fortino, G., Guerrieri, A., Spezzano, G., Vinci, A.: Metamodeling of smart environments: from design to implementation. Adv. Eng. Inform. **33**, 274–284 (2017)
22. Gamboa, M.A., Syriani, E.: Using workflows to automate activities in MDE tools. In: Communications in Computer and Information Science (2017)
23. Neumayr, B., Schuetz, C.G., Jeusfeld, M.A., Schrefl, M.: Dual deep modeling: multi-level modeling with dual potencies and its formalization in F-logic. Softw. Syst. Model. **17**, 233–268 (2018)
24. Carré, B., Vanwormhoudt, G., Caron, O.: On submodels and submetamodels with their relation: a uniform formalization through inclusion properties. Softw. Syst. Model. **17**, 1105–1137 (2018)
25. Carvalho, V.A., Almeida, J.P.A.: Toward a well-founded theory for multi-level conceptual modeling. Softw. Syst. Model. **17**, 205–231 (2018)
26. Durisic, D., Staron, M., Tichy, M., Hansson, J.: Addressing the need for strict meta-modeling in practice—a case study of AUTOSAR. In: 2016 4th International Conference on Model-Driven Engineering and Software Development (MODELSWARD), pp. 317–322 (2016)
27. Tolvanen, J.-P.: MetaEdit + for collaborative language engineering and language use (tool demo). In: Proceedings of the 2016 ACM SIGPLAN International Conference on Software Language Engineering, pp. 41–45 (2016)
28. Perrouin, G., Amrani, M., Acher, M., Combemale, B., Legay, A., Schobbens, P.-Y.: Featured model types: towards systematic reuse in modelling language engineering. In: Proceedings of the 8th International Workshop on Modeling in Software Engineering (2016)
29. Kalnins, A., Barzdins, J.: Metamodel specialization for graphical modeling language support. In: Proceedings—19th ACM/IEEE International Conference on Model Driven Engineering Languages and Systems, MODELS 2016 (2016)

30. De La Vara, J.L., et al.: Model-based specification of safety compliance needs for critical systems: a holistic generic metamodel. Inf. Softw. Technol. **72**, 16–30 (2016)
31. Şutîi, A.M., Verhoeff, T., van den Brand, M.: Modular multilevel metamodeling with MetaMod. In: Companion Proceedings of the 15th International Conference on Modularity, pp. 212–217 (2016)
32. Theisz, Z., Mezei, G.: Multi-level dynamic instantiation for resolving node-edge dichotomy. In: 2016 4th International Conference on Model-Driven Engineering and Software Development (MODELSWARD), pp. 274–281. IEEE (2016)
33. Hamid, B., Gürgens, S., Fuchs, A.: Security patterns modeling and formalization for pattern-based development of secure software systems. Innov. Syst. Softw. Eng. **12**, 109–140 (2016)
34. Larrucea, X., Gonzalez-Perez, C., McBride, T.: Standards-based metamodel for the management of goals, risks and evidences in critical systems development. Comput. Stand. Interfaces **48**, 71–79 (2016)
35. Nastov, B., Chapurlat, V., Dony, C., Pfister, F.: Towards semantical DSMLs for complex or cyber-physical systems. In: ENASE: Evaluation of Novel Software Approaches to Software Engineering, pp. 115–123 (2016)
36. Chaari, M., Ecker, W., Kruse, T., Novello, C., Tabacaru, B.A.: Transformation of failure propagation models into fault trees for safety evaluation purposes. In: Proceedings—46th Annual IEEE/IFIP International Conference on Dependable Systems and Networks, DSN-W 2016 (2016)
37. Balaji, B., et al.: Brick: towards a unified metadata schema for buildings. In: Proceedings of the 3rd ACM International Conference on Systems for Energy-Efficient Built Environments—BuildSys '16 (2016)
38. Mezhuyev, V., Samet, R.: Metamodeling methodology for modeling cyber-physical systems. Cybern. Syst. **47**, 277–289 (2016)
39. Sarjoughian, H.S., Alshareef, A., Lei, Y.: Behavioral DEVS metamodeling. In: Proceedings—Winter Simulation Conference (2016)
40. Rosen, S.L., Slater, D., Beeker, E., Guharay, S., Jacyna, G.: Critical infrastructure network analysis enabled by simulation metamodeling. In: Proceedings—Winter Simulation Conference (2016)
41. Davies, J., Gibbons, J., Milward, A., Milward, D., Shah, S., Solanki, M., Welch, J.: Domain specific modelling for clinical research. In: Proceedings of the Workshop on Domain-Specific Modeling, pp. 1–8 (2015, October)
42. Zhang, X., Zou, L.: Simulation metamodeling in the presence of model inadequacy. In: Proceedings of the 2016 Winter Simulation Conference, pp. 566–577 (2016)
43. Karagiannis, D.: Agile modeling method engineering. In: Proceedings of the 19th Panhellenic Conference on Informatics, pp. 5–10 (2015, October)
44. Rosen, S.L., Ramsey, J., Harvey, C.E., Guharay, S.K.: Efficient analysis for emergency management using simulation metamodeling: a case study for a medical trauma center. In: 47th Summer Computer Simulation Conference, SCSC 2015, Part of the 2015 Summer Simulation Multi-Conference, SummerSim 2015 (2015)
45. Tolvanen, J.P., Djukić, V., Popovic, A.: Metamodeling for medical devices: code generation, model-debugging and run-time synchronization. Procedia Comput. Sci. **63**, 539–544 (2015)
46. Henderson-Sellers, B., Eriksson, O., Gonzalez-Perez, C., Ågerfalk, P.J., Walkerden, G.: Software modelling languages: a wish list. In: Proceedings—7th International Workshop on Modeling in Software Engineering, MiSE 2015 (2015)
47. Latombe, F., Crégut, X., Combemale, B., Deantoni, J., Pantel, M.: Weaving concurrency in executable domain-specific modeling languages. In: SLE 2015—Proceedings of the 2015 ACM SIGPLAN International Conference on Software Language Engineering (2015)
48. De Lara, J., Guerra, E., Cuadrado, J.S.: When and how to use multilevel modelling. ACM Trans. Softw. Eng. Methodol. **24**, 1–46 (2015)
49. Hamid, B., Percebois, C.: A modeling and formal approach for the precise specification of security patterns. In: Lecture Notes in Computer Science (including subseries Lecture Notes in Artificial Intelligence and Lecture Notes in Bioinformatics) (2014)

50. Kiwelekar, A.W., Joshi, R.K.: An ontological framework for architecture model integration. In: Proceedings of the 4th International Workshop on Twin Peaks of Requirements and Architecture, pp. 24–27 (2014, June)
51. Piho, G., Tepandi, J., Thompson, D., Tammer, T., Parman, M., Puusep, V.: Archetypes based meta-modeling towards evolutionary, dependable and interoperable healthcare information systems. Procedia Comput. Sci. **37**, 457–464 (2014)
52. Kim, S.H.: Automating building energy system modeling and analysis: an approach based on SysML and model transformations. Autom. Constr. **41**, 119–138 (2014)
53. Goknil, A., Kurtev, I., Van Den Berg, K., Spijkerman, W.: Change impact analysis for requirements: a metamodeling approach. Inf. Softw. Technol. **56**, 950–972 (2014)
54. Rabbi, F., Lamo, Y., MacCaull, W.: Co-ordination of multiple metamodels, with application to healthcare systems. Procedia Comput. Sci. **37**, 473–480 (2014)
55. Bucaioni, A., Cicchetti, A., Sjödin, M.: Towards a metamodel for the Rubus component model. In: CEUR Workshop Proceedings (2014)
56. Frank, U.: Multilevel modeling: toward a new paradigm of conceptual modeling and information systems design. Bus. Inf. Syst. Eng. **6**, 319–337 (2014)
57. Goknil, A., Kurtev, I., Millo, J.V.: A metamodeling approach for reasoning on multiple requirements models. In: Proceedings—IEEE International Enterprise Distributed Object Computing Workshop, EDOC (2013)
58. Othman, S.H., Beydoun, G.: Model-driven disaster management. Inf. Manag. **50**, 218–228 (2013)
59. El Kouhen, A., Gérard, S., Dumoulin, C., Boulet, P.: A component-based approach for specifying DSML's concrete syntax. In: Proceedings of the Second Workshop on Graphical Modeling Language Development, pp. 3–11 (2013, July)
60. Piriou, P.Y., Faure, J.M., Deleuze, G.: A meta-model for integrating safety concerns into systems engineering processes. In: 2013 IEEE International Systems Conference (SysCon), pp. 298–304 (2013)
61. Tekinerdogan, B., Demirli, E.: Evaluation framework for software architecture viewpoint languages. In: Proceedings of the 9th International ACM Sigsoft Conference on Quality of Software Architectures, pp. 89–98 (2013, June)
62. Rouillé, E., Combemale, B., Barais, O., Touzet, D., Jézéquel, J.M.: Integrating software process reuse and automation. In: Proceedings—Asia-Pacific Software Engineering Conference, APSEC (2013)
63. MacIel, R.S.P., Gomes, R.A., Magalhães, A.P., Silva, B.C., Queiroz, J.P.B.: Supporting model-driven development using a process-centered software engineering environment. Autom. Softw. Eng. **20**, 427–461 (2013)
64. De Lara, J., Guerra, E., Sánchez Cuadrado, J.: Reusable abstractions for modeling languages. Inf. Syst. **38**, 1128–1149 (2013)
65. Ristić, S., Aleksić, S., Čeliković, M., Luković, I.: Meta-modeling of inclusion dependency constraints. In: Proceedings of the 6th Balkan Conference in Informatics, pp. 114–121 (2013)
66. Dermeval, D., Castro, J., Silva, C., Pimentel, J., Bittencourt, I. I., Brito, P., … Pedro, A.: On the use of metamodeling for relating requirements and architectural design decisions. In: Proceedings of the 28th Annual ACM Symposium on Applied Computing, pp. 1278–1283 (2013, March)
67. Kuzenkova, A., Deripaska, A., Bryksin, T., Litvinov, Y., Polyakov, V.: QReal DSM platform—an environment for creation of specific visual IDEs. In: ENASE, pp. 205–211 (2013)
68. Wuest, D., Seyff, N., Glinz, M.: Semi-automatic generation of metamodels from model sketches. In: 2013 28th IEEE/ACM International Conference on Automated Software Engineering, ASE 2013—Proceedings (2013)
69. Spacek, P., Dony, C., Tibermacine, C., Fabresse, L.: Wringing out objects for programming and modeling component-based systems. In: Proceedings of the Second International Workshop on Combined Object-Oriented Modelling and Programming Languages (2013)
70. Jackson, E.K., Levendovszky, T., Balasubramanian, D.: Automatically reasoning about metamodeling. Softw. Syst. Model. **14**, 271–285 (2013)

71. Sousa, G.C.M., Costa, F.M., Clarke, P.J., Allen, A.A.: Model-driven development of DSML execution engines. In: Proceedings of the 7th Workshop on Models@ run. time, pp. 10–15 (2012, October)
72. Rutle, A., MacCaull, W., Wang, H., Lamo, Y.: A metamodelling approach to behavioural modeling. In: Proceedings of the Fourth Workshop on Behaviour Modelling-Foundations and Applications (2012)
73. Nogueras-Iso, J., Latre, M.Á., Béjar, R., Muro-Medrano, P.R., Zarazaga-Soria, F.J.: A model driven approach for the development of metadata editors, applicability to the annotation of geographic information resources. Data Knowl. Eng. **81**, 118–139 (2012)
74. Demuth, A., Lopez-Herrejon, R.E., Egyed, A.: Automatically generating and adapting model constraints to support co-evolution of design models. In: 2012 Proceedings of the 27th IEEE/ACM International Conference on Automated Software Engineering, pp. 302–305. IEEE (2012, September)
75. Ionita, A.D., Radulescu, S.A.: Metamodeling for assigning specific roles in the migration to service-oriented architecture. In: Proceedings—3rd International Conference on Emerging Intelligent Data and Web Technologies, EIDWT 2012 (2012)
76. Gascueña, J.M., Navarro, E., Fernández-Caballero, A.: Model-driven engineering techniques for the development of multi-agent systems. Eng. Appl. Artif. Intell. **25**, 159–173 (2012)
77. Brüning, J., Kunert, M., Lantow, B.: Modeling and executing ConcurTaskTrees using a UML and SOIL-based metamodel. In: Proceedings of the 12th Workshop on OCL and Textual Modelling, pp. 43–48 (2012, September)
78. Lucrédio, D., Renata, R.P., Whittle, J.: MOOGLE: a metamodel-based model search engine. Softw. Syst. Model. **11**, 183–208 (2012)
79. Schütz, C., Schrefl, M., Delcambre, L.M.L.: Multilevel business process modeling: motivation, approach, design issues, and applications. In: International Conference on Information and Knowledge Management, Proceedings (2012)
80. Cho, H., Gray, J., Syriani, E.: Syntax map: a modeling language for capturing requirements of graphical DSML. In: Proceedings—Asia-Pacific Software Engineering Conference, APSEC (2012)

A Systematic Review of the Technological Factors Affecting the Adoption of Advanced IT with Specific Emphasis on Building Information Modeling

Mohamed Ghayth Elghdban, Nurhidayah Binti Azmy, Adnan Bin Zulkiple, and Mohammed A. Al-Sharafi

Abstract Despite the sensitivity of the architecture, engineering, and construction (AEC) industry to the changes in demographic factors, economic activity, and social development, the industry is consistently progressing and becoming more successful. Organizations in the AEC industry are lagging in the adoption of advanced IT. For instance, Building Information Modelling (BIM) is among the top technologies utilized by the industry. BIM is extensively known as one of the innovations of IT to have emerged within the AEC industry. Although BIM processes require organization-wide adoption, few studies focused on the technological factors influencing BIM adoption at the organizational level in the AEC industry. Therefore, the present study aims to further enrich the literature of such studies through a systematic literature review (SLR). The main objective of this SLR is to analyze the current studies that involved the technological factors that influence organizations to adopt and use advanced IT, such as BIM. Therein, 78 up-to-date studies retrieved from two primary databases (Scopus and Web of Science) were critically analyzed in the period between 2015 and October 2019. The review identified 42 technological factors that may affect BIM adoption in the AEC industry. The outcome of this SLR will add significantly to the existing literature in IT adoption in general, and BIM adoption practically.

M. G. Elghdban · N. B. Azmy · A. B. Zulkiple
Faculty of Engineering Technology, Universiti Malaysia Pahang, Lebuhraya Tun Razak, 26300 Gambang, Pahang, Malaysia
e-mail: jawal_75@yahoo.com

N. B. Azmy
e-mail: hidayahba@ump.edu.my

A. B. Zulkiple
e-mail: adnanz@ump.edu.my

M. A. Al-Sharafi (✉)
Faculty of Computing, College of Computing and Applied Sciences, Universiti Malaysia Pahang, Lebuhraya Tun Razak, 26300 Gambang, Pahang, Malaysia
e-mail: alsharafi@ieee.org

M. Al-Emran et al. (eds.), *Recent Advances in Intelligent Systems and Smart Applications*, Studies in Systems, Decision and Control 295,
https://doi.org/10.1007/978-3-030-47411-9_2

Keywords IT adoption · Technological factors · Building information modeling · Architecture · Engineering · And construction (AEC) industry

1 Introduction

The architecture, engineering, and construction (AEC) industry is always seen as lagging behind the other sectors in terms of increasing efficiency and productivity [1]. Previous researches have shown that the lack of adoption of IT technology is the reason behind this stand [2]. On the other side, the studies have shown that the need to accelerate the innovation adoption rate in the construction industry has been well reported [3]. The AEC industry must improve in their IT utilization as it can increase their chances of reaching their expectations and improve their credibility. BIM is commonly regarded as one of the IT innovations to have evolved within the AEC industry over the past few decades [4]. It provides a novel design, construction, and facility management approach by enabling the presentation of design information in 3D and improving project and design coordination [5]. This implies that BIM technology has taken the place of the conventional paper-based drawings as the major source of design concept, and as a collaborative working tool [6].

Despite the significant benefit of BIM and massive attention within the construction industry, the increasing utilization of BIM among construction players in the developing country is still at the lower level as they perceive BIM as a new technology [7]. The low adoption rate is not just in the developing countries, but also in some of the developed countries. The 2015 Canadian BIM Survey has shown that BIM adoption in Canada is at a low rate as it has only increased by 3% over the last 2 years (against the expected 20% +) [8]. In this regard, various attempts have been made to understand and explain the critical factors that have impact on BIM adoption from the perspective of the user [9, 10].

Although BIM processes require organization-wide adoption, few studies have focused on the technological factors influencing BIM adoption at the organizational level [11, 12].

Due to the lack of empirical studies that examined the element of the TOE in BIM adoption field, this study seeks to review the literature regarding IT innovation adoption to determine the technological factors that have been used as a part of a developed framework. In this regard, this study intends to answer the following research question in order to fulfil the aim of this study:

RQ1: What are the technological factors that influence organizations to adopt and use advanced IT such as Building Information Modeling in their activities?

This paper is organized as follows: Sect. 2 presents the BIM adoption concepts and studies while Sect. 3 presents the employed method in this study. Section 4 presents the findings and discussion while Sect. 5 presents the conclusion and future works.

2 BIM Adoption

BIM is an advanced IT-enabled tool with integral digital representation (data repository) for different phases of a project lifecycle [12]. It is evolving as a promising tool in building and construction management [4]. The methods and tools for risk control, fragmentation, and collaboration improvement in construction projects can be improved by using BIM tools. As a framework, BIM was designed to guarantee sustainability, reduce poor quality, drive the integration of disjointed practices, as well as stimulate the need for changes in the business process [5, 13]. BIM is revolutionary and has transformed the manner of conceiving, designing, constructing, and operating buildings [10].

Several studies have focused on the understanding and explanation of the concept of advanced IT adoption by relying mainly on the common theories such as Technology Acceptance Model (TAM) [14, 15], Theory of Planned Behavior (TPB) [16, 17] Unified Theory of Acceptance and Use of Technology (UTAUT) [18], Diffusion of Innovation (DOI) [19].

The previous studies have tried modifying and developing these theories by integrating two or more of these theories to understand the users' opinions [9, 11, 20, 21].

To enhance the predictive power of those theories in the organizational level, it has been combined with theories of innovation adoption at the organizational level, such as diffusion of innovation (DOI) and using the factors that have defined the characteristics of the technology, such as relative advantage, compatibility, complexity, trialability and observability, to examine their effect on user perception [10, 12]. However, studies on BIM adoption at the organization level have remained lacking. Very few studies have been carried out at organizational level using the TOE framework for BIM adoption [17, 18] which is considered an important theory in explaining and understanding the determinants of IT innovation at the organizational level, including BIM.

3 Methodology

This study employed a systematic literature review (SLR) methodology because it aims to identify the essential scientific contribution to the field of IT adoption with specific emphasis on BIM by providing a solid structured overview of the current knowledge and identifying gaps. Thus, the research question that guided the systematic review was, "What are the technological factors that influence organizational adoption and use of advanced IT, such as Building Information Modeling in their activities?" Applying the SLR method would help in mitigating the gaps of traditional narrative reviews of IT adoption literature, thereby limiting bias, reducing chance effects, enhancing the authority and legitimacy of the resulting evidence, and providing reliable results upon which decisions can be made for subsequent conclusions [22].

The SLR represents a precise and reproducible method knowledge gathering and integration. SLR can help in the identification of the existing gaps in the existing literature and in suggesting the possible areas for improvement in future [23].

The stages of a well-defined SLR protocol includes the following: review question formulation, identification of the pertinent studies, selection and evaluation of the identified studies, analysis and synthesis of the identified studies, presentation of the reports [22, 23].

3.1 Data Sources and Search Strategies

In this study, the reviewed articles were retrieved from two notable databases, including Scopus and Web of Science. The reason for selecting only these two databases was that they host high-quality publications from many good publishers and ranked journals. The first round of search in the selected databases was based on the strings: "Technology-organization-environment" OR "TOE framework" OR "Information system", OR "Information technology", AND "Building Information Modelling", AND "adoption" OR "implementation", "diffusion, of innovation". Unfortunately, no articles appeared in the search results because no studies had been conducted using TOE framework and BIM adoption. The second round of search required a change in the search string from AND "Building Information Modelling" to OR "Building Information Modelling", thereby becoming "Technology-organization-environment" OR "TOE framework" OR "Information system", OR "Information technology", OR "Building Information Modelling", AND "adoption" OR "implementation", "diffusion, of innovation".

This search yielded a total of 281 published empirical studies in the initial search but after removing the duplicated reports, a total of 122 was retained. After applying the criteria shown in Table 1, the search came up with 81 studies. The search for these articles was conducted between 2015 and October 2019.

3.2 Inclusion and Exclusion Criteria

The inclusion and exclusion criteria were defined based on sources, language, research questions, the theory used, type of publication, date, and research design. The scholars employed the inclusion and exclusion criteria to ensure the selection of only the relevant articles for the SLR process. Before selecting any of the articles for the analysis, it must be critically analyzed to ensure it meets the conditions as described in Table 1.

Table 1 Inclusion and exclusion criteria

Criteria	Inclusion criteria	Exclusion criteria
Sources	Journal articles or conference	Studies in other-than-English language
Language	Written in English	Studies that are irrelevant to the research questions
Research questions	Primary studies related to the research questions	Studies that are un-related to BIM/IT innovation adoption
Theory used	Studies that have reported the use of TOE framework to study IT adoption	Duplicate materials (i.e. same studies that resulted from the application of different search string or retrieved from different online databases)
Type of publication	Stated the technological factors	Master dissertations, books chapters, conference review, prefaces and opinions
Date	2015–October 2019	Before 2015
Research design	Empirical study	

3.3 Selecting and Evaluating Studies

Having sourced the articles, the selection and exclusion criteria were applied to remove all the unrelated articles to the studied concept. First, studies that are mainly theoretical or review were excluded while studies that focused on TOE framework to study IT adoption were selected for the study. Applying these criteria, we ended up with 78 studies. Figure 1 shows the SLR execution process.

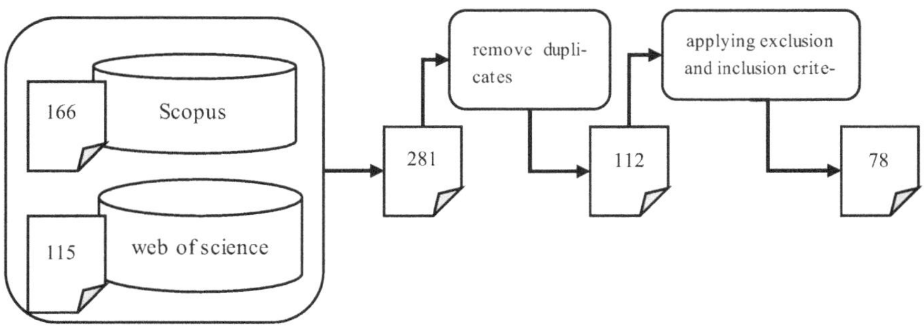

Fig. 1 SLR execution process

4 Results and Discussion

The current SLR examined 78 studies that involved the technological factors that influence organizational adoption and use of advanced IT. These articles were analyzed to answer the defined research question. To identify the technological factors that influence adoption and use of advanced IT, this study counted the frequency for each used factor and its significance in the collected studies. Table 2 showed the frequencies of the significant and insignificance factors that were derived from the analyzed studies.

Table 2 showed that several studies have employed several terminologies in describing the same factor. For example, the advantages of using advanced IT, such as Relative advantage [24, 25, 27–30, 32, 33, 60–63, 70], 'Perceived Usefulness' [54, 98], Perceived direct benefits [53, 93], 'Perceived Benefits' [31, 40, 57, 75, 79–86], and 'Perceived values' [56] have been described by several researchers. Other studies have employed the term 'complexity' in describing the perceived level of ambiguity in the understanding of the use of advanced IT in order to portray different terminologies [7, 34, 35, 39, 42, 44, 45, 76]. Terminologies such as 'Perceived Ease of use' have also been used [63, 98]. Prior IT innovation adoption studies also frequently found that Complexity (Perceived Ease of use) is an important technology characteristic that influences organizational adoption.

This SLR demonstrated that the specific characteristics of IT mentioned in DOI theory had been shown in the extracted factors, with Compatibility, Relative Advantage, Complexity, and Trialability being the most common factors in the literature. In general, the most frequent factors included in the technological context are Compatibility, Relative advantage (Perceived Benefits), Complexity, IT infrastructure, Security and Privacy concern, Financial costs, Trialability, Cost savings, Technology competence, and Technology readiness.

5 Conclusion and Future Work

The implementation of BIM can improve project performance and increase productivity, reduce wastes, improve business process flow, decrease uncertainties and complexities, minimize conflicts and fragmentations. The need for BIM adoption is becoming obvious for the AEC industry, especially for the integration of construction process, as well as in addressing problems in the building lifecycles. Although BIM processes require organization-wide adoption, few studies focused on the factors influencing BIM adoption at the organizational level, and very few studies have been carried out at the organizational level using the TOE framework for BIM adoption. Due to the absence of studies on BIM adoption, this SLR analyzed the current studies that involved the technological factors that influence organizational adoption and use of advanced IT and utilized those factors for the BIM adoption case. Therein, 78 up-to-date studies retrieved from two primary databases (Scopus and Web of

Table 2 Factors analysis derived from the analyzed studies

Technological factors	Significant	Frequency	Insignificant	Frequency	Total
Compatibility	[24–59]	37	[60–69]	10	47
Relative advantage	[24, 25, 27–30, 32–39, 41, 42, 44, 45, 48, 50, 52, 55, 58, 60–63, 65, 67, 69–73]	35	[26, 43, 46, 59, 64, 66, 71, 74]	8	43
Complexity	[7, 25, 27, 29–35, 39, 42–46, 48, 50, 51, 54, 59–62, 67, 73, 75–78]	30	[24, 26, 36, 38, 47, 53, 55, 58, 64, 69, 74]	11	41
Perceived benefits	[31, 40, 57, 75, 79–86]	12	[68, 87]	2	14
IT infrastructure	[28, 47, 49, 50, 57, 61, 74, 79, 81, 88]	11	[89, 90]	2	13
Security and privacy concern	[29, 32, 37, 49, 53, 63, 65, 70, 75, 91, 92]	11	[26, 66]	2	13
Financial costs	[37, 50, 53, 54, 63, 68, 93, 94]	8	[70, 78, 87, 88]	4	12
Trialability	[29, 30, 42, 51, 52, 54]	7	[59, 61, 63, 64, 73]	5	12
Cost savings	[7, 8, 32, 44, 52, 54, 91]	9	[71]	1	10
Technology competence	[24, 27, 37, 40, 54, 62, 72, 75, 85]	9	[95]	1	10
Technology readiness	[25, 41, 44, 65, 67, 96]	6	[26, 62, 97]	3	9
Observability	[29, 59, 61, 63]	4	[64]	1	5
Risks	[43, 83, 86]	3	[73]	1	4
Technical barriers	[26, 81, 82]	3	[47]	1	4
Reliability	[92, 94]	2	[54]	1	3
Perceived direct benefits	[78]	1	[53, 93]	2	3
Perceived usefulness	[98]	1	[54]	1	2

(continued)

Table 2 (continued)

Technological factors	Significant	Frequency	Insignificant	Frequency	Total
Perceived indirect benefits	[78]	1	[93]	1	2
Perceived ease of use		0	[63, 98]	2	2
Applicability to data management	[98]	1		0	1
Business concerns	[80]	1		0	1
Data quality and integration	[99]	1		0	1
IT support	[100]	1		0	1
IT effectiveness	[100]	1		0	1
Interoperability	[96]	1		0	1
Legal concern	[94]	1		0	1
Network externality	[58]	1		0	1
Organisational fit	[82]	1		0	1
Perceived availability	[101]	1		0	1
Perceived simplicity	[56]	1		0	1
Perceived values	[56]	1		0	1

(continued)

Table 2 (continued)

Technological factors	Significant	Frequency	Insignificant	Frequency	Total
Strategic flexibility	[54]	1		0	1
Systems quality	[101]	1		0	1
Technology integration	[90]	1		0	1
Technology maturity	[77]	1		0	1
Use of standards and platforms	[47]	1		0	1
Ubiquity	[53]	1		0	1
Uncertainty	[30]	1		0	1
Value creation	[26]	1		0	1
Data size		0	[73]	1	1
License concern		0	[94]	1	1
Industry-wide technology readiness	[96]	1		0	1

Science) were critically analyzed in the period between 2015 and October 2019. A total of 42 technological factors were extracted and the most common extracted technological factors were Compatibility, Relative advantage (Perceived Benefits), Complexity, IT infrastructure, Security and Privacy concern, Financial costs, Trialability, Cost savings, Technology competence, and Technology readiness. Identifying the technological factors that affect the adoption and use of advanced IT is crucial for the decision-makers in the AEC industry to adopt BIM properly and overcome the challenges of BIM implementation.

There are insufficient studies related to the adoption of BIM technology in the AEC industry. Therefore, the outcome of this study will contribute significantly to the existing literature on IT adoption and will be helpful to the decision-makers in the AEC industry during the formulation of the grand strategies for BIM technology adoption. This study also contributes by providing the most commonly extracted technological factors used in the previous works which can be used by experts and professionals in determining the most important factors affecting BIM adoption. Besides, the factors that will be chosen and validated by the experts will help in building future conceptual models for BIM adoption in the AEC industry.

References

1. World Economic Forum: Shaping the Future of Construction: A Breakthrough in Mindset and Technology (2016)
2. Murray, M.: Rethinking construction: the egan report (1998), pp. 178–195. Blackwell Science, Oxford, UK (2003)
3. Mitropoulos, P., Tatum, C.B.: Technology adoption decisions in construction organizations. J. Prof. Nurs. **30**(4), 292–299 (1999)
4. Lee, H.W., Oh, H., Kim, Y., Choi, K.: Quantitative analysis of warnings in building information modeling (BIM). Autom. Constr. **51**(C), 23–31 (2015)
5. Eastman, C.M.: BIM Handbook: A Guide to Building Information Modeling for Owners, Managers, Designers, Engineers and Contractors, vol. 12, no. 3 (2011)
6. Baharuddin, H.E.A., Othman, A.F., Adnan, H., Ismail, W.N.W.: BIM training: the impact on BIM adoption among quantity surveyors in government agencies. In: IOP Conference Series: Earth and Environmental Science, vol. 233, no. 2, p. 022036. IOP Publishing (2019)
7. Gerges, M., Austin, S., Mayouf, M., Ahiakwo, O., Jaeger, M., Saad, A.: An investigation into the implementation of building information modeling in the Middle East. J. Inf. Technol. Constr. **22**(2), 1–15 (2017)
8. NBS: International BIM Report 2016—The International Picture, p. 24. NBS (2016)
9. Howard, R., Restrepo, L., Chang, C.-Y.: Addressing individual perceptions: an application of the unified theory of acceptance and use of technology to building information modelling. Int. J. Proj. Manage. **35**(2), 107–120 (2017)
10. Kim, S., Park, C.H., Chin, S.: Assessment of BIM acceptance degree of Korean AEC participants. KSCE J. Civ. Eng. **20**(4), 1163–1177 (2016)
11. Acquah, R., Eyiah, A.K., Oteng, D.: Acceptance of building information modelling: a survey of professionals in the construction industry in Ghana. J. Inf. Technol. Constr. **23**, 75–91 (2018)
12. Xu, H., Feng, J., Li, S.: Users-orientated evaluation of building information model in the Chinese construction industry. Autom. Constr. **39**, 32–46 (2014)

13. Tsai, M.-H., Kang, S.-C., Hsieh, S.-H.: Lessons learnt from customization of a BIM tool for a design-build company. J. Chin. Inst. Eng. **37**(2), 189–199 (2014)
14. Davis, F.D.: A technology acceptance model for empirically testing new end-user information systems: theory and results (1986)
15. Davis, F.D.: Perceived usefulness, perceived ease of use, and user acceptance of information technology. MIS Q. **13**(3), 319 (1989)
16. Schifter, D.E., Ajzen, I.: Intention, perceived control, and weight loss: an application of the theory of planned behavior. J. Pers. Soc. Psychol. **49**(3), 843 (1985)
17. Ajzen, I.: The theory of planned behavior. Organ. Behav. Hum. Decis. Process. **50**(2), 179–211 (1991)
18. Venkatesh, V., Morris, M.G., Davis, G.B., Davis, F.D.: User acceptance of information technology: toward a unified view. MIS Q. 425–478 (2003)
19. Rogers, E.M.: Diffusion of Innovations. The Free Press, New York (1995)
20. Wu, Y.W., Hsu, I.T., Lin, H.Y.: Using TAM to explore vocational students' willingness to adopt a web-based BIM cost estimating system. Adv. Mater. Res. **1079**(1080), 1098–1102 (2015)
21. Batarseh, S., Kamardeen, I.: The impact of individual beliefs and expectations on BIM adoption in the AEC industry, vol. 1, pp. 466–475 (2017)
22. Al-Emran, M., Mezhuyev, V., Kamaludin, A.: Technology acceptance model in m-learning context: a systematic review. Comput. Educ. **125**, 389–412 (2018)
23. Al-Emran, M., Mezhuyev, V., Kamaludin, A., Shaalan, K.: The impact of knowledge management processes on information systems: a systematic review. Int. J. Inf. Manage. **43**, 173–187 (2018)
24. Junior, C.H., Oliveira, T., Yanaze, M.: The adoption stages (evaluation, adoption, and routinisation) of ERP systems with business analytics functionality in the context of farms. Comput. Electron. Agric. **156**, 334–348 (2019)
25. Bhuyan, S., Dash, M.: Exploring cloud computing adoption in private hospitals in India: an investigation of DOI and TOE model. J. Adv. Res. Dyn. Control Syst. **10**(8), 443–451 (2018)
26. AL-Shboul, M.A.: Towards better understanding of determinants logistical factors in SMEs for cloud ERP adoption in developing economies. Bus. Process Manag. J. **25**(5), 889–907 (2018). https://doi.org/10.1108/BPMJ-01-2018-0004
27. Martins, R., Oliveira, T., Thomas, M.A.: An empirical analysis to assess the determinants of SaaS diffusion in firms. Comput. Hum. Behav. **62**, 19–33 (2016)
28. Yang, Z., Sun, J., Zhang, Y., Wang, Y.: Understanding SaaS adoption from the perspective of organizational users: a tripod readiness model. Comput. Hum. Behav. **45**, 254–264 (2015)
29. Safari, F., et al.: The adoption of software-as-a-service (SaaS): ranking the determinants. J. Enterp. Inf. Manage. **28**(3), 400–422 (2015)
30. Alshamaila, Y., Papagiannidis, S., Li, F.: Cloud computing adoption by SMEs in the north east of England: a multi-perspective framework. J. Enterp. Inf. Manage. **26**(3), 250–275 (2013)
31. Rosli, K., Yeow, P.H.P., Siew, E.-G.: Adoption of audit technology among audit firms. In: 24th Australasian Conference on Information Systems (ACIS) (2013)
32. Chong, A.Y.L., Chan, F.T.S.: Structural equation modeling for multi-stage analysis on Radio Frequency Identification (RFID) diffusion in the health care industry. Expert Syst. Appl. **39**(10), 8645–8654 (2012)
33. Henderson, D., Sheetz, S.D., Trinkle, B.S.: The determinants of inter-organizational and internal in-house adoption of XBRL: a structural equation model. Int. J. Account. Inf. Syst. **13**(2), 109–140 (2012)
34. Ifinedo, P.: An empirical analysis of factors influencing internet/e-business technologies adoption by SMEs in Canada. Int. J. Inf. Technol. Decis. Making **10**(04), 731–766 (2011)
35. Wang, Y.M., Wang, Y.S., Yang, Y.F.: Understanding the determinants of RFID adoption in the manufacturing industry. Technol. Forecast. Soc. Change **77**(5), 803–815 (2010)
36. Doolin, B., Al Haj Ali, E.: Adoption of mobile technology in the supply chain. Int. J. eB. Res. **4**(4), 1–15 (2008)

37. Zhu, K., Dong, S., Xu, S.X., Kraemer, K.L.: Innovation diffusion in global contexts: determinants of post-adoption digital transformation of European companies. Eur. J. Inf. Syst. **15**(6), 601–616 (2006)
38. Hassan, H., Tretiakov, A., Whiddett, D.: Factors affecting the breadth and depth of e-procurement use in small and medium enterprises. J. Organ. Comput. Electron. Commer. **27**(4), 304–324 (2017)
39. Azmi, A., Sapiei, N.S., Mustapha, M.Z., Abdullah, M.: SMEs' tax compliance costs and IT adoption: the case of a value-added tax. Int. J. Account. Inf. Syst. **23**, 1–13 (2016)
40. Wang, Y.M., Wang, Y.C.: Determinants of firms' knowledge management system implementation: an empirical study. Comput. Hum. Behav. **64**, 829–842 (2016)
41. Alharbi, F., Atkins, A., Stanier, C.: Understanding the determinants of cloud computing adoption in Saudi healthcare organisations. Complex Intell. Syst. **2**(3), 155–171 (2016)
42. Gangwar, H., Date, H., Ramaswamy, R.: Developing a cloud-computing adoption framework. Glob. Bus. Rev. **16**(4), 632–651 (2015)
43. Van Huy, L., Rowe, F., Truex, D., Huynh, M.Q.: An empirical study of determinants of e-commerce adoption in SMEs in Vietnam. J. Glob. Inf. Manage. **20**(3), 23–54 (2012)
44. Haberli, C., Oliveira, T., Yanaze, M.: Understanding the determinants of adoption of enterprise resource planning (ERP) technology within the agrifood context: the case of the Midwest of Brazil. Int. Food Agribusiness Manage. Rev. **20**(5), 729–746 (2017)
45. Ali, O., Soar, J., Yong, J., McClymont, H., Angus, D.: Collaborative cloud computing adoption in Australian regional municipal government: an exploratory study. In: 2015 IEEE 19th International Conference on Computer Supported Cooperative Work in Design (CSCWD), pp. 540–548. IEEE (2015)
46. Agrawal, K.P.: Investigating the determinants of Big Data Analytics (BDA) adoption in emerging economies. Acad. Manage. Proc. **2015**(1), 11290 (2016)
47. MacLennan, E., Van Belle, J.P.: Factors affecting the organizational adoption of service-oriented architecture (SOA). Inf. Syst. eB. Manage. **12**(1), 71–100 (2014)
48. Hwang, B.N., Huang, C.Y., Wu, C.H.: A TOE approach to establish a green supply chain adoption decision model in the semiconductor industry. Sustainability **8**(2), 168 (2016)
49. Awa, H.O., Ojiabo, O.U.: A model of adoption determinants of ERP within T-O-E framework. Inf. Technol. People **29**(4), 901–930 (2016)
50. Alam, M.G.R., Masum, A.K.M., Beh, L.S., Hong, C.S.: Critical factors influencing decision to adopt human resource information system (HRIS) in hospitals. PLoS One **11**(8) (2016)
51. Ahuja, R., Jain, M., Sawhney, A., Arif, M.: Adoption of BIM by architectural firms in India: technology–organization–environment perspective. Archit. Eng. Des. Manage. **12**(4), 311–330 (2016)
52. Al Isma'ili, S., Li, M., Shen, J., He, Q.: Cloud computing adoption determinants: an analysis of Australian SMEs. In: Pacific Asia Conference on Information Systems 2016 Proceedings, pp. 1–17 (2016)
53. Lai, H.M., Lin, I., Tseng, L.T.: High-Level Managers' Considerations for RFID Adoption in Hospitals: An Empirical Study in Taiwan. J Med Syst **38**(2), 1–17 (2014)
54. Chauhan, S., Jaiswal, M., Rai, S., Motiwalla, L., Pipino, L.: Determinants of adoption for open-source office applications: a plural investigation. Inf. Syst. Manage. **35**(2), 80–97 (2018)
55. Gangwar, H.: Understanding the determinants of big data adoption in India. Inf. Resour. Manage. J. **31**(4), 1–22 (2018)
56. Awa, H.O., Ojiabo, O.U., Orokor, L.E.: Integrated technology-organization-environment (T-O-E) taxonomies for technology adoption. J. Enterp. Inf. Manage. **30**(6), 893–921 (2017)
57. Lin, H.F., Lin, S.M.: Determinants of e-business diffusion: a test of the technology diffusion perspective. Technovation **28**(3), 135–145 (2008)
58. Zhai, C.: Research on post-adoption behavior of B2B e-marketplace in China. In: 2010 International Conference on Management and Service Science, MASS 2010, no. 1 (2010)
59. Mangula, I.S., Van De Weerd, I., Brinkkemper, S.: The adoption of software-as-a-service: an Indonesian case study. In: Proceedings—Pacific Asia Conference on Information Systems, PACIS 2014 (2014)

60. Chen,Y.,Yin,Y., Browne, G.J., Li, D.: Adoption of building informationmodeling in Chinese construction industry: The technology organization environment framework. Eng. Constr. Archit. Manage. **26**(9), 1878–1898 (2019)
61. AlBar, A.M., Hoque, M.R.: Factors affecting cloud ERP adoption in Saudi Arabia: an empirical study. Inf. Dev. **35**(1), 150–164 (2019)
62. Khan, M.J., Mahmood, S.: Assessing the determinants of adopting component-based development in a global context: a client-vendor analysis. IEEE Access **6**, 79060–79073 (2018)
63. Hsu, C.L., Lin, J.C.C.: Factors affecting the adoption of cloud services in enterprises. Inf. Syst. eB. Manage. **14**(4), 791–822 (2016)
64. Simamora, B.H., Sarmedy, J.: Improving services through adoption of cloud computing at PT XYZ in Indonesia. J. Theor. Appl. Inf. Technol. **73**(3), 395–404 (2015)
65. Senyo, P.K., Effah, J., Addae, E.: Preliminary insight into cloud computing adoption in a developing country. J. Enterp. Inf. Manage. **29**(4), 505–524 (2016)
66. Yoon, T.E., George, J.F.: Why aren't organizations adopting virtual worlds? Comput. Hum. Behav. **29**(3), 772–790 (2013)
67. Tarhini, A., Al-Gharbi, K., Al-Badi, A., AlHinai, Y.S.: An analysis of the factors affecting the adoption of cloud computing in higher educational institutions. Int. J. Cloud Appl. Comput. **8**(4), 49–71 (2018)
68. Ajjan, H., Kumar, R.L., Subramaniam, C.: Understanding differences between adopters and nonadopters of information technology project portfolio management. Int. J. Inf. Technol. Decis. Making **12**(06), 1151–1174 (2013)
69. Xu, W., Ou, P., Fan, W.: Antecedents of ERP assimilation and its impact on ERP value: a TOE-based model and empirical test. Inf. Syst. Front. **19**(1), 13–30 (2017)
70. Ilin, V., Ivetić, J., Simić, D.: Understanding the determinants of e-business adoption in ERP-enabled firms and non-ERP-enabled firms: a case study of the Western Balkan Peninsula. Technol. Forecast. Soc. Change **125**, 206–223 (2017)
71. Puklavec, B., Oliveira, T., Popovič, A.: Understanding the determinants of business intelligence system adoption stages an empirical study of SMEs. Ind. Manage. Data Syst. **118**(1), 236–261 (2018)
72. Chandra, S., Kumar, K.N.K.N., Road, H., Kumar, K.N.K.N., Road, H.: Exploring factors influencing organizational adoption of augmented reality in e-commerce: empirical analysis using technology–organization–environment model. J. Electron. Commer. Res. **19**(3), 237–265 (2018)
73. Alkhalil, A., Sahandi, R., John, D.: An exploration of the determinants for decision to migrate existing resources to cloud computing using an integrated TOE-DOI model. J. Cloud Comput. **6**(1) (2017)
74. Wei, J., Lowry, P.B., Seedorf, S.: The assimilation of RFID technology by Chinese companies: a technology diffusion perspective. Inf. Manage. **52**(6), 628–642 (2015)
75. Chana, F.T.S., Chong, A.Y.L.: Determinants of mobile supply chain management system diffusion: a structural equation analysis of manufacturing firms. Int. J. Prod. Res. **51**(4), 1196–1213 (2013)
76. Sila, I., Dobni, D.: Patterns of B2B e-commerce usage in SMEs. Ind. Manage. Data Syst. **112**(8), 1255–1271 (2012)
77. Wu, X., Subramaniam, C.: Understanding and predicting radio frequency identification (RFID) adoption in supply chains. J. Organ. Comput. Electron. Commer. **21**(4), 348–367 (2011)
78. Rouhani, S., Ashrafi, A., Ravasan, A.Z., Afshari, S.: Business intelligence systems adoption model. J. Organ. End User Comput. **30**(2), 43–70 (2018)
79. Ammar, A., Ahmed, E.M.: Factors influencing Sudanese microfinance intention to adopt mobile banking. Cogent Bus. Manage. **3**(1), 1–20 (2016)
80. Hsu, P.F., Ray, S., Li-Hsieh, Y.Y.: Examining cloud computing adoption intention, pricing mechanism, and deployment model. Int. J. Inf. Manage. **34**(4), 474–488 (2014)

81. Cao, Q., Baker, J., Wetherbe, J., Gu, V.: Organizational adoption of innovation: identifying factors that influence RFID adoption in the healthcare industry. In: European Conference on Information Systems 2012, pp. 5–15 (2012)
82. Troshani, I., Rampersad, G., Plewa, C.: Organisational adoption of e-business: the case of an innovation management tool at a university and technology transfer office. Int. J. Netw. Virtual Organ. **9**(3), 265 (2011)
83. Cao, Y., Ajjan, H., Hong, P., Le, T.: Using social media for competitive business outcomes: an empirical study of companies in China. J. Adv. Manage. Res. **15**(2), 211–235 (2018)
84. Ifinedo, P.: Internet/e-business technologies acceptance in Canada's SMEs: an exploratory investigation. Internet Res. **21**(3), 255–281 (2011)
85. Zhang, H., Xiao, J.: Assimilation of social media in local government: an examination of key drivers. Electron. Libr. **35**(3), 427–444 (2017)
86. Shim, S., Lee, B., Kim, S.L.: Rival precedence and open platform adoption: an empirical analysis. Int. J. Inf. Manage. **38**(1), 217–231 (2018)
87. Lin, H.F.: Understanding the determinants of electronic supply chain management system adoption: using the technology-organization-environment framework. Technol. Forecast. Soc. Change **86**, 80–92 (2014)
88. Maditinos, D., Chatzoudes, D., Sarigiannidis, L.: Factors affecting e-business successful implementation. Int. J. Commer. Manage. **24**(4), 300–320 (2016)
89. Rondović, B., Djuričković, T., Kašćelan, L.: Drivers of e-business diffusion in tourism: a decision tree approach. J. Theor. Appl. Electron. Commer. Res. **14**(1), 30–50 (2019)
90. Wolf, M., Beck, R., König, W.: Environmental dynamics as driver of on-demand computing infrastructures—empirical insights from the financial services industry in UK. ECIS 1–14 (2012)
91. Ali, O., Soar, J., Shrestha, A.: Perceived potential for value creation from cloud computing: a study of the Australian regional government sector. Behav. Inf. Technol. **37**(12), 1157–1176 (2018)
92. Sulaiman, H., Magaireh, A., Ramli, R.: Adoption of cloud-based e-health record through the technology, organization and environment perspective. Int. J. Eng. Technol. **7**(4.35), 609 (2018)
93. Nam, D.W., Kang, D.W., Kim, S.: Process of big data analysis adoption: defining big data as a new IS innovation and examining factors affecting the process. In: Proceedings of Annual Hawaii International Conference on System Sciences, pp. 4792–4801. IEEE (2015)
94. Ramanathan, L., Krishnan, S.: An empirical investigation into the adoption of open source software in information technology outsourcing organizations. J. Syst. Inf. Technol. **17**(2), 167–192 (2015)
95. Martins, R., Oliveira, T., Thomas, M., Tomás, S.: Firms' continuance intention on SaaS use—an empirical study. Inf. Technol. People **32**(1), 189–216 (2019)
96. Hossain, M.A., Standing, C., Chan, C.: The development and validation of a two-staged adoption model of RFID technology in livestock businesses. Inf. Technol. People **30**(4), 785–808 (2017)
97. Indriasari, E., Wayan, S., Gaol, F.L.: Intelligent Information and Database Systems, vol. 7803. Springer International Publishing, Cham (2013)
98. Kim, D.J., Hebeler, J., Yoon, V., Davis, F.: Exploring determinants of semantic web technology adoption from IT professionals' perspective: industry competition, organization innovativeness, and data management capability. Comput. Hum. Behav. **86**, 18–33 (2018)
99. Cruz-Jesus, F., Pinheiro, A., Oliveira, T.: Understanding CRM adoption stages: empirical analysis building on the TOE framework. Comput. Ind. **109**, 1–13 (2019)
100. Lin, H.F.: Contextual factors affecting knowledge management diffusion in SMEs. Ind. Manage. Data Syst. **114**(9), 1415–1437 (2014)
101. Schwarz, C., Schwarz, A.: To adopt or not to adopt: a perception-based model of the EMR technology adoption decision utilizing the technology-organization-environment framework. J. Organ. End User Comput. **26**(4), 57–79 (2014)

A Study on Software Testing Standard Using ISO/IEC/IEEE 29119-2: 2013

Cristiano Patrício, Rui Pinto, and Gonçalo Marques

Abstract ISO/IEC/IEEE 29119 is an internationally agreed set of standards for software testing that must be adopted during all software development processes and incorporated by every software company when conducting every software testing development. This article presents a study on ISO/IEC/IEEE 29119 standard, focusing on the description of a three-layer process model that covers: (1) organizational test specifications; (2) test management and (3) dynamic testing. Therefore, a comprehensive survey has been done to summarise and analyse the intended purpose of the implementation and adoption of this software testing standard. Furthermore, this paper also states the added value related to the implementation of this standard for any organisation in terms of software quality assurance. The adoption of software testing processes is an essential part of the software development that must be included in every development project, in particular, considering the artificial intelligence software development. The costs associated are relatively high. However, the cost associated with the treatment of bugs and software problems is even higher.

Keywords ISO/IEC/IEEE 29119-2 · ISO standard · Software testing

1 Introduction

Software testing is a crucial component in the development cycle done right. The demand for a standard and a baseline for testing discipline motivated the release of new standards of software testing. The IEEE 29119-2:2013 presents a group of globally agreed standards for software testing that can be used by any software developer

C. Patrício · R. Pinto
Universidade da Beira Interior, 6201-001 Covilhã, Portugal
e-mail: cristiano.ppatricio@gmail.com

R. Pinto
e-mail: srpruipedro@gmail.com

G. Marques (✉)
Instituto de Telecomunicações, Universidade da Beira Interior, 6201-001 Covilhã, Portugal
e-mail: goncalosantosmarques@gmail.com

M. Al-Emran et al. (eds.), *Recent Advances in Intelligent Systems and Smart Applications*, Studies in Systems, Decision and Control 295,
https://doi.org/10.1007/978-3-030-47411-9_3

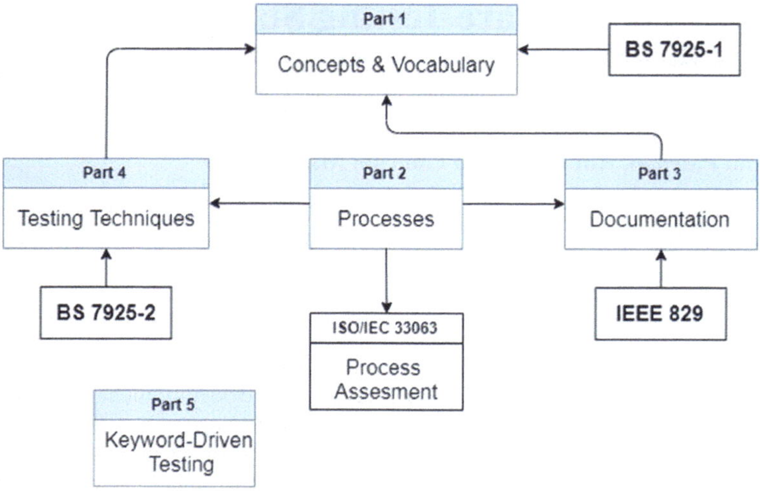

Fig. 1 ISO/IEC 29119 structure

or software testing organization within any step of the software development [1]. As illustrated in Fig. 1, the standard is structured in five parts in conjunction with glossaries and other criteria.

This document presents a comprehensive survey of the second part of the ISO/IEC/IEEE 29119, Testing Processes. The model describes the test processes for management, government and implementation of software testing actions. This standard specifies a process model with three-layer that includes, organizational test specifications, test management and dynamic Testing.

The successive technological improvements in numerous domains of computer science, in particular with artificial intelligence (AI), Internet of Things and Big Data, software testing assumes an even more critical role for the efficient and useful application of these systems. These technologies and related research fields influence people's daily routine and offer numerous opportunities for the development of smart or intelligent systems in many applications domains.

The main contribution of this paper is to present a comprehensive review to summarise and analyse the purpose of the implementation and adoption of this software testing standard. The main objective is to describe the elements and concepts associated with the ISO/IEC/IEEE 29119 standard.

This study aims to answer the following research questions: (RQ1) What is the ISO/IEC/IEEE 29119 standard, and which practices include regarding software testing? (RQ2) Which companies or groups must incorporate this standard? (RQ3) Which are the limitations associated with the implementation of this standard? (RQ4) Which are the outcomes of the application of this software testing method? (RQ5) Which advantages this standard brings for software developers? and (RQ6) The effort in terms of time and cost of adopting this kind of criteria for testing are compensatory?

This paper is organized as follows: Sect. 2 presents the related work and Sect. 3 presents the test process model. The tree-layer process model: (1) organizational test process; (2) test management process and (3) dynamic testing process are described and analyzed in Sects. 4–6, respectively. Finally, the conclusions are discussed in Sect. 7.

2 Related Work

Several studies on software testing are available in the literature. This section presents some research initiatives on software testing. According to the authors of [2], software testing methods can be divided into white-box testing and black-box testing. Black-box methods are used without information on the software source code, and their tests are based on the software's and user's requirements or specifications. These methods include boundary value analysis, equivalence partitioning, all-pairs testing, and state transition tables. White-box methods can be applied with knowledge of the source code, and their tests are based on the software's original architecture or code.

The authors of [3] state that software testing is a crucial activity of the software development life cycle as well as the testing never ends. Testing aims to discover errors in software using innovative techniques and methodologies which are necessary to ensure software quality. In addition, it is also introduced a new type of software testing method technique denominated by grey-box testing, which is a mix of black-box and white-box testing on which there is limited knowledge of the internal architecture (data structure and algorithms) and design of the application.

The authors of [4] compare manual and automated testing. In this study, the authors state the difference between these two types of testing is that one is done manually by humans, and the other is performed by testing tools. Moreover, the authors of [4] state that manual testing cannot be accurate due to human errors. Therefore, automated testing is more reliable and less time-consuming.

The authors of [5] discuss the benefits and drawbacks of 17 software process models. The review shows that the adoption of a testing model has advantages and disadvantages in specific circumstances and enterprise contexts. Some of the existing models are identified as generally applicable, whereas others are customized to satisfy a target domain such as automated testing or military systems. However, existing many testing models with a variety of characteristics, it is essential to choose the appropriate based on the required goals.

Despite being mentioned in [4] that automated testing is reliable, it is not perfect, since the authors of [6] state that a testing tool can produce fail negatives and positives when we are talking of accessibility web applications. These failures need to be reduced by using a combination of results provided by these tools to mitigate this scenario. The proposed method for testing these types of applications includes a model proposed by the International Software Testing Qualifications Board, which serves as a base for the approach proposed by the authors of [6]. In this study, the test planning, test analysis and design, test implementation and execution, evaluating exit criteria and test incident reporting is presented by the authors.

The comparison between three software testing strategies, such as unit testing, integration testing, and system testing, is presented in [7]. On the one hand, in-unit testing, individual units of software are tested. On the other hand, in integration testing, particular units are combined and tested as a group. System testing is referred to check the system as a whole; that is when the system becomes complete and meets the specified requirements.

The importance of having the most significant interaction between who designs the specifications and testers is underlined presented by the authors of [8]. The authors advice to provide an early review of the requirements that means saving cost and time when exists problems in software to fix. The testers should be clear when they write the specifications and provide a specific type of test model to developers, so they make sure that the original specifications are met before the project is sent to the official testing.

3 Test Process Model

The testing model is shown in Fig. 2. This model is based on a primary approach where processes related to testing are in the center of the model surrounded with more three entities: (1) concepts and definitions; (2) techniques, and (3) documentation.

The documentation entity is the closing outcome of process execution in test cases that provide a report of the output. The methods come from the request analysis and are used to plan test cases [9]. The concepts and definitions are related to other terminologies used by other models. The standard test process model illustrated in Fig. 3 and is a multi-layered model with a top-level named "organizational test process", a middle level named "test management processes", and a bottom level labelled as "dynamic test processes".

Fig. 2 Testing model

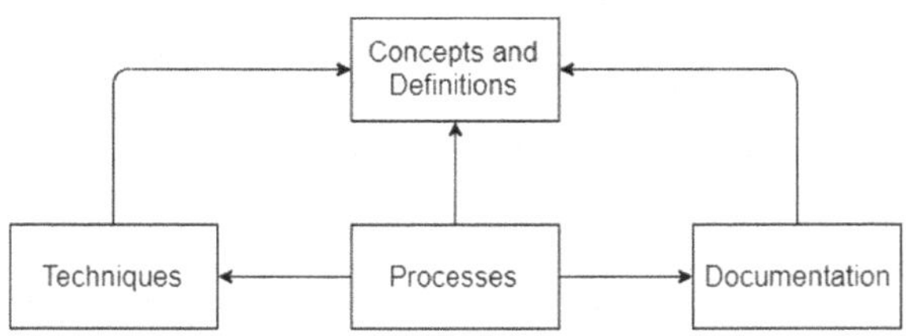

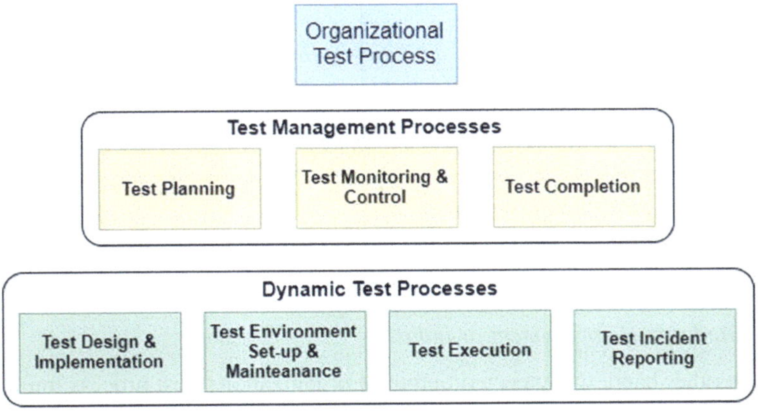

Fig. 3 Test process model detailed view

4 Organizational Test Process

On the top-level of the pyramid, is the organizational test process, developed to control and administer the organizational test specifications [10]. The organizational test strategy, as well as organizational test policy, are typical specifications situated at this level. As illustrated in Fig. 4, the organizational test strategy and organizational test policy correspond between them and with the test management process.

The purpose of the organizational test process is to design, maintain and manage the test requirements: the organizational test strategy and organizational test policy [11]. Below, the examples of inputs to activities in this process:

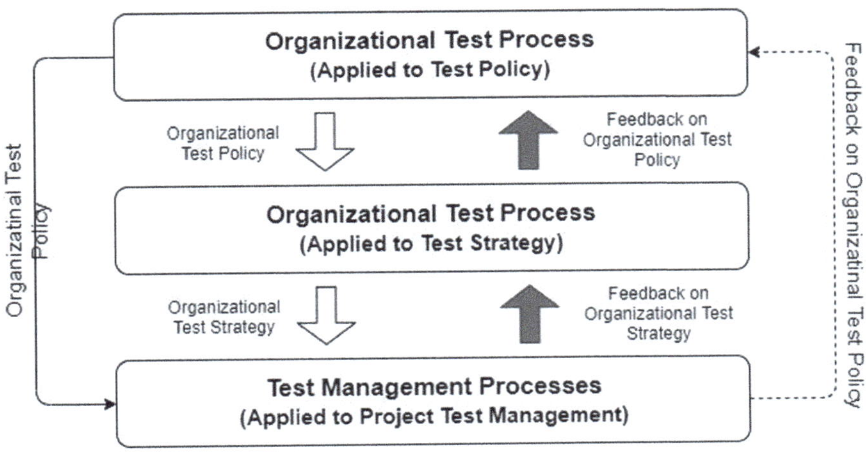

Fig. 4 Communication between the test management processes and organizational test processes

- *The insights of stakeholders*;
- *The knowledge of the actual test activities in the company*;
- *The company mission and visualization*;
- *The information technology policy*;
- *The information technology project management policy*;
- *The quality policy*;
- *The feedback on test requirements*;
- *The current organizational test policy*;
- *The current organizational test strategy*;
- *The samples of test plans from the company*;
- *The standards of government or industry*.

On the other hand, the expected output of organizational test process implementation includes:

- *The identified requirements for the organizational test specifications*;
- *The developed organizational test specifications*;
- *The agreed organizational test specifications*;
- *The availability of organizational test specifications*;
- *The coherence to the organizational test specifications*;
- *The updates decided by stakeholders and organizational test requirements*;
- *The creation of updates to the organizational test requirements*.

4.1 Graphic Overview of the Organizational Test Process

The organizational test process can be graphically explained, as illustrated in Fig. 5 [12]. The diagram shows three activities of the organizational test process and the relationships between them:

- Develop organizational test specification (OT1)
- Monitor and control of organizational test specification (OT2)
- Update organizational test specification (OT3).

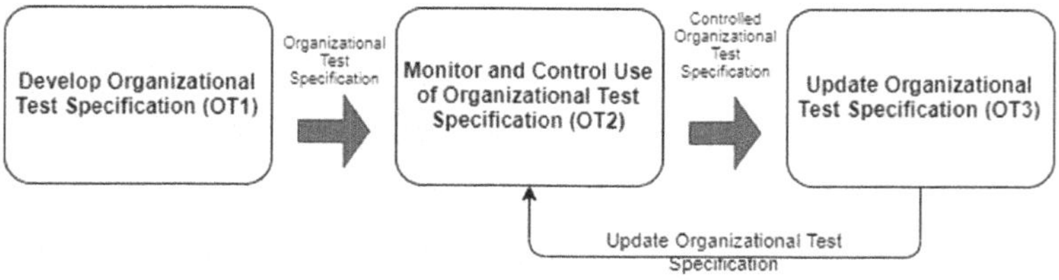

Fig. 5 Graphic schema of the organizational test process

(OT1): The main tasks involved in this activity are related to:

1. Identify the needs for the organizational test requirements from stakeholders or by other parts and from the testing practices established in the organization;
2. Use the requirements of organizational test requirements to create the organizational test requirements;
3. Report if the organizational test requirements are available to the involved stakeholders.

(OT2): This activity aims to determine if the organizational test specification is being used effectively within the organization. It is also essential to take decisions to align the stakeholders with the organizational test specification.

(OT3): In the last activity, the reaction on the use of the organizational test requirements should be reviewed. The management and the use of the organizational test specification shall be taken into consideration, as well as any changes and feedback to improve their effectiveness. Where changes are identified and approved in the organizational test specification, these changes must be implemented and reported across the organization, including all stakeholders.

5 Test Management Processes

As illustrated in Fig. 3, there are three types of test management processes:

(1) Test planning processes;
(2) Test monitoring and control processes;
(3) Test completion processes.

These test management processes are general and broad processes making it possible for them to be used and applied during different levels of testing while using different methods of testing [13, 14].

Furthermore, the test management processes can be applied to various test types gather from the product quality model stated in *ISO/IEC 25010* that includes eight quality features: (1) functional suitability, (2) performance efficiency; (3) compatibility; (4) usability; (5) reliability; (6) security; (7) maintainability and (8) portability. Figure 6 illustrates the relationship between the organizational test process, the dynamic test processes and other applications of the test management processes [15, 16].

This process requires an organizational test strategy and an organizational test policy produced by the organization test process, and also the test management processes need to provide feedback on organizational test policy and strategy [17].

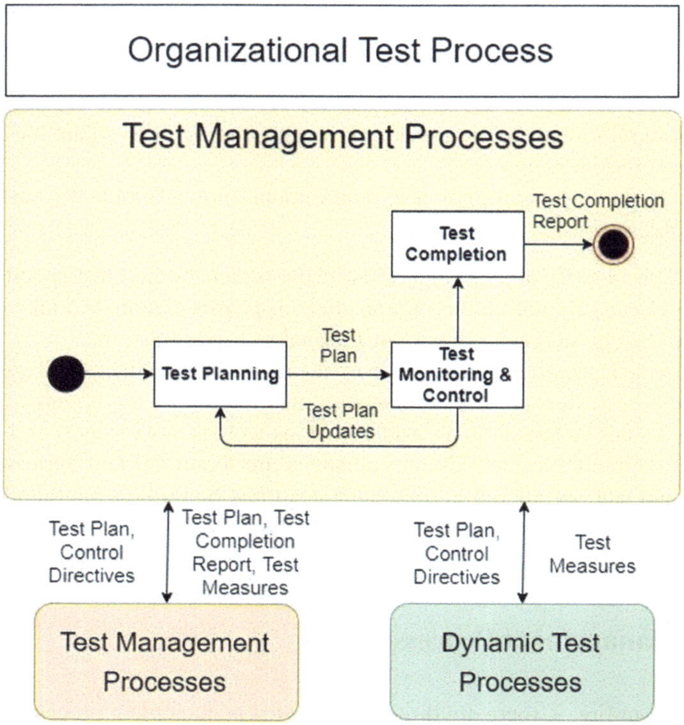

Fig. 6 Relationship between the organizational test process, test management process and dynamic test processes

5.1 *Test Planning Processes*

The test planning process is a process used to create a test plan. The creation of a test plan is a cyclical process, as illustrated in Fig. 7, and some of the activities might need to be re-done to make a final version of a test plan. Table 1 shows the list of tasks necessary to complete in each activity from the test planning process.

5.2 *Test Monitoring and Control Process*

The test monitoring and control processes are used to verify and validate if the testing being done is following the test plan from the test planning process and the organizational specifications [18, 19]. This process can be applied during different test levels, but usually, this process is referred to be able to manage the entire test project [20, 21]. The graphic overview of a test monitoring and control process is illustrated in Fig. 8. In practice, some of the activities represented in Fig. 8 are done

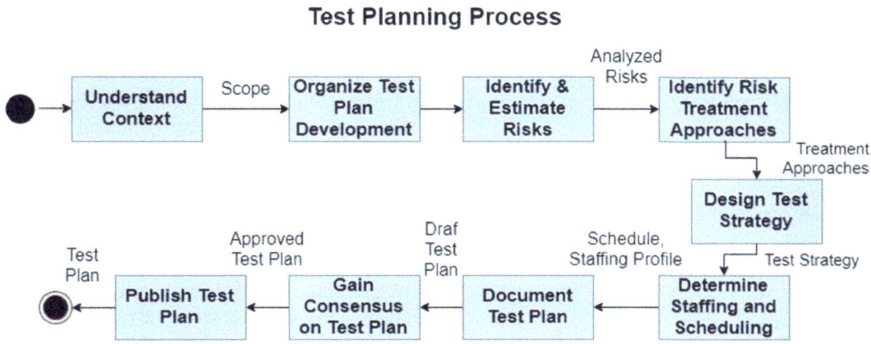

Fig. 7 Test planning process

in an iterative manner, making it possible to revisit some events more than once. Table 2 gives details about the various tasks in need of completion for each activity.

5.3 *Test Completion Processes*

The test completion process is the last process of the test management processes [22]. This process archives the test assets for a future purpose, cleans up the test environment, records the test results and reports the test completion results to the stakeholders [23, 24]. These activities are all illustrated in Fig. 9. Table 3 gives further details about the various tasks in need of completion for each action.

6 Dynamic Test Processes

As illustrated in Fig. 10, the dynamic testing process includes four different activities:

- Test design and implementation;
- Test environment set-up and maintenance;
- Test execution;
- Test incident reporting.

Some activities of the dynamic testing process, in practice, might need to be revisited multiple times [25]. This process has various iterations that will show if the completion criteria defined in the test process plan have not been met. The dynamic test process will receive the test plan, and the control directives from the test management processes and will send test measures for the test supervision, and control monitors test progress.

Table 1 Tasks and activities from the test planning process

Activity	Tasks
Understand context	(a) Understand the context and the software testing requirements by identifying and interacting with the relevant stakeholders (b) Initiate a communication plan
Organize test plan development	(a) Identify and schedule the activities that need to be performed to complete test planning (b) Identify the stakeholders required (c) Obtain the approval from the relevant stakeholders of the activities, schedules, and participants (d) Organize stakeholder involvement
Identify and analyse risks	(a) Review previously identified risks (b) Identify additional risks (c) Classify risks (d) Assign levels of exposure to each risk (e) Obtain approval for the results of the risk assessments from stakeholders (f) Record the results of the risk assessment
Identify risk mitigation approaches	(a) Identify the appropriate means of treating the risk (b) Record the results of the risk mitigation
Design test strategy	(a) Produce initial estimate of resources required, defined by the organizational test strategy and policy (b) Produce an initial assessment of the resources needed to perform the individual mitigation actions (c) Design a test strategy (d) Recognize metrics to be used for test monitoring and control (e) Recognize test information (f) Identify test environment conditions and test tool requirements (g) Identify test deliverables (h) Produce an initial estimate of the required resources to perform the complete set of activities from the test strategy (i) Record the test strategy (j) Obtain the approval on the test strategy from the stakeholders
Determine staffing and scheduling	(a) Identify the roles and skills of staff (b) Schedule each required test activity (c) Obtain approval on staffing scheduling from relevant stakeholders

(continued)

Table 1 (continued)

Activity	Tasks
Record test plan	(a) Calculate final estimates for the testing (b) Incorporate the test strategy, staffing profile, and schedule in the test plan
Gain consensus on the test plan	(a) Gather the views of the stakeholders on the test plan (b) Resolve conflicts between the test plan and stakeholders' views (c) Update test plan to take into account the feedback from stakeholders (d) Obtain approval on the test plan from the stakeholders
Make the test plan available and communicate	(a) Make available the test plan (b) Communicate to the stakeholders the availability of the test plan

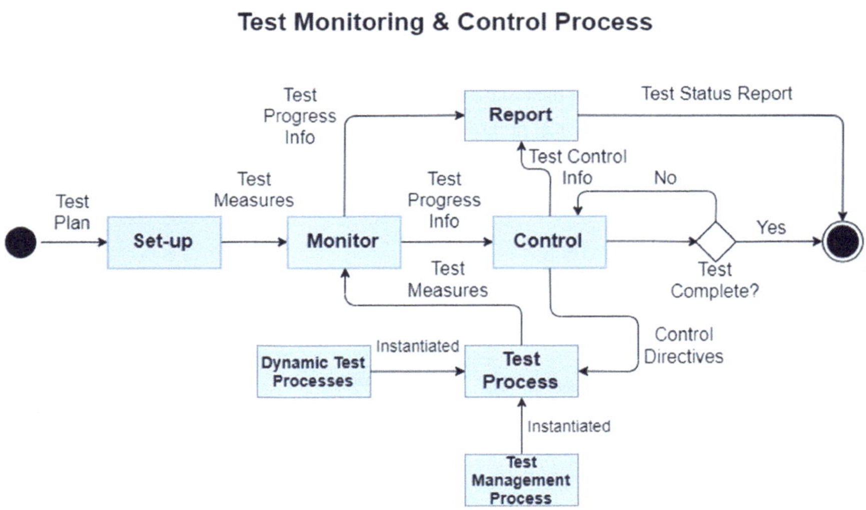

Fig. 8 Test control and monitoring process

6.1 *Test Design and Implementation Process*

Test design and implementation processes are used to document test conditions, test cases and test procedures for execution in the test execution process [12]. This method is iterative because it usually is revisited various times throughout the test project.

Table 2 Tasks and activities from the test monitoring and control process

Activity	Tasks
Set-up	(a) Identify a suitable measure for monitoring progress (b) Identify appropriate means for identifying new and changing risks (c) Start monitoring activities (test status reporting and test metrics collection)
Monitor	(a) Record and collect test measures (b) Start monitoring the progress against the test plan (c) Identify and record divergences from planned testing activities (d) Identify and analyze new risks (e) Start monitoring and identifying the changes to know risks
Control	(a) Perform the necessary actions to implement the test plan (b) Perform the steps required to achieve control directives from higher-level management processes (c) Identify the actions needed to manage the divergence of planned testing to actual testing (d) Identify means of treating for newly identified and changed risks (e) Issue control directives to change the way measurement is performed, any change to the test plan should be in the form of test plan updates, communicate recommended changes to the relevant stakeholders (f) Establish each selected test action before starting that exercise (g) Obtain approval for completion of assigned test activities (h) Obtain consent for the test completion when the test has met is completion criteria
Report	(a) Communicate any test progress against the test plan to stakeholders in a test status report during the reporting period (b) Update the new risks and changes to existing uncertainty in the risk register and communicate them with the relevant stakeholders

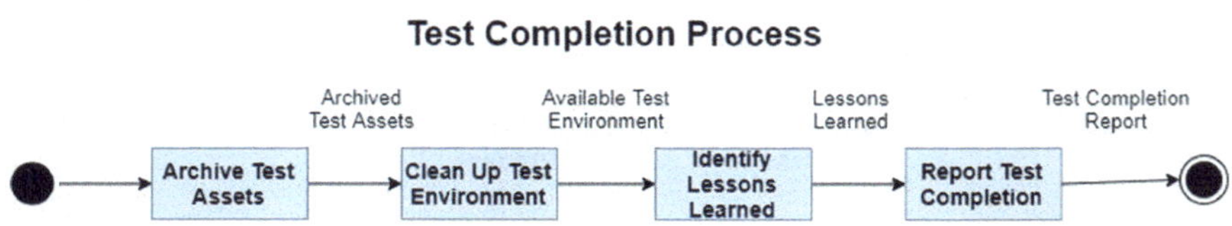

Fig. 9 Test completion process

Figure 11 illustrates a graphic overview of the various activities of the test design and implementation process. Table 4 gives further details about the multiple tasks in need of completion for each action.

6.2 *Test Environment Set-up and Maintenance Process*

The test environment set-up and maintenance processes are needed to create an environment in which a test is executed and to maintain the said environment [12]. This process is also required to support interaction with all relevant stakeholders

Table 3 Tasks and activities from the test completion process

Activity	Tasks
Archive test assets	(a) Identify and make available the test assets using appropriate means (b) Identify and archive the test assets that may be reused on other projects (c) Record the availability of reusable test assets in the test completion report and communicate them to the relevant stakeholders
Clean up the test environment	(a) Restore the test environment to a predefined state upon conclusion of all testing actions
Identify lessons learned	(a) Record the lessons learned during the project execution (what went well and not so well) (b) Record the outcome in the test completion report and communicate with the relevant stakeholders
Report test completion	(a) Collect relevant information from, but not limited to test plans, test results, test status reports, test completion; reports and incident reports (b) Evaluate and summarize the collected information (c) Obtain approval for the test completion report from the responsible stakeholders (d) Distribute the approved test completion report to the relevant stakeholders

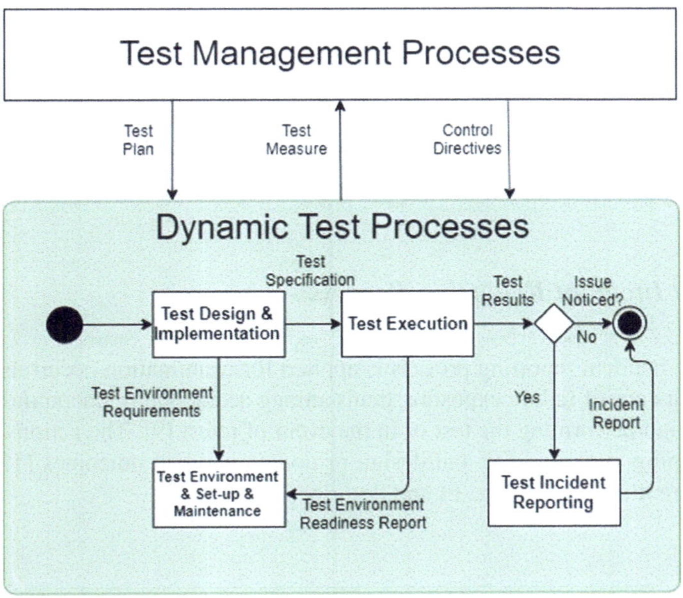

Fig. 10 Dynamic test processes activities

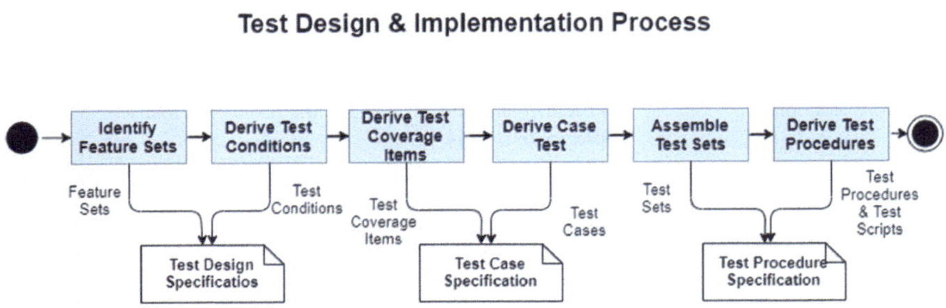

Fig. 11 Test design and implementation process activities

about the status of the test environment. Figure 12 illustrates a graphic overview of the various activities of the test design and implementation process. Table 5 gives further details about the multiple tasks in need of completion for each action.

6.3 Test Execution Process

The test execution process is applied to manage the test methods that are created as a consequence of the development process. The test execution demands running numerous times, considering all feasible test procedures which cannot be performed a specific iteration. Moreover, if a problem is solved, the test execution must be conducted again [12]. The actions included in this workflow are the test procedure execution, test results comparison, and test execution registration [1]. Figure 13 illustrates the overview of the test execution process graphically. Table 6 summarizes the actions and tasks of this method.

6.4 Test Incident Reporting Process

The testing incident reporting process is applied for examination occurrence report as a consequence of failure exposure, items among accidental or uncommon performance though performing the test or in the event of retest [9]. The actions required are: examining test sequence and design or update incident outcomes [12]. These activities are illustrated in Fig. 14 and described in Table 7.

7 Discussion and Conclusions

This article presents a study on ISO/IEC/IEEE 29119 standard, and the findings are summarized as follows:

Table 4 Tasks and activities from the test design and implementation

Activity	Tasks
Feature sets identification	(a) Analyse the test data to know the conditions for the test (b) Combine the elements to be examined into feature settings (c) Prioritize the examination of the feature settings (d) Obtain agreement with the stakeholder concerning the composition and prioritization of feature sets (e) Document feature sets in the test designation (f) Record the traceability within the test data and the feature sets
Derive test conditions	(a) Determine the test requirements for every feature (b) Prioritize the test requirements using risk levels (c) Record the test conditions in the test designation (d) Record the traceability among the test data, features sets and test requirements (e) Obtain approval for test design specifications by the stakeholders
Derive test coverage items	(a) Derive the test coverage details to be handled during the test (b) Prioritize the test coverage items considering exposure levels (c) Record the test coverage item in the test designation (d) Record the traceability among the test data, feature settings, test requirements, and test coverage items
Derive test cases	(a) Derive individual or group test cases by defining pre-conditions, choosing input values and activities to execute the decided test coverage items (b) Record test cases in the test designation (c) Record the traceability within the test data, feature sets, test conditions, test coverage items, and test cases (d) Obtain approval from the stakeholders for the test case specification
Assemble test sets	(a) Distribute the test cases into several test settings according to the execution constraints (b) Record test sets in the test procedure specification (c) Record the traceability among the test basis, feature settings, test description, test coverage, test cases and test settings
Derive test procedures	(a) Derive test procedures by organising test cases inside a test set (b) Identify each test data and test conditions that are not incorporated in the plan (c) Prioritize the test methods using risk exposure levels (d) Record the test procedures in the test designation (e) Record the traceability among the test data, feature sets, test requirements, test coverage items, test samples, test sets, and test methods (f) Obtain approval on the test procedures specification with the stakeholders

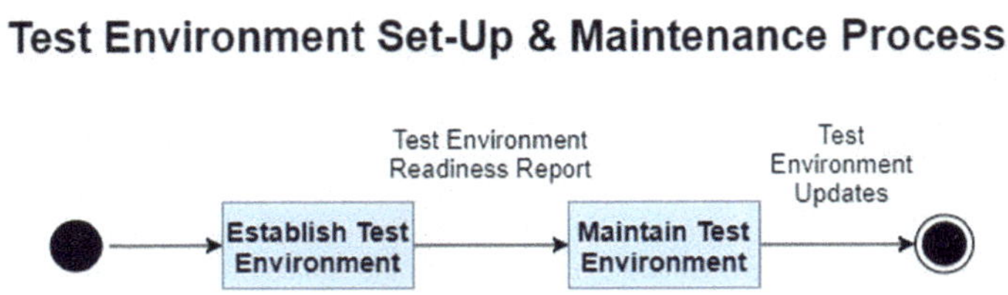

Fig. 12 Test environment set-up and maintenance

Table 5 Tasks and activities from the test environment set-up and maintenance process

Activity	Tasks
Establish a test environment	(a) Based on the test plan, perform the following:
	1. Prepare the set-up of the test settings
	2. Plan the test conditions
	3. Define the configuration management
	4. Implement the test environment
	5. Prepare the test data that will be used in the testing process
	6. Prepare the test tools to assist the examination
	7. Configure the test object
	8. Check if the test environment fulfils the specification
	9. Assure that the test environment meets the defined requirements
	(b) Record the test environment status and data and communicate it to the relevant stakeholders
	(c) Include the details of the disparities among the test and the operation environment
Maintain test environment	(a) Maintain the test environment as defined
	(b) Communicate the updates to the status regarding the test environment to the stakeholders

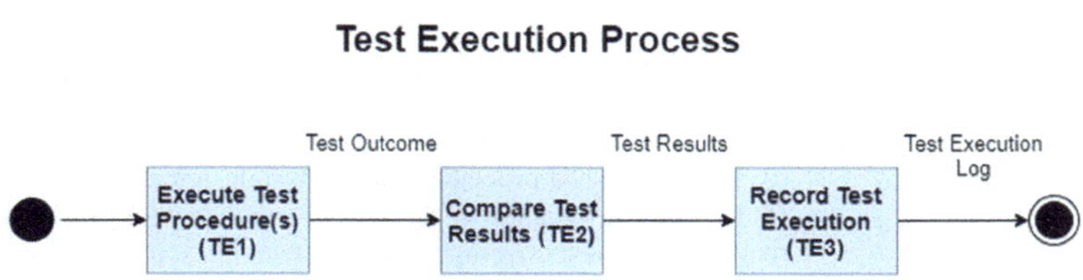

Fig. 13 Graphic overview of the test execution process

Table 6 Tasks and activities from the test execution process

Activity	Tasks
Execute test procedure(s)	(a) Perform individual or group test methods in the qualified test environment (b) Supervise the real results for every test case (c) Record the actual outcomes
Compare test results	(a) Compare the actual and expected results for each test case (b) Determine the test outcome of performing the test samples in the inspection procedure. If the rate is reached, this order needs an update to an event report by the test process
Record test execution plan	(a) recording test performance, as defined in the standard

Fig. 14 Graphic overview of the test incident reporting process

Table 7 Activity and tasks from the test reporting phase

Activity	Tasks
Analyse test results	(a) Analyze the test result and update the incident details when the test is associated with a previous incident (b) Analyze the test result when it is related to a newly identified bug. This analysis will be conducted whether it is an occurrence that needs reporting, an operation item that will be solved without conflict reporting, or needs no additional action to be considered (c) Assign the activity details to a proper person for decision
Create/update incident report	(a) Identify and report/update the data that requires to be registered concerning the occurrence (b) Communicate the status of new or updated episodes to the relevant stakeholders

(RQ1) The ISO/IEC/IEEE 29119 is a group of globally agreed practices for software testing guidelines. It must be used by any software developer or software testing organization within any step of the application development [1].

(RQ2) The ISO/IEC/IEEE 29119 Standard implementation is a rigorous process for small and medium-sized companies and scenarios.

(RQ3) The main limitation associated with the implementation of this standard is related to the standard architecture. This standard is divided into sixteen outputs, which range from planning, system requirements, project execution to the application reports [10]. The implementation of this standard in the software development scenario incorporates several limitations regarding its complexity and scalability.

(RQ4) The adoption of this standard makes it possible to decrease the number of documents which promote the acceptability in small and medium scale organizations [10]. The incorporation of the new methods and models, the testing process can be improved which leads to several improvements in different sectors of the companies, in particular associated with the development, analysis, and team management.

(RQ5) This standard allows the testing team to centralize their efforts on the development processes by abstracting from details, which prevents effects such as the discovery of bugs during the testing phase. Therefore, developers can focus on their essential tasks. Furthermore, the testing and management team can focus on the development of high-quality software and avoid losing their time fixing bugs [11]. The developers and managers should understand that by adopting significant procedures in the testing phase does not depend on the technology used and must be taking into consideration in terms of capability assessment [20–25].

(RQ6) The effort in terms of time and cost of adopting this kind of standard for testing is relevant. However, the final price can be higher in the case of bugs during the product installation phase, especially in enterprise and safety-critical systems.

In sum, software testing is a relevant element of the code development which can not be avoided in any situation. Nowadays, software testing must be done in line with all the software development processes for the design of reliable applications. Software testing is currently particularly critical regarding the new advances in AI applications. These smart or intelligent systems will change our daily routine, and therefore software testing practices must be adopted to ensure high-quality services and applications. Is should be noted that the adoption of a standard test process model is an asset for any organization since both software and product quality is guaranteed by the established metrics of quality of software and the test process models followed.

References

1. Matalonga, S., Rodrigues, F., Travassos, G.H.: Matching context aware software testing design techniques to ISO/IEC/IEEE 29119. In: Rout, T., O'Connor, R.V., Dorling, A. (eds.) Software Process Improvement and Capability Determination, pp. 33–44. Springer International Publishing, Cham (2015). https://doi.org/10.1007/978-3-319-19860-6_4
2. Thakur, M.S.: Review on structural software testing coverage approaches. Int. J. Adv. Res. Ideas Innovations Technol. 281–286 (2017)
3. Anwar, N., Kar, S.: Review paper on various software testing techniques & strategies. Glob. J. Comput. Sci. Technol. [S.l.], May 2019. ISSN: 0975-4172. Available at: https://computerresearch.org/index.php/computer/article/view/1873. Date accessed: 25 Feb 2020

4. Rana, I., Goswami, P., Maheshwari, H.: A review of tools and techniques used in software testing. Int. J. Emerg. Technol. Innovative Res. **6**(4), 262–266. www.jetir.org. ISSN:2349-5162, April 2019. Available: http://www.jetir.org/papers/JETIR1904Q46.pdf
5. Hrabovská, K., Rossi, B., Pitner, T.: Software testing process models benefits & drawbacks: a systematic literature review. arXiv preprint arXiv:1901.01450 (2019)
6. Sanchez-Gordon, S., Luján-Mora, S.: A method for accessibility testing of web applications in agile environments. In: Proceedings of the 7th World Congress for Software Quality (WCSQ), pp. 13, 15 (85) (2017)
7. Jan, S.R., Shah, S.T.U., Johar, Z.U., Shah, Y., Khan, F.: An innovative approach to investigate various software testing techniques and strategies. Int. J. Sci. Res. Sci. Eng. Technol. (IJSRSET). Print ISSN: 2395-1990 (2016)
8. Ali, S., Yue, T.: Formalizing the ISO/IEC/IEEE 29119 software testing standard. In: 2015 ACM/IEEE 18th International Conference on Model Driven Engineering Languages and Systems (MODELS), pp. 396–405. IEEE, Ottawa, ON, Canada (2015). https://doi.org/10.1109/MODELS.2015.7338271
9. Jamil, M.A., Arif, M., Abubakar, N.S.A., Ahmad, A.: Software testing techniques: a literature review. In: 2016 6th International Conference on Information and Communication Technology for the Muslim World (ICT4M), pp. 177–182. IEEE (2016)
10. Eira, P., Guimaraes, P., Melo, M., Brito, M.A., Silva, A., Machado, R.J.: Tailoring ISO/IEC/IEEE 29119-3 standard for small and medium-sized enterprises. In: 2018 IEEE International Conference on Software Testing, Verification and Validation Workshops (ICSTW), pp. 380–389. IEEE, Vasteras (2018). https://doi.org/10.1109/ICSTW.2018.00077
11. Pröll, R., Bauer, B.: Toward a consistent and strictly model-based interpretation of the ISO/IEC/IEEE 29119 for early testing activities: In: Proceedings of the 6th International Conference on Model-Driven Engineering and Software Development, pp. 699–706. SCITEPRESS—Science and Technology Publications, Funchal, Madeira, Portugal (2018). https://doi.org/10.5220/0006749606990706
12. Departamento de Ingeniería, Pontificia Universidad Católica del Perú, Lima, Lima 32, Perú, Dávila, A., García, C., Departamento de Ingeniería, Pontificia Universidad Católica del Perú, Lima, Lima 32, Perú, Cóndor, S., Escuela Profesional de Ingeniería de Software Universidad Nacional Mayor de San Marcos, Lima, Lima 1, Perú: Análisis exploratorio en la adopción de prácticas de pruebas de software de la ISO/IEC 29119-2 en organizaciones de Lima, Perú. risti. 1–17 (2017). https://doi.org/10.17013/risti.21.1-17
13. Park, B.H., Seo, Y.G.: Process improvement for quality increase of weapon system software based on ISO/IEC/IEEE 29119 test method. 한국컴퓨터정보학회논문지 **23**, 115–122 (2018). https://doi.org/10.9708/JKSCI.2018.23.12.115
14. Sánchez-Gordón, M.-L., Colomo-Palacios, R.: From certifications to international standards in software testing: mapping from ISQTB to ISO/IEC/IEEE 29119-2. In: Larrucea, X., Santamaria, I., O'Connor, R.V., Messnarz, R. (eds.) Systems, Software and Services Process Improvement, pp. 43–55. Springer International Publishing, Cham (2018). https://doi.org/10.1007/978-3-319-97925-0_4
15. Henderson-Sellers, B., Gonzalez-Perez, C., McBride, T., Low, G.: An ontology for ISO software engineering standards: creating the infrastructure. Comput. Stand. Interfaces **36**, 563–576 (2014). https://doi.org/10.1016/j.csi.2013.11.001
16. Condor, S., Garcia, C., Davila, A.: Adoption of ISO/IEC 29119-2 software testing practices: an exploratory analysis in organizations in Lima, Perú. In: 2016 International Conference on Software Process Improvement (CIMPS), pp. 1–8. IEEE, Aguascalientes, Mexico (2016). https://doi.org/10.1109/CIMPS.2016.7802802
17. Munir, H., Runeson, P.: Software testing in open innovation: an exploratory case study of the acceptance test harness for jenkins. In: Proceedings of the 2015 International Conference on Software and System Process—ICSSP 2015, pp. 187–191. ACM Press, Tallinn, Estonia (2015). https://doi.org/10.1145/2785592.2795365
18. Felderer, M., Wendland, M.-F., Schieferdecker, I.: Risk-based testing. In: Margaria, T., Steffen, B. (eds.) Leveraging Applications of Formal Methods, Verification and Validation. Specialized

Techniques and Applications, pp. 274–276. Springer, Berlin (2014). https://doi.org/10.1007/978-3-662-45231-8_19
19. Kawaguchi, S.: Trial of organizing software test strategy via software test perspectives. In: 2014 IEEE Seventh International Conference on Software Testing, Verification and Validation Workshops, pp. 360–360. IEEE, OH, USA (2014). https://doi.org/10.1109/ICSTW.2014.42
20. Ruy, F.B., Falbo, R.A., Barcellos, M.P., Guizzardi, G., Quirino, G.K.S.: An ISO-based software process ontology pattern language and its application for harmonizing standards. SIGAPP Appl. Comput. Rev. **15**, 27–40 (2015). https://doi.org/10.1145/2815169.2815172
21. Dussa-Zieger, K., Ekssir-Monfared, M., Schweigert, T., Philipp, M., Blaschke, M.: The current status of the TestSPICE® project. In: Stolfa, J., Stolfa, S., O'Connor, R.V., Messnarz, R. (eds.) Systems, Software and Services Process Improvement, pp. 589–598. Springer International Publishing, Cham (2017). https://doi.org/10.1007/978-3-319-64218-5_49
22. Garcia, C., Dávila, A., Pessoa, M.: Test process models: systematic literature review. In: Mitasiunas, A., Rout, T., O'Connor, R.V., Dorling, A. (eds.) Software Process Improvement and Capability Determination, pp. 84–93. Springer International Publishing, Cham (2014). https://doi.org/10.1007/978-3-319-13036-1_8
23. Siegl, S., Russer, M.: Systematic use case driven environmental modeling for early validation of automated driving functionalities. In: Gühmann, C., Riese, J., von Rüden, K. (eds.) Simulation and Testing for Vehicle Technology, pp. 383–392. Springer International Publishing, Cham (2016). https://doi.org/10.1007/978-3-319-32345-9_26
24. Großmann, J., Seehusen, F.: Combining security risk assessment and security testing based on standards. In: Seehusen, F., Felderer, M., Großmann, J., Wendland, M.-F. (eds.) Risk Assessment and Risk-Driven Testing, pp. 18–33. Springer International Publishing, Cham (2015). https://doi.org/10.1007/978-3-319-26416-5_2
25. Adlemo, A., Tan, H., Tarasov, V.: Test case quality as perceived in Sweden. In: Proceedings of the 5th International Workshop on Requirements Engineering and Testing—RET '18, pp. 9–12. ACM Press, Gothenburg, Sweden (2018). https://doi.org/10.1145/3195538.3195541

Towards the Development of a Comprehensive Theoretical Model for Examining the Cloud Computing Adoption at the Organizational Level

Yousef A. M. Qasem, Rusli Abdullah, Yusmadi Yah, Rodziah Atan, Mohammed A. Al-Sharafi, and Mostafa Al-Emran

Abstract Cloud computing (CC) is a new computing paradigm in higher educational institutions (HEIs). CC allows educational services to be delivered at anytime anywhere settings. Despite all its advantages, the adoption of CC in general and cloud-based education as a service (CEaaS) in specific is still not yet clearly understood in HEIs. Besides, there is a scarce of knowledge regarding the adoption of CC at the organizational level. Therefore, this study aims to develop a comprehensive theoretical model through the integration of four dominant theories, including the technology-organization-environment framework (TOE), fit viability model (FVM), diffusion of innovations (DOI), and institutional theory (INT). The primary purpose of the developed model is to understand the factors affecting the CC adoption at the organizational level of HEIs. It is believed that the developed model would assist the decision-makers in HEIs to make informed decisions concerning the future implementation of CC. Theoretical contributions and practical implications were also discussed.

Y. A. M. Qasem · R. Abdullah · Y. Yah · R. Atan
Faculty of Computer Science and Information Technology, Universiti Putra Malaysia, 43400 Serdang, Malaysia
e-mail: y.alsharaei@gmail.com

R. Abdullah
e-mail: rusli@upm.edu.my

Y. Yah
e-mail: yusmadi@upm.edu.my

R. Atan
e-mail: rodziah@upm.edu.my

M. A. Al-Sharafi (✉)
Faculty of Computing, Universiti Malaysia Pahang, Gambang, Malaysia
e-mail: alsharafi@ieee.org

M. Al-Emran
Department of Information Technology, Al Buraimi University College, Al Buraimi, Oman
e-mail: mustafa.n.alemran@gmail.com

M. Al-Emran et al. (eds.), *Recent Advances in Intelligent Systems and Smart Applications*, Studies in Systems, Decision and Control 295,
https://doi.org/10.1007/978-3-030-47411-9_4

Keywords Cloud computing · Adoption · TOE framework · FVM · DOI · INT theory · Higher education

1 Introduction

As a model, cloud computing (CC) is seen to benefit from fast and easy network access, and its convenience is enhanced by the fact that it allows rapid access with minimal efforts from the company that provides the service [1]. A revolution in technology has been created through the relationship between information technology and XaaS. This stems from the reason that services can be accessed from anywhere in the world, and their cost-effective benefits can be enjoyed by individuals and companies of all sizes as it operates on the simple basis of pay-per-use [2, 3].

In the education sector, there is an expectation that the institutions that educate students for degree levels, Tertiary, and higher education (HEI), providers should keep pace with technology [4]. In the past, this has created a problem for such institutions due to the expensive resources of IT in investment. Therefore, educational services should be affordable while maintaining excellent quality [3]. In addition, HEIs must become more efficient while focusing on the delivery of excellent services and looking for ways in which they can maximize their resources [3]. In addition to providing good standards of education delivered through practical skills, HEIs also have a unique opportunity to enable students to become professional members of society and capable of handling problem-solving activities [4].

For HEIs, CC represents an ideal opportunity to lower their IT costs while increasing efficiency, which would have a positive impact on their long-term sustainability. As suggested by Thomas [5], CC is not only a learning tool for HEIs but also an essential platform for educators as it enables them to improve their practices, encourage team collaboration, and enhance their productivity. Furthermore, CC would be able to save both costs and energy outputs as the same cloud infrastructure can be utilized across different activities, including teaching, learning, and research [6]. Despite all these advantages, the adoption of CC in general and cloud-based education as a service (CEaaS) in specific is still not yet clearly understood in HEIs [7].

Using constructs from the technology acceptance model (TAM) and DOI theory, Ref. [8] developed a model for CC adoption in education. Using the TOE theory, Ref. [9] developed a framework that explained the migration from the traditional system into a cloud-based system in HEIs. It has been observed from the literature that there is a lack of understanding the adoption of CC at the organizational level of HEIs. Therefore, this research contributes to the existing literature by developing a research model through the integration of four well-established models, namely the technology-organization-environment framework (TOE) [10], fit viability model (FVM) [11], diffusion of innovations (DOI) [12], and institutional theory (INT) [13]. This research intends to establish a starting point through which the decision-makers of HEIs can determine the suitability of CC services adoption in their institutions and enable them to make informed decisions before engaging in the process.

2 Cloud Computing in HEIs

In education, cloud computing has the ability to provide both teachers and students with numerous advantages. Whether in education or research, the ability to store large amounts of data, collaborate on projects, and share materials is an attractive proposition. Due to its remote access ability, users can take this advantage to access their materials at anytime anywhere settings. HEIs have decided to overcome the old style of systems by adopting CC due to its efficiency and rapid implementation [14]. The collaborative approaches to learning are one of the key benefits that CC offers, which in turn makes it an ideal choice for those institutions that are looking for computer-based technologies to enhance the cooperative learning styles [15]. CC also facilitates e-learning in HEIs as it is capable of utilizing facilities such as data access monitoring and storage through a cloud platform [16]. Although it is still in its early stages of adoption, the popularity of CC is increasing in HEIs [17]. Rather than being a choice, CC becomes an essential part of the HEIs [18].

Table 1 provides a summary of the CC adoption studies in higher education. Despite the tremendous amount of research conducted on the CC adoption, there is a scarce of knowledge regarding the adoption of CC at the organizational level [8, 19].

Table 1 A summary of the previous cloud-based education as a service (CEaaS) adoption studies

Source	Theory	Methodology	Country
[19]	DOI and TAM	Survey	Universities in sub-Saharan Africa
[8]	DOI and TAM	Conceptual model	Universities in sub-Saharan Africa
[20]	N/A	Semi-structured interviews	HEIs in Oman
[9]	Success factors based on literature	Survey	Universities in Saudi Arabia
[16]	LR	–	–
[21]	N/A	Survey	University in Southeast Michigan in USA
[4]	FVM and DOI	Conceptual model	HEIs in Oman
[22]	TAM	Survey	University Politehnica of Bucharest in Romania
[23]	TAM	Survey	Universities in Thailand
[24]	TAM3	Survey	Saudi Arabia
[25]	DOI	Focus groups and DEMATEL	Science and technology institutions in Taiwan
[26]	TOE	Survey	HEIs in Saudi Arabia
[27]	TAM3	Focus groups and interviews	Rural and urban community colleges in USA

This lack of evidence at the organizational level suggests to conduct further research regarding the adoption of CC in HEIs and to explore the factors affecting its adoption in such institutions.

3 Mapping Matrix for Model Development

A review of the existing relevant literature suggested that the TOE framework and the DOI theory could be used in combination to increase the efficacy of the model with regard to IT adoption [28]. The enrichment of DOI with TOE (i.e., technological context) [29] is mirrored by the combination of the INT theory and TOE framework (i.e., environmental context) [30].

When a new technology is used to implement a system, a degree of risk is usually involved, so there is an advantage to develop a model that is able to predict how the new technology will be applied within a particular context. The context readiness and the characteristics of the technology would affect the adoption of a new technological innovation [31], and the environmental characteristics would also exert their influence. The FVM is useful in establishing whether or not CC is appropriate in facilitating the delivery of CEaaS in HEIs. In that, the fit is used to describe how far CC is appropriate for the delivery of CEaaS, and it can be measured by defining the tasks related to CEaaS. This is combined with the use of DOI factors to establish the effect of CC on the CEaaS adoption [32–37]. Additionally, the viability describes how much added-value that CC might bring to the CEaaS delivery process and to what extent are the HEIs ready for the adoption of CC technology. These theoretical perspectives provide a theoretical basis through which to assess the factors affecting the adoption of CC in HEIs by considering different characteristics, including task, organization, technology, and environment. This approach has received a great deal of empirical support in prior research [29, 31, 38–44].

Ideas from the previous studies were considered as part of the literature review, and the information from these studies was filtered and consolidated so that the broad range of factors derived from past studies could be incorporated [45]. Through this process of filtering, the factors that occurred most often were established and selected for use. This has resulted in a mapping matrix of the salient adoption theories derived from the previous literature (see Table 2). In this research, we proposed a conceptual model by integrating four well-known dominant adoption theories used at the organizational level, namely FVM Model, TOE framework, DOI theory, and INT theory (see Fig. 1). The proposed conceptual model is believed to offer a better understanding of the factors affecting the CC adoption in HEIs.

Table 2 Mapping matrix for deriving the model constructs from TOE, FVM, DOI, and INT

Model/theory	Technology	Source	Factors/constructs											
			Fit context					Viability context				Environmental		
			Task	Technological				Organizational						
				RA	Comp	CX	SC	TR	CS	CK	TMS	CP	NP	MP
TOE	Open systems	[46]				×								
TOE	Electronic data interchange	[47]						×	×					
TOE	E-business	[48]						×						
DOI and TOE	E-business	[49]		×	×	×	×	×	×					
TOE	E-business	[50]						×						
INT	Enterprise System	[51]										×	×	×
TOE	E-business	[52]			×									
TOE	Knowledge management and enterprise systems	[53]		×	×	×					×			
DOI and TOE	Collaborative commerce	[40]		×	×	×					×			
DOI, TOE, and others	Internet utilization	[54]						×	×		×			
DOI and TOE	Benchmarking	[55]		×	×	×								
DOI	RFID	[56]		×		×					×			
TOE and others	E-business	[57]						×						
TOE	E-commerce	[58]		×	×				×		×			
TOE	E-business	[59]		×	×	×					×			
TOE	RFID	[60]			×	×			×		×			
DOI and TOE	Cloud computing	[61]		×	×	×		×	×		×			
INT	Green IS	[62]										×	×	×

(continued)

Table 2 (continued)

Model/theory	Technology	Source	Factors/constructs											
			Fit context					Viability context				Environmental		
			Task	Technological				Organizational						
				RA	Comp	CX	SC	TR	CS	CK	TMS	CP	NP	MP
DOI	Internet-based purchasing application	[63]			×	×								
DOI	Cloud computing	[64]		×	×	×								
TOE	Cloud computing	[65]			×	×	×		×					
TOE	Cloud computing	[66]		×	×	×		×	×		×			
DOI and others	Cloud computing	[67]			×	×								
INT	Electronic human resource management	[68]											×	×
TOE and DOI	Cloud computing	[42]		×	×	×	×	×	×		×			
TOE, DOI, and INT	SaaS	[29]		×	×	×	×	×	×		×	×	×	×
FVM and DOI	Cloud computing	[31]	×	×	×	×	×	×	×	×	×			
FVM, TOE, and DOI	Cloud ERP	[69]	×	×	×		×		×	×	×			
This study	CEaaS	–	×	×	×	×	×	×	×	×	×	×	×	×

RA Relative advantage, *Comp* Compatibility, *CX* Complexity, *SC* Security concerns, *TR* Technology readiness, *CS* Cost savings, *CK* Cloud knowledge, *TMS* Top management support, *CP* Coercive pressures, *NP* Normative pressures, and *MP* Mimetic pressures

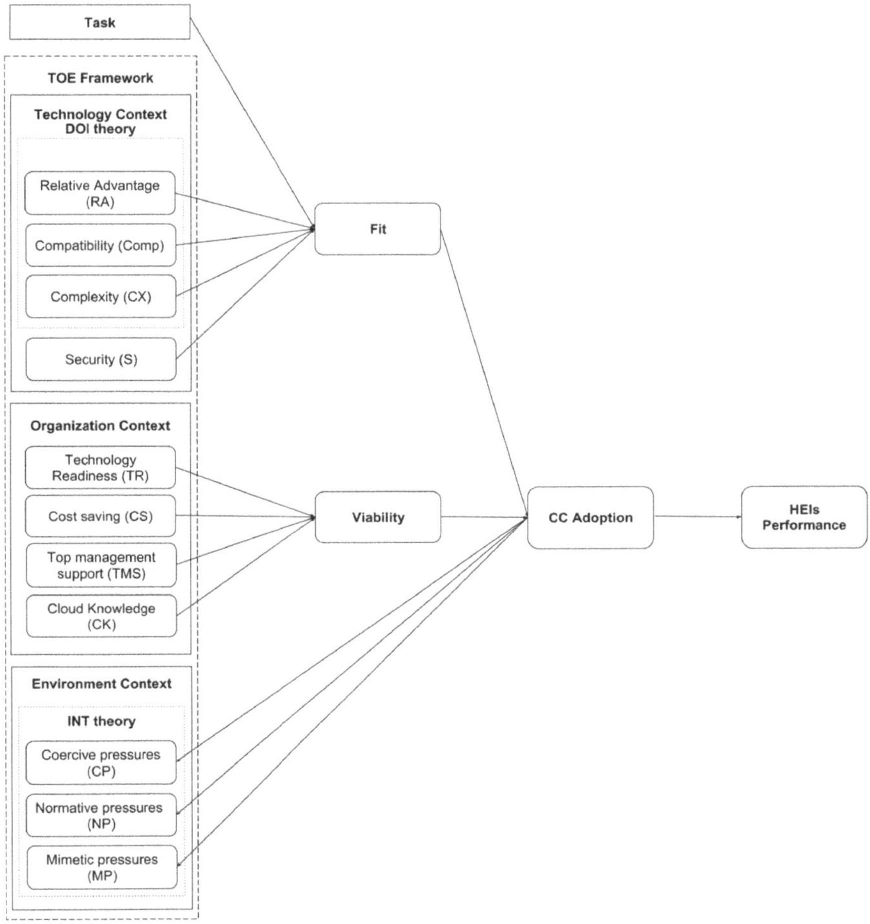

Fig. 1 Research model

4 Discussion

4.1 Theoretical Contributions and Practical Implications

This study throws up a number of theoretical contributions and practical implications. First, this research develops a new conceptual model through the integration of four well-known established theories, including FVM, TOE, DOI, and INT to understand the factors affecting the CC adoption at the organizational level of HEIs. This, in turn, is believed to add a significant contribution to the existing organizational theories in general and the CC adoption in particular. Second, the developed conceptual model could be used to explain the adoption of other technologies in higher education or other sectors. Third, this study provides a starting point for further research concerning the adoption of CC in HEIs. Fourth, understanding the factors affecting the CC adoption in HEIs would assist the decision-makers in these institutions to make informed decisions concerning the future implementation of CC.

4.2 Limitations and Future Research

Although the present study developed a conceptual model based on four well-known established theories, the main limitation is that it did not validate the proposed model in an empirical study. Therefore, it is highly recommended that further research should validate the proposed model in an empirical study in order to understand the factors affecting the adoption of CC in HEIs. In addition, it was unfortunate that the present study did not consider the role of any moderating variables in understanding the adoption of CC in HEIs. Further research could consider this point by extending the proposed conceptual model with some moderating factors.

5 Conclusion

CC offers a number of opportunities in terms of scalability and cost-efficiency. The unique features of CC allow the HEIs to make CEaaS as a reality in the educational community at a rapid speed and low costs. Despite these significant features, understanding the factors affecting the CC adoption at the organizational level of HEIs is still scarce and requires further research. Therefore, this research developed a conceptual model based on the integration of four well-known established theories, including TOE, FVM, DOI, and INT in order to understand the factors affecting the adoption of CC at the organizational level of HEIs.

The proposed model provides a theoretical basis through which to assess the factors affecting the adoption of CC in HEIs by considering different characteristics, including task, organization, technology, and environment. With regard to the provision of CEaaS initiatives in HEIs, it is considered that this research was able to make an original contribution through its focus on the effect of CC adoption. It is believed that the conclusions derived from this study would add a significant contribution to the existing literature on one hand, and the decision-makers in the HEIs on the other hand.

Acknowledgements The authors would like to thank Universiti Putra Malaysia (UPM)—RMC—for supporting and funding this research under Grant No. 95223100.

References

1. Mell, P., Grance, T.: The NIST definition of cloud computing (2011)
2. Armbrust, M., et al.: A view of cloud computing. Commun. ACM **53**(4), 50–58 (2010)
3. Abdullah, R., Eri, Z.D., Talib, A.M.: A model of knowledge management system for facilitating knowledge as a service (KaaS) in cloud computing environment. In: 2011 International Conference on Research and Innovation in Information Systems (ICRIIS), pp. 1–4. IEEE (2011)
4. AlAjmi, Q., Arshah, R.A., Kamaludin, A., Sadiq, A.S., Al-Sharafi, M.A.: A conceptual model of e-learning based on cloud computing adoption in higher education institutions. In: 2017 International Conference on Electrical and Computing Technologies and Applications (ICECTA), 21–23 Nov 2017, pp. 1–6 (2017). https://doi.org/10.1109/icecta.2017.8252013
5. Thomas, P.Y.: Cloud computing: a potential paradigm for practising the scholarship of teaching and learning. Electron. Libr. **29**(2), 214–224 (2011). https://doi.org/10.1108/02640471111125177
6. Razak, S.F.A.: Cloud computing in Malaysia universities. In: 2009 Innovative Technologies in Intelligent Systems and Industrial Applications, 25–26 July 2009, pp. 101–106 (2009). https://doi.org/10.1109/citisia.2009.5224231
7. Abdullah, R., Alsharaei, Y.A.: A mobile knowledge as a service (mKaaS) model of knowledge management system in facilitating knowledge sharing of cloud education community environment. In: 2016 Third International Conference on Information Retrieval and Knowledge Management (CAMP). IEEE, pp. 143–148 (2016)
8. Sabi, H.M., Uzoka, F.M.E., Langmia, K., Njeh, F.N.: Conceptualizing a model for adoption of cloud computing in education (in English). Int. J. Inf. Manage. **36**(2), 183–191 (2016). https://doi.org/10.1016/j.ijinfomgt.2015.11.010
9. Alharthi, A., Alassafi, M.O., Walters, R.J., Wills, G.B.: An exploratory study for investigating the critical success factors for cloud migration in the Saudi Arabian higher education context. Telematics Inform. **34**(2), 664–678 (2017). http://doi.org/10.1016/j.tele.2016.10.008
10. Tornatzky, L.G., Fleischer, M., Chakrabarti, A.K.: Processes of Technological Innovation. Lexington Books, Lexington (1990)
11. Tjan, A.K.: Finally, a way to put your internet portfolio in order. Harvard Bus. Rev. **79**(2), 76–85, 156 (2001)
12. Rogers, E.M.: Diffusion of Innovations, vol. 12. Free Press, New York (1995)
13. Chatterjee, D., Grewal, R., Sambamurthy, V.: Shaping up for e-commerce: institutional enablers of the organizational assimilation of web technologies. MIS Q. 65–89 (2002)
14. Sultan, N.: Cloud computing for education: a new dawn? Int. J. Inf. Manage. **30**(2), 109–116 (2010). https://doi.org/10.1016/j.ijinfomgt.2009.09.004
15. Thorsteinsson, G., Page, T., Niculescu, A.: Using virtual reality for developing design communication. Stud. Inform. Control **19**(1), 93–106 (2010)
16. Pocatilu, P., Alecu, F., Vetrici, M.: Using cloud computing for E-learning systems. In: Proceedings of the 8th WSEAS International Conference on Data Networks, Communications, Computers, pp. 54–59. Citeseer (2009)
17. Katz, R., Goldstein, P., Yanosky, R., Rushlo, B.: Cloud computing in higher education. In: EDUCAUSE [Online], vol. 10. Retrieved October 5, 2010. http://net.educause.edu/section_params/conf/CCW (2010)
18. Sasikala, S., Prema, S.: Massive centralized cloud computing (MCCC) exploration in higher education (2011)
19. Sabi, H.M., Uzoka, F.-M.E., Langmia, K., Njeh, F.N., Tsuma, C.K.: A cross-country model of contextual factors impacting cloud computing adoption at universities in sub-Saharan Africa. Inform. Syst. Front. 1–24 (2017). https://doi.org/10.1007/s10796-017-9739-1
20. Alajmi, Q.A., Kamaludin, A., Arshah, R.A., Al-Sharafi, M.A.: The effectiveness of cloud-based e-learning towards quality of academic services: an Omanis. Expert view. Int. J. Adv. Comput. Sci. Appl. (IJACSA) **9**(4), 158–164 (2018). https://doi.org/10.14569/ijacsa.2018.090425
21. Ashtari, S., Eydgahi, A.: Student perceptions of cloud applications effectiveness in higher education. J. Comput. Sci. **23**, 173–180 (2017)

22. Militaru, G., Purcărea, A.A., Negoiţă, O.D., Niculescu, A.: Examining cloud computing adoption intention in higher education: exploratory study. In: International Conference on Exploring Services Science, pp. 732–741. Springer, Berlin (2016)
23. Bhatiasevi, V., Naglis, M.: Investigating the structural relationship for the determinants of cloud computing adoption in education. Educ. Inf. Technol. **21**(5), 1197–1223 (2016)
24. Almazroi, A.A., Shen, H., Teoh, K.-K., Babar, M.A.: Cloud for e-learning: determinants of its adoption by university students in a developing country. In: 2016 IEEE 13th International Conference on E-Business Engineering (ICEBE), pp. 71–78. IEEE (2016)
25. Hwang, B.-N., Huang, C.-Y., Yang, C.-L.: Determinants and their causal relationships affecting the adoption of cloud computing in science and technology institutions. Innovation **18**(2), 164–190 (2016)
26. Tashkandi, A.N., Al-Jabri, I.M.: Cloud computing adoption by higher education institutions in Saudi Arabia: an exploratory study. Cluster Comput. **18**(4), 1527–1537 (2015)
27. Behrend, T.S., Wiebe, E.N., London, J.E., Johnson, E.C.: Cloud computing adoption and usage in community colleges. Behav. Inf. Technol. **30**(2), 231–240 (2011)
28. Hsu, P.-F., Kraemer, K.L., Dunkle, D.: Determinants of e-business use in US firms. Int. J. Electron. Commer. **10**(4), 9–45 (2006)
29. Martins, R., Oliveira, T., Thomas, M.A.: An empirical analysis to assess the determinants of SaaS diffusion in firms. Comput. Hum. Behav. **62**, 19–33 (2016)
30. Oliveira, T., Martins, M.F.: Literature review of information technology adoption models at firm level. Electron. J. Inf. Syst. Eval. **14**(1), 110–121 (2011)
31. Mohammed, F., Ibrahim, O., Nilashi, M., Alzurqa, E.: Cloud computing adoption model for e-government implementation. Inf. Dev. **33**(3), 303–323 (2017)
32. Goodhue, D.L., Thompson, R.L.: Task-technology fit and individual performance. MIS Q. 213–236 (1995)
33. Lee, C.-C., Cheng, H.K., Cheng, H.-H.: An empirical study of mobile commerce in insurance industry: task–technology fit and individual differences. Decis. Support Syst. **43**(1), 95–110 (2007)
34. Dishaw, M.T., Strong, D.M.: Extending the technology acceptance model with task–technology fit constructs. Inf. Manage. **36**(1), 9–21 (1999)
35. Teo, T.S., Men, B.: Knowledge portals in Chinese consulting firms: a task–technology fit perspective. Eur. J. Inf. Syst. **17**(6), 557–574 (2008)
36. Nance, W.D., Straub, D.W.: An investigation of task/technology fit and information technology choices in knowledge work. J. Inf. Technol. Manage. **7**, 1–14 (1996)
37. Lam, T., Cho, V., Qu, H.: A study of hotel employee behavioral intentions towards adoption of information technology. Int. J. Hospitality Manage. **26**(1), 49–65 (2007)
38. Yadegaridehkordi, E., Iahad, N.A., Ahmad, N.: Task-technology fit assessment of cloud-based collaborative learning technologies. In: Remote Work and Collaboration: Breakthroughs in Research and Practice, p. 371. IGI Global, USA (2017)
39. Chan, F.T., Chong, A.Y.-L.: Determinants of mobile supply chain management system diffusion: a structural equation analysis of manufacturing firms. Int. J. Prod. Res. **51**(4), 1196–1213 (2013)
40. Chong, A.Y.-L., Lin, B., Ooi, K.-B., Raman, M.: Factors affecting the adoption level of c-commerce: an empirical study. J. Comput. Inf. Syst. **50**(2), 13–22 (2009)
41. Ciganek, A.P., Haseman, W., Ramamurthy, K.: Time to decision: the drivers of innovation adoption decisions. Enterp. Inf. Syst. **8**(2), 279–308 (2014)
42. Oliveira, T., Thomas, M., Espadanal, M.: Assessing the determinants of cloud computing adoption: an analysis of the manufacturing and services sectors. Inf. Manag. **51**(5), 497–510 (2014)
43. Wang, Y.-M., Wang, Y.-S., Yang, Y.-F.: Understanding the determinants of RFID adoption in the manufacturing industry. Technol. Forecast. Soc. Change **77**(5), 803–815 (2010)
44. Yoon, T.E., George, J.F.: Why aren't organizations adopting virtual worlds? Comput. Hum. Behav. **29**(3), 772–790 (2013)

45. Wymer, S.A., Regan, E.A.: Factors influencing e-commerce adoption and use by small and medium businesses. Electron. Markets **15**(4), 438–453 (2005)
46. Chau, P.Y., Tam, K.Y.: Factors affecting the adoption of open systems: an exploratory study. MIS Q. 1–24 (1997)
47. Kuan, K.K., Chau, P.Y.: A perception-based model for EDI adoption in small businesses using a technology–organization–environment framework. Inf. Manage. **38**(8), 507–521 (2001)
48. Zhu, K., Kraemer, K.L.: Post-adoption variations in usage and value of e-business by organizations: cross-country evidence from the retail industry. Inf. Syst. Res. **16**(1), 61–84 (2005)
49. Zhu, K., Dong, S., Xu, S.X., Kraemer, K.L.: Innovation diffusion in global contexts: determinants of post-adoption digital transformation of European companies. Eur. J. Inf. Syst. **15**(6), 601–616 (2006)
50. Zhu, K., Kraemer, K.L., Xu, S.: The process of innovation assimilation by firms in different countries: a technology diffusion perspective on e-business. Manage. Sci. **52**(10), 1557–1576 (2006)
51. Liang, H., Saraf, N., Hu, Q., Xue, Y.: Assimilation of enterprise systems: the effect of institutional pressures and the mediating role of top management. MIS Q. 59–87 (2007)
52. Lin, H.-F., Lin, S.-M.: Determinants of e-business diffusion: a test of the technology diffusion perspective. Technovation **28**(3), 135–145 (2008)
53. Ramdani, B., Kawalek, P., Lorenzo, O.: Predicting SMEs' adoption of enterprise systems. J. Enterp. Inf. Manage. **22**(1/2), 10–24 (2009)
54. Shah Alam, S.: Adoption of internet in Malaysian SMEs. J. Small Bus. Enterp. Dev. **16**(2), 240–255 (2009)
55. Azadegan, A., Teich, J.: Effective benchmarking of innovation adoptions: a theoretical framework for e-procurement technologies. Benchmarking Int. J. **17**(4), 472–490 (2010)
56. Tsai, M.-C., Lee, W., Wu, H.-C.: Determinants of RFID adoption intention: evidence from Taiwanese retail chains. Inf. Manage. **47**(5), 255–261 (2010)
57. Oliveira, T., Martins, M.F.: Understanding e-business adoption across industries in European countries. Ind. Manage. Data Syst. **110**(9), 1337–1354 (2010)
58. Ghobakhloo, M., Arias-Aranda, D., Benitez-Amado, J.: Adoption of e-commerce applications in SMEs. Ind. Manage. Data Syst. **111**(8), 1238–1269 (2011)
59. Ifinedo, P.: An empirical analysis of factors influencing Internet/e-business technologies adoption by SMEs in Canada. Int. J. Inf. Technol. Decis. Making **10**(04), 731–766 (2011)
60. Thiesse, F., Staake, T., Schmitt, P., Fleisch, E.: The rise of the "next-generation bar code": an international RFID adoption study. Supply Chain Manage. Int. J. **16**(5), 328–345 (2011)
61. Low, C., Chen, Y., Wu, M.: Understanding the determinants of cloud computing adoption. Ind. Manage. Data Syst. **111**(7), 1006–1023 (2011). https://doi.org/10.1108/02635571111161262
62. Butler, T.: Compliance with institutional imperatives on environmental sustainability: building theory on the role of Green IS. J. Strateg. Inf. Syst. **20**(1), 6–26 (2011)
63. Klein, R.: Assimilation of internet-based purchasing applications within medical practices. Inf. Manage. **49**(3), 135–141 (2012)
64. Lin, A., Chen, N.-C.: Cloud computing as an innovation: perception, attitude, and adoption. Int. J. Inf. Manage. **32**(6), 533–540 (2012). http://dx.doi.org/10.1016/j.ijinfomgt.2012.04.001
65. Nkhoma, M.Z., Dang, D.P., De Souza-Daw, A.: Contributing factors of cloud computing adoption: a technology-organisation-environment framework approach. In: Proceedings of the European Conference on Information Management & Evaluation, pp. 180–189 (2013)
66. Abdollahzadegan, A., Hussin, C., Razak, A., Moshfegh Gohary, M., Amini, M.: The organizational critical success factors for adopting cloud computing in SMEs (2013)
67. Wu, Y., Cegielski, C.G., Hazen, B.T., Hall, D.J.: Cloud computing in support of supply chain information system infrastructure: understanding when to go to the cloud. J. Supply Chain Manage. **49**(3), 25–41 (2013)

68. Heikkilä, J.-P.: An institutional theory perspective on e-HRM's strategic potential in MNC subsidiaries. J. Strateg. Inf. Syst. **22**(3), 238–251 (2013)
69. Salum, K.H., Rozan, M.Z.A.: Conceptual model for cloud ERP adoption for SMEs. J. Theor. Appl. Inf. Technol. **95**(4), 743 (2017)

Factors Affecting Online Shopping Intention Through Verified Webpages: A Case Study from the Gulf Region

Mohammed Alnaseri, Müge Örs, Mustefa Sheker, Mohanaad Shakir, and Ahmed KH. Muttar

Abstract Increasing sales is the main objective of any business. In the past, the sellers displayed their products, and the customer came to try and buy it. But after the new invention (the Internet), there is a new shopping platform called online shopping. Each society has its culture and specificities that change or influence their intentions. This research examined the Gulf States to find out the basic factors that affect their intentions to buy from the Internet. Several factors were chosen in this study. These factors were adopted through the results obtained through the initial questionnaire published in the study community. The study population was specific to the GCC (Gulf Cooperation Council) countries (citizens and residents) to determine their priorities for online shopping. The data collected for this research was carried out through the publication of the questionnaire in some universities in the targeted countries of the study society in addition to real customers who already shopped through online stores. To analyze these data, different methods of analysis (descriptive analysis, multiple regression analysis, dummy variables, and ANOVA test for hypothesis testing) were used. The results found that customers in the study community do care about some factors such as payment methods, discounts, ease to use, ordering method, and verified page rather than other factors chosen through this study like security, product quality, shipping, warranty, and order status. In this study, one of the factors was studied carefully because it has a unique characteristic, and it is the verified pages because it entered the market in recent years and there

M. Alnaseri · M. Örs
Department of Business Administration/Istanbul Aydin University, Istanbul, Turkey

M. Sheker
Department of the Sewage of Fallujah, Ministry of Municipalities and Public Works, Al Fallujah, Iraq

M. Shakir (✉)
Department of Business Administration and Accounting/Al Buraimi University College, Al Buraimi, Oman
e-mail: mohanaad@buc.edu.om

A. KH. Muttar
Applied Science University, College of Administrative Sciences, Al Eker, Bahrain

M. Al-Emran et al. (eds.), *Recent Advances in Intelligent Systems and Smart Applications*, Studies in Systems, Decision and Control 295,
https://doi.org/10.1007/978-3-030-47411-9_5

is not a lot of research written on this topic. Thus, in this research, we will find its value on the customer's intention to make new purchasing through new websites.

Keywords Online shopping · Customer intention · E-Commerce

1 Introductıon

Since the invention of computers, people have been started to use it in a variety of sectors in business [19]. The computer is very easy to get or use. Many homes started to get it, especially with the wide use of the internet in various aspects of life [7]. In the last 20 years, the internet has become very popular to use in various sections of business, which is generated the need of creating a model for the purpose of assessing ease of use and user acceptance in any new technology. Accordingly, Davis et al. suggested the technology acceptance model (TAM) for explaining the behavioral intention towards any technology [9, 11].

Nowadays, the internet has been applied in many different areas, such as communication and television broadcasting. It has also been applied in communication, listening, dissemination knowledge, social communication, health, security, education, and E-commerce [5]. E-commerce term appeared after the discovery of the Internet. It means the operations of online transactions by electronic devices through the internet, such as computers and mobile phones. The huge increase in internet users builds the importance of E-commerce. Via the internet, users can find much information about companies and their business easily when they just make a small search. All information will be listed for free. This feature leads companies to reach many customers easily and assure their products or services to customers faster and cheaper. In addition, the Internet decreases the distances between e-vendor and customers [10]. E-commerce is a very wide platform, including many sectors. One of these sectors is the online shopping [18].

Online shopping, as it is known, Michael Aldrich created (E-commerce) in 1979 to make a new market, which is easier than the traditional one (Özsurunc 2017). It allows making transactions directly between customers and business or business to business. In the beginning, it was not very common because the internet was not easy to have access, but nowadays, the internet, as we mentioned before, became one of the most important things in human life for searching information, learning, communicate with others or even trading. The internet has made a new platform for trading and for the new marketing environment [23, 25]. This prospective study was designed to investigate the factors effecting on customer's intention to improve online shopping in the study area since [1] found that the online shopping in Arab countries not developed because the customers in this area prefer to touch the products before buying. Thus, in this study, factors affecting customers were indicated to find out which factors have a high priority for customers in the study area and to see how we can improve online shopping in the study community. To find which factor affects positively on customer's intention or has high effect on their intention 10 factors have been chosen carefully to find the best way for online stores wanted to cover the study community in order to melt the ice between the consumer and the websites providing this service to develop this field in the study area. In this study, a new component that has recently been used by some e-shopping sites, verified pages, has been studied to find out the true impact of this factor on the intentions of consumers in the study area. In addition to selecting some factors that have taken on the high importance of consumer interest in the study area through the initial questionnaire published in the same region.

In this study, we analyzed the factors that affect the intentions of the consumers in the Gulf countries (the study society). This study aims to find a shopping platform that meets the needs of customers and provides a safe environment for online shopping, which leads to the development of this field to serve all parties online sites and customers. On the other hand, to find these factors, we conducted an initial study to identify the most important factors of the customers in the study society. Purchasing online has many risks, such as customers cannot use what they bought at the moment of purchasing [2]. In another study made only for UAE customers, Saxena found that customers in the UAE are using online shopping for some products. Also, they have the intention to rely more on online shopping just in case they find a safe platform meets with their expectations [24]. In 2019, the research made by [4] researched the reasons that prevented the development of online shopping in Saudi Arabia, despite having all the ingredients that help in the development of this field. So, this study focused on how we can build a platform that leads to develop online shopping in the study area.

2 Advantages and Disadvantages of E-Commerce

2.1 Advantages of E-Commerce

Appropriateness: All information on the internet is very easy to get. Also, you can compare between different sites to determine whether the products they found makes them fully satisfied their needs. Among the appropriate things on the internet, the consumer can search for a product on all websites, whether covering its region or in the various parts of the world, creating a broader space and more information about the product around the world [29, 10].

Saving time: Through internet users can get more time because they will not spend their time to go to the traditional market and check. This is one of the most important advantages offered to consumers because most consumers work in a field and they cannot go to the markets to buy their needs. And if they want to compare prices between the various traditional stores, it may take those days, but by using online shopping, they can obtain the necessary information about the product within seconds [15, 32].

Alternatives: Users can easily compare between these sites in a very short time, unlike traditional shopping way. Through online shopping, consumers can search for alternatives, whether it is a seller or a product. It means that the best product can be found at the best price, and the scope of the search will be on a regional or globally [29].

Sharing the opinion is very easy: users can easily write their comments to express their experience through the site. In traditional markets, the consumers cannot know the opinion of customers of these markets, because the nature of the business does not allow knowledge of such information, but through electronic shopping, the consumer can know the opinion of people about a product or an electronic shopping site by viewing the page dedicated to customer opinions [21, 29].

Cheaper price: In online shops, you can find the best deal by comparing the prices between different sites. In addition, websites do not need to rent a place to present the goods, which is extra expenses will reflect [21].

2.2 *Disadvantages of E-Commerce*

Privacy and security: One of the most important reasons will face users. They have to be sure that all information must be protected by the site. Security is related to a lot of information related to the customer's contact information, such as phone number, email, and address. In addition, the information on their bank card, it scares consumers the most through online shopping [12].

Quality. In this type of shopping, customers cannot tach the products until they receive it at home, so users must check the return policy before to protect themselves. Through online shopping, the consumer can see the picture of the product as they want, but the real product will be totally different event that the product does not meet their expectations. And this point is one of the most important points that negatively affect the reputation of the website [10, 21].

Hidden cost: All aspects of cost must be aware to customers such as (product cost, shipping fee, and other costs) because sometimes hidden costs charged automatically without informing the users. These hidden costs can be, for example, the amount paid to the customs upon purchase from out of the country, an additional cost to be paid to the bank for every purchase abroad, etc. [29].

Internet access: For shopping, online users must spend money to have access to the internet, so it will cost them extra money [21].

3 Hypothesis and Research Model Development

To create a research model, the researchers took some theories into consideration, such as TAM, TPB, and TRA. According to these theories and the result got from the initial questionnaire, the research model was created [9, 13, 16] (Fig. 1).

Fig. 1 Research model

H1*—Product quality has a positive effect on online shopping intention.* To determine the acceptability of this hypothesis, a set of questions were adopted and studied in a good way to determine the possibility of questions in determining the acceptability of this hypothesis. Liying and Cai found that providing the customers with all documents that explain the quality of their product directly impacts on customers through online shopping activity. By using the previous result of many studies, this hypothesis was created [14, 34].

H2*—Providing customers with discount increase the ability to buy products they normally do not buy.* To determine the acceptability of this hypothesis, a set of questions were adopted and studied in a good way to determine the possibility of questions in determining the acceptability of this hypothesis. Many studies adopted the discount as the main topic of their studies as one of the most effective factors that influence online shopping activities because of that H2 were established [30].

H3*—Shipping process has a positive effect on online shopping intention.* To determine the acceptability of this hypothesis, a set of questions were adopted and studied in a good way to determine the possibility of questions in determining the acceptability of this hypothesis. Therefore, Liying, found that the shipping process has many details and providing customers with all delivery terms information will increase the ability to make orders [34]. On the other hand, Yingxi, found that track term has a direct and significant effect on customers through online shopping activities [31]. According to these studies, H3 was created to check the impact of shipping term on online shopping intention.

H4*—Website security has a positive effect on online shopping intention.* To determine the acceptability of this hypothesis, a set of questions were adopted and studied in a good way to determine the possibility of questions in determining the acceptability of this hypothesis. Liying suggested that guarantee influence on online shopping and increase the ability to buy through online web store this guarantee term including personal information and credit card information [34]. In a study conducted by Recep. found that financial risk one of the main factors that affect online shopping intention [23]. According to these results, H4 was established to measure the security factor, including financial and personal information, and its effect on customers' intention.

H5—*Easy and not complicated websites design have a positive effect on online shopping intention*. To determine the acceptability of this hypothesis, a set of questions were adopted and studied in a good way to determine the possibility of questions in determining the acceptability of this hypothesis. In some studies, made approved that the website structure very important to influence customers to make their order through online store these studies found that making the website very sample and put all information related with using the website increased the business value more than the online store with a difficult design (Humaira 2008; Turan 2011) [23].

H6—*Being verified pages has a positive effect on online shopping intention*. To determine the acceptability of this hypothesis, a set of questions were adopted and studied in a good way to determine the possibility of questions in determining the acceptability of this hypothesis. Therefore, the following two questions were used to determine this hypothesis. (1—Do you find effective to make first purchasing from a website has a verified page on Facebook, Instagram, and Twitter? 2—I feel more confident when I see the online store has a verified page?) (Köse et al. 2016) [8, 27, 22].

H7—*Explaining ordering method in the website has a positive effect on online shopping intention*. To determine the acceptability of this hypothesis, a set of questions were adopted and studied in a good way to determine the possibility of questions in determining the acceptability of this hypothesis. In a study made by Liying found that ordering method may affect negatively or positively on customers' attention concluding of this study the researcher found that explaining the ordering method has a significant effect on customer's attitude. By conduct, this study with some studies made before H7 has been created to check the real effect of the ordering method on GCC customers [34].

H8—*Warranty-related information on online shopping sites will positively influence consumers' intention*. To determine the acceptability of this hypothesis, a set of questions were adopted and studied in a good way to determine the possibility of questions in determining the acceptability of this hypothesis Lying found that ISO certificate and another term of approving the website effected positively on customers to purchase online (Lying et al. 2018). On the other hand, Cai states that customers' comments increase trust and motivate new customers to purchase through online stores [14]. Relying on these studies, H8 has been created to recognize the warranty term effect on GCC customers.

H9—*Providing order status has a positive effect on consumer intention*. To determine the acceptability of this hypothesis, a set of questions were adopted and studied in a good way to determine the possibility of questions in determining the acceptability of this hypothesis. To measure this hypothesis some studies were included in this sector in order to understand which result they got to complete from the point other researchers have already reached. Liying suggests that putting the order status gives the customers more confidence and influence the customer's attitude of making purchasing through the online store. For Recep, he found that adding the cancellation step to the store helps customers to be more familiar with online shopping activity [34, 23].

H10*—Having a different way of Payments have a positive effect on online shopping intention.* To determine the acceptability of this hypothesis, a set of questions were adopted and studied in a good way to determine the possibility of questions in determining the acceptability of this hypothesis. Liying Found that payment method has a positive effect on online shopping activities such as credit card, cash on delivery, PayPal, and others were will give the customer better alternatives according to their priority [34].

4 Methodology

The results of the initial questionnaire showed that product quality, offers and discounts, easy to use, shipping, and customer service. Besides, the researcher added a new factor which is verified pages to know whether it has an effect or not on online shopping intention and to know the extent of their impact on the intentions of consumers since this factor used in recent years by websites and there are not many researchers have researched in this field.

The data collected through e-mail, through the establishment of a Google form questionnaire in addition to the self-administrated questionnaires to university students in Oman, the Kingdom of Bahrain, and Saudi Arabia. The questionnaires used to collect data from some of the sample population, preferably identified as general respondents. The Appendix illustrates a sample questionnaire designed mainly according to the objectives of the study. All the scales adopted have been used in previous studies, which have already been adopted as basic scales in many of the questionnaires conducted over the last 10 years. Questionnaires altogether were prepared and administered. The questionnaires divided into several sections of the first section. It consists of 10 questions, which are aimed at determining the demographic factors. The second part of the questionnaire is the section on the factors adopted in this study. It consists of 36 questions to measure all existing variables and to determine the extent of each component's influence on the intentions of consumers. The last part is related to the pages documented and recently adopted by some sites such as Facebook, Instagram, and Twitter. It consists of two questions, and the answer was determined as either YES or NO because this type of pages is new. The purpose of this part of this study is to know whether to own this type of pages will increase the intentions of consumers and their desire to shop online or not. The questionnaires contained close-ended questions that used to collect quantifiable data relevant for precise and effective correlation of research variables. The Likert scale was used for some of the closed-ended questions one of the most widely and successfully used techniques to measure attitudes toward a study variable based on answer item such as; (1-strongly disagree), (2-disagree), (3-Not sure) (4-Agree) or (5-strongly agree). The questionnaires were developed from the scales of the previous authors [34, 31, 6, 28, 14].

Therefore, we studied the most important factors of the study carefully based on previous studies conducted in the same field and was built research model as shown in the first figure based on the theories that study the intentions and behavior of online shopping customers such as (TAM, TPB, and TRA) to determine the priorities [9, 3, 13]. The total number of samples for this study was divided into two parts. The first sample was 25 responds. It was only for the first questionnaire (initial questionnaire) and to determine the study community priorities. The second sample was 250 responds participants from all the Gulf Cooperation Council countries. In this research, the sample size was calculated according to Machin's formula [16].

$$n = \frac{(zs)^2}{d} n = \frac{z^2 p\,(1-p)}{d^2}$$

d = Confidence Interval which is considered to be 10% in this research.

The common confidence levels 99% Z = 99%—Z score = 2.32

In this research Confidence Level is considered 99% so, Z = 2.32

P = standard deviation = 0.5

$$n = \left(\frac{(2.32)(1-0.5)}{0.1}\right)^2$$

$$n = 134.56$$

According to the above formula, 135 respondents will be the minimum sample rate. So, 250 were used in this study to increase reliability. The survey was published on a wide platform that included many universities in the Gulf countries. As well as the questionnaire sent to the real customer who has a real experience by using the customer data from the website www.vipbrands.com which cover the study area to get real information depending on their experience through this website in order to obtain realistic results that increase the importance of this research as it is a examine the problem with real customers and they are not hypothetical In addition, many statistical analyses technic used to know the real effect of each factor chosen in this study. The aim of this study is to find the most effective factors influencing consumer intentions to make first purchasing in the GCC.

The research methodology adopted in this study was summarized in Fig. 2 from the first step to reaching the final results.

Most of the participants in this study have previous (real) experience in the field of online shopping, and they have already purchased through websites so that the results of the questionnaire are credible, and based on actual experiences. A database was used at the site VIPBrands. Also, including university students to determine the percentage of online purchases in the Gulf countries overall.

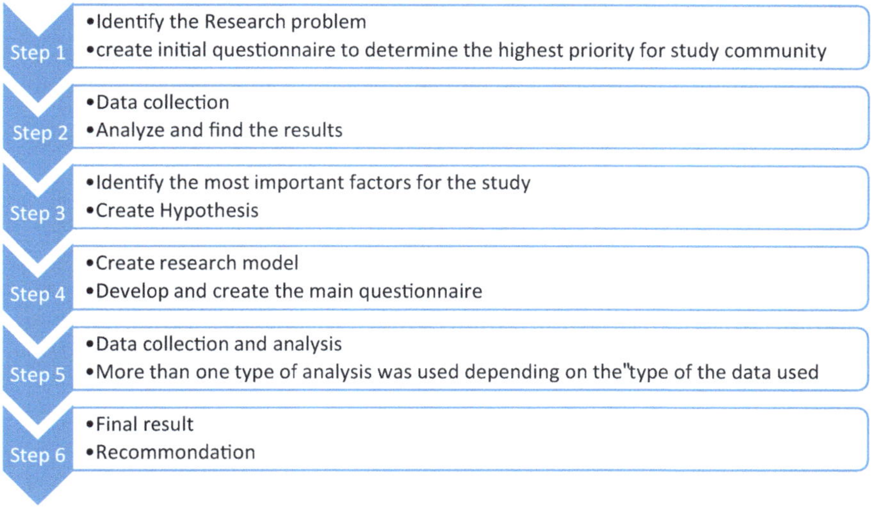

Fig. 2 Research methodology design

5 Results

The results of the questionnaire showed that the proportion of women is higher than the percentage of men in the Gulf countries, which is higher than 60%, which means that the society of women in the study society is more inclined to shop online. It can also be observed that most of the responses were between ages 19 and 34, Higher than 79%, which means that the youth category is the most used for the Internet or shopping through different shopping platforms, in addition to the educational level, which shows that only 4% of the respondents had limited education and that more than 70% of these with good education Bachelor Degree. It can be concluded that the study population (i.e., the Gulf Countries) has a good education and have the ability to understand the instructions and policies that are explained by websites.

In terms of income, it shows that 50% of responses with a limited income of less than \$500. Also, about 25% of responders' income was between \$500 and \$1000, which means that 75% of the response interested in online shopping have limited income. So, it is necessary to design a marketing policy commensurate with the purchasing power of the study community or designing a discounts campaign that will give them a high desire to buy from the website as they are low-income. It is clear from the results collected that the responses received included all six countries included in this study in addition to the residents in varying percentages, which gives the possibility of applying this study result to any of the six countries (GCC countries).

Here, it is important to clarify the important point that over 65% have already purchased online, which means that the countries of the study community are from countries with high rates of shopping online, which creates high opportunities for

Table 1 Demographic analysis

		Frequency	Percent			Frequency	Percent
Gender	Male	99	39.60	Nationality	Kingdome of Saudi Arabia	16	6.40
	Female	151	60.40		United Arab Emirates	21	8.40
Age	Less than 18	2	0.80		Qatar	25	10
	19–24 Years	111	44.40		Kuwait	18	7.20
	25–34	88	35.20		Oman	120	48
	35–44	33	13.20		Bahrain	11	4.40
	Above 44	16	6.40		Other country	39	15.60
Educational level	No schooling completed	4	1.60	Internet use per day	Less than 3 h	57	22.80
	Elementary School	8	3.20		4–6 h	116	46.40
	High school	27	10.80		7–9 h	36	14.40
	Bachelor's Degree	176	70.40		Above 9 h	41	16.40
	Post Graduate	35	14	Purpose of using internet	Education	60	24
Employment	Unemployed	49	19.60		Entertainment	46	18.40
	Public job	42	16.80		Information	93	37.20
	Private sector	56	22.40		Shopping	16	6.40
	Retired	9	3.60		Listening to Music	29	11.60
	Student	94	37.60		Other reasons	6	2.40
Income	<500$	125	50	Purchasing online	Yes	164	65.60
	501$–1000$	64	25.60		No	86	34.40
	1001$–2000	35	14	Marital status	Single, Never Married	137	54.80
	2001$ +	26	10.40		Married	113	45.20

online websites to attract large numbers and get new customers. The remaining 34% did not shop through the websites, so some factors that positively affect the customers in the Gulf countries should be developed in order to increase the overall percentage of online shopping in the study community (Table 1).

6 Model Summary

Table 2 gives the overall model parameters (Beta values and the significance of these values which tell about the individual contribution of the variables in the fit model of online purchase intentions, the beta values and significance values for product quality (−0.017, sig value of 0.842), negative relationship, discounts had (0.201, Sig

Table 2 Model summary

Model	R	R Square	Adjusted R Square	Std. Error of the Estimate
1	0.738[a]	0.544	0.517	2.99239

Dependent Variable online shopping intention
[a]Predictors: (Constant), Product Quality, Discount, warranty, Website security, Ease to use, ordering method, order status, Shipping Process

ANOVA[a]

Model		Sum of Squares	df	Mean Square	F	Sig.
1	Regression	1645.464	9	182.829	20.418	.000[b]
	Residual	1378.945	154	8.954		
	Total	**3024.439**	**163**			

[a]*Dependent Variable* online shopping intention
[b]*Predictors* (Constant), Product Quality, Discount, warranty, Website security, Ease to use, Ordering method, order status, Shipping Process

Coefficients[a]

Model		Unstandardized Coefficients		Standardized Coefficients	t	Sig.	Collinearity Statistics	
		B	Std. Error	Beta			Tolerance	VIF
1	(Constant)	0.927	0.989		0.937	0.35		
	Discount	0.42	0.161	0.201	2.614	0.01	0.501	1.996
	Easy to Use	0.5	0.11	0.39	4.551	0	0.404	2.475
	Shipping	0.015	0.087	0.018	0.169	0.866	0.254	3.938
	Ordering Method	0.351	0.14	0.24	2.503	0.013	0.323	3.095
	Order Status	0.213	0.123	0.187	1.735	0.085	0.256	3.904
	Warranty	0.182	0.1	0.162	1.819	0.071	0.373	2.679
	Payment	0.347	0.128	0.249	2.71	0.007	0.35	2.855

(continued)

Table 2 (continued)

Model	R	R Square		Adjusted R Square		Std. Error of the Estimate		
	Privacy and security	−0.2	0.134	−0.132	−1.488	0.139	0.378	2.646
	Product quality	−0.024	0.108	−0.017	−0.222	0.824	0.497	2.011

[a] *Dependent Variable* Online Shopping Intention

value of 0.010), positive relationship, shipping process had a positive relationship Beta = 0.018, Sig = 0.866), positive. Website security had beta values of (−0.132, Sig = 0.139), ease accessibility had a positive relationship with beta = 0.390, Sig = 0.000), Ordering method had a negative effect on online purchasing intentions with (Beta, −0.240, sig value of 0.013), provision of warranty had a beta value of (Beta = 0.162, Sig value of 0.071), positive relationship. For order status had Beta = 0.187, Sig. value = 0.085). Payments had a beta value of 0.249, Sig of 0.007, the results were positive. For verified pages, as seen in the Table 3 the value of beta and sig. (Beta = 3.053 and sig. value .000) which means having verified pages will affect positively on consumer's intention to increase the possibility of purchasing through online sites.

By testing the assumption of multicollinearity, the test used here is VIF Variance Inflation Factors used to determine how much the independent variable is linearly related to one another. The model presented demonstrates that the variables of product quality had the VIF of 2.011, discounts had 1.996, shipping process had 3.938, website security had 2.646, ease had 2.475, being verified pages had 1.061, Ordering method had 3.095, the warranty had 2.679, Order status had 3.904, and payment had 2.855. The diagnostic tests for the regression model show that there exist no instances of collinearity as the VIF statistics associated with each of the independent variables in the model were within the acceptable range.

7 Hypotheses Testing Results

In this study, many factors were indicated as the main subject of study in order to understand the GCC customers thought about online shopping platform also to improve this sector according to customers' needs. The essential model was depicted in the methodology (Fig. 1). Also, some of the factors have been rejected according to the data collected. Thus, for the final model, which already accepted by analyzing the data.

The researcher assumed that the quality of the product had a positive impact on the intentions of consumers, but the results proved the opposite. It is because consumers are concerned that the product does not match what they expect to receive. From this, we find that the relationship between product quality and consumer intentions to buy online is negative. In the second factor, the researcher assumed that the relationship between the discount and consumer intentions is positive, and the results proved this hypothesis. It means creating discount campaigns from one period to another, increasing the customer's intention to buy online.

In the next factor, shipping, the researcher assumed that shipping has a positive impact on the intentions of consumers to buy through websites, but the results have proven that shipping has a negative impact as it turns out that most consumers are afraid that the product will not be delivered on time. So, we know that shipping must be developed to increase the intentions of consumers to purchase through the websites. Safety and security are one of the factors studied in this research, where the

Table 3 Model Summary of Verified Pages

Model	R	R square	Adjusted R square	Std. error of the estimate
1	0.435[a]	0.189	0.184	3.89036

Dependent Variable online shopping intention
[a]verified pages

ANOVA[a]

Model		Sum of Squares	Df	Mean Square	F	Sig.
1	Regression	572.586	1	572.586	37.832	0.000[b]
	Residual	2451.853	162	15.135		
	Total	**3024.439**	**163**			

[a]*Dependent Variable* online shopping intention
[b]verified pages.

Coefficients[a]

Model		Unstandardized Coefficients		Standardized Coefficients			Collinearity Statistics	
		B	Std. Error	Beta	t	Sig.	Tolerance	VIF
1	(constant) Verified page	11.869 3.053	0.364 .496	0.435	32.646 6.151	0.000 0.000	1.000	1.000

[a]Dependent Variable: online shopping intention

researchers assumed that the relationship between safety and security with consumer intentions was positive, and the results proved to be negative. Therefore, this factor must be improved to create a safety platform that helps to increase online purchasing.

The easy-to-use designs are one of the factors that have been identified and studied in this research, which have been assumed to have a positive impact on the intentions of consumers and the results have proven that this factor is important and has a positive impact on the consumer.

Having verified pages, which are among the most important features of this research, because this factor has not been researched a lot, so it was focused on in this research to know the true impact on the intentions of consumers. As the relationship between verified pages and consumer, intentions were to be positive, and the results approved this hypothesis. This type of page appeared in recent years, and the study proved that having this type of page increases the confidence of consumers on the site, which leads to an increase in the intentions of consumers to purchase through the Internet.

The ordering method is one of the factors that have been studied in this research, where the results have proven that the relationship between explaining the ordering method and the intentions of consumers is positive, therefore putting simple points explaining the way of making the order on the site increases the possibility of purchasing.

The researcher assumed that the warrantee had a positive impact on the consumers' intentions, but the results proved the opposite. As the results showed that consumers in the Gulf countries do not give this factor much attention compared to the factors that were adopted in this study. Order status, which is one of the factors that the researcher assumed had a positive relationship with the intentions of consumers, but the results obtained by the researcher proved to be unimportant through the study conducted in this research and it has a negative effect on customer's intention.

The payment method, one of the important factors in this research, has a relationship in safety, as explained previously. Therefore, the researchers assumed that the presence having more than one method of payment has a positive effect on the intentions of consumers in the study community, and the results are identical to this hypothesis (Fig. 3 and Table 4).

8 Conclusion

The study was set to investigate factors effect on online shopping intention to make first purchasing in terms of being verified web Pages in the Gulf countries. The study results indicate that five of the study factors had a negative effect on online shopping intention, which is clear according to the availability of the hypothesis. All hypotheses established to investigate the positive effect of the factors chosen, which means the hypothesis rejected according to this study has a negative effect on online shopping intention. These were product quality, website security, shipping proses, warranty, and order status. The researcher concluded that these factors

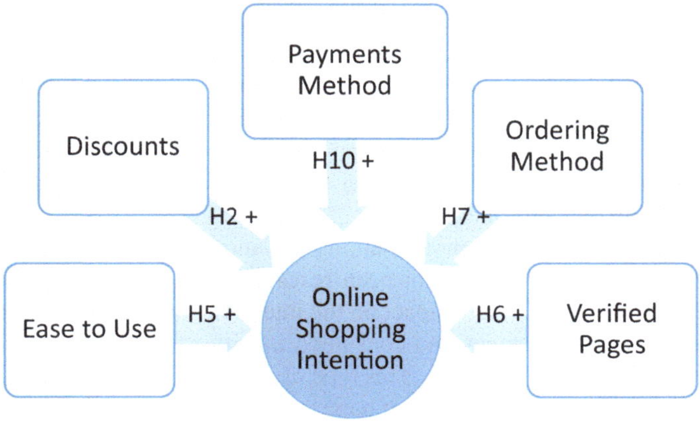

Fig. 3 Final accepted models

Table 4 Availability of hypothesis

	Hypothesis	Availability
H1	Product quality has a positive effect on online shopping intention	Rejected
H2	Discounts have positive effect on online shopping intention	Accepted
H3	Shipping process has a positive effect on online shopping intention	Rejected
H4	Website security has a positive effect on online shopping intention	Rejected
H5	Easy and not complicated websites design have a positive effect on online shopping	Accepted
H6	Being verified pages have a positive effect on online shopping intention	Accepted
H7	Explaining the way of making the order in website has a positive effect on online shopping intention	Accepted
H8	Warranty has positive effect on online shopping intention	Rejected
H9	Providing information of order status has a positive effect on consumer intention	Rejected
H10	Payments have a positive effect on online shopping intentions	Accepted

reduced online purchasing intention by the customers. We can understand that any misunderstanding in any factors above will lead to refuse the customer to take any action of making a new order. On the positive side, discounts, easy accessibility, Payments method, ordering method, and verified pages, have a positive impact on online shopping intentions. The study concluded that providing discounts offers, explaining the payments availability methods, the easy structure of the website (easy to get the product data- maximum 3 ticks according to the researchers studied) explaining the way of making order and being verified or having verified pages on (Facebook, Instagram, and other social media channels) will influence online shopping intention positively. The study indicates that websites that want to work in Gulf countries should explain the factors which have high value to the customers in this area; those factors were accepted in this study. Another factor also important, but each country has different cultures different thought.

The study based on the findings recommends that there is a need for increasing the product quality term since customers afraid to get a low-quality product to affect customer's attitudes on online shopping activities. Website security also needs enhancement through developing strict. So, those factors must be developed to increase the rate of affecting online shopping intention. So, the quality is one of the highest required of GCC customers. All websites want to access the gulf countries market should focus on this factor. Also, website security and customers' information one of the factors that affect online shopping intention (Negatively if it cannot be assured or positively if it assured by the website).

The shipping process and accessibility to the websites also contribute to the performance of online purchasing. There is a need to enhance the accessibility by possibly ensuring easy accessibility to the internet that can enhance the accessibility of the products to the customers online.

The study findings on the verified pages, warranty, contribute positively to the online shopping intention, improving verification of the web pages for the customers is fundamental since it will increase the customers' trust. On the other hand, having this feature will improve the image of the company in the eyes of customers. Providing and improving the warranty and information accessibility to online purchasing will lead to increase the customer intention according to this study because the GCC customers already accepted that warranty can influence their intention by increasing the ability to buy through online stores. Payment systems can be developed by establishing the form of payment method to generate performance for the organization. On the other hand, websites have different payment method will give the users more advantages. For instance, if the website has only credit/ debit card payment, they will lose customers who want to pay cash on delivery. So that having more than one payment method affecting positively on customer's intention.

This study found that customers need more development for the factors that were rejected, such as shipping, product quality, website security, warranty, and order status, because the customers are scared about having some problems related to those terms [26, 33, 20].

Survey Items

	Questions	Sources
1	I will strongly recommend online shopping to others	[17, 34]
2	I have strong possibility to purchase from online shops	

(continued)

(continued)

	Questions	Sources
3	I will likely visit online stores to shop for my needs	
4	I have high intention to purchase products through online shops	
5	When I hear a price discount announcement I immediately buy that product	[30]
6	I am more likely to buy brands that on sale	
7	It is extremely easy to prove the size, numbers, measure of products on online shopping	[23]
8	It is extremely easy to find the right products on online shopping	
9	It is extremely easy to shop online	Humaira (2008)
10	Webpages contain all necessary information about products and processes (product information, purchasing, return, cancelation, delivery etc)	
11	The website described the delivery method of the products in details	[34]
12	The website informed customers in advance of expected delivery time and charges involved	
13	The websites provided the carrier's mane and tracking mode of shipping	
14	The ability to track my shipment with my mobile devices	[31]
15	It is easy to make return/ exchanges through online shops	
16	The online stores have clear return policy	
17	The website indicated how products can be ordered step by step	[34]
18	The website offered instructions for the whole process of ordering	
19	The websites described the way to order the product in detail	
20	The websites provided customers with detail instructions on how to check their order	[34]
21	The websites provided customers a link to check their order status	
22	Customers could receive an e-mail to confirm whether or not the order process was successful	
23	It is extremely easy to cancel and return the order on online shopping	
24	The websites explained the return policy for the products it sold in detail	[34]
25	The website provided a detailed instruction of product return procedure	
26	The website showed whether or not the money back guarantee is available	
27	The websites provided the service policy for its customers such as installation and maintenance service	
28	The website explained various ways to pay for the product	[34]
29	The websites showed detailed instructions on how to pay for the product	
30	The websites explained each payment method available in detail	

(continued)

(continued)

	Questions	Sources
31	The website provided the guarantee that customers' personal information is protected online	[34]
32	The website showed that it could protect credit card information by encryption during a transaction	
33	The websites provided third-party authentication, Hacker safe, Trustee, and Verisign Secured to show that consumers' personal information is safety	
34	The website provided quality certification of the product (e.g. ISO quality certification)	[34]
35	The websites provided other users' comments on product quality	[14]
36	The websites presented expert comments on product quality	

References

1. Abu-Shamaa, R., Abu-Shanab, E.: Factors influencing the intention to buy from online stores: An empirical study in Jordan. In: 2015 IEEE 8th GCC Conference & Exhibition. https://doi.org/10.1109/ieeegcc.2015.7060022 (2015)
2. Aldhmour, F., Sarayrah, I.: An investigation of factors influencing consumers'intention to use online shopping: An empirical study in south of Jordan. J. Internet Banking Commerce. **21**(2) (2016)
3. Al-hashimi, M., Al-nidawi, W. J., Othman, M., Shakir, M., Sulaiman, H.: Evaluate information security governance frameworks in cloud computing environment using main and sub criteria. **16**(xx). https://doi.org/10.1166/jctn.2019.7989 (2019)
4. Alqahtani, A.S., Goodwin, R., de Vries, D.: Structural equation modelling of the factors influencing the adoption of E-Commerce in Saudi Arabia. J. Electr. Commerce Organ. **17**(4), 58–78 (2019). https://doi.org/10.4018/jeco.2019100105
5. Al-Qaysi, N., Mohamad-Nordin, N., & Al-Emran, M.: A systematic review of social media acceptance from the perspective of educational and information systems theories and models. J. Educ. Comput. Res. 073563311881787. https://doi.org/10.1177/0735633118817879 (2020)
6. Brengman, M., Geuens, M., Weijters, B., Smith, S.M., Swinyard, W.R.: Segmenting internet shoppers based on their Web-usage-related lifestyle: a cross-cultural validation. J. Bus. Res. **58**(1), 79–88 (2005)
7. Buğra, C.: Türkiye'de yeni kapitalizm: siyaset, din ve iş dünyası: new capitalism in Turkey: the relationship between politics, religion and business. İletişim (2015)
8. Correa, D., Silva, L.A., Mondal, M., Benevenuto, F., Gummad, K.P.: The many shades of anonymity: Characterizing anonymous social media content. In: ICWSM, (2015)
9. Davis, R., Stretton, A.: Signaling properties of Ascaris motoneurons: Graded active responses, graded synaptic transmission, and tonic transmitter release. J. Neurosci. **9**(2), 415–425. https://doi.org/10.1523/jneurosci.09-02-00415.1989 (1989)
10. Dumičić, K., Žmuk, B.: Forecasting of the internet usage in the world regions. Adv. Sci. Technol. Innov. 155–165. https://doi.org/10.1007/978-3-030-01662-3_18 (2019)
11. Eastin, M.S., Brinson, N.H., Doorey, A., Wilcox, G.: Living in a big data world: Predicting mobile commerce activity through privacy concerns. Comput. Human Behav. **58**, 214–220. https://doi.org/10.1016/j.chb.2015.12.050 (2016)

12. Elliot, S., Fowell, S.: Expectations versus reality: A snapshot of consumer experiences with Internet retailing. Int. J. Inf. Manage. **20**(5), 323–336. https://doi.org/10.1016/s0268-4012(00)00026-8 (2000)
13. Fishbein, M., Ajzen, I.: Belief, attitude, intention, and behavior: An introduction to theory and research. Addison-Wesley, Reading, MA (1975)
14. Galloway, J.N., Townsend, A.R., Erisman, J.W., Bekunda, M., Cai, Z., Freney, J.R., … Sutton, M.A.: Transformation of the Nitrogen Cycle: Recent trends, questions, and potential solutions. Science. **320**(5878), 889–892. https://doi.org/10.1126/science.1136674 (2008)
15. Granados, N., Gupta, A., Kauffman, R.J.: Online strategies for the information age: Information transparency as a dimension of information systems design. Info. Sys. Res. (2008)
16. Hajdu, J., Machin, P.A., Campbell, J.W., Greenhough, T.J., Clifton, I.J., Zurek, S., … Elder, M.: Millisecond X-ray diffraction and the first electron density map from Laue photographs of a protein crystal. Nature **329**(6135), 178–181. https://doi.org/10.1038/329178a0 (1987)
17. Hung, K., Pen, N., Hackley, C., Tiwsakul, R.A., Chou, C.: Antecedents of luxury brand purchase intention. J. Prod. Brand Manage. **20**(6), 457–467 (2011)
18. Huseynov, F., Yıldırım, S.Ö.: Internet users' attitudes toward business-to-consumer online shopping. Inf. Dev. **32**(3), 452–465. https://doi.org/10.1177/0266666914554812 (2014)
19. Martens, J., Grosse, R.: Optimizing neural networks with kronecker-factored approximate curvature. In: International conference on machine learning, pp. 2408–2417. (2015)
20. Monaco, G., Falconi, S., Crichton, W. A., Mezouar, M.: Nature of the first-order phase transition in fluid Phosphorus at high temperature and pressure. Phys. Rev. Lett. **90**(25). https://doi.org/10.1103/physrevlett.90.255701 (2003)
21. Niranjanamurthy, M., Kavyashree, N.: Analysis of e-commerce and m-commerce: advantages, limitations and security issues. Int. J. Adv. Res. Comput. Commun. Eng. **2**, 2360–2370 (2013)
22. Pinheiro, A., Cappelli, C., Maciel, C.: Building up a verified page on facebook Using Information Transparency Guidelines. Lecture Notes in Computer Science, 125–137 (2017). https://doi.org/10.1007/978-3-319-58562-8_10
23. Recep, Ö.: The effect of perceived risk on online. Marmara University, (2017)
24. Saxena, R.P.: Shoppers' attitude towards online retailing/an exploratory study performed in the context of United Arab Emirates (UAE). ravindrasaxena@ uowdubai. ac. ae (2016)
25. Sherif, A.: Challenges of attracting new customers to online shopping in Istanbul, Turkey (2016)
26. Siddiqui, H.N.: Investigation of intention to use e-commerce in the Arab countries: A comparison of self-efficacy, usefulness, culture, gender, and socioeconomic status in Saudi Arabia and the United Arab Emirates. Nova Southeastern University, (2008)
27. Singh, R.K., Rastogi, S., Aggarwal, M.: Analyzing the factors for implementation of green supply chain management. Competitiveness Review, (2016)
28. Swinyard, W.R., Smith, S.M.: Why people (don't) shop online: A lifestyle study of the internet consumer. Psychol. Mark. **20**(7), 567–597 (2003)
29. Wang, W., Qiu, L., Kim, D., and Benbasat, I.: Effects of rational and social appeals of online recommendation agents on cognition- and affect-based trust. Decision support systems (2016)
30. Yildirim, Y., Aydin, O.: Investigation of the effects of discount announcements on consumers' purchase decisions: A case study in supermarket. Proc. Soc. Behav. Sci. **62**, 1235–1244 (2012). https://doi.org/10.1016/j.sbspro.2012.09.212
31. Yingxia, C., Haya, A., Paul, H.: Post-purchase shipping and customer service experiences in online shopping and their impact on customer satisfaction: An empirical study with comparison. Asia Pacific J. Mark. Logist. **30**(2), 400–416. https://doi.org/10.1108/apjml-04-2017-0071 (2018)
32. y Monsuwé, T.P., Dellaert, B.G., De Ruyter, K.: What drives consumers to shop online? A literature review. Int. J. Serv. Ind. Manage. (2004)

33. Zhang, H., Zhao, L., Gupta, S.: The role of online product recommendations on customer decision making and loyalty in social shopping communities. Int. J. Inf. Manage. **38**(1), 150–166 (2018). https://doi.org/10.1016/j.ijinfomgt.2017.07.006
34. Zhou, L., Wang, W., Xu, J., Liu, T., Gu, J.: Perceived information transparency in B2C e-commerce: An empirical investigation, information and management. https://doi.org/10.1016/j.im.2018.04.005 (2018)

Knowledge Management

Critical Review of Knowledge Management in Healthcare

Afrah Almansoori, Mohammed AlShamsi, Said A. Salloum, and Khaled Shaalan

Abstract Healthcare is considered a pillar for keeping humans alive, thus managing their health through diagnosing and treating the disease or illness is a must, in addition to preventing it. Knowledge management (KM) in healthcare domain plays a substantial role in the implementation of different processes to ensure the existence of good and outstanding healthcare systems. In this study, a systematic review is carried out among previous studies to observe KM processes studies with respect to healthcare. Based on a search strategy, the systematic review offers a broad analysis of 10 quality assessed research articles published in peer-reviewed journals from 2009 till 2019. The results provide findings regarding the most KM processes studied, which are knowledge sharing and applications, and the least KM processes studied, which are knowledge acquisition and protection. Surveys and interviews were used as research methods to avail information in the context of KM processes. Taiwan, Finland, United States, Ghana, and Pakistan were the most active countries in the surveyed studies. Furthermore, the contributors to these studies were healthcare staff. This review study attempts to determine and feature KM processes research as an increase considering healthcare systems.

Keywords Healthcare · Knowledge management · Knowledge acquisition

1 Introduction

Providing healthcare considered a complicated effort that depends on data [1]. Healthcare organizations depend on information on curing, patients, providing care and health satisfaction [1–3]. Additionally, it relies on the performance to deliver,

A. Almansoori · M. AlShamsi · S. A. Salloum (✉) · K. Shaalan
Faculty of Engineering and IT, The British University in Dubai, Dubai, UAE
e-mail: ssalloum@sharjah.ac.ae

A. Almansoori
General Department of Forensic Science and Criminology, Dubai Police, Dubai, UAE

S. A. Salloum
Research Institute of Sciences and Engineering, University of Sharjah, Sharjah, UAE

M. Al-Emran et al. (eds.), *Recent Advances in Intelligent Systems and Smart Applications*, Studies in Systems, Decision and Control 295,
https://doi.org/10.1007/978-3-030-47411-9_6

coordinate and incorporate services [4]. Also, improving today's hospitals clinically and operationally are part of KM principles, which in turn positively impact healthcare [5].

A definition of knowledge management (KM) can be simplified as getting the most out of knowledge resources by doing what is needed [6]. Another way to define KM as a what it does: "comprises the identification and analysis of available and required knowledge, and the subsequent planning and control of actions to develop knowledge assets to fulfill individual and/or organizational objectives" [7].

In the 1990s, the research field of KM was developed by researchers from different disciplines such as organizational behaviour, strategic technology administration, organizational learning, Computer Science and Artificial Intelligence [8, 9]. Recently, organizations also adopt KM along with individuals [6]. In addition, a large number of studies were conducted to examine the role of KM and its processes on adopting different educational technologies [10–15]. Knowledge classifications is divided into tacit knowledge and explicit knowledge. Tacit knowledge cannot be transferred to others because it only resides in an individual mind as it is a result of personal experience [16]. Furthermore, it has an individual quality, thus it is to formalize and communicate [17]. Certainly, sharing "tacit knowledge among multiple individuals with different backgrounds, perspectives, and motivations becomes a critical step for organizational knowledge creation to take place" [18]. On the other hand, Explicit knowledge is easy to share or communicate in different forms [8]. Additionally, explicit knowledge can be created, written, documented, transferred through a communication mode such as telephone, emails or information systems [19].

KM has few processes as create, acquire, capture, share, use and re-use of knowledge. It also includes tacit and explicit knowledge [20]. Alternatively, KM is a significant discipline as it endorses the creation, sharing, and influencing of the corporation's knowledge [6]. Then again, "Traditional knowledge processes associated with KM are knowledge creation, knowledge acquisition, storage, transfer, sharing and application (or re-use) of knowledge. Traditionally, KM is associated with knowledge processes, various methodologies, and the use of knowledge technologies or KM systems that guarantee that knowledge assets are improved and effectively employed within organizations" [8]. The KM cycle model that includes all KM processes are needed to increase the performance of an organization [21]. Therefore, the purpose of this study is to produce a systematic review of previous studies on KM in healthcare to provide a comprehensive analysis. The study aims to answer five research questions. According to Table 1, five research questions are possessed for this study.

2 Literature Review

KM is considered the core of the successful business as it includes physical, employee, supplier, customer, and organization assets [22]. On the other hand, the foundation of creating a maintainable advantage is KM, as it plays a role of helping

Table 1 Research questions

Question ID	Research question
Q1	Considering the relationship with healthcare, what are the principle KM processes studied?
Q2	What are the fundamental research methods and findings tended to in the gathered studies?
Q3	What sorts of the healthcare system are primarily considered involving KM processes, and what are the sorts of participants within the collected studies?
Q4	What is the year of publication and what are the countries of implementation to KM processes studies in relation to healthcare systems?
Q5	What are the databases provided studies covering KM processes?

the organization to keep, develop, shape and use their knowledge. Globalization is the reason of an organization to survive by maintaining its knowledge resources [23]. KM strategies ensure easy access, custom-made, beleaguered, spread, and sharing of information among knowledge users [24]. In addition, KM aims to improve the effectiveness and performance of an organization by emphasizing the significance of knowledge creation and development [25].

Knowledge sharing is considered a pillar for the organization to improve its performance; as in healthcare, it is considered as a source of contribution to increasing patient safety [26]. The integration of people, technology and processes in healthcare involving KM is to ensure the improvements and availability of ideas and knowledge, expertise and experience to improve the delivery of healthcare [25]. Adding that KM in healthcare can be considered as the union of practices and methods to enable the creation, identification, acquisition, development, preservation, distribution and utilization of the various sides of a healthcare enterprise's knowledge assets [27].

KM provides advantages to healthcare delivery through reducing medical error, cooperation, and innovation, quality of care, cost reduction, organizing knowledge and organizational learning [28]. By using various techniques KM can define enablers in the healthcare sector, such as policy incentives as the key enabler, long-term strategic planning, alignment of KM efforts with business strategy and IT for KM [29]. On the other hand, barriers can affect applying KM, such as the non-existence of KM integration, doubt about the efficiency of KM, lack of common definition and discomfiture for sharing incorrect information [23]. In addition, lack of trust, deficiency of technological services, shortage of organizational policies, network failure, culture, and the unwanted task or extra responsibility are challenges facing knowledge sharing in healthcare [30]. Thus, there are three models of KM in healthcare organizations exist: KM conceptualization, total KM and KM infrastructure [31].

3 Research Methodology

Various reasons are considered for conducting a literature review, including eliciting information for creating and developing policies, in addition to evidence-based care [32]. Moreover, conducting literature reviews familiarize the researcher with his research topic; therefore, high-quality research is a reflection of the used sources [33]. As different type of research reviews exists and systematic review is one of it, [32], p. 39, states: "In contrast to the traditional or narrative review, systematic reviews use a more rigorous and well-defined approach to reviewing the literature in a specific subject area. Systematic reviews are used to answer well-focused questions about clinical practice". Alternatively, "A review of a formulated question that uses systematic and explicit methods to identify, select, and critically appraise relevant research, and to collect and analyze data from the studies that are included in the review. Statistical methods (meta-analysis) may or may not be used to analyze and summarise the results of the included studies" [34].

This study aims to provide a systematic review of the previous studies' methods for knowledge management in healthcare and to answer the research above questions. Consequently, we followed the guidelines of conducting systematic literature review developed by Ref. [35] and other systematic reviews [36–39]. In that, we identified the inclusion and exclusion criteria, sources of collected data, search strategy, quality assessment of the collected data, data coding, and analysis.

3.1 Search Criteria

The study accomplished by relating an inclusion-exclusion criterion on collected data to determine what paper should be included, as shown in Table 2.

The inclusion criteria cover papers of a period of 10 years from 2009 to 2019, written in the English language, and must focus on knowledge management and healthcare. On the other hand, the exclusions ruled out papers for more than ten years, other languages than English also papers not focused on knowledge management and healthcare.

Table 2 Inclusion and exclusion criteria

Criteria Inclusion	Criteria exclusion
Ten years (2009–2019)	more than ten years (2008 and less)
English language	Other languages
Must include knowledge management	Not including knowledge management
Must include healthcare	Not including healthcare

3.2 Search Strategy

The conducting mapping study was based on online databases of papers such as; BMJ, DIT, ScienceDirect, Emerald, IEEE, IGI, BioMed Central, Springer, MDPI, ECRI, Informs, Wiley Periodicals, Aspen, and Taylor & Francis. The mentioned databases covered specific conference proceedings and journals for the topic of our interest, Knowledge Management in healthcare. In addition to the selected online databases, the search strategy followed terms included the queries separated with a combination of the Boolean operator "AND", "OR": (("Knowledge Management" OR "KM") AND ("in healthcare" OR "health")) as showed in Table 3. Selecting related keywords for this study is a crucial step because it determines the retrieval of appropriate papers from databases [40].

Using the above search strategies, (74) relevant papers were found in total. As per Table 4, (N = 2) from BMJ, (N = 5) from DIT, (N = 15) from ScienceDirect,

Table 3 Searching queries

#	Search query
1	"Knowledge Management" and "in healthcare"
2	"KM" and "in healthcare"
3	"Knowledge Management" and "health"
4	"KM" and "health"

Table 4 Total number of papers found in a database

#	Database Name	Total number of papers found (N)
1	BMJ	2
2	DIT	5
3	ScienceDirect	15
4	Emerald	15
5	IEEE	5
6	IGI	2
7	BioMed Central	2
8	Springer	6
9	MDPI	2
10	ECRI	1
11	Informs	3
12	Wiley Periodicals	2
13	Aspen	5
14	Taylor & Francis	9
Total	74	

(N = 15) from Emerald, (N = 5) from IEEE, (N = 2) from IGI, (N = 2) from BioMed Central, (N = 6) from Springer, (N = 2) from MDPI, (N = 1) from ECRI, (N = 3) from Informs, (N = 2) from Wiley Periodicals, (N = 5) from Aspen, and (N = 9) from Taylor & Francis.

3.3 Processing and Selecting

A total of 74 articles included these keywords. Among these, it was found that 3 articles were repeated, and so they were eliminated. A total of 71 papers were now remaining for the review. The way these articles are distributed with respect to the databases they are a part of is shown in Fig. 1. As a second step, the title and abstract were screened based on the previously defined criteria, which made the total number of papers (N = 25). By checking the eligibility of papers after entire text reading against the search queries, the papers left for this study are (N = 10).

3.4 Quality Assessment

Quality assessment is an essential factor that must be examined along with the inclusion-exclusion criterion [41–43]. A checklist with nine criteria (Table 5) for a quality assortment was identified to evaluate the 10 studies and to prepare it for further analysis. This checklist was adapted from those proposed by [35]. The checklist was not planned to be a form of criticism of any researchers' work [35]. Each question in the checklist was scored according to the three-point scale. An answer of "Yes" being worth 1 point, an answer of "No" being worth 0 point, and an answer of "Partially" being worth 0.5 point. Therefore, each study could score between 0 and 9, with the higher the total score a study attains, the higher the degree to which this study addresses the research questions.

As a result, Table 6 presents the quality assessment results of all 10 studies. As all of the studies passed the quality assessment, hence, it is eligible for further analysis.

3.5 Data Coding and Analysis

Several interrelated features to the research methodology quality were tilted including (a) author, (b) year of publication, (c) research aim, (d) type KM processes, (e) research method, (f) research findings, (g) country of which the study was about, (h) research participants and (i) database (Table 8). Some of the studies didn't clarify impact of KM process on healthcare systems, therefore it got excluded from the synthesis.

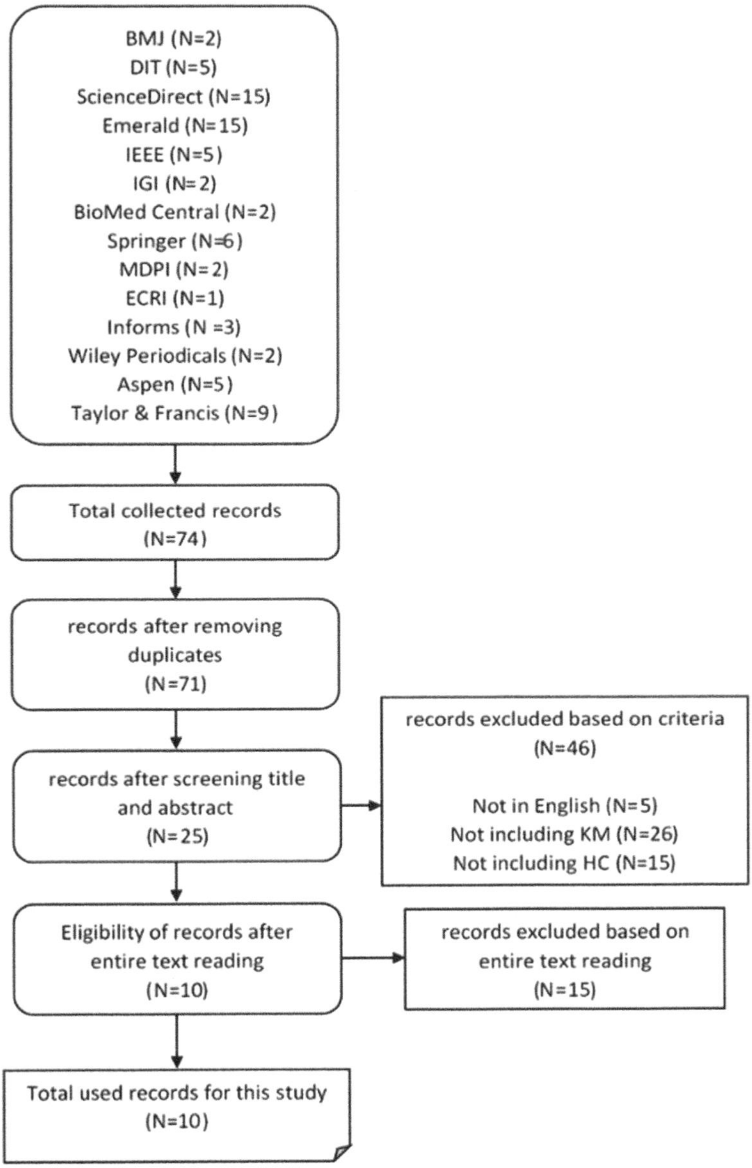

Fig. 1 Literature search flowchart

Table 5 Quality assessment checklist

#	Questions
1	Is the research aim specified clearly?
2	Did the study achieve its aim?
3	Are the variables considered by the study clearly indicated?
4	Is the context/discipline of the study clearly defined?
5	Are the method of collecting data sufficiently detailed?
6	Is the study describing the measure's reliability/validity?
7	Are the statistical techniques used to analyze the data sufficiently described?
8	Do the findings add to the literature?
9	Does the study add to your knowledge or understanding?

4 Results

RQ1. Considering the relationship with healthcare, what are the principle KM processes studied?
Numerous research studies were conducted to examine the effect and adoption of KM in healthcare sector. As 10 studies were used for this research, Table 7 presents the most frequent KM processes studied by the authors. It is noticeable that knowledge sharing is the most studied KM process (N = 7). Followed by knowledge application and healthcare knowledge (N = 6), then knowledge creation (N = 2), Knowledge storage (N = 1) and last Knowledge acquisition and protection to be (N = 0). With reference to Table 7, it is obvious that knowledge sharing was affecting the acceptance and implementation of healthcare systems [22, 24–26, 28, 30]. On the other hand, Knowledge storage, acquisition, and protection was less studied.

RQ2: What are the fundamental research methods and findings tended to in the gathered studies?
Figure 2 shows that 60% of the analyzed studies did not specify the research methods used for the study (N = 6). Interviews and surveys were found as research methods in total (N = 4), divided equally (N = 2) for Interviews and (N = 2) for surveys. The distribution of research outcomes, as shown in Fig. 3, is revealing that (N = 6) of the analyzed studies research outcomes as positive (N = 3) as neutral and (N = 1) as negative. As a percentage calculation, 60% are positive, 30% are neutral, and 10% are negative. The positive outcomes of the studies were relaying on questionnaire surveys as a method of data collection.

RQ3: What sorts of healthcare systems are primarily considered involving KM processes, and what are the sorts of participants within the collected studies?
The distribution of the analyzed studies with regards to types of healthcare systems considering KM processes. It is observed that healthcare systems were focusing on studying the impact of KM processes on the implementation of the healthcare system

Table 6 Quality assessment results

Question study	Q1	Q2	Q3	Q4	Q5	Q6	Q7	Q8	Q9	Total	Percentage (%)
S1	1	1	1	0.5	0.5	0.5	0.5	1	1	7	78
S2	1	0	0.5	0.5	1	1	1	0	1	6	67
S3	1	1	1	1	1	0.5	0	0.5	1	7	78
S4	1	1	0.5	1	0.5	0.5	0.5	1	1	7	78
S5	0.5	1	0.5	0.5	0	0.5	0	0.5	1	4.5	50
S6	1	0.5	1	0.5	0.5	0.5	0.5	0.5	1	6	67
S7	1	1	1	1	1	0.5	0.5	0.5	1	7.5	83
S8	1	1	0.5	1	1	1	1	1	1	8.5	94
S9	1	1	1	0.5	1	0.5	1	0	1	7	78
S10	1	0.5	0.5	0.5	1	0.5	1	0	1	6	67

Table 7 KM processes classification across analysed studies

#	Source	Knowledge acquisition	Knowledge sharing	Knowledge application	Knowledge protection	Knowledge storage	Healthcare knowledge	Knowledge creation
1	[28]		*	*			*	
2	[27]			*			*	
3	[25]		*				*	
4	[26]		*				*	
5	[22]		*					
6	[24]		*	*				*
7	[31]		*	*		*		*
8	[30]		*					
9	[23]			*			*	
10	[23]			*			*	

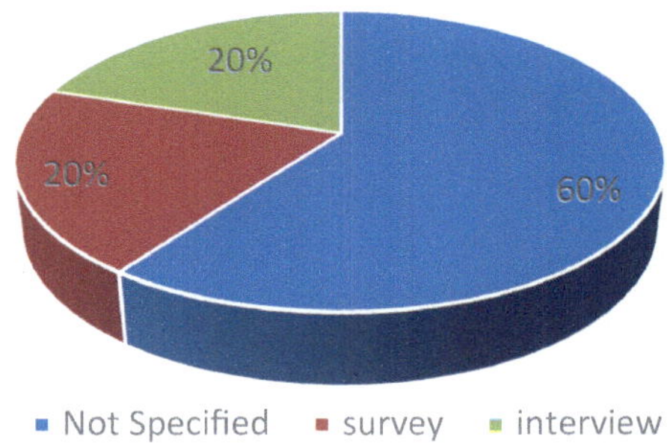

Fig. 2 Distribution of studies by research methods

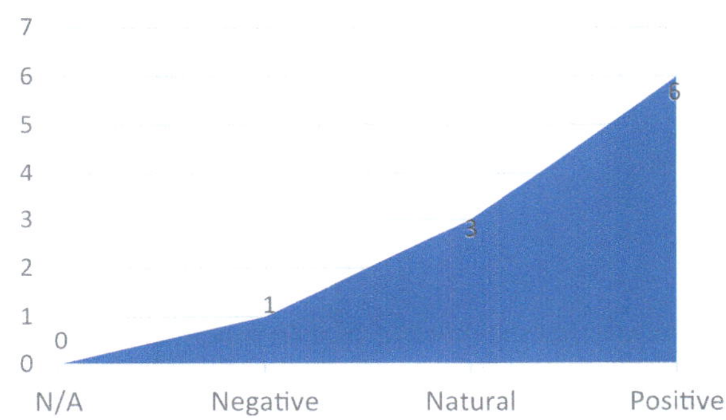

Fig. 3 Distribution of studies in terms of research outcomes

(N = 7). Studies didn't specify the exact numbers of participants of each category, therefore it only showed the total number (X = 494) of participants of mixed doctors, nurses, patients, and technical staff.

RQ4: What is the year of publication and what are the countries of implementation to KM processes studies in relation to healthcare systems?

This section presents a year of publication and its distribution over the research studies. Hence this study was conducted on published studies over the last ten years from 2009 until 2019, Fig. 4 presents the distribution of the analyzed studies. It is

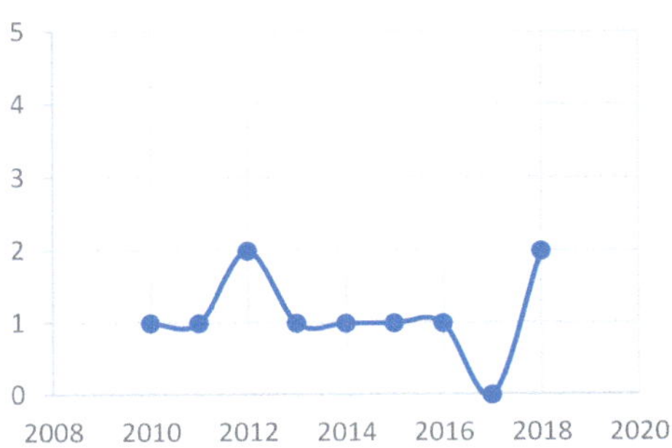

Fig. 4 Distribution of studies by publication year

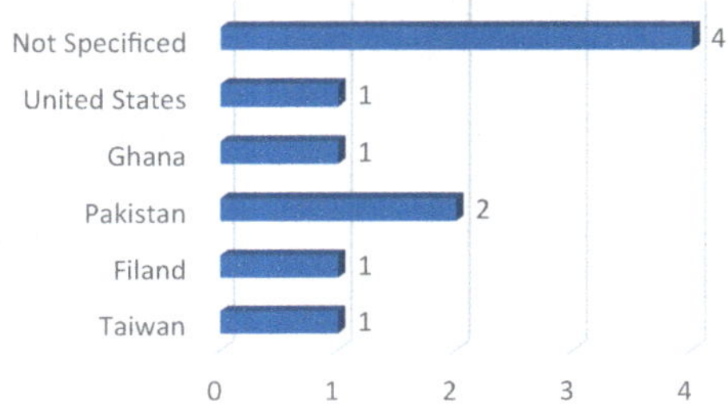

Fig. 5 Distribution of studies by country of implementation

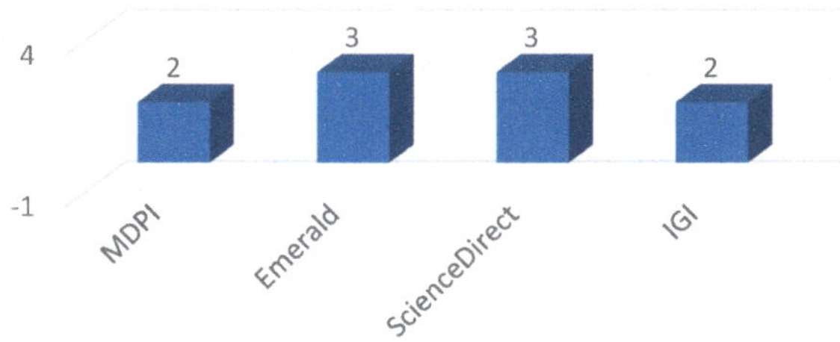

Fig. 6 Distribution of studies in terms of database

noticeable that the number of yearly publications is increasing. In the year of 2009 and 2017, there were no published studies. However, between 2010 and 2018, a good number of publications was available in the database, and at a rate of 2 in the year 2012 and 2018.

In relation to the country that implemented KM processes and its distribution over the research, Fig. 5 shows the distribution of the analyzed studies over the countries in which these studies were carried out. It is noticeable that two of the studies were carried out in Pakistan, followed by whoever United Stated, Ghana, Taiwan, and Finland.

RQ5: What are the databases provided studies covering KM processes?
Here it is showing the database that had an influence of publishing studies in relation to KM processes and healthcare systems as a relationship. As Fig. 6, Emerald and ScienceDirect are the databases that mostly supported publishing studies (N = 3) each, hence the productive database among the other, followed by MDPI and IGI (N = 2) each.

5 Discussion

The combination of KM with healthcare allows organizations to use the information as a good and effective decision-making tool. Accepting and implementing of different kinds of healthcare systems are affected by whether KM processes play a significant role in it. What this study aims to is reviewing systematically the studies published on the topic and synthesize it as an attempt to improve the understanding of the related features of KM processes and its relationships to successful healthcare systems.

Table 7 shows the classification of KM processes among the ten analyzed studies. It is noticeable that knowledge sharing is studied frequently, followed by knowledge application, creation, storage, acquisition, and protection, respectively (Table 8).

Figure 3 shows that 40% of the analyzed studies used interviews and surveys as research methods; however, 60% of the studies did not specify the research methods used for the study. 40% emphasizes that interviews and surveys are suitable methods for such research. In addition, Fig. 4 indicates that 60% of the analyzed studies research outcomes are positive, 30% are neutral, and 10% are negative. The adoption, acceptance, and implementation of healthcare systems are usually affected by KM processes. Another indication is that questionnaire surveys as a method for data collection is the reason for positive outcomes results. These findings point out that questionnaire surveys are considered as great methods for data collection, which could achieve positive outcomes based on the context, participants, and sample tool. In terms of participants, the studies generally focused on whoever affecting the implementation of KM in healthcare. They were doctors, nurses, patients, and technical staff. This means that those participants were eligible to evaluate healthcare systems from adoption, acceptance, and implementation aspects. With respect to country, Fig. 5 clearly shows that a good number of KM processes studies were carried out in Pakistan, followed by whoever United Stated, Ghana, Taiwan, and Finland. These findings could be qualified for the reason that the country is investing in the healthcare field rather than other research fields.

In relation to the year of publication, Fig. 4 indicates that the number of research articles was improved from one study in 2010 to two studies in the last year. This increase can be a contribution to improving KM processes that study influenced these processes on healthcare systems. With respect to the database, Fig. 6 presents that Emerald and ScienceDirect are considered the most supportive and productive database that stores many KM processes research in healthcare, followed by MDPI and IGI respectively among the other databases. This result could be useful for future research as these databases can be used for collecting the KM articles from.

6 Conclusion and Future Work

Earlier KM review studies gave an appreciated vision about the research directions of KM. These review studies neglected the effect of KM processes on healthcare. In this study, a systematic review of the literature of KM studies related to healthcare was carried out, in an attempt to deliver a complete analysis of the existing studies and to deliberate the effects of analysis results. This study shows that 7 new findings were revealed. Starting at the knowledge sharing process is the most KM process studied, followed by knowledge application and healthcare knowledge, creation, storage, acquisition, and protection, respectively. Second, interviews and questionnaire

Table 8 Analysis of KM processes studies related to different healthcare systems

#	Source	Year	Aim of study	Methods	Country	Participants	Database	Findings
1	[28]	2010	(a) An overview methods and techniques of KM (b) Challenges and needs of KM (c) KM application healthcare and KM future	–	–	–	IGI	(a) By adopting KM ways complexness challenges facing the healthcare sector are often resolved. (b) Using KM in healthcare promises to support the quality of patient care by ensuring continuity of care (c) Implementation of KM system for health can change health partners to conduct evidence based follow and to collaborate depend rely upon the supply of best knowledge
2	[27]	2011	(a) Define KM and its impact which can support in decision making (b) The importance of using real data in the study of the research base for the implementation of knowledge management in the healthcare industry	Survey	Taiwan	227 doctors, nurses and technicians	Science Direct	(a) Many Taiwanese organizations supported the management of classified knowledge, and the government has created numerous governing mechanisms for this (b) The development and implementation of KM in the healthcare sector is not noticeable, especially in the area of infection control

(continued)

Table 8 (continued)

#	Source	Year	Aim of study	Methods	Country	Participants	Database	Findings
3	[25]	2012	(a) Improve considerate about role of knowledge and information in processing of healthcare (b) To support service provision its need to build basis for practices	Interview	Finland	32 patient and different level personnel	Emerald	(a) For analyzing various aspects of knowledge management in the health process a systematic framework need to be Provide (b) Provides useful ideas for analysing knowledge-depend on the basics of medical services
4	[26]	2012	(a) Find out to what extent institutional structures contribute to the sharing of knowledge and its impact on organizational performance	Survey	United States	220 medical staff	Emerald	(a) Institutions have severely compromised knowledge sharing practices and together dramatically improves patient safety (b) Institutional structures that had a major impact on leadership roles and also the limitation of repressing behaviour, successively together considerably improve its patient care
5	[22]	2013	(a) Explain role of knowledge management in healthcare sector	–	–	–	Science Direct	(a) New ways of operating were needed when the number of players and the exponential increased (b) A new business model need to be built around the sector's most prized resource

(continued)

Table 8 (continued)

#	Source	Year	Aim of study	Methods	Country	Participants	Database	Findings
6	[24]	2014	(c) What are some common features of managing public health knowledge sharing portals? (d) How can KEP help manage knowledge for healthcare professionals? (e) Do KEPs effectively apply them and facilitate knowledge management in public health practices?	Peer-reviewed and grey literature	–	–	Science Direct	(a) Accessing resources in one place, sharing and distributing information, and bringing people together to exchange the knowledge is done by using knowledge exchange portals as a platform
7	[31]	2016	(a) Discuss KM models (b) Development of proposed framework	–	–	–	IGI	(a) Lacking of expansive and detailed focus on the role of the upper management in knowledge sharing and creation (b) Processes linked to the transfer of knowledge creation and application inside the health organization may vary depending on the type of work of each group of workers

(continued)

Table 8 (continued)

#	Source	Year	Aim of study	Methods	Country	Participants	Database	Findings
8	[30]	2015	(a) Status of healthcare professionals share knowledge in the Ghanaian healthcare sector (b) Challenges that healthcare professionals in Ghana face in sharing knowledge	Research paper	Ghana	15 doctors, nurses, pharmacists, physician assistants and biomedical scientists	Emerald	(a) No formal knowledge management systems for healthcare (b) Healthcare professionals depend on informal conversations and seminars to share knowledge
9	[23]	2018	(a) Identify knowledge management enablers (b) Identify and analyse the relationships between the enablers using ISM technique (c) Improve a hierarchy of enablers within Pakistan healthcare sector	–	Pakistan	–	MDPI	(a) The base enabler, policy incentive is focused on, it will lead toward more facilitators that will support for successful implementation of KM

(continued)

Table 8 (continued)

#	Source	Year	Aim of study	Methods	Country	Participants	Database	Findings
10	[29]	2018	(a) Identify and analyse the barriers (b) How the different barriers support and influence each other	–	Pakistan	–	MDPI	(a) The ISM method does not statistically check relationships; ISM is a tool that assistances in developing original models (b) SEM can statistically verify them and propose for this case

surveys as a method for data collection is the reason for positive outcomes results. Third, 60% of the analyzed studies registered positive research outcomes, followed by 30% as neutral outcomes. Forth, in terms of participants, most of the analyzed studies were primarily focused on doctors, nurses, patients, and technical staff in terms of data collection. Fifth, most of the analyzed studies were carried out in Pakistan, followed by whoever the United Stated, Ghana, Taiwan, and Finland. Sixth, the number of publications concerned with KM related to healthcare are increasing. This study witnessed a remarkable increase in published articles in 2012 and 2018.

As a limitation, specific databases restricted collecting research studies for this systematic review, such as Emerald, ScienceDirect, MDPI, and IGI. Those databases could not provide a good illustration of the relationship between articles published on KM and healthcare. Further research could be extended in the future to include studies from other databases such as Web of Science, Scopus, and Sage.

Acknowledgements This work is a part of a project undertaken at the British University in Dubai.

References

1. Alhashmi, S.F.S., Salloum, S.A., Abdallah, S.,: Critical success factors for implementing Artificial Intelligence (AI) projects in Dubai Government United Arab Emirates (UAE) health sector: Applying the extended technology acceptance model (TAM). In: International Conference on Advanced Intelligent Systems and Informatics, pp. 393–405 (2019)
2. Alshurideh, M.: Pharmaceutical promotion tools effect on physician's adoption of medicine prescribing: Evidence from Jordan. Mod. Appl. Sci. **12**(11) (2018)
3. Alhashmi, S.F.S., Salloum, S.A., Mhamdi, C.: Implementing Artificial Intelligence in the United Arab Emirates healthcare sector: An extended technology acceptance model. Int. J. Inf. Technol. Lang. Stud. **3**(3) (2019)
4. Bose, R.: Knowledge management-enabled health care management systems: Capabilities, infrastructure, and decision-support. Expert Syst. Appl. **24**(1), 59–71 (2003)
5. Guptill, J.: Knowledge management in health care. J. Health Care Finance **31**(3), 10–14 (2005)
6. Becerra-Fernandez, I., Sabherwal, R.: Knowledge management systems and processes. M. E. Sharpe, Inc. (2010)
7. Macintosh, A.: Position paper on knowledge management. Artifitial Intell. Appl. Institute, Univ. Adinburgh **2** (1997)
8. Razmerita, L., Wren, G., Jain, L.C.: Innovations in knowledge management: The impact of social media, semantic web and cloud computing (2016)
9. Salloum, S.A., Al-Emran, M., Shaalan, K.: The impact of knowledge sharing on information systems: A review. In: International Conference on Knowledge Management in Organizations, pp. 94–106 (2018)
10. Al-Emran, M., Mezhuyev, V., Kamaludin, A.: Towards a conceptual model for examining the impact of knowledge management factors on mobile learning acceptance. Technol. Soc. (2020)
11. Al-Emran, M., Mezhuyev, V., Kamaludin, A.: An innovative approach of applying knowledge management in M-Learning application development: A Pilot study. Int. J. Inf. Commun. Technol. Educ. **15**(4), 94–112 (2019)
12. Al-Emran, M., Mezhuyev, V.: Examining the effect of knowledge management factors on mobile learning adoption through the use of importance-performance map analysis (IPMA). In: International Conference on Advanced Intelligent Systems and Informatics, pp. 449–458 (2019)

13. Al-Emran, M., Teo, T.: Do knowledge acquisition and knowledge sharing really affect e-learning adoption? An empirical study. Educ. Inf. Technol. (2019)
14. Al-Emran, M., Mezhuyev, V., Kamaludin, A., AlSinani, M.: Development of M-learning application based on knowledge management processes. In: 2018 7th International conference on Software and Computer Applications (ICSCA 2018), pp. 248–253 (2018)
15. Al-Emran, M., Mezhuyev, V., Kamaludin, A.: Students' perceptions towards the integration of knowledge management processes in M-learning systems: A preliminary study. Int. J. Eng. Educ. **34**(2), 371–380 (2018)
16. Sensky, T.: Knowledge management strategy (2002)
17. Nonaka, I.: A dynamic theory knowledge of organizational creation. Organ. Sci. **5**(1), 14–37 (1994)
18. Nonaka, I., Takeuchi, H.: The knowledge-creating company: How Japanese companies create the dynamics of innovation (1995)
19. Choi, B., Lee, H.: An empirical investigation of KM styles and their effect on corporate performance. Inf. Manag. **40**(5), 403–417 (2003)
20. Jain, K.K., Sandhu, M., Sidhu, G.: Knowledge sharing among academic staff: a case study of business schools in Klang Valley, Malaysia. J. Adv. Sci. Arts **2**, 23–29 (2007)
21. King, W.R., Rachel Chung, T., Haney, M.H.: Knowledge management and organizational learning. Omega **36**, 6068–6078 (2006)
22. Hasoneh, A.: The role of knowledge management in the healthcare sector. Int. J. Adv. Res. **1**(6), 436–442 (2013)
23. Karamat, J., Shurong, T., Ahmad, N., Waheed, A., Khan, S.: Barriers to knowledge management in the health sector of Pakistan. Sustain. **10**(11) (2018)
24. Quinn, E., Huckel-Schneider, C., Campbell, D., Seale, H., Milat, A.J.: How can knowledge exchange portals assist in knowledge management for evidence-informed decision making in public health? BMC Public Health **14**(1) (2014)
25. Myllärniemi, J., Laihonen, H., Karppinen, H., Seppänen, K.: Knowledge management practices in healthcare services. Meas. Bus. Excell. **16**(4), 54–65 (2012)
26. Kim, Y.M., Newby-Bennett, D., Song, H.J.: Knowledge sharing and institutionalism in the healthcare industry. J. Knowl. Manag. **16**(3), 480–494 (2012)
27. Chen, Y.H., Liu, C.F., Hwang, H.G.: Key factors affecting healthcare professionals to adopt knowledge management: The case of infection control departments of Taiwanese hospitals. Expert Syst. Appl. **38**(1), 450–457 (2011)
28. El Morr, C., Subercaze, J.: Knowledge management in healthcare. no. May, pp. 490–510 (2010)
29. Karamat, J., Shurong, T., Ahmad, N., Waheed, A., Mahmood, K.: Enablers supporting the implementation of knowledge management in the healthcare of Pakistan. Int. J. Environ. Res. Public Health. **15**(12) (2018)
30. Assem, P.B., Pabbi, K.A.: Knowledge sharing among healthcare professionals in Ghana. VINE J. Inf. Knowl. Manag. Syst. **46**(4), 479–491 (2016)
31. Sabeeh, Z.A., Mustapha, S.S., Mohamad, R.: A review of three models for knowledge management in healthcare organizations. IGI (2016)
32. Aleshire, R.A.: Planning and citizen participation. Urban Aff. Q. **5**(4), 369–393 (1970)
33. Dawidowicz, P.: Literature reviews made easy: A quick guide to success (2010)
34. Collaboration The Cochrane.: Glossary of Terms in the Cochrane Collaboration. Cochrane Handb. Syst. Rev., no. May, p. 51 (2005)
35. Kitchenham, B, Charters, S.: Guidelines for performing systematic literature reviews in software engineering. Softw. Eng. Group, Sch. Comput. Sci. Math. Keele Univ. pp. 1–57 (2007)
36. Al-Saedi, K., Al-Emran, M., Abusham, E., El_Rahman, S.A.: Mobile payment adoption: A systematic review of the UTAUT Model. In: International Conference on Fourth Industrial Revolution (2019)
37. Saa, A.A., Al-Emran, M., Shaalan, K.: Factors affecting students' performance in higher education: A systematic review of predictive Data Mining techniques. Technol. Knowl. Learn. (2019)

38. Al-Qaysi, N., Mohamad-Nordin, N., Al-Emran, M.: A systematic review of social media acceptance from the perspective of educational and information systems theories and models. J. Educ. Comput. Res. **57**(8), 2085–2109 (2020)
39. Al-Emran, M., Mezhuyev, V., Kamaludin, A., Shaalan, K.: The impact of knowledge management processes on information systems: A systematic review. Int. J. Inf. Manage. **43**, 173–187 (2018)
40. Costa, V., Monteiro, S.: Key knowledge management processes for innovation: A systematic literature review. VINE J. Inf. Knowl. Manag. Syst. **46**(3), 386–410 (2016)
41. Al-Emran, M., Mezhuyev, V., Kamaludin, A.: Technology acceptance model in M-learning context: A systematic review. Comput. Educ. **125**, 389–412 (2018)
42. Salloum, S.A., Alhamad, A.Q.M., Al-Emran, M., Monem, A.A., Shaalan, K.: Exploring students' acceptance of E-learning through the development of a comprehensive technology acceptance model. IEEE Access **7**, 128445–128462 (2019)
43. Salloum, S.A.S., Shaalan, K.: Investigating students' acceptance of E-learning system in Higher Educational Environments in the UAE: Applying the Extended Technology Acceptance Model (TAM). The British University in Dubai (2018)

Knowledge Sharing Challenges and Solutions Within Software Development Team: A Systematic Review

Orabi Habeh, Firas Thekrallah, Said A. Salloum, and Khaled Shaalan

Abstract Effective and efficient management of project teams and effective knowledge sharing policy imposed by organization have significant positive impact on the knowledge sharing among software development team. Thus, organizations should enhance and improve its culture in order to offer an effective knowledge sharing process that leads to project success; thus, address challenges. This paper conducts a systematic literature review of the knowledge sharing challenges within software development team and evaluate the suggested solutions based on the methods used for the selected studies. This systematic literature review emphasizes on all knowledge sharing challenges and recommend the optimum knowledge sharing solutions based on comparing and contrasting all challenges and barriers gathered from the literature. Data collection is performed, and 9 research papers were found related to the stated objective, which can answer the stated research questions. In our methodology, 37 challenges explored, and 19 suggested solutions were gathered, compared, and discussed in the systematic literature review. Generally, challenges and suggested solutions across all selected studies were structured under four main pillars: individual, technology, project, and organization. Recommended framework and solutions have been addressed in this paper based on the results of this systematic review.

Keywords Knowledge sharing barriers in software development · Knowledge sharing challenges in software development · Knowledge sharing drivers among software development members · Impact of knowledge sharing among software development teams

O. Habeh · F. Thekrallah · S. A. Salloum (✉) · K. Shaalan
Faculty of Engineering and IT, The British University in Dubai, Dubai, UAE
e-mail: ssalloum@sharjah.ac.ae

S. A. Salloum
Research Institute of Sciences and Engineering, University of Sharjah, Sharjah, UAE

M. Al-Emran et al. (eds.), *Recent Advances in Intelligent Systems and Smart Applications*, Studies in Systems, Decision and Control 295,
https://doi.org/10.1007/978-3-030-47411-9_7

1 Introduction

Software development is considered one of the most essential and critical tasks in any organization. The known steps of software development task are defined as analysing, designing and implementing the desired information system application to enhance and improve organization activities and procedures. Thus, it can achieve organization goals, sustain and increase its profit [1]. However, and due to the technological issues, time constraint, among other challenges, are considered a complex, lengthy and critical task in any organization [2]. Many factors could affect the acceptance of software engineering techniques [3]. Due to these factors and challenges rationale organizations used to appoint specialized project teams which may consist of, for instance, specialized developers, designers, network specialists, project managers and others [4, 5]. The purpose is to overcome the challenges; thus, seamless improvement and aim for successful software development projects [6].

Basically, software development team in many critical industries are considered one of the most important teams in the organization, for instance, in public transportation industry software development teams are fully responsible to develop, update, maintain and implement all transportations systems, including fare deductions, command control, road signals, ...etc. Therefore, any system failure, delays, or failure to perform scheduled maintenance will cause significant negative impacts to the whole chain of the involved organizations which will impacts its profit and publicity [7, 8]. Thus, they are considered critical to the most important teams in the organization. Similar case to banks [9–11]. Based on the known fact "Knowledge is power" and knowledge sharing is part of spreading this power, the key element of software development organization success is to be able to allow transferring the knowledge among members of the project team effectively and efficiently; thus, spreading knowledge power among team members [12, 13]. Moreover, and due to the fact that many organizations are using project teams in its software development projects, several studies have shed the light on identifying successful elements of the project team as well as key challenges of sharing knowledge among members of a software development team in order to support project success factors [14].

2 Literature Review

Knowledge sharing was found to affect the adoption of several technologies in the past [15–20]. A large and growing body of the literature has investigated knowledge sharing challenges and suggested different solutions, modules in order to eliminate stated challenges. According to the literature, challenges identified on the selected studies vary, which are based on different categories and sub-categories. [14] categorized the challenges into three different categories and each category consists of different sub-categories: Knowledge-related category which consists of (value of knowledge, knowledge as power and nature of knowledge), and individual-related category which consists of (absorptive capacity, relationship between senders and recipients and different mentalities), and organization-related category which consists of (culture, restraining forces in the organization). Similarly, and according to [21], barriers of sharing knowledge among software development teams categorized into individual, organizational, but he added the technological levels into it. However, the result of the survey conducted in this research shows most effective challenges under each level identified as below [21].

- Individual level: Low awareness of others and its values, lack of social networking, and less power in terms of relations and time.
- Organization level: across distance, challenges of networking across organization, communication of the benefits of knowledge sharing by management, comment culture and attitude.
- Technological level: incompatible technology.

Many scholars such as [22–24] claimed that the knowledge sharing challenges between software developers are categorized based on: Team diversity, team perceptions, team capabilities, project communication, project organization, project technology, project setting. Total of 37 challenges highlighted under these categories; similarly, [25] claimed perceptions on knowledge as a challenges, but they pointed out time availability, the gained values and power of knowledge as well as distance of employees are considered important challenges which require organization attentions. Additionally, they argued that relationships between the two challenges "knowledge and distance" and "gain value for the of employees" resulted in negative relationship. On the one hand, the perceptions and time availability challenges shows positive and strong relationships which leads to significant consideration when proposing any mitigated solutions for these particular challenges [25]. on the other hand, [23] claims that when offering more generalized conceptualization of knowledge sharing barriers specifically to agile software development teams it will support future research to explore the relationship between these barriers.

Furthermore, [26] pointed out that task-oriented, social, and institutional context between developers consider key factors which impact sharing knowledge between developers; thus, massively impacts its performance. Moreover, they claimed a strong relationship between knowledge sharing in software development and developer's performance which they prove it through five theories. However, the argued conceptual model mainly focuses on four variables only (social, task, formal and Informal Institutional Dimension). Other variables are not considered; thus, stated theory and conceptual model can be considered but in narrow fields, environment and specific situations. Moreover, it is argued that one of the most essential challenges of knowledge sharing is the lack of awareness of the knowledge sharing policies adopted in the organization; however, extrinsic and intrinsic motivations consider one the core fact which can influence knowledge sharing awareness; thus, increase developers' performance. Additionally, the argued module which was based on the survey results shows significant impact of the extrinsic and intrinsic knowledge sharing motivations on a team's performance. [27]; therefore, we can conclude that lack of motivations for sharing knowledge can be considered one of the knowledge sharing challenges among members of the software development team which impact team's performance.

On the other hand, [26] claimed different challenges (task-oriented, social, and institutional context) which impacts teams performance; However, other selected studies claimed the same challenges but at a higher level. For examples, [14] stated that individual related category can be considered as challenge which consists of motivations challenge, but the major differences from the other selected studies are that it focuses on in depth with prof and deep understanding of how motivations in knowledge sharing aspect impacted from extrinsic and intrinsic factors and how it positively impact a team performance. In addition, [28, 29] claimed that trust–dependence between developers are considered a major challenge of knowledge sharing within the team. They argued that if this challenge eliminated, it will result in significant contributions to the project success [28]. Nevertheless, and according to [30], other challenges which are not mentioned in the selected studies are highlighted and requires significant considerations. For instance, consider time constraints, job security, resistance to change, communication, physiological factors and rewards [30]. However, [30] shows more precise results of the stated challenges (33 challenges) due to the fact that sampling size was the largest among selected studies (218 concerned participant); thus, it strengthens the claimed solutions for the claimed challenges.

Finally, social interaction factor is considered one of the challenges of tacit knowledge sharing between members of the software development team; thus, the more quality and improvement of social interactions the better the communication. However, efficient sharing of tacit knowledge within teams has significant impact on teams performance and efficiency [31]. It is argued that the more efficient and detailed explicit knowledge, such as process, procedures, and policy's, the better adoption of the formal knowledge sharing approach between team members; thus, efficiency depends on explicit knowledge [31].

3 Method

According to [32], "a systematic literature review is a means of evaluating and interpreting all available research relevant to a particular research question, topic area, or phenomenon of interest". Based on that, the method used in this research follows the guidelines provided by Kitchenham and Charters [32] and other relevant systematic reviews [33–36]. Systematic literature review steps are subsequently followed: planning followed by collection, analysis, evaluation and selection, then reporting the findings of the review.

3.1 Planning Stage

Planning phase concludes research objectives, propose research questions, search strategy and criteria, apply inclusion/exclusion criteria, perform data extraction process and assure quality as illustrated below in more in details.

3.1.1 Objectives and Research Questions

Nowadays, developing a software becomes highly dependent on a team not an individual; therefore, researchers are more interested to identify success criteria's in order to complete the required software development tasks successfully; thus, massively contribute to the project success [14]. For example, within the software developer teams the leadership and teamwork characteristics [37], team communications [38–40], continuous learning among team members [41, 42], and self-confidence [43–45] are all considered important key factors for a project success.

In fact, communication and knowledge sharing are considered essential aspects for project success. Therefore, and due to that fact that the main objective of this research is to identify, compare and contrast all challenges/barriers and the solutions of knowledge sharing among members of a software development team based on the selected research studies. Then, answer following research questions thoroughly.

1. What are the studies that exhibit main challenges or barriers of sharing knowledge among the software development team?
2. What are the studies that provide solutions for knowledge sharing challenges among members of the software development team?
3. What are the research methods used for addressing challenges and provide solution for knowledge sharing between members of the software development team?

3.1.2 Search Strategy

Based on the research objectives and research questions, a search keywords and search database sources were identified [46]. For example, electronic databases, Google Schooler and the library of the British university in Dubai.

3.1.3 Search Criteria

Two set of criteria have been adopted and followed in this research as follows:

- First keyword string: "knowledge sharing impact", "sharing knowledge challenges", and "Knowledge sharing challenges solutions".

Table 1 Inclusion and exclusion criteria

Inclusion criteria	Exclusion criteria
Should include knowledge sharing challenges/barriers among software development team	Knowledge sharing challenges not used in software development industry
Should include knowledge sharing drivers among software development members	Knowledge sharing drivers not used in software development industry
Should include challenges in software development originations	Knowledge sharing used in context other than in software development aspects
Can include knowledge challenges sharing in information systems	Studies used languages other than English
Should be written in English language	

- Second keyword string: "challenges" with "software development teams", "barriers" with "software development teams", "challenges in software development", and "drivers in software development teams".

Based on the above string, the research plan executed on the selected database resources; for example, WorldCat.org, ABI/INFORM Global, ScienceDirect, Directory of Open Access Journals, ERIC, Taylor and Francis Journals, Emerald, IEEE and Google Scholar. Companied keywords used in this research based on the 1st and 2nd strings targeting abstract AND title, abstract OR title, or full text; for example, (ab: knowledge sharing barriers in software development", OR ti:"knowledge sharing challenges in software development", OR"ab: knowledge sharing drivers among software development teams", AND "ab: impact of knowledge sharing among software development teams".

3.1.4 Inclusion/Exclusion Criteria

The main purpose of the inclusion/exclusion criteria is to be more precise and focus on the selected articles based on the research objectives and questions [32]. The selected studies were analysed in depth based on the inclusion and exclusion criteria mentioned in Table 1.

3.2 Selection and Evaluation Phase

3.2.1 Quality Assessment

The quality assessment is considered as an essential factor which support the inclusion and exclusion criteria [36]. The quality checklist were followed to evaluate each

Table 2 Quality assessment checklist

#	Questions
1.	Are drivers of knowledge sharing among software development team are clearly stated?
2.	Did the study use scientific statistical approach to state the experiment outcomes?
3.	Are challenges of knowledge sharing among software development team clearly stated?
4.	Are the methods used to collect data (sample) mentioned in details?
5.	Did the recommended solution/conclusion resolve the stated gaps or challenges in the study?

study based on its relevance against research questions [32]. Each research question scored based on three-point scale where 1 = 100% fulfil the stated question "Yes", 0.5 = 50% fulfil the stated question" Partially" and 0 does not fulfil the stated question "No". Total scores of all questions related to each study will demonstrate its relevance, fulfilment for this systematic literature review and its eligibility for further analyses [36]. Table 2 shows the quality assessment checklist.

3.3 Data Filtration and Extraction

The selection filtration details where described thoroughly in Fig. 1. The search results indicate N = 360 studies considering abstract filtration and excluding studies published before 2015. After removing duplicate studies, total result indicates N = 181. After performing quick scanning of the titles and abstracts and excluding the ones its full text is not available, N = 97 concluded as relevant studies. All studies don't meet the inclusion criteria has been excluded; for example, studies explain the knowledge sharing challenges in context and not focused on software development were excluded. Indeed, total studies meet the inclusion and exclusion criteria N = 35. After reading full text of the selected studies, verifying references, evaluating and scoring each study against research questions. Indeed, high score given to the study which fulfil all research questions criteria. For example, any paper doesn't score in total 80% and above it is excluded; thus, final relevant studies aim to answer research questions N = 9. Figure 1 illustrates the filtration process used for the purpose of this systematic review and Table 3 illustrates the evaluation and selection of studies based on the research questions.

3.4 Data Coding and Analysis

The research methodology quality features were coded and outlines as per the following:

1. Research methods

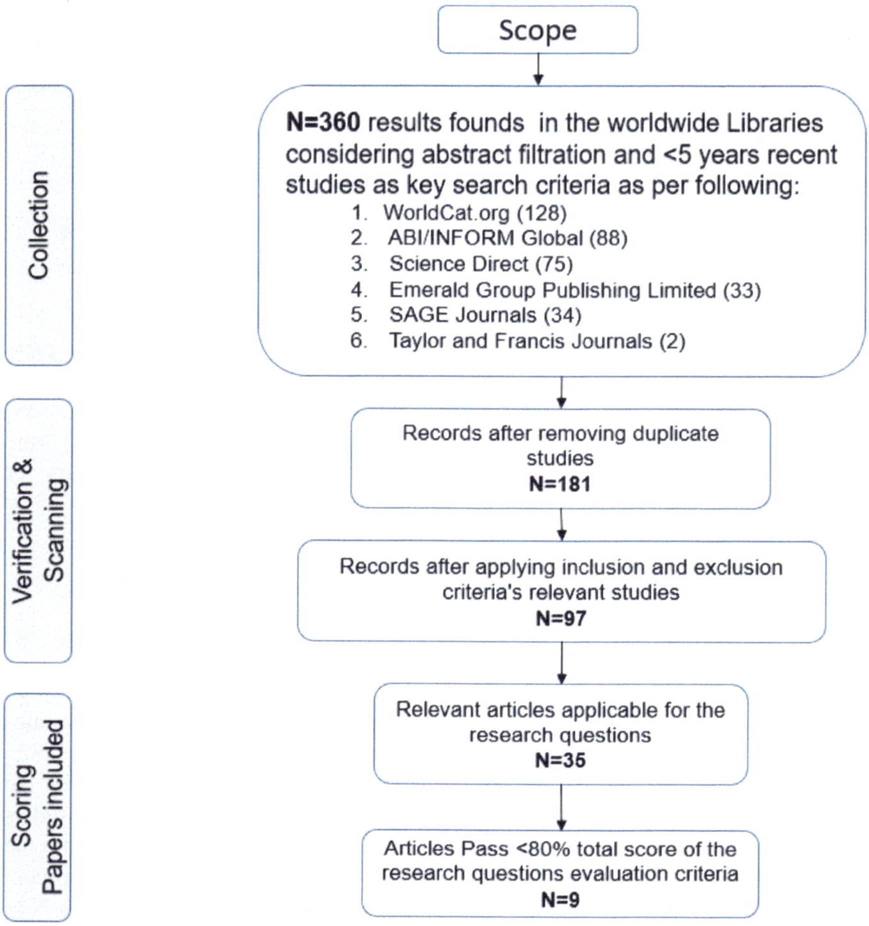

Fig. 1 Filtration process

2. Data Analysis Types
3. Data Collection Method
4. Country
5. Sample size
6. Citation
7. Research outcomes and recommendations
8. Research coverage within context.

While analysing data phase, any article doesn't satisfy the research questions with total of 80% and more was excluded as per Table 3, While coding the extracted data, following table illustrate the selected papers based on the predefined coded criteria.

Table 3 Evaluating and selecting of research papers

Study	Study title	Q1 challenges in context	Q2 solutions in context	Q3 methods in context	Total	Percentage	Status
S1	Factors negatively influencing knowledge sharing in software development	1	1	0.5	2.5	83%	In
S2	An Investigation of the Acquisition and Sharing of Tacit Knowledge in Software Development Teams	1	1	0.5	2.5	83%	In
S3	Knowledge Sharing Barriers of Acquisitioned Growth: A Case Study from a Software Company	0.5	1	1	2.5	83%	In
S4	Contextualized Relationship Between Knowledge Sharing and Performance in Software Development	1	1	0.5	2.5	83%	In
S5	Extrinsic and intrinsic motivation for experience grounded tacit knowledge sharing in Chinese software organizations	1	0.5	1	2.5	83%	In
S6	Knowledge sharing barriers in software development teams: a multiple case study in Turkey	1	0.5	1	2.5	83%	In
S7	Perceived barriers to effective knowledge sharing in agile software teams	1	1	0.5	2.5	83%	In

(continued)

Table 3 (continued)

Study	Study title	Q1 challenges in context	Q2 solutions in context	Q3 methods in context	Total	Percentage	Status
S8	Propensity of Knowledge Sharing in Software Development: Case in Sri Lanka	1	1	0.5	2.5	83%	In
S9	Knowledge sharing in information systems development projects: Explicating the role of dependence and trust	0.5	1	1	2.5	83%	In
S10	10 years of software architecture knowledge management-Practice	0	0	0.5	0.5	17%	Out
S11	A systematic examination of knowledge loss in open source software projects	1	1	0	2	67%	Out
S12	systematic review of knowledge sharing challenges and practices in global software development	1	1	0	2	67%	Out
S13	Biomedical Big Data Training Collaborative (BBDTC): An effort to bridge the talent gap in biomedical science and research	0	1	1	2	67%	Out
S14	Challenges and recommended practices for software architecting in global software development	1	0.5	0.5	2	67%	Out
S15	RE challenges in multi-site software development organizations	1	0	1	2	67%	Out

(continued)

Table 3 (continued)

Study	Study title	Q1 challenges in context	Q2 solutions in context	Q3 methods in context	Total	Percentage	Status
S16	Communities of practice as an initiative for knowledge sharing in business organizations: a literature review	0	0	1	1	33%	Out
S17	Effect of knowledge management on software product-case study in India	0	1	0.5	1.5	50%	Out
S18	Examining the impacts of organizational culture and top management support of knowledge sharing on the success of software process improvement	0	0.5	1	1.5	50%	Out
S19	Examining the impacts of organizational culture and top management	0.5	0.5	0.5	1.5	50%	Out
S20	Impact of supply chain collaboration and knowledge sharing on organizational outcomes in pharmaceutical industry of Bangladesh	0.5	0.5	1	2	67%	Out
S21	In- and extra-role knowledge sharing among information technology professionals-The five-factor model perspective	0	1	1	2	67%	Out

(continued)

Table 3 (continued)

Study	Study title	Q1 challenges in context	Q2 solutions in context	Q3 methods in context	Total	Percentage	Status
S22	Integration by communication: knowledge exchange in global outsourcing of product software development	0	1	0	1	33%	Out
S23	Investigating the impact of information systems on knowledge sharing	1	0.5	0	1.5	50%	Out
S24	Knowledge construction for efficient man-month estimation in software development	1	0.5	0.5	2	67%	Out
S25	Knowledge sharing in open source software communities: motivations and management	0.5	0.5	1	2	67%	Out
S26	Knowledge strategy, sharing behavior and performance	0	0	1	1	33%	Out
S27	Knowledge management and business strategy	0	0.5	0.5	1	33%	Out
S28	Knowledge sharing, Knowledge management strategy and performance A knowledge based view	0	0	0.5	0.5	17%	Out
S29	Knowledge sharing via social media in software development: a systematic literature review	0.5	0.5	1	2	67%	Out

(continued)

Table 3 (continued)

Study	Study title	Q1 challenges in context	Q2 solutions in context	Q3 methods in context	Total	Percentage	Status
S30	"Old" theories, "New" technologies: Understanding knowledge sharing and learning in Brazilian software development companies	0.5	0.5	0.5	1.5	50%	Out
S31	PBL Framework for Enhancing Software Development Skills: An Empirical Study for Information Technology Students	0.5	0	1	1.5	50%	Out
S32	The effect of governance on global software development: An empirical research in transactive memory systems	0	0.5	0.5	1	33%	Out
S33	The impact of knowledge management processes on information systems: A systematic review	0	0.5	0.5	1	33%	Out
S34	The Relationship between Management Support and Knowledge Sharing: An Exploratory Study of Manufacturing Firms	0	0.5	0.5	1	33%	Out
S35	What drives knowledge sharing in software development teams: A literature review and classification framework	1	1	0	2	67%	Out

4 Results

Based on the selected studies published concerning knowledge sharing challenges in the context of software development teams from 2013 to 2018, the below results shall answer the research questions as an outcome of the systematic literature review:

5 What are the studies that exhibit main challenges or barriers of sharing knowledge among the software development team?

Based on the literature review all selected articles demonstrate the challenges of knowledge sharing between software development teams, but each study illustrates the challenges from different categories, insights and focus. However, all challenges discussed falls within these categorizes: Knowledge-related, individual-related, organization-related, and project related. The following are most important challenges claimed and focused from the selected studies:

1. [23] identify 37 theoretical general challenges that falls under the mentioned categorise, but its sample size was very small to prove the mentioned challenges.
2. Time availability, the gained values and power of knowledge as well as distance of employees consider an essential challenge categorized under individual-related [25].
3. Task-oriented, social, and institutional context are considered challenges which impacts knowledge sharing; thus, impacts team performance [26].
4. Lack of awareness of the knowledge and less motivations are considered challenges which impact team performance [27].
5. Trust–dependence is considered one of the most essential knowledge challenges among team members [28].
6. Time constraints, job security, and resistance to change, communication, physiological factors and rewards are considered a challenge [30].
7. Social interaction factor can be considered one of the main challenges of tacit knowledge sharing between software developments teams [31].
8. Value of knowledge, knowledge as power, nature of knowledge, absorptive capacity, relationship between senders and recipients, different mentalities, culture, restraining forces in the organization are considered a challenge [14].
9. Low awareness of others and their values, lack of social networks, power of the relations, time, distance, challenges of networking across organization, communication of the benefits of knowledge sharing by management, comment culture and attitude, incompatible technology [21].

5.1 What Are the Studies that Provide Solutions for Knowledge Sharing Challenges Among Members of the Software Development Team?

All selected studies suggest different solutions based on their methods, case studies and hypotheses adopted to achieve the desired solutions. The following are most significant conclusions mentioned in the selected studies:

1. Improve the process and procedures of the software development through enhancing and improving efficient utilization of the knowledge sharing process [26].
2. Improve the awareness of the benefits associated with knowledge sharing by offering training and enhancing organization policy towards awareness of knowledge sharing of an individual soft skills instead of focusing on technical trainings [27].
3. Improve organization culture in order to enhance knowledge sharing process towards project success. Moreover, offer more time to project team members in order to share knowledge [30].
4. Improve project leadership approach and consider members' emotions [14].
5. Identifying, understanding and analysing existing knowledge sharing challenges among software development teams will better allow organizations to be ready for managing growth [21].
6. Enhance the trust among team members. Effective selection of the project team considering its similar values and goals. Additionally, provide training to the team; for example, on problem-solving and teamwork programs [28].
7. Organization should distinguish between explicit and tacit knowledge and enhance its tacit knowledge sharing process by encouraging interpersonal and social networks communications among teams and team members [23].
8. Implement causal morning meetings with/within the team(s) in order to improve knowledge sharing process. Eliminate team distance challenge by allowing job, project or team rotations [25].
9. Increase Social interactions by allowing team(s) to interact via social channels and facilitate team members family gathering to increase social interaction; thus, improve tacit knowledge sharing and increase trust [31].

5.2 What Are the Research Methods Used for Addressing Challenges and Provide Solution for Knowledge Sharing Between Members of the Software Development Team?

Based on Table 4 most of the studies used th'e survey, interview and questioner methods to obtain required outcomes of the claimed challenges and solutions; however, [14, 23] sample size was small comparing to other studies. Additionally, [21]

Table 4 Selected Papers

Research method	Factors negatively influencing knowledge sharing in software development	An investigation of the acquisition and sharing of tacit knowledge in software development teams	Knowledge sharing barriers of acquisitioned growth: a case study from a software company	Contextualized relationship between knowledge sharing and performance in software development	Extrinsic and intrinsic motivation for experience grounded tacit knowledge sharing in Chinese software organizations	Knowledge sharing barriers in software development teams: a multiple case study in Turkey	Perceived barriers to effective knowledge sharing in agile software teams	Propensity of knowledge sharing in software development: case in Sri Lanka	Knowledge sharing in information systems development projects: explicating the role of dependence and trust
Study number	S1	S2	S3	S4	S5	S6	S7	S8	S9
Case study		√	√		√				
Comparative causal mapping							√		
Controlled experiment		√							
Survey	√	√		√	√	√	√	√	√
Focus group									
Discussion			√						
Unclear									
Data analysis types									
Qualitative				√	√	√	√		√
Quantitative			√					√	

(continued)

Table 4 (continued)

Research method	Factors negatively influencing knowledge sharing in software development	An investigation of the acquisition and sharing of tacit knowledge in software development teams	Knowledge sharing barriers of acquisitioned growth: a case study from a software company	Contextualized relationship between knowledge sharing and performance in software development	Extrinsic and intrinsic motivation for experience grounded tacit knowledge sharing in Chinese software organizations	Knowledge sharing barriers in software development teams: a multiple case study in Turkey	Perceived barriers to effective knowledge sharing in agile software teams	Propensity of knowledge sharing in software development: case in Sri Lanka	Knowledge sharing in information systems development projects: explicating the role of dependence and trust
Mixed	√	√							
Unclear									
Study Setting			√	√					
Academic		√	√						
Industry	√		√	√	√	√			
Data collection method									
Interview					√	√	√	√	√
Archival record		√							
Observation		√	√						
Questionnaire	√	√		√				√	√
Workshop									
Focus groups									

(continued)

Table 4 (continued)

Research method	Factors negatively influencing knowledge sharing in software development	An investigation of the acquisition and sharing of tacit knowledge in software development teams	Knowledge sharing barriers of acquisitioned growth: a case study from a software company	Contextualized relationship between knowledge sharing and performance in software development	Extrinsic and intrinsic motivation for experience grounded tacit knowledge sharing in Chinese software organizations	Knowledge sharing barriers in software development teams: a multiple case study in Turkey	Perceived barriers to effective knowledge sharing in agile software teams	Propensity of knowledge sharing in software development: case in Sri Lanka	Knowledge sharing in information systems development projects: explicating the role of dependence and trust
Unclear									
Country	South Africa	Ireland	Finland	Hong Kong	China	Turkey	Australia	Sri Lanka	South Korea
Year of publication	2017	2013	2013	2015	2018	2017	2016	2014	2014
Sample size	218	181		11	44	18	8	46	195
Citation	10	11	9	30	14685	28	88	20	215

conducted a case study from the industry and concluded its results based on the analyses of the case study results and author observation. Finally, and due to the rapid changes in the world, results and recommendations of [27, 30] can be more adopted than others due to the reasonable sample size (218,44) as well as its recent publications (2017, 2018), respectively.

6 Conclusion and Recommendations

In fact, enriching knowledge sharing culture and adoption comes from top management by motivating its team's rewards program or salary increments [30]. However, all selected studies emphasize mostly on the knowledge sharing challenges among members of the software development team(s) as well as they provide great contributions to resolve each mentioned challenge. Due to the fact that knowledge sharing challenges are categorized into different categories, organizations should take serious consideration to identify and analyse its knowledge sharing challenges and adopt the optimum solutions based on their own environment, culture, policy, processes and its desire goals.

To conclude, we recommend to follow solutions adopted by [30] as shown in Table 4 due to the facts mentioned in third research question, i.e. What are the research methods used for addressing challenges and provide solution for knowledge sharing between members of the software development team? In other words, organizations should enhance and improve its culture in order to enhance knowledge sharing process and mentality towards project success. Nevertheless, we recommend to use the claimed module by [23] which shown in order to explore the relationship between knowledge sharing challenges for future research's specifically when agile software development teams is adopted.

Acknowledgements This work is a part of a project undertaken at the British University in Dubai.

References

1. Yang, T.-H., Ku, C.-Y., Liu, M.-N.: Case study: Application of enhanced Delphi method for software development and evaluation in medical institutes. Kybernetes **45**(4), 637–649 (2016)
2. Xiang, C., Lu, Y., Gupta, S.: Knowledge sharing in information system development teams: Examining the impact of shared mental model from a social capital theory perspective. Behav. Inf. Technol. **32**(10), 1024–1040 (2013)
3. Mezhuyev, V., Al-Emran, M., Ismail, M.A., Benedicenti, L., Chandran, D.A.: The acceptance of search-based software engineering techniques: An empirical evaluation using the technology acceptance model. IEEE Access (2019)
4. ELSamen, A., Alshurideh, M.: The impact of internal marketing on internal service quality: A case study in a Jordanian pharmaceutical company. Int. J. Bus. Manag. **7**(19), 84 (2012)
5. Ghannajeh, A., et al.: A qualitative analysis of product innovation in Jordan's Pharmaceutical sector. Eur. Sci. J. **11**(4), 474–503 (2015)

6. Zhang, X., Vogel, D.R., Zhou, Z.: Effects of information technologies, department characteristics and individual roles on improving knowledge sharing visibility: a qualitative case study. Behav. Inf. Technol. **31**(11), 1117–1131 (2012)
7. Alkalha, Z., Al-Zu'bi, Z., Al-Dmour, H., Alshurideh, M., Masa'deh, R.: Investigating the effects of human resource policies on organizational performance: An empirical study on commercial banks operating in Jordan. Eur. J. Econ. Financ. Adm. Sci. **51**(1), 44–64 (2012)
8. Al-dweeri, R.M., Obeidat, Z.M., Al-dwiry, M.A., Alshurideh, M.T., Alhorani, A.M.: The impact of e-service quality and e-loyalty on online shopping: moderating effect of e-satisfaction and e-trust. Int. J. Mark. Stud. **9**(2), 92 (2017)
9. Alshurideh et al.: The impact of Islamic Bank's service quality perception on Jordanian customer's loyalty. J. Manag. Res. **9** (2017)
10. Obeidat, B., Sweis, R., Zyod, D., Alshurideh, M.: The effect of perceived service quality on customer loyalty in internet service providers in Jordan. J. Manag. Res. **4**(4), 224–242 (2012)
11. Regaliza, J.C.P.: Extending the viable system model scope on ICT-sector software projects in Castilla y León. Kybernetes (2014)
12. Ghobadi, S.: What drives knowledge sharing in software development teams: A literature review and classification framework. Inf. Manag. **52**(1), 82–97 (2015)
13. Salloum, S.A., Al-Emran, M., Shaalan, K.: The impact of knowledge sharing on information systems: A review. In: International Conference on Knowledge Management in Organizations, pp. 94–106 (2018)
14. Akgün, A.E., Keskin, H., Ayar, H., Okunakol, Z.: Knowledge sharing barriers in software development teams: A multiple case study in Turkey. Kybernetes **46**(4), 603–620 (2017)
15. Al-Emran, M., Mezhuyev, V., Kamaludin, A.: Towards a conceptual model for examining the impact of knowledge management factors on mobile learning acceptance. Technol. Soc. (2020)
16. Al-Emran, M., Mezhuyev, V., Kamaludin, A.: An innovative approach of applying knowledge management in M-Learning application development: A pilot study. Int. J. Inf. Commun. Technol. Educ. **15**(4), 94–112 (2019)
17. Al-Emran, M., Mezhuyev, V.: Examining the effect of knowledge management factors on mobile learning adoption through the use of importance-performance map analysis (IPMA). In: International Conference on Advanced Intelligent Systems and Informatics, pp. 449–458 (2019)
18. Al-Emran, M., Teo, T.: Do knowledge acquisition and knowledge sharing really affect e-learning adoption? An empirical study. Educ. Inf. Technol. (2019)
19. Al-Emran, M., Mezhuyev, V., Kamaludin, A., AlSinani, M.: Development of M-learning application based on knowledge management processes. In: 2018 7th International conference on Software and Computer Applications (ICSCA 2018), pp. 248–253 (2018)
20. Al-Emran, M., Mezhuyev, V., Kamaludin, A.: Students' perceptions towards the integration of knowledge management processes in M-Learning systems: A preliminary study. Int. J. Eng. Educ. **34**(2), 371–380 (2018)
21. Kukko, M.: Knowledge sharing barriers of acquisitioned growth: A case study from a software company. Int. J. Eng. Bus. Manag. **5**(1), 1–12 (2013)
22. Altamony, H., Alshurideh, M., Obeidat, B.: Information systems for competitive advantage: Implementation of an organisational strategic management process. In: Proceedings of the 18th IBIMA conference on innovation and sustainable economic competitive advantage: From regional development to world economic, Istanbul, Turkey, 9–10 May 2012
23. Ghobadi, S., Mathiassen, L.: Perceived barriers to effective knowledge sharing in agile software teams. Inf. Syst. J. **26**(2), 95–125 (2016)
24. Shannak, R., Masa'deh, R., Al-Zu'bi, Z., Obeidat, B., Alshurideh, M., Altamony, H.: A theoretical perspective on the relationship between knowledge management systems, customer knowledge management, and firm competitive advantage. Eur. J. Soc. Sci. **32**(4), 520–532 (2012)
25. Weerakoon, S., Wijayanayake, J.: Propensity of knowledge sharing in software development: Case in Sri Lanka. Int. J. Multidiscip. Stud. **1**(2), 87 (2014)

26. Ozer, M., Vogel, D.: Contextualized relationship between knowledge sharing and performance in software development. J. Manag. Inf. Syst. **32**(2), 134–161 (2015)
27. Chen, H., Baptista Nunes, M., Ragsdell, G., An, X.: Extrinsic and intrinsic motivation for experience grounded tacit knowledge sharing in Chinese software organisations. J. Knowl. Manag. **22**(2), 478–498 (2018)
28. Park, J.G., Lee, J.: Knowledge sharing in information systems development projects: Explicating the role of dependence and trust. Int. J. Proj. Manag. **32**(1), 153–165 (2014)
29. Alshurideh, M.: Customer service retention–a behavioural perspective of the UK mobile market. Durham University (2010)
30. Khoza, L.T., Pretorius, A.B.: Factors negatively influencing knowledge sharing in software development. SA J. Inf. Manag. **19**(1), 1–10 (2017)
31. Ryan, S., O'Connor, R.V.: An investigation of the acquisition and sharing of tacit knowledge in software development teams: an empirical study. Inf. Softw. Technol. **55**(9), 1614–1624 (2013)
32. Kitchenham, B., Charters, S.: Guidelines for performing systematic literature reviews in software engineering. Softw. Eng. Group, Sch. Comput. Sci. Math. Keele Univ. 1–57 (2007)
33. Al-Saedi, K., Al-Emran, M., Abusham, E., El_Rahman, S.A.: Mobile payment adoption: a systematic review of the UTAUT model. In: International Conference on Fourth Industrial Revolution (2019)
34. Saa, A.A., Al-Emran, M., Shaalan, K.: Factors affecting students' performance in higher education: A systematic review of predictive data mining techniques. Technol. Knowl. Learn. (2019)
35. Al-Qaysi, N., Mohamad-Nordin, N., Al-Emran, M.: A Systematic Review of Social Media Acceptance from the Perspective of Educational and Information Systems Theories and Models. J. Educ. Comput. Res. **57**(8), 2085–2109 (2020)
36. Al-Emran, M., Mezhuyev, V., Kamaludin, A., Shaalan, K.: The impact of knowledge management processes on information systems: A systematic review. Int. J. Inf. Manage. **43**, 173–187 (2018)
37. Hasani, K., Sheikhesmaeili, S.: Knowledge management and employee empowerment: A study of higher education institutions. Kybernetes **45**(2), 337–355 (2016)
38. Alshurideh, M., Shaltoni, A., Hijawi, D.: Marketing communications role in shaping consumer awareness of cause-related marketing campaigns. Int. J. Mark. Stud. **6**(2), 163 (2014)
39. Ammari, G., Al kurdi, B., Alshurideh, M., Alrowwad, A.: Investigating the impact of communication satisfaction on organizational commitment: a practical approach to increase employees' loyalty. Int. J. Mark. Stud. **9**(2), 113–133 (2017)
40. Aubert, B., Hooper, V., Schnepel, A.: Revisiting the role of communication quality in ERP project success. Am. J. Bus. **28**(1), 64–85 (2013)
41. Al Kurdi, B., Alshurideh, M., Salloum, S.A., Obeidat, Z.M., Al-dweeri, R.M.: An empirical investigation into examination of factors influencing university students' behavior towards elearning acceptance using SEM approach. Int. J. Interact. Mob. Technol. **14**(02), 19–41 (2020)
42. Chan, C.C.A., Pearson, C., Entrekin, L.: Examining the effects of internal and external team learning on team performance. Team Perform. Manag. Int. J. **9**(7/8), 174–181 (2003)
43. Alshurideh, M., Al Kurdi, B., Salloum, S.A.: Examining the main mobile learning system drivers' effects: a mix empirical examination of both the expectation-confirmation model (ECM) and the technology acceptance model (TAM). In: International Conference on Advanced Intelligent Systems and Informatics, pp. 406–417 (2019)
44. Alshurideh, M.T., Salloum, S.A., Al Kurdi, B., Monem, A.A., Shaalan, K.: Understanding the quality determinants that influence the intention to use the mobile learning platforms: A practical study. Int. J. Interact. Mob. Technol. **13**(11), 157–183 (2019)
45. Porter, T.W., Lilly, B.S.: The effects of conflict, trust, and task commitment on project team performance. Int. J. Confl. Manag. **7**(4), 361–376 (1996)
46. Zahedi, M., Shahin, M., Babar, M.A.: A systematic review of knowledge sharing challenges and practices in global software development. Int. J. Inf. Manage. **36**(6), 995–1019 (2016)

The Role of Knowledge Management Processes for Enhancing and Supporting Innovative Organizations: A Systematic Review

Sufyan Areed, Said A. Salloum, and Khaled Shaalan

Abstract The rapidly advancing technology and the globalizing market have espoused the notion of knowledge management (KM) as an appropriate plan in the highly competitive commercial contexts. The importance of knowledge is evident in multidimensional society and the corporate domain. The use of acquaintance and information has currently increased in every type of business to simplify all the operations and improve the decision-making procedure. As such, the KM systems are among the fastest developing parts of the corporate sector. Likewise, the innovative competency offers mechanisms and unique insights that practitioners should consider in building up a creative business to raise performance. The business environment has dramatically transformed, and thereby, forcing companies to innovate in the ways they function and learn other new concepts. The present research reviews and sheds the light on innovation-related KM processes to provide a thorough assessment of 50 research articles published in peer-reviewed journals. The main findings of this study indicate that multiple KM processes can play an important role in an innovative organization. However, the most essential process for innovation was found to be knowledge sharing. Besides, KM was found to be involved in all of the innovation types.

Keywords Knowledge management (KM) · Innovation · Knowledge management processes · Knowledge sharing

1 Introduction

Organizations can make more accurate decisions on significant problems and enhance their knowledge practices based on superior knowledge [1, 2]. Therefore, managing this knowledge is regarded to be more significant than the knowledge itself, and many

S. Areed · S. A. Salloum (✉) · K. Shaalan
Faculty of Engineering and IT, The British University in Dubai, Dubai, UAE
e-mail: ssalloum@sharjah.ac.ae

S. A. Salloum
Research Institute of Sciences and Engineering, University of Sharjah, Sharjah, UAE

M. Al-Emran et al. (eds.), *Recent Advances in Intelligent Systems and Smart Applications*, Studies in Systems, Decision and Control 295,
https://doi.org/10.1007/978-3-030-47411-9_8

organizations try to explain how information and knowledge can be transformed into knowledge and skills in individuals and groups [3]. On the other hand, businesses have experienced several difficulties within the present economic climate as a result of the multiplicity of aspects, including globalization and intensified competition between corporations [4]. Additionally, the underlying circumstance encourages firms to apply new management paradigms and concepts, such as KM and Cross-Functional Teams [1, 5], which is an essential tool for enhancing the effectiveness and success of firms [6]. Mostly, the knowledge creation and learning are crucial processes in the modern world since they drive innovation, which in turn generates to competitive advantage. Successful executives outperform others by building an atmosphere without fear based on confidence that will make employees share their knowledge, an environment that maximizes the development of knowledge leads expertise towards innovation. Comparatively, innovation entails the introduction of a different combination of vital aspects of production in the manufacturing system. Novelty has become the strength of every company in the contemporary fast shifting business domain [7]. Accomplishing a competitive advantage requires enterprises to be proactive, innovative, and responsive. Additionally, companies with creative proficiency can change the ideas and knowledge of their personnel into goods and services that have been tailored to attain the needs of clients and innovations in the original manufacturer of products and services. Unfortunately, academics have conducted limited systematic researches to investigate the influence of KM as the major precursor of innovation on several organizational members. As well, the invention has increasingly become multifaceted as a result of changing consumer needs, widespread competitive pressure, and fast technological variation. The presence of progression in the level of acquaintance to organizations as a base for novelty has contributed to the complexity of innovation. Origination highly depends on the presence of knowledge; therefore, corporations must identify and manage the complication built by the upsurge of richness and reach of experience to ensure prosperous innovation. However, minimal intelligibility exists in contemporary literature, and thereby, the current review aims at exploring the research inquiry through a systematic review strategy. The gap of information offers the prospects for the exploration of how KM processes affect innovative companies and improve the understanding of KM and innovation. The main aim of this exploration is to provide a comprehensive systematic review of the influence of KM processes in enhancing and supporting innovative organizations. In addition, this research addresses the following primary questions:

1. RQ1. What are the main KM processes that influence innovation?
2. RQ2. What types of innovation that involves KM?
3. RQ3. What are the main research outcomes addressed in KM that are related to innovation?

The paper is organized as follows. First, the literature review of previous studies about the KM process and organizational innovation, in addition to organizational learning. Then, the research methodology including data sources and papers collection, the validity and reliability of the selected studies. It is then followed by the results section, which represents the finding of the proposed research questions. Finally, the discussion and conclusion sections are provided and presented.

2 Literature Review

KM is considered to be a fast-growing academic discipline [8]. While innovation can be defined as a transdisciplinary strategy in order to develop the synergies of wholeness, improving the organization's competitiveness through integral, competitive and complementary relationships in a particular setting between innovation members [9, 10]. Distinct definitions about KM have shown various thoughts on its significance. There are a large number of definitions with more or less prevalent characteristics that highlight different elements of the concept [11]. It is beyond the scope of this research to preview all definitions of KM. However, it would be useful to select some of these definitions. For instance, Lee [12] has defined KM as "the process of capturing, storing, sharing, and using knowledge". Gottschalk [13] emphasizes knowledge ownership and defines KM to release and leverage expertise so that it becomes an organizational resource that is not dependent on the same people, considering it as the intellectual capital of the organization. Mårtensson [14] describes KM as a competitive advantage for intellectual capital management controlled by the organization. Accordingly, Dzenopoljac et al. [15] defines KM as systematic management of all operations and procedures involving the generation and growth, codification and storage, transfer and share of information for the competitive edge of the organization.

2.1 Knowledge Management Processes

Hesamamiri et al. [16] defines KM processes as discovering, capturing, sharing, and applying knowledge. KM is the development of a mechanism that can find specific data more effectively and organize information for fast retrieval and reuse. The aforementioned process is a vital resource in modern organizations since it sustains organizational learning, victory, growth, and innovation. According to Dave et al. [17], KM entails the systematic running of the knowledge assets of an institution to generate value-enhanced output. Lee [12] contends that businesses tend to lose grip of their significant emphasis not due to the absence of knowledge, but as a result of poor KM. Therefore, the system of data implies a robust connection to corporate strategy, comprehension of the form of knowledge existence, creating procedures that span company functions and making sure members of a corporation accept and support initiatives. Hashemi et al. [18] states that the general objective of KM is to provide value, leverage, and refine the knowledge assets of an enterprise to attain organizational goals. The two significant aspects of KM include information control and people management. Mostly, the information management links with the controlling of knowledge related to elements that are determined and handled with data systems. Chang et al. [19] explains that the practice of managing info became broadly accepted when managers realized that data is an essential corporate resource that businesses should accomplish to enhance their competitiveness. The primary source of prosperity and wealth in the modern information economy is the production and supply of knowledge and information. According to Rusly et al. [20], KM has become an essential theme in most large corporations since executives have realized that much of their company's value relies on the enterprise's capacity to generate and manage knowledge. Tubigi and Alshawi [21] reports that business leaders must ensure efficient KM practices to acquire a sustainable competitive edge, which, in turn, facilitates innovation and improves organizational performance. The information-based procedures are the critical processes for enhancing the capabilities of a specific technology, and the fruitful adoption and execution of the technology greatly relies on the effective use of the progressions. Mostly, KM entails the identification and mapping of knowledgeable assets within a company, for instance, determining the various expert roles in an organization. As indicated by Zaim et al. [22], people management, which is a component of KM, basically encompasses the controlling of implicit information. Moreover, it involves the control of the information that is present alongside company processes, including a composite set of dynamic skills, expertise, and other knowledge-associated capabilities. Sangari et al. [23] reports that the knowledge related resources in an organization can include information in the form of manuals and patents, the reserved experience in electronic sources, and workers' understanding concerning the best way to perform their jobs. The overall responsibility by teams that work on intensive problems and facts are embedded in products, relationships, and processes of institutions. Therefore, the effective management of the human resources that possess the required tacit knowledge requires companies to take into consideration the individuals' social and cultural values, aspirations, attitudes, likes, and dislikes. According to Yusr et al. [24], organizations should accurately execute the internal procedures to ensure proper creation, storage, and utilization of knowledge. The KM processes support the transformation of tacit information to explicit instruction and vice versa. Storey and Barnett [25] emphasizes that the effectiveness and efficiency of managing the knowledge is a core component to develop key competencies for organizations in the present vibrant worldwide companies. This was endorsed by Zaim [26], who stated that KM is the only way to achieve long-term sustainable competitive benefits for organizations. KM processes were found to affect the adoption of several technologies [27–32]. Organizations therefore increasingly consider KM to be their most important strategic asset and strive to discover new methods of improving business efficiency through the development of efficient KM procedures [33]. Various scholars have introduced different KM processes. For instance, according to Sangari et al. [23], KM procedures encompass the creation, transfer, and application of knowledge. On the other hand, [22] indicates that KM processes entails information identification, capture, development, sharing, dissemination, application, and storage of knowledge. The goals of KM include the leverage and enhancement of the knowledge assets of an organization to effectuate better awareness practices, enhanced institutional behaviors, and superior decisions as well as upgraded organizational performance. Tubigi and Alshawi [21] postulates that although people can individually perform every KM process, KM is substantially a regulatory activity that emphasizes on what executives can do to enable the achievement of KM goals, how managers can motivate personnel to participate in attaining the objectives and how the leaders can create social procedures that will expedite the success of KM. As stated by Gunasekera and Chong, Al Dmour et al. [34, 35], KM systems include applications of a company's computer-based communications and data systems to back several KM practices.

2.2 *Organizational Innovation*

Busaibe et al. [36] defines innovation as the utilization of an originative strategy to enhance the firm's operations. Organizational advancement includes the execution of new company techniques in the undertaking's business performs, external relations or workstation organization to improve the use of knowledge, quality of products and services, and workflow efficiency. The four significant examples of innovation as provided by Leovaridis and Popescu [37], include architectural, radical, disruptive, and incremental novelty. A change can encompass a different good or service, modern production practice technology, new administrative system or structure, different program or plan concerning organizational members. Alves et al. [38] reports that innovation is a way of altering an organization, whether as a reaction to variations in its external or internal surroundings or as a preventive action to impact an environment. Chaubey and Sahoo [39] highlights that institutional innovation has enabled pioneering companies in the world to create structures that eliminate structural barriers virtually. Additionally, the businesses have launched an innovation culture and upgraded their invention capability. According to Ho [40], organizational innovation is a complex, multifaceted process, where the executives that deal with the revolution management must be aware of the extent of complexity contained in the process. Organizational learning plays an essential role in creating a creative enterprise [41]. Moreover, institutions cannot endure innovation and competitive benefits without consistent learning and the development of new knowledge [42]. The importance of company innovation as provided by Goby and Alhadhrami [43], including solving problems, adapting to the transformation, maximizing on globalization, evolving workplace dynamics, facing up the competition and consumers changing preferences and tastes. Most innovative ideas result from attempts to resolve present problems. The adoption to improve is apparent in the technological world, whereby rapid transformations define the business. Change is inevitable, and therefore, innovation assists organizations to keep their businesses afloat and ensure the commerce remains profitable and relevant. Li et al. [44] clarify that markets have become internationally interlinked, and the prospects have presented more excellent opportunities that emerge in the new marketplaces. However, it is imperative to note that opportunities present unique challenges and needs. Therefore, tapping into the new market share requires corporations to embrace innovation to capitalize on emerging changes. The business world is usually very competitive, and with the various prospective enterprises, the highest position in the business is no longer a reservation of a few. The establishment or the retention of the cutting edge of a firm requires organizations to compete strategically through a dynamic business with the ability to make innovative moves. Guisado-González et al. [45] explains that the workplace demographics are continuously varying, and new market trends evolve due to the new generation that has joined the marketplace. Moreover, the current clients have a wide range of goods and services, and they are well informed of their selections than previous cycles [46]. Innovation is therefore vital, and businesses must be up-to-date with the evolving tastes and implement new ways of gratifying customers.

2.3 Organization Learning

Learning entails how individuals enhance and create new knowledge across a wide range of technologies [47–52]. Nery-Kjerfve and McLean [53] describes organizational learning (OL) as a constant process that improves the collective capability to make sense of, accept, and reply to the interior and exterior modification. The concept of OL has experienced essential growth over the last few years in both the business and academic arenas. Sony and Naik [54] highlights that the reasons behind the development include the new features of the business world, the widespread analytical value of OL in adding to the enhancement of the comprehension of companies and their operations. According to Snell and Morris, Alshurideh et al. [55, 56], organizational learning theory entails the procedure of evolving, retaining, and conveying knowledge within a company. OL happens due to experience, and a corporation is said to have learned from an encounter when there exists a change in the behavior and performance of the corporation. Cacioppe [57] characterizes organizational learning as an intricate three-phase process that consists of knowledge acquisition, distribution, and shared execution. Individuals can obtain knowledge from the direct encounter, the experience of other people, or organizational memory. Progressively, King et al. [58] indicates that organizational learning occurs when a company attains its intended achievements and when a corporation identifies and corrects a mismatch between intentions and results. The outcomes of OL contribute to the growth of a company's resource capability. Therefore, it is necessary for firms to invest in persons to grow the intellectual capital required by the corporation and thereby, raise the stock of skills and knowledge. According to Law and Chuah [59], maintaining viability in the current business setting that is characterized by uncertainty and transformation requires individuals and companies to rely on the ability to learn. The ability of a company to learn relies on various organizational features. For instance, as reported by Cacioppe [57], the organizational features that influence OL include an enterprise's learning intent, approaches supporting capability development or innovation, liberal leadership and dispersed authority, belief systems and norms supporting backing learning, decision-making forums and the tools and processes that allow the transfer of information between groups and individuals. According to

Nery-Kjerfve and McLean [53], knowledge is a challenging notion to describe and measure, particularly at the organizational level of evaluation. Some scholars measure OL by estimating the cognitions of company members. Other various authors emphasize that knowledge entrenches on the practices or routines and view variations in them as reflective of modifications in awareness and therefore symbolic that OL occurred. Correspondingly, businesses can estimate OL by measuring the deviations in the features of performance, including its speed or accuracy, as indicative that personnel acquired knowledge and OL transpired. Individual learning frameworks exist in three groups, which include humanists, behaviorist, and cognitivist perspective. Sony and Naik [54] contends that the cognitive perspective assumes OL as an effective technique to process, interpret, and answer both internal and external info of a primarily explicit nature. The present research findings emphasize on cognitive development that has determined several sub-processes that allow the acquisition of different insights, including acquirement methods, perceptual and interpretation procedures and degrees of learning [55]. Companies can seek knowledge using active search practices such as pilot studies, shadowing, benchmarking, formal experiments, and action research initiatives. According to the social-process viewpoint, learning no longer involves appropriation or attainment of pieces of information, but it entails the development of positioned identities depending on involvement in a community of practice [58]. The social outlook of OL assumes that organizational knowledge arises from communal interactions, usually happening in the workplace. Moreover, according to this viewpoint, OL focuses on how employees make sense or interpret their encounters at work. Cacioppe [57] reports that the social perspective provides an alternative to the dominant cognitive concept, which comprehends the learning person to an individual who processes facts and adjusts his or her mental structures. Snell and Morris [55] reports that the social perspective of OL comprises three viewpoints, which include the concept of learning as a social construction, cultural process, and administrative procedure. The societal outlook of OL implies that persons are social beings who collectively construct a comprehension of the things they have around them and study from public relations within social systems like companies. Individuals can acquire knowledge through active participation based on the social practice of institutional life, which organizations continuously change.

3 Research Methodology

The detailed review is based on findings obtained from research articles published in a variety of online journals and databases that focus on knowledge management. Subsequent sections explain the collection of data as well as the modes used for appraisal of information obtained during the search. The selected publications broadly elaborate on the various concepts related to the subject of knowledge management. The areas highlighted include the processes involved in KM, its influence on innovations at different organizational levels as well as the types of changes that are associated with knowledge management. This research follows the systematic review procedures provided by Kitchenham and Charters [60] and other systematic review studies [61–64].

3.1 Data Sources and Search Strategies

A variety of articles were selected to incorporate diverse approaches towards the topic of discussion. In particular, 50 materials emanated from credible online journals available on various databases used for the uses included Emerald Insight, Research gate, Sage journals, Science Direct, and Google Scholar.

The search algorithms included searching for the presence of the required elements in the research objectives, as indicated in the sorted sources. Additionally, the exclusion aided to sort the sources with undesirable and far-fetched content was sidelined, leaving only those with the required data. The keywords used to locate information from the journals were ((“knowledge management processes” OR “knowledge management”) AND (“innovation” OR “organizational innovation” OR “organizational learning”)). The research and refining phases in this review were performed according to the Preferred Reporting Items for Systematic Reviews and Meta-Analysis (PRISMA) as shown in Fig. 1.

3.2 Inclusion/Exclusion Criteria

The choice of a source was dependent on the accuracy of the information available about the guiding research questions. Additionally, articles with abstracts were preferred since it preempted the contents of the entire study and its relevance to the objectives in question. However, other selected items were, disregarded due to time frame considerations, the depth of research articulated in the materials, and depending on whether they were peer-reviewed or not. Subsequent sections of the paper will provide a detailed analysis of the information obtained from the sources identified. The papers critically analyzed in this systematic review research should fulfill Table 1 requirements for inclusion and exclusion.

The essence of the keywords was to narrow down the research, such that the data obtained answered the questions sufficiently. Four hundred ninety-eight papers were found using the above keywords. 152 papers were discovered and filtered out as duplicates. Thus, the total amount of papers left was reduced to 346. Every research was verified according to the requirements. Therefore, 50 study papers met the inclusion/exclusion criteria and were thus incorporated into this systematic review. As a result, the information collected through the study is credible in explaining the necessity of KM in the success of an organization in terms of recent innovations.

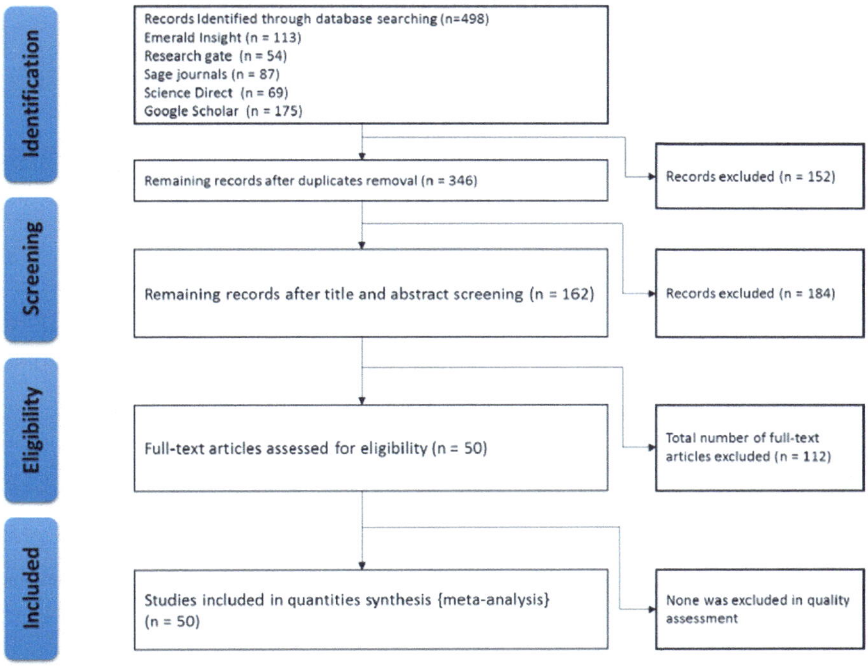

Fig. 1 PRISMA flowchart

Table 1 Inclusion/exclusion criteria

Inclusion criteria	Exclusion criteria
Accuracy relating to research questions	The credibility of the databases where articles are found
Articles ought to have an abstract. Articles should have information on knowledge management and innovation	Information about KM not relating to innovation. Time frames for the publishing of the articles

3.3 Quality Assessment

Quality assessment is a crucial factor to be examined in conjunction with inclusion and exclusion criteria. A quality evaluation checklist of 6 criteria was prepared and used to assess the quality of the study papers (N = 50). The criteria checklist is presented in Table 2.

The scale of each question answers was based on three-point. "Yes" has a value of 1 point, "No" has a value of 0 point, and "Partially" has a value of 0.5 point. Each study can, therefore, score from 0 to 6 and, as the score goes higher, the more research questions this study answers. The quality assessment results for all 50 research are shown in Table 3. In this context, it is evident that the quality assessment has been

Table 2 Quality assessment checklist

#	Question
1	Are the research objectives outlined?
2	Does the study explain the KM processes relation with innovation clearly?
3	Are the data collection methods indicated and well detailed?
4	Does the study explain the KM processes relation with innovation clearly?
5	Are the statistical techniques for data analysis described and illustrated?
6	Do the results support the literature review?

passed by all selected researches, which shows that all the selected papers are eligible for further analysis.

4 Results

The findings from the research indicate a detailed analysis of the research questions proposed in the study. The significant issues in the study involved the influence of knowledge management in innovation, the processes that are involved in KM as well as other changes that require the input of knowledge management. In essence, the various articles covered the topic based on the specific sections outlined in the research objectives.

4.1 RQ1. What Are the Main KM Processes that Influence Innovation?

The first research question entailed the examination of the main KM processes that influence innovation. Primarily, the research identifies the integration of the various methods into the concept of KM. According to [16, 34, 65–69], knowledge management involves steps such as awareness creation or discovery, acquisition, sharing, and application. These steps are critical in ensuring that the ideas conceived by the workforce are carefully considered and put into the appropriate use. Acquiring skills involve borrowing from external sources as well as developing them within the organization. The procedure of knowledge acquisition enables organizations to enlarge their mental models and therefore creating the conditions necessary to adapt and innovate processes, products, and services. Sharing ensures that all employees are acquainted with the developments taking place in the company, which in turn builds team spirit and co-operation.

Similarly, the application provides the utilization of the technical, as well as technological know-how to enhance the generation of new goods and services of the enterprise. The application procedure can influence innovation via practice since

Table 3 Quality assessment results

Study	Q1	Q2	Q3	Q4	Q5	Q6	Total	Percentage
S1	1	0	1	0.5	1	1	4.5	75
S2	1	1	1	1	1	1	6	100
S3	1	0.5	1	0	1	1	4.5	75
S4	1	1	0.5	1	1	1	5.5	91.6
S5	0.5	0.5	1	0	1	0.5	3.5	58.3
S6	1	0.5	1	0.5	0.5	0.5	4	66.67
S7	1	1	1	0	1	1	5	83.3
S8	1	1	1	1	1	1	6	100
S9	1	1	1	0.5	1	1	5.5	91.6
S10	1	1	1	0.5	1	1	5.5	91.6
S11	1	1	1	1	1	1	6	100
S12	1	0.5	0.5	1	1	1	5	83.3
S13	1	1	0.5	1	0.5	0.5	4.5	75
S14	1	1	1	1	1	1	6	100
S15	1	1	1	0.5	1	1	5.5	91.6
S16	1	1	0.5	1	1	1	5.5	91.6
S17	1	1	1	1	1	1	6	100
S18	1	0.5	1	1	1	1	5.5	91.6
S19	1	0.5	0.5	1	1	0.5	4.5	75
S20	1	1	1	1	1	0.5	5.5	91.6
S21	1	1	1	1	1	0.5	5.5	91.6
S22	1	0.5	0.5	1	1	1	5.5	91.6
S23	1	1	1	0	1	1	5	83.3
S24	1	1	1	1	1	1	6	100
S25	1	1	0.5	1	1	0.5	5	91.6
S26	1	1	0	0	1	0.5	3.5	58.3
S27	1	1	0.5	0.5	1	0.5	4.5	75
S28	1	1	0.5	0.5	1	1	5	83.3
S29	1	0.5	0.5	1	1	1	4	66.67
S30	1	0	0.5	0.5	1	0.5	3	60
S31	1	1	1	0.5	1	1	5.5	91.6
S32	1	1	1	0.5	1	1	5.5	91.6
S33	1	1	1	1	1	1	6	100
S34	1	1	1	1	1	1	6	100
S35	1	1	1	0	1	1	5	83.3
S36	1	0.5	1	1	1	1	5.5	91.6

(continued)

Table 3 (continued)

Study	Q1	Q2	Q3	Q4	Q5	Q6	Total	Percentage
S37	1	1	1	0	1	1	5	83.3
S38	1	1	1	0	1	1	5	83.3
S39	1	1	1	1	1	1	6	100
S40	1	0	1	1	1	1	5	83.3
S41	1	0	1	1	1	1	5	83.3
S42	1	1	0.5	0	1	1	4.5	75
S43	1	1	1	0	1	1	5	83.3
S44	1	1	1	0	1	0.5	5.5	91.6
S45	1	1	0.5	1	1	1	5.5	91.6
S46	1	1	0.5	1	0.5	0	4	66.67
S47	1	1	1	0	1	0.5	4.5	75
S48	1	1	1	0.5	1	1	5.5	91.6
S49	1	1	1	0	1	1	5	83.3
S50	1	0.5	1	1	1	1	5.5	91.6

frequent interaction with a piece of info provides people with extra insight into its likely applications. All the operations discussed are fundamental elements of innovation. However, the essential method for innovation includes the knowledge sharing procedure. The practice of sharing knowledge improves the ability of individuals to hasten adaptation to change and the growth of innovation. Moreover, when organizational members endorse information flows among themselves, it triggers new interpretations and fosters novelty.

4.2 RQ2. What Types of Innovation that Involves KM?

The second research question entailed the evaluation of the various types of innovation that involves KM. The four significant examples of innovation, as provided by Leovaridis and Popescu [37], include architectural, radical, disruptive, and incremental novelty. The disruptive novelty includes developing new goods or services to substitute present technologies and acquire a competitive edge. For instance, in a usual innovative high technology commerce, disruptive invention shakes up a market when introduced externally, and it needs a creative internal approach toward the promotion process and product development. Architectural innovation involves taking lessons, general technology, and skills and using them within a diverse market. Furthermore, the aspect alters the nature of interactions between major components, while strengthening the main design concepts. Fundamental novelty entails producing new industries and the creation of revolutionary technology. The radical invention introduces a different meaning, possibly a paradigm shift. Incremental novelty is the most prevalent form of innovation that utilizes the available technology and increases the value to clients through changes, features, and design within the existing market. Moreover, this kind of invention introduces quality enhancements in the core components. As explained by [23, 37, 39, 69–72], the fundamental categories are focused on products, process and organizational innovations. Process innovation deals with the improvement of the existing technologies and resources in production.

On the other hand, corporate change involves using the competencies amongst the workforce to actualize ideas on new inventions. Additionally, the activities involved in the advancement of existing goods in the market require the input of experience and abilities such that the outcome meets the needs of the consumers. In essence, boosting business activities involves input from all stakeholders in a company. Organizations cannot sustain the different types of innovation and obtain competitive benefit without consistent learning, and the development of new knowledge, and therefore, all the categories of novelty involve KM.

4.3 RQ3. What Are the Main Research Outcomes Addressed in KM that Are Related to Innovation?

Various authors have tackled the topic of the influence of KM on organizational changes. Notably, the previous studies [23, 24, 34, 39, 66, 67, 73–78], have explored the topic, providing detailed explanations on the link between knowledge management and the introduction of innovations in firms. The writers assert that creative renovations require the transformation of virtual knowledge possessed by the workers in the implementation of the tactical skills in the process of production. Additionally, the publications indicate that organizational collaboration, as well as teamwork, are enhanced through the monitoring of available expertise. On the other hand, managing the employees' skills ensures that firms attain a competitive advantage over rivals in the market. Fundamentally, innovations in a company require the input of expert skills to make the process viable and profitable in the long run.

5 Discussion

The principal aim of the present analysis entails the systematic evaluation and synthesis of the influence of KM processes in innovative organizations. As explained in the results section, creative renovations require the transformation of virtual knowledge possessed by the workers in the implementation of the tactical skills in the process of production. Innovations in a company need the input of expert skills to make the process viable and profitable in the long run. As explained in the literature review part of the exploration, the primary source of prosperity and wealth in the modern information economy is the production and supply of knowledge and information.

KM is a strategy that helps businesses to use understanding for envisaging, making, and controlling the entire decision-making procedure. KM procedures are the critical processes for enhancing the capabilities of a specific technology, and the fruitful adoption and execution of the technology dramatically relies on the effective use of the progressions. The findings by Andreeva and Kianto [79] indicate that the current market demands of business processes mandate that manual labor should be replaced by automation, which necessitates the need for knowledge and expert skills.

The first goal of the investigation included the aim to determine the significant KM processes that impact innovation. As indicated in the results section, knowledge management involves steps such as creation, acquisition, sharing, and application. In the literature review chapter, the various scholars highlight that the processes of KM include acquisition, generation, refinement, storage, transmission, sharing, and application of knowledge. Although people can individually perform every KM process, KM is substantially an organizational activity that emphasizes on what executives can do to enable the achievement of KM goals, how managers can motivate personnel to participate in attaining the objectives and how the leaders can create standard procedures that will expedite the success of KM. Based on the study results, there exists a positive association between the practical application of KM processes and successful innovation. For instance, the findings postulate that acquiring skills entails borrowing from external sources and developing them within the organization. Sharing ensures that all staffs are familiar with the developments occurring in the company, which in turn builds team spirit and co-operation. Likewise, the application ensures that technical and theoretical know-how is utilized to improve the generation of new goods and services of the enterprise.

The second aim of the research involved the investigation of the various types of innovation that require KM. According to Rahimi et al. [80], knowledge management is an essential antecedent of embracing and executing various types of innovation. KM augments engagement in novelty by producing, using, and sharing unique concepts as well as the exploitation of the thinking power of the firm. Based on the findings from the systematic review, the significant innovation categories emphasize on products, processes, and organizational inventions. A change can encompass a different good or service, production practice technology, new administrative system or structure, different program or plan concerning corporate members. In the results section, the activities involved in the advancement of existing goods in the market requires the input of experience and abilities such that the outcome meets the needs of the consumers. In essence, boosting business activities involves input from all stakeholders in a company.

The last research objective entailed the exploration of the significant research outcomes addressed in KM that relate to innovation. As seen in the study findings based on the systematic review, various authors have tackled the topic of the influence of KM on organizational changes. For instance, Rahimi et al. [80] explains the link between knowledge management and the introduction of innovations in firms. Renovations require the transformation of virtual knowledge possessed by the workers into tactic skills in the process of production. Changes in a company need the input of expert skills to make the process viable and profitable in the long run.

6 Conclusion and Future Work

The above systematic review presented an analytical discussion of the present researches concerning the impact of KM processes for innovative companies. The analysis sought to respond to the three research questions that achieve the study objective. As highlighted in the preceding sections of the study, there exists a positive and significant link between various KM processes and organizational innovation. The literature review section of the research paper and the results chapter formulates the basis of the consolidation of findings to attain the study goals. Based on the investigation purpose and the results collected from various articles, the findings postulate a positive relationship between the KM practice such as processes such as creation, acquisition, sharing, and application influence organizational innovation in multiple ways. Moreover, as explained in the study, the variables above are critical in ensuring that the ideas conceived by the workforce are carefully considered and put into the appropriate use. While it has been shown that KM processes have a significant impact on innovation, there is still an open door for some further work to examine the influence of these processes on other forms of innovation. Future work should take into account this gap and seek to explore the effect of these KM processes on different types of innovation. In conclusion, the results of this review study offer insight into the current trend of research involving innovation studies in KM processes and provide a valuable guide for future studies in innovation-based KM processes.

Acknowledgements This work is a part of a project undertaken at the British University in Dubai.

References

1. Salloum, S.A., Al-Emran, M., Shaalan, K.: The impact of knowledge sharing on information systems: A review. In: International Conference on Knowledge Management in Organizations, pp. 94–106 (2018)
2. Altamony, H., Alshurideh, M., Obeidat, B.: Information systems for competitive advantage: Implementation of an organisational strategic management process. In: Proceedings of the 18th IBIMA conference on innovation and sustainable economic competitive advantage: From regional development to world economic, Istanbul, Turkey, 9–10 May 2012
3. Shannak, R., Masa'deh, R., Al-Zu'bi, Z., Obeidat, B., Alshurideh, M., Altamony, H.: A theoretical perspective on the relationship between knowledge management systems, customer knowledge management, and firm competitive advantage. Eur. J. Soc. Sci. **32**(4), 520–532 (2012)
4. Zu'bi, Z., Al-Lozi, M., Dahiyat, S., Alshurideh, M., Al Majali, A.: Examining the effects of quality management practices on product variety. Eur. J. Econ. Financ. Adm. Sci. **51**(1), 123–139 (2012)
5. Ghannajeh, A., et al.: A qualitative analysis of product innovation in Jordan's pharmaceutical sector. Eur. Sci. J. **11**(4), 474–503 (2015)
6. Alshraideh, A., Al-Lozi, M., Alshurideh, M.: The Impact of training strategy on organizational loyalty via the mediating variables of organizational satisfaction and organizational performance: An empirical study on Jordanian agricultural credit corporation staff. J. Soc. Sci. **6**, 383–394 (2017)

7. Abu Zayyad, H.M., Obeidat, Z.M., Alshurideh, M.T., Abuhashesh, M., Maqableh, M., Masa'deh, R.: Corporate social responsibility and patronage intentions: The mediating effect of brand credibility. J. Mark. Commun. pp. 1–24 (2020)
8. Serenko, A., Bontis, N.: Global ranking of knowledge management and intellectual capital academic journals: 2017 update. J. Knowl. Manag. **21**(3), 675–692 (2017)
9. Swink, M.: Building collaborative innovation capability. Res. Technol. Manag. **49**(2), 37–47 (2006)
10. Bommert, B.: Collaborative innovation in the public sector. Int. Public Manag. Rev. **11**(1), 301–316 (2010)
11. Lytras, M.D., Pouloudi, A., Poulymenakou, A.: Knowledge management convergence—Expanding learning frontiers. J. Knowl. Manag. **6**(1), 40–51 (2002)
12. Lee, J.N.: The impact of knowledge sharing, organizational capability and partnership quality on IS outsourcing success. Inf. Manag. **38**(5), 323–335 (2001)
13. Gottschalk, P.: Strategic knowledge networks: The case of IT support for Eurojuris Law firms in Norway. Int. Rev. Law, Comput. Technol. **14**(1), 115–129 (2002)
14. Mårtensson, M.: Emerald Article: A critical review of knowledge management as a management tool. J. Knowl. Manag. **4**, 204–216 (2000)
15. Dzenopoljac, V., Alasadi, R., Zaim, H., Bontis, N.: Impact of knowledge management processes on business performance: Evidence from Kuwait. Knowl. Process Manag. **25**(2), 77–87 (2018)
16. Hesamamiri, R., Mazdeh, M.M., Jafari, M., Shahanaghi, K.: Knowledge management reliability assessment: An empirical investigation (2015)
17. Dave, M., Singh, K., Kumar, A., Kumar, S.: Knowledge management measures: An empirical investigation of Indian cement industry. Benchmarking **26**(3), 734–752 (2019)
18. Hashemi, P., Khadivar, A., Shamizanjani, M.: Developing a domain ontology for knowledge management technologies Parvin (2017)
19. Chang, C.M., Hsu, M.H., Yen, C.H.: Factors affecting knowledge management success: The fit perspective. J. Knowl. Manag. **16**(6), 847–861 (2012)
20. Rusly, F.H., Corner, J.L., Sun, P.: Positioning change readiness in knowledge management research. J. Knowl. Manag. **16**(2), 329–355 (2012)
21. Tubigi, M., Alshawi, S.: The impact of knowledge management processes on organisational performance: The case of the airline industry. J. Enterp. Inf. Manag. **28**(2), 167–185 (2015)
22. Zaim, H., Keceli, Y., Jaradat, A., Kastrati, S.: The effects of knowledge management processes on human resource management. J. Sci. Technol. Policy Manag. **9**(3), 310–328 (2018)
23. Sangari, M.S., Hosnavi, R., Zahedi, M.R.: The impact of knowledge management processes on supply chain performance an empirical study (2014)
24. Yusr, M.M., Mokhtar, S.S.M., Othman, A.R., Sulaiman, Y.: Does interaction between TQM practices and knowledge management processes enhance the innovation performance? Int. J. Qual. Reliab. Manag. **34**(7), 955–974 (2017)
25. Storey, J., Barnett, E.: Knowledge management initiatives: Learning from failure. J. Knowl. Manag. **4**(2), 145–156 (2000)
26. Zaim, H.: Knowledge management implementation in IZGAZ. J. Econ. Soc. Res. **8**(2), 1–25 (2006)
27. Al-Emran, M., Mezhuyev, V., Kamaludin, A.: Towards a conceptual model for examining the impact of knowledge management factors on mobile learning acceptance. Technol. Soc. (2020)
28. Al-Emran, M., Mezhuyev, V., Kamaludin, A.: An innovative approach of applying knowledge management in M-Learning application development: A pilot study. Int. J. Inf. Commun. Technol. Educ. **15**(4), 94–112 (2019)
29. Al-Emran, M., Mezhuyev, V.: Examining the effect of knowledge management factors on mobile learning adoption through the use of importance-performance map analysis (IPMA). In: International Conference on Advanced Intelligent Systems and Informatics, pp. 449–458 (2019)
30. Al-Emran, M., Teo, T.: Do knowledge acquisition and knowledge sharing really affect e-learning adoption? An empirical study. Educ. Inf. Technol. (2019)

31. Al-Emran, M., Mezhuyev, V., Kamaludin, A., AlSinani, M.: Development of M-learning application based on knowledge management processes. In: 2018 7th International conference on Software and Computer Applications (ICSCA 2018), pp. 248–253 (2018)
32. Al-Emran, M., Mezhuyev, V., Kamaludin, A.: Students' perceptions towards the Integration of knowledge management processes in M-learning systems: A preliminary study. Int. J. Eng. Educ. **34**(2), 371–380 (2018)
33. Alzubi, H.M., Hanandeh, R., AL ALI, R.: Investigating the relationship between knowledge management processes and organizational performance the mediating effect of organizational innovation. Int. Rev. Manag. Bus. Res. (2015)
34. Gunasekera, V.S., Chong, S.C.: Knowledge management for construction organisations: A research agenda. Kybernetes **47**(9), 1778–1800 (2018)
35. Al Dmour, H., Alshurideh, M., Shishan, F.: The influence of mobile application quality and attributes on the continuance intention of mobile shopping. Life Sci. J. **11**(10), 172–181 (2014)
36. Busaibe, L., Singh, S.K., Ahmad, S.Z., Gaur, S.S.: Determinants of organizational innovation: A framework. Gend. Manag. **32**(8), 578–589 (2017)
37. Leovaridis, C., Popescu, G.: Organizational innovation—A means to enhance quality of life for employees in knowledge economy. Manag. Dyn. Knowl. Econ. **3**(1), 25–43 (2015)
38. Alves, M.F.R., Galina, S.V.R., Dobelin, S.: Literature on organizational innovation: Past and future. Innov. Manag. Rev. **15**(1), 2–19 (2018)
39. Chaubey, A., Sahoo, C.K.: Enhancing organizational innovation in Indian automobile industry. Int. J. Innov. Sci. **11**(1), 82–101 (2019)
40. Ho, L.A.: Meditation, learning, organizational innovation and performance. Ind. Manag. Data Syst. **111**(1), 113–131 (2011)
41. Jia, X., Chen, J., Mei, L., Wu, Q.: How leadership matters in organizational innovation: A perspective of openness Introduction (2017)
42. Al Kurdi, B., Alshurideh, M., Salloum, S.A., Obeidat, Z.M., Al-dweeri, R.M.: An empirical investigation into examination of factors influencing university students' behavior towards elearning acceptance using SEM approach. Int. J. Interact. Mob. Technol. **14**(02), 19–41 (2020)
43. Goby, V.P., Alhadhrami, A.: National citizenship as a mediator of organizational innovation. Leadersh. Organ. Dev. J. **39**(4), 482–494 (2018)
44. Li, W., Bhutto, T.A., Nasiri, A.R., Shaikh, H.A., Samo, F.A.: Organizational innovation: The role of leadership and organizational culture. Int. J. Public Leadersh. **14**(1), 33–47 (2017)
45. Guisado-González, M., González-Blanco, J., Coca-Pérez, J.L.: Analyzing the relationship between exploration, exploitation and organizational innovation. PAL Driven Organ. Learn. Theory Pract. A Light Learn. Journey Organ. pp. 1–112 (2017)
46. Pauget, B., Wald, A.: Creating and implementing organizational innovation: The role of professional identity and network embeddedness in healthcare organizations. Eur. J. Innov. Manag. **21**(3), 384–401 (2018)
47. Alshurideh, M., Al Kurdi, B., Salloum, S.A.: Examining the main Mobile Learning system drivers' effects: A mix empirical examination of both the expectation-confirmation model (ECM) and the technology acceptance model (TAM). In: International Conference on Advanced Intelligent Systems and Informatics, pp. 406–417 (2019)
48. Salloum, S.A., Al-Emran, M., Shaalan, K., Tarhini, A.: Factors affecting the E-learning acceptance: A case study from UAE. Educ. Inf. Technol. **24**(1), 509–530 (2019)
49. Salloum, S.A., Maqableh, W., Mhamdi, C., Al Kurdi, B., Shaalan, K.: Studying the social media adoption by university students in the United Arab Emirates. Int. J. Inf. Technol. Lang. Stud. **2**(3), 83–95 (2018)
50. Salloum, S.A., Shaalan, K.: Adoption of e-book for university students. In: International Conference on Advanced Intelligent Systems and Informatics, pp. 481–494 (2018)
51. Alshurideh, M.T., Salloum, S.A., Al Kurdi, B., Monem, A.A., Shaalan, K.: Understanding the quality determinants that influence the intention to use the mobile learning platforms: A practical study. Int. J. Interact. Mob. Technol. **13**(11), 157–183 (2019)
52. Salloum, S.A., Alhamad, A.Q.M., Al-Emran, M., Monem, A.A., Shaalan, K.: Exploring students' acceptance of E-Learning through the development of a comprehensive technology acceptance model. IEEE Access **7**, 128445–128462 (2019)

53. Nery-Kjerfve, T., McLean, G.N.: Repatriation of expatriate employees, knowledge transfer, and organizational learning: What do we know? Eur. J. Train. Dev. **36**(6), 614–629 (2012)
54. Sony, M., Naik, S.: Six Sigma, organizational learning and innovation: An integration and empirical examination. Int. J. Qual. Reliab. Manag. **29**(7), 797–818 (2012)
55. Snell, S.A., Morris, S.S.: Building dynamic capabilities around organizational learning challenges. J. Organ. Eff. **1**(3), 214–239 (2014)
56. Alshurideh, M., Nicholson, M., Xiao, S.: The effect of previous experience on mobile subscribers' repeat purchase behaviour. Eur. J. Soc. Sci. **30**(3), 366–376 (2012)
57. Cacioppe, R.L.: Integral mindflow. Learn. Organ. **24**(6), 408–417 (2017)
58. King, W.R., Qureshi, S., Kamal, M.: Knowledge management and organizational learning. pp. 215–234 (2009)
59. Law, K.M., Chuah, K.B.: Organizational learning as a continuous. DELO (2015)
60. Kitchenham, B., Charters, S.: Guidelines for performing systematic literature reviews in software engineering. Softw. Eng. Group, Sch. Comput. Sci. Math. Keele Univ. 1–57 (2007)
61. Al-Emran, M., Mezhuyev, V., Kamaludin, A., Shaalan, K.: The impact of knowledge management processes on information systems: A systematic review. Int. J. Inf. Manage. **43**, 173–187 (2018)
62. Al-Qaysi, N., Mohamad-Nordin, N., Al-Emran, M.: A systematic review of social media acceptance from the perspective of educational and information systems theories and models. J. Educ. Comput. Res. **57**(8), 2085–2109 (2020)
63. Saa, A.A., Al-Emran, M., Shaalan, K.: Factors affecting students' performance in higher education: A systematic review of predictive Data Mining techniques. Technol. Knowl. Learn. (2019)
64. Al-Saedi, K., Al-Emran, M., Abusham, E., El_Rahman, S.A.: Mobile payment adoption: A systematic review of the utaut model. In: International Conference on Fourth Industrial Revolution (2019)
65. Bharadwaj, S.S., Chauhan, S., Raman, A.: Impact of knowledge management capabilities on knowledge management effectiveness in Indian organizations. Vikalpa **40**(4), 421–434 (2015)
66. Costa, V., Monteiro, S.: Key knowledge management processes for innovation: A systematic literature review. VINE J. Inf. Knowl. Manag. Syst. **46**(3), 386–410 (2016)
67. Teixeira, E.K., Oliveira, M., Curado, C.M.M.: Knowledge management process arrangements and their impact on innovation. Bus. Inf. Rev. **35**(1), 29–38 (2018)
68. Latilla, V.M., Frattini, F., Petruzzelli, A.M., Berner, M.: Knowledge management, knowledge transfer and organizational performance in the arts and crafts industry: A literature review. J. Knowl. Manag. **22**(6), 1310–1331 (2018)
69. Chang, D.K.S.: How does knowledge management influence the innovation management process? J. Knowl. Manag. **4**(2), 385–386 (2007)
70. Costa, V., Monteiro, S.: Knowledge processes, absorptive capacity and innovation: A mediation analysis. Knowl. Process Manag. **23**(3), 207–218 (2016)
71. Tsinopoulos, C., Sousa, C.M.P., Yan, J.: Process innovation: Open innovation and the moderating role of the motivation to achieve legitimacy. J. Prod. Innov. Manag. **35**(1), 27–48 (2018)
72. Lousã, E.P., dos, L., Mónico, S.M.: How can leadership and organizational culture predict innovation in small, medium and large enterprises? J. Organ. Manag. Stud. **2018** (2018)
73. Omotayo, F.O.: Knowledge management as an important tool in organisational management : A review of literature. Libr. Philos. Pract. (e-journal). **1238**, 1–23 (2015)
74. du Plessis, M.: The role of knowledge management in innovation. J. Knowl. Manag. **11**(4), 20–29 (2007)
75. Akram, K., Siddiqui, S.H., Atif Nawaz, M., Ghauri, T.A., Khawar, A., Cheema, H.: Role of knowledge management to bring innovation: An integrated approach. Int. Bull. Bus. Adm. **11**(11), 1451–243 (2011)
76. Gonzalez, R.V.D., Martins, M.F.: Knowledge management: An analysis from the organizational development. J. Technol. Manag. Innov. **9**(1), 131–147 (2014)

77. Mohajan, H.: The impact of knowledge management models for the development of organizations the impact of knowledge management models for the development of organizations. J. Environ. Treat. Tech. **5**(1), 12–33 (2017)
78. Oke, A.: Innovation types and innovation management practices in service companies. Int. J. Oper. Prod. Manag. **27**(6), 564–587 (2007)
79. Andreeva, T., Kianto, A.: Knowledge processes, knowledge-intensity and innovation: A moderated mediation analysis. J. Knowl. Manag. **15**(6), 1016–1034 (2011)
80. Rahimi, E., Rostami, N., Shad, F., Vafaei, V.: The importance of knowledge management on innovation. Appl. Math. Eng. Manag. Technol. **5**(1), 68–73 (2017)

The Impact of Artificial Intelligence and Information Technologies on the Efficiency of Knowledge Management at Modern Organizations: A Systematic Review

Saeed Al Mansoori, Said A. Salloum, and Khaled Shaalan

Abstract In the twenty-first century, with the evolution of Information Technology and Artificial Intelligence, Knowledge Management seems to be more advanced and sophisticated. Knowledge management has been studied and discussed for a long time by several researchers from academia and business sector due to its vitality for the success of organizations. Additionally, many leading companies around the world have adopted several knowledge management practices to ensure that they remain ahead of their rival companies in today's competitive business world. Hence, companies continue to look for ways to improve these knowledge management practices. Several studies were conducted to have more profound understanding of the recent research trend of knowledge management processes and its best practices at organizations. However, this subject requires further investigation from other perspectives. Most of the previous studies neglect exploring the impact of applying artificial intelligence (AI) and information technology (IT) concepts and techniques on the efficiency of knowledge management at modern organizations in particular. This research systematically reviews and highlights the current knowledge management practices that have relied on IT and AI mechanisms and their impacts on modern organizations along with their challenges and limitations. The current research aims to provide a profound analysis of 15 (out of 844) research papers published in highly standard journals and conference proceedings throughout the past 10 years, from 2009 to 2019. This review study can be considered a pivotal reference for scholars as it fills in some gaps in knowledge management especially in IT and AI related studies.

Keywords Knowledge management (KM) · Information technology (IT) · Artificial intelligence (AI) · Modern organizations

---

S. Al Mansoori · S. A. Salloum (✉) · K. Shaalan
Faculty of Engineering and IT, The British University in Dubai, Dubai, UAE
e-mail: ssalloum@sharjah.ac.ae

S. Al Mansoori · S. A. Salloum
Research Institute of Sciences and Engineering, University of Sharjah, Sharjah, UAE

M. Al-Emran et al. (eds.), *Recent Advances in Intelligent Systems and Smart Applications*, Studies in Systems, Decision and Control 295,
https://doi.org/10.1007/978-3-030-47411-9_9

1 Introduction

There is no doubt on how several technologies have changed the landscape of several sectors [1–4]. In the past, human beings have been preserving knowledge since the dawn of history as they began expressing their ideas through painting on the walls of caves using primitive tools [5]. This was the first stage of publicizing individual knowledge and transferring it to the public [6]. Later on, humanity took a further step to save and spread knowledge by evolving various writing systems through introducing letters written on scattered papers and in books [7]. The twentieth century witnessed a roof hitting revolution in information management through the appearance of digital documents that are credited for the invention of knowledge management tools [8–11]. In the twenty-first century, Artificial Intelligence has been introduced as a cornerstone in knowledge management since it has evolved knowledge gaining, developing and sharing as well as effective utilization within organizations [6, 12]. Several studies were conducted to explore the up-to-date progress in knowledge management systems, processes and best practices along with their repercussions on organizations [6, 13, 14]. In addition, knowledge management processes have proved its effectiveness in adopting educational technologies [15–20].

The term "*Knowledge Management*" is not an easy concept to define since it is multifaceted and applied in different contexts by different individuals. According to [21], for instance, knowledge management includes the processes used to identify, gather and reinforce knowledge. By contrast, [22] defines knowledge management as a field that systematically enriches expertise and content to boost competency, responsiveness, efficacy and innovation in an organization. Moreover [23] adds that knowledge management is all about managing the flow of information, which ensures that the right people get the right information at the right time. The main aspects that can be derived from these definitions of knowledge management according to [24, 25] are that knowledge is embedded in processes, services as well as products and it is presented in databases and documents.

Generally, knowledge management has three major elements that are people, technology and process [26]. People (human factor) are considered the core component of knowledge management as they account 70% of its success. This is due to the fact that people are the source of knowledge since they evolve and promulgate it. However, processes which weighs 20%, shapes all the actions taking place in the tangible management of knowledge including creating, storing, sharing, transferring and utilizing knowledge. Technology is the third component that enables people to effectuate the processes and make knowledge accessible everywhere at any time.

Due to recent advancements in technology, knowledge management has become less costly, standardized, ubiquitous and more effective in responding to individuals' needs [5]. The capabilities of the Artificial Intelligence tools contribute effectively to taking knowledge management to the next level. It helps employees to get the targeted information in a fraction of a second and boost real-time decision making. Also, it aids organizations to improve their ability to track processes and documents in an efficient manner. Artificial Intelligence can go farther than that as it can be utilized to create

Table 1 The research questions of this study

#	Research question
RQ1	What is the role of IT in knowledge management with respect to modern organizations?
RQ2	What are the common AI techniques used to generate knowledge at modern organizations?
RQ3	What are the main domains utilizing such technology in their daily business?
RQ4	Which countries are more engaged in the researches related to AI and KM?
RQ5	Which year(s) witnessed the highest rate of researches on the impact of AI on KM?
RQ6	What are the active repositories in the context of AI and KM?

and extract knowledge from a big amount of structured and unstructured data along with predicting the future trends of business. Accordingly, Artificial Intelligence technologies have been widely applied to boost the knowledge management systems at the majority of modern organizations to improve their performance [27].

As per literature, it has been noticed that there are some research gaps when it comes to the outcomes, challenges and limitations of applying IT and AI concepts in knowledge management at modern organizations. Thus, the objective of this research is to systematically review the knowledge management studies related to IT and AI in order to give a comprehensive analysis of the shortlisted studies. More precisely, the present study aims to add to the body of knowledge in this domain by responding to the six research questions illustrated in Table 1.

2 Literature Review

This section highlights the contributions of the recent researches that focus on the topic of this study. It discusses IT and AI best practices, challenges, and proposed approaches to foster knowledge management activities at modern organizations.

2.1 Integration of Technology and Knowledge Management

According to [28], knowledge management has been significantly affected by technology in various ways. For instance, IT and AI provide support to activities of knowledge management including database decision support systems, management information systems, expert systems, resource planning systems, lessons learned systems, among others. Additionally, there are various social mechanisms that support knowledge management, such as on-job training, learning through observation, employees' rotations and monitoring knowledge sharing.

2.2 The Transition from Organizational Learning to Knowledge Management

As put forward by Webb [29], organizational learning entails incorporating new members who can add knowledge that an organization lacks. This learning process is at the organizational level, and it requires learning from the results obtained through employees' behaviours. The convergence of knowledge management and organizational learning began during the 1960s, and it was mainly introduced in a similar study by Wang et al. [30].

Some researchers consider knowledge management as a part of organizational learning; however, others see it broader than this frame. The shift from organizational learning to knowledge management took place in three steps [31]. Firstly, organizational learning mainly focuses on creating knowledge and transferring it, while knowledge management mainly involves capturing knowledge and getting the best use of it. Secondly, organizational learning mainly emphasizes the maximization of employees' potentials through the process of self-development, while knowledge management entails a focus on networks and individuals to help the organization grow and expand [5]. Lastly, organizational learning focuses on social and structural aspects, whereas knowledge management identifies the importance of technology in creating, sharing and applying the obtained knowledge.

2.3 Increased Implementation of Knowledge Management Across Organizational Processes

Knowledge management is vital at modern organizations since it facilitates the creation of an effective organizational structure and it helps in performing various organizational tasks and processes that potentially lead to the discovery of new knowledge. [13] state that the increased dependence on knowledge management practices at modern organizations has been observed in shaping organizational behaviours.

Additionally, the current model knowledge-intensive organization is required to have more advanced communication and computing equipment that ease the integration of multimedia for capturing, processing, displaying, and storing information [32, 33]. In contrast to hierarchical organizations, knowledge communities at organizations that value knowledge management are interconnected via groupware [7]. These organizations use groupware facilities to conduct meetings, create multi-author documents in addition to access data and knowledge from their storage locations or devices.

2.4 Enablers of Knowledge Management

A knowledge management system entails the integration of IT to capture, organize, transfer, and distribute knowledge. The process of implementation is not easy as it requires proper planning to enable accessibility, timeliness and availability. Below are the IT technologies and the roles they play in knowledge management:

1. *Intranets*: this helps in sharing dynamic information at organizations. The intranet is divided into two components, namely: the web server and technology infrastructure. The technology entails HTML, IP networks and universal web server; while the webserver is the respiratory for content.

2. *Document management systems*: These are respiratory for organizational documents and explicit knowledge. They are mainly used to create, process and review organizational documents. These systems are vital knowledge respiratory and thus must be integrated into the organization's knowledge structure.

3. *Information retrieval engines*: These are part of corporate text respiratory and they exist at most organizations as a knowledge silo. New features are continuously being added as relevancy ranking, natural language, summarization, preferential searching and others for satisfying the information seekers' needs and ensuring the precision of finding information.

4. *Workflow systems and groupware*: These are used for communication and collaboration among departments and teams in formal Ad hoc conversations when users cannot communicate in real time. As such, this is a vital technology for exchanging and enhancing tacit knowledge.

Other IT tools include brainstorming applications as well as data warehousing and mining tools. These tools support knowledge management practices at modern organizations to increase their efficiency and effectiveness in business.

2.5 Role of AI in Knowledge Management

Nowadays, people have stopped considering businesses as industries and they viewed them from knowledge perspective. According to [30], in today's business world, knowledge is used as a resource. This is extremely true because the successful organizations are those that have managed to identify, evaluate, create and transform the knowledge they possess into assets. The life cycle of knowledge, which begins from data to information, then to knowledge, should be reduced to ensure that there is more value to satisfy the needs of organizations and individuals. Currently, ICT is accompanied by AI capabilities to ensure that an organization is getting the most out of what these technologies offer.

Artificial Intelligence tools and systems are applied widely at most modern organizations. These tools are used to recognize patterns, search heuristics and mathematical logic. In recent years, knowledge management has attracted an increasing attention among the fields where AI advances can be applied [34]. Various technologies under the umbrella of Artificial Intelligence including Neural Networks, Intelligent Agents and Genetic Algorithms provide organizations with intelligent agents for different operations such as users' profiling, matching patterns, text mining and semantic analysis of texts.

Artificial Intelligence technologies enable organizations to support and enhance practices of knowledge management. From the AI perspective, the representation of knowledge stipulates familiarizing the organizational processes with this knowledge [35]. This leads to the automation of knowledge management processes which allow computer systems to derive information and draw conclusions from knowledge in a form that can be interpreted by machines. The extent to which Artificial Intelligence can play a role in knowledge management has been discussed widely. The main categories of Artificial Intelligence that enhance knowledge management practices include Artificial Neural Networks and Intelligent Agents.

Artificial Neural Networks have much potential in knowledge management at modern organizations since they operate in a manner closer to the human brain. The main advantage of these neural networks is their ability to complete tasks and operate using data that is incomplete [7, 36]. This technology can be utilized in knowledge management to advance the distribution and sharing of knowledge.

However, intelligent search is used to ease the process of searching to form information on different search programs. For example, this happens when an employee types a keyword on the search engine, the intelligent search will reduce the number of results to only the relevant information, thus making it easier for the employee to find the information they are looking for [37]. After training the search engine using examples, it can be used to search for information only based on the keywords provided, as well as the content, context, and meaning. Many activities in knowledge management need intelligent support that can be implemented as an intelligent agent.

3 Method

A critical literature review is a paramount phase before carrying out any research [7]. It creates the foundation for gathering information which empowers theories' expansions and developments, fills in the gaps in the related research field and reveals the domains which previous researches have not probed thoroughly [38]. A literature review can be considered a systematic literary study only when it is grounded on explicit research questions, investigates and analyses other related research studies and assesses their quality based on specified standards. This research follows the guidelines provided by Kitchenham and Charters [39] for conducting systematic

Table 2 Inclusion and exclusion standards

Inclusion criteria	Exclusion criteria
Should include the processes of KM	Articles not focusing on KM concepts
Should include AI and IT technology	KM references highlighting other areas other than AI and IT
Should focus on modern organizations	Articles investigating traditional organizations
Should be published in the recent 10 years	Articles published prior to 10 years
Should be written in English	Articles not written in English

review studies, and other systematic reviews [40–42]. Accordingly, this systematic review is carried out in three distinct stages: the definition of inclusion and exclusion standards, sources of data and search strategies as well as quality evaluation. These stages are briefly discussed in the following sub-sections.

3.1 Selection Criteria (Inclusion and Exclusion Standards)

The aim of setting such criteria is to ensure that only the studies related to the scope of the research are shortlisted for further analysis. This was attained through implementing the inclusion-exclusion criterion on the mapping study to decide if a certain article should be selected for critical investigation. The research articles targeted for the critical analysis in this study shall meet both the inclusion and exclusion standards defined in Table 2.

3.2 Search Strategy

This systematic review was conducted based on highly reputed digital repositories such as Emerald, IEEE Xplore, Springer, Taylor & Francis, Wiley and ScienceDirect. The initial phase of the search strategy was to identify the main keywords forming the basis of this study: ("Knowledge Management" AND "Artificial Intelligence") OR ("Knowledge Management" AND "modern organizations"). At this stage of the systematic review, it is essential to select the proper keywords as they affect shortlisting the articles to be investigated. According to the above-mentioned keywords, 844 articles were retrieved. 150 articles were ruled out because of duplication. As a result, the total number of articles becomes 694. Each of these articles are analyzed and filtered based on the inclusion and exclusion standards. Hence, 15 articles were incorporated in the analytical process as they met the selection criteria. Figure 1 shows a flowchart summarizing the entire review process including number of research articles excluded at each phase.

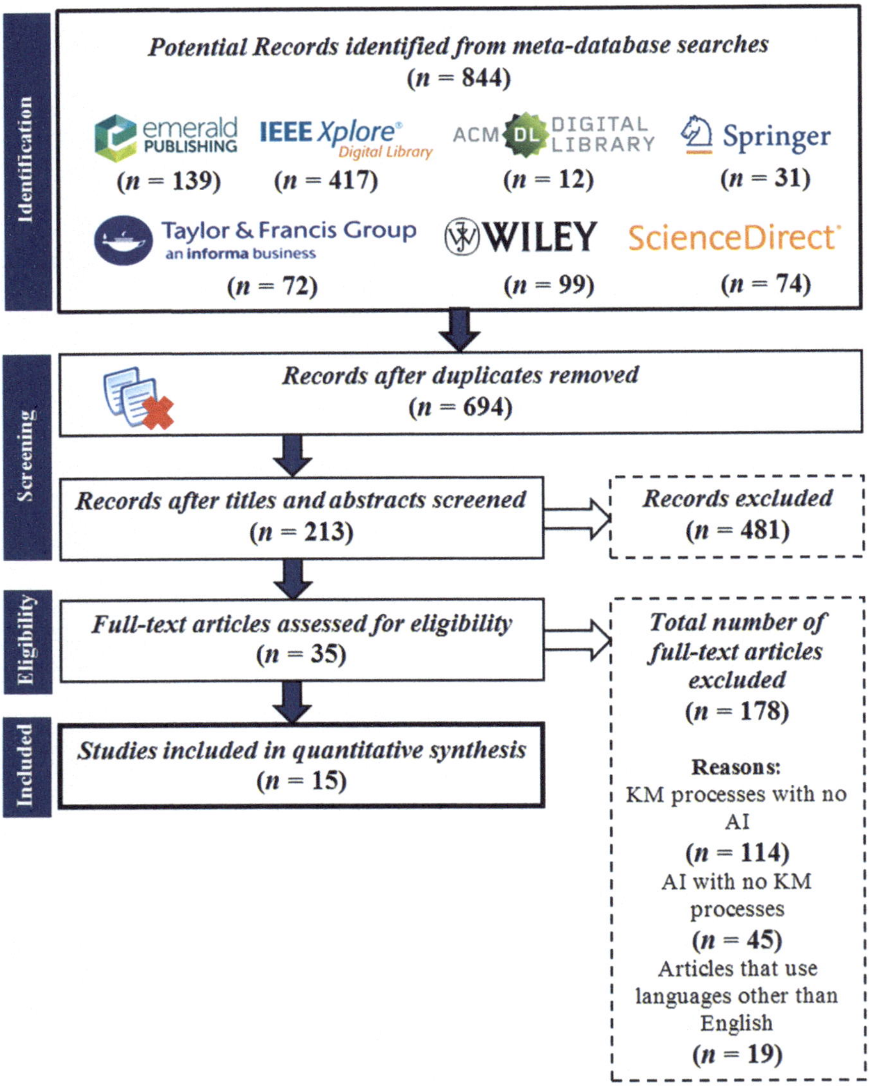

Fig. 1 A flowchart illustrating the literature search process

Table 3 Quality assessment checklist

#	Question
1	Are the study objectives precisely identified?
2	Are the KM processes investigated by the research precisely identified?
3	Are the AI techniques investigated by the research precisely identified?
4	Are the methods of data collection clearly stated?
5	Does the study explicate the validity of the measures?
6	Are the numerical methods utilized in analyzing the data clearly depicted?
7	Do the outcomes of the study add to the body of knowledge?
8	Does the study enrich your knowledge and add to your expertise?

3.3 Quality Assessment

One of the essential elements that requires to be investigated together with the inclusion and exclusion standards is quality assessment. In this systematic review, a checklist with eight criteria is utilized to appraise the quality of the articles and shortlist those that meet most of the criteria for further analysis ($N = 15$). Table 3 displays the quality assessment checklist used in this study. Each question was rated based on three-point scale that are 1 if the answer is "Yes", 0 if the answer is "No" and 0.5 in case the answer is "Partially". Therefore, each study could be rated from 0 to 8. The higher the study scores, the higher its potentiality to address the research questions. Table 4 illustrates the results of the quality assessment for the entire 15 studies. It can be observed that all the studies passed the quality check and they are eligible for further analysis.

4 Results

With reference to the shortlisted 15 research articles, published between 2009 and 2019, on the impact of IT and AI on knowledge management at modern organizations, the outcomes of this review are reported in accordance with six research questions.

4.1 RQ1: What Is the Role of IT in Knowledge Management with Respect to Modern Organizations?

There are several studies conducted to examine the role of IT and AI in knowledge management and how they affect modern organizations. A multilayer fundamental question regarding knowledge management is on the role of information technology

Table 4 Quality assessment results

Study	Q1	Q2	Q3	Q4	Q5	Q6	Q7	Q8	Total	Percentage (%)
S1	1	1	1	1	1	1	1	1	8	100
S2	1	1	1	1	1	1	0.5	1	7.5	93.75
S3	1	1	1	1	0.5	1	0.5	1	7	87.5
S4	1	1	1	0	1	1	0.5	0.5	6	75
S5	1	1	1	1	0	1	0.5	0.5	6	75
S6	1	1	1	1	0.5	1	1	0.5	7	87.5
S7	1	1	1	1	1	0.5	0.5	0.5	6.5	81.25
S8	1	1	1	0	1	1	0.5	0.5	6	75
S9	1	1	1	1	0	1	1	0.5	6.5	81.25
S10	1	1	1	0	0	1	0.5	0.5	5	62.5
S11	1	1	1	1	0.5	0	1	0.5	6	75
S12	1	1	1	0	1	0.5	0.5	0.5	5.5	68.75
S13	1	1	1	1	1	1	0.5	0.5	7	87.5
S14	1	1	1	1	1	0.5	1	0.5	7	87.5
S15	1	1	1	1	0.5	1	1	0.5	7	87.5

at organizations that have applied knowledge management practices. This question has attracted responses from various researchers concerning whether IT is an evitable or inevitable element and whether it is a key component of knowledge management. The success of knowledge management practices requires an appropriate balance between technology, people and processes, which should be supported by proper programs and strategies [35]. Organizations must clearly understand knowledge management and its importance. Different researchers hold different views with regard to the role of IT and AI in knowledge management. For instance, [43] state that IT plays a vital role by managing, storing and enabling access to documents and databases. Nonetheless, for any knowledge management practice to be successful, IT professionals must be aware of its various implementation processes. When IT is integrated into the knowledge management process, it becomes a major player. Additionally, a conceptual framework built by Wong et al. [7], reveals that information technology is a principal key player in knowledge management, along with other organizational pillars such as learning, leadership and organizational structure.

The debate on the role of IT in knowledge management is never-ending. On one hand, IT is used as a medium for the flow of knowledge to support communication, collaboration, knowledge-seeking and to enable collaborative learning. One the other hand, researchers of knowledge management warn individuals against the huge investment in IT at the expense of the human capital [21, 44]. This can lead to converting IT strategies and practices in knowledge management into static and inert information. Thus, failing to regard the role played by tacit knowledge, in turn, may bring back the earlier organizational motto "More IT, fewer people."

Integrating IT tools into knowledge management at modern organizations is an unavoidable move [43]. In this case, the best solution is to apply information technology by being aware of its limits that can give the most optimal results for knowledge management. One fact to mention is that the deployment of information technology in knowledge management does not guarantee success until it is accompanied by a change in the organization's culture towards values of knowledge.

Accordingly, knowledge management needs support from technology, and thus its projects have more chances of success when the organization adopts a sophisticated technology structure. Some IT infrastructure includes data warehousing, decision support systems, document management and workflow software as well as E-mails [5]. Knowledge management is not useful in a competitive business environment unless its applications and communication systems support various business processes in addition to creating, seeking and disseminating knowledge.

Table. 5 Classification of AI techniques used by modern organizations across the analysed studies

Source	Fuzzy logic	Artificial neural network (ANN)	Support vector machine (SVM)	Genetic algorithm (GA)	Knowledge-based systems (KBS)	Application domain
[30]		×			×	Finance
[45]		×				Software development
[46]		×				Energy
[47]		×	×	×		Energy
[48]		×				Manufacturing
[49]			×	×		Construction
[50]	×					Gas
[51]		×				Manufacturing
[52]		×	×			Telecommunications
[13]		×		×		Health
[53]	×	×				Health
[54]			×	×		Health
[23]		×			×	Finance
[55]		×	×			Health

4.2 *RQ2: What Are the Common AI Techniques Used to Generate Knowledge at Modern Organizations?*

Several studies are conducted to analyse the performance of applying Artificial Intelligence techniques such as Fuzzy Logic, Artificial Neural Network (ANN), Support Vector Machine (SVM), Genetic Algorithm (GA), Knowledge-Based Systems (KBS), etc. in generating knowledge in various business domains. Table 5 demonstrates the classification of AI techniques used by modern organizations stated in the analysed studies. It can be observed that ANN is the most frequently used AI technique ($N = 12$), followed by SVM and GA ($N = 5$), KBS ($N = 3$) and Fuzzy Logic ($N = 2$), respectively, in the analysed studies.

4.3 *RQ3: What Are the Main Domains Utilizing Such Technology in Their Daily Business?*

AI technology plays an essential role in improving the performance of a wide range of organizations. Based on literature, it can be noticed that AI has more contributions in some domains than others. Table 5 shows that AI is widely implemented at health care organizations as it supports many functionalities such as diagnosing some common diseases and providing medication prescriptions. Moreover, they can precisely predict some diseases that a patient may suffer from in the future along with the percentage of the surgical operations success. IT and AI related technologies also help in detecting a variety of chronic diseases at early stages which alleviates the financial cost and the degree of their severity. In addition, AI introduces a variety of intelligent tools and automated systems such as Brain-Computer Interfaces (BCIs), ASL-MRI, biomarkers, Natural Language Processing (NLP), etc. that helps the reduction of human errors to the minimal level and the control of disease progression. From another perspective, some domains such as finance, energy, construction and manufacturing experienced almost the same level of AI growth.

4.4 RQ4: Which Countries Are More Engaged in the Researches Related to AI and KM?

There have been many countries interested in exploring various AI technologies to boost knowledge management practices in their organizations. Figure 2 illustrates that Taiwan ($N = 3$) was the most involved country in enriching the body of knowledge in this domain followed by Spain ($N = 2$), Malaysia ($N = 2$) and Denmark ($N = 1$), respectively.

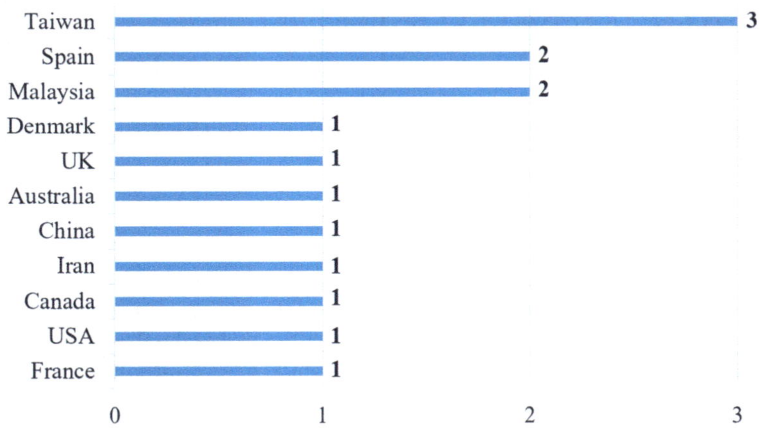

Fig. 2 Number of studies per the country of implementation

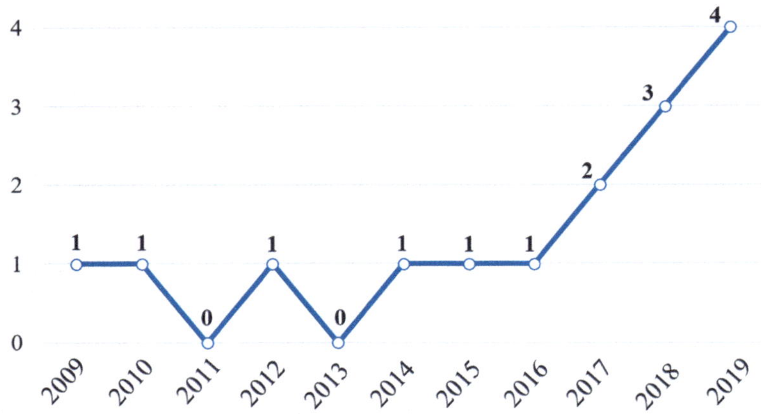

Fig. 3 Number of studies per the publishing year

4.5 RQ5: Which Year(s) Witnessed the Highest Rate of Researches on the Impact of AI on KM?

According to the findings of this study (Fig. 3), the year 2019 witnessed the highest number of the published studies in this domain ($N = 4$). Also, it can be noticed that there was a gradual increase in the number of the published articles related to the role of AI in KM in the recent three years from 2016 to 2019.

4.6 RQ6: What Are the Active Repositories in the Context of AI and KM?

This section is devoted to identify the ultimate repositories that contain valuable research articles on the impact of AI on knowledge management at modern organizations. Figure 4 shows the researches distribution according to repositories. It can be observed that ScienceDirect ($N = 6$) is the most valuable database as it contains the majority of knowledge management related studies especially those relevant to IT and AI, followed by IEEE ($N = 4$), Taylor & Francis ($N = 2$), Springer ($N = 1$), Wiley ($N = 1$) and Emerald ($N = 1$) respectively among the other databases.

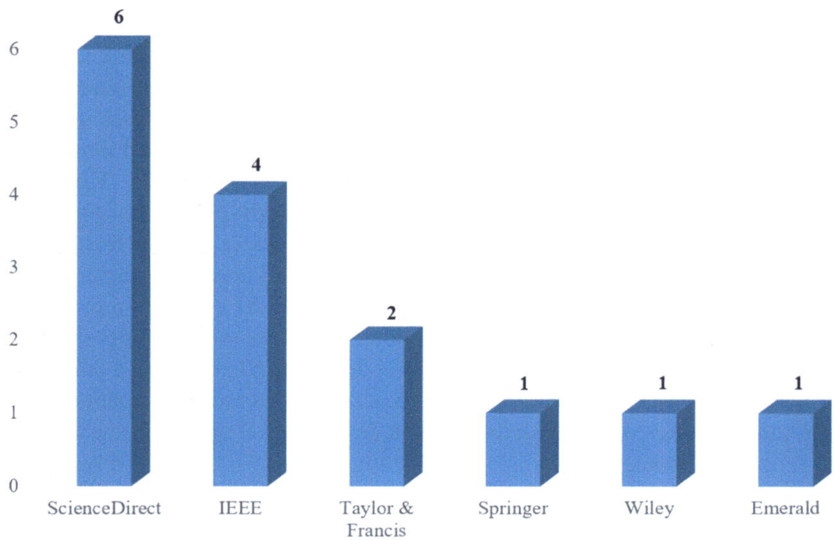

Fig. 4 Researches distribution according to repositories

5 Discussion and Conclusion

5.1 Discussion

The integration of IT and AI concepts in knowledge management practices enable entities to get the required information for a more efficient decision-making process. The ultimate goal of this study is to systematically review and investigate the articles published on this topic to fill in the research gaps and enrich the understanding of the aspects related to IT and AI and their impacts on knowledge management at modern organizations in particular. Among many researches carried out in knowledge management domain, few studies only emphasize on IT and AI influences on knowledge management at modern organizations. The majority of these studies stated that data warehousing and mining tools are of vital necessity for helping managers to make informed decisions based on complex sets of data. More specifically [35] confirmed that data mining is used to acquire the targeted data from large databases during investigations, which saves time and effort. Other studies revealed that IT tools, such as brainstorming applications, are used to convert tacit knowledge to explicit knowledge. According to [5, 56], these tools are useful for creating knowledge, and they help in categorizing, organizing and identifying resource of knowledge and its applications.

From another perspective, the recent advancements made in Artificial Intelligence play a vital role in knowledge management at modern organizations, as they foster processes and facilitate tasks. There are various Artificial Intelligence tools and techniques that can be of a great practical value to support different business domains and organizations such as Fuzzy Logic, Artificial Neural Network (ANN), Genetic Algorithm (GA), Support Vector Machine (SVM), Knowledge-Based Systems (KBS), etc. Table 5 illustrates the classification of the AI techniques used by modern organizations based on the analyzed studies. It can be observed that ANN is the most frequently used AI technique, followed by SVM and GA, KBS and Fuzzy Logic, respectively, in the analyzed studies. These findings are almost aligned with [57] who pointed out that ANN and SVM are the widely used AI techniques by organizations concerned with automated processes and knowledge management best practices. In addition, it can be noticed that IT and AI concepts are widely implemented in health care, compared with other domains, as they support many functionalities such as diagnosing some common diseases and providing medication prescriptions. Moreover, they can precisely predict some diseases that a patient may suffer from in the future along with the percentage of surgical operation success. IT and AI related technologies also help in detecting a variety of chronic diseases at early stages which alleviates the financial cost and the degree of its severity.

With regard to the countries of the published studies, Fig. 2 evidently indicates that the majority of studies related to AI in knowledge management were conducted in Taiwan followed by Spain and Malaysia, respectively, among the other countries. With respect to research findings, most of the analysed articles that attained worth outcomes took place in Taiwan and Malaysia. The reason behind achieving such outcomes could be due to the incentives and high motivation that encourage the researchers of these countries to carry out studies targeting AI and IT technologies implementations in knowledge management rather than other research domains. This could possibly clarify the reason behind carrying out the majority of these studies in the pre-mentioned countries in comparison with the others.

Concerning the publishing year, Fig. 3 points out that the number of IT and AI related studies in knowledge management domain witnessed a remarkable gradual increase in the recent three years from 2016 to 2019. There was only one study in this domain in 2016; however, the number of studies increased four times in 2019. This increase might be an indication of AI technology growing popularity in various fields especially throughout the past few years. With respect to the search repositories, Fig. 4 demonstrates that ScienceDirect is the most valuable database as it contains the majority of knowledge management related studies especially those relevant to IT and AI, followed by IEEE, Taylor & Francis, Springer, Wiley and Emerald respectively among the other databases.

6 Conclusion

Knowledge management is one of the hottest research topics in academia and business sectors. However, scholars did not pay enough attention to explore the impact of IT and AI on knowledge management at modern organizations. In this study, a systematic review was conducted to give a comprehensive analysis of the shortlisted research articles and to debate the implications of analysis findings. This review study came up with five main outcomes. First, IT tools and systems such as data warehousing, brainstorming applications, decision support systems, document management systems and information retrieval engines are used to enhance knowledge management practices by easing access to information, creating, organizing and disseminating relevant knowledge and information within the organization, to enhance organizational performance. Second, there are several AI techniques such as Fuzzy Logic, Artificial Neural Network (ANN), Support Vector Machine (SVM), Genetic Algorithm (GA), Knowledge-Based Systems (KBS), etc. that are successfully implemented in a variety of business domains. ANN was found to be the most frequently used AI technique, followed by SVM and GA, KBS and fuzzy logic respectively. With respect to business domains, AI was found to be widely applied at health care organizations followed by other sectors such as construction, finance, energy, manufacturing, telecommunications, etc. Third, most of the analysed studies were conducted in Taiwan, followed by Spain and Malaysia, respectively, among the other countries. Fourth, regarding the publishing year, knowledge management studies related to AI have witnessed a noticeable gradual increase in the recent three years from 2016 to 2019. Fifth, ScienceDirect is the most valuable database as it contains the majority of knowledge management related studies especially those relevant to IT and AI, followed by IEEE, Taylor & Francis, Springer, Wiley and Emerald respectively among the other databases. In conclusion, the outcomes of this systematic review provide a profound insight in the role of IT and AI in shaping the future trend of knowledge management. These outcomes also form a comprehensive reference for the upcoming knowledge management studies' especially those related to IT and AI.

As a limitation, this review is relied on certain repositories for gathering the research articles that are Emerald, IEEE Xplore, Springer, Taylor & Francis, Wiley and ScienceDirect. Hence, these repositories may not provide all the valuable published researches on the impact of IT and AI on knowledge management at modern organizations. This study can be extended to cover more articles from other repositories such as Scopus, Sage, and Google Scholar, along with others, to add more value and dimensions to the research outcomes.

Acknowledgements This work is a part of a project undertaken at the British University in Dubai.

References

1. Al-Maroof, R.S., Salloum, S.A., AlHamadand, A.Q., Shaalan, K.: Understanding an extension technology acceptance model of google translation: a multi-cultural study in United Arab Emirates. Int. J. Interact. Mob. Technol. **14**(03), 157–178 (2020)
2. Al Kurdi, B., Alshurideh, M., Salloum, S.A., Obeidat, Z.M., Al-dweeri, R.M.: An empirical investigation into examination of factors influencing university students' behavior towards e-learning acceptance using SEM approach. Int. J. Interact. Mob. Technol. **14**(02), 19–41 (2020)

3. Al-Emran, M., Arpaci, I., Salloum, S.A.: An empirical examination of continuous intention to use m-learning: an integrated model. Educ. Inf. Technol. (2020)
4. Alghizzawi, M., Habes, M., Salloum, S.A., Ghani, M.A., Mhamdi, C., Shaalan, K.: The effect of social media usage on students' e-learning acceptance in higher education: a case study from the United Arab Emirates. Int. J. Inf. Technol. Lang. Stud. **3**(3) (2019)
5. Lee, K.-W., Lanting, M.C.L., Rojdamrongratana, M.: Managing customer life cycle through knowledge management capability: a contextual role of information technology. Total Qual. Manag. Bus. Excell. **28**(13–14), 1559–1583 (2017)
6. Salloum, S.A., Al-Emran, M., Shaalan, K.: The Impact of knowledge sharing on information systems: a review. In: International Conference on Knowledge Management in Organizations, pp. 94–106 (2018)
7. Wong, K.Y., Tan, L.P., Lee, C.S., Wong, W.P.: Knowledge management performance measurement: measures, approaches, trends and future directions. Inf. Dev. **31**(3), 239–257 (2015)
8. Salloum, S.A., Al-Emran, M., Shaalan, K.: Mining text in news channels: a case study from facebook. Int. J. Inf. Technol. Lang. Stud. **1**(1), 1–9 (2017)
9. Mhamdi, C., Al-Emran, M., Salloum, S.A.: Text mining and analytics: a case study from news channels posts on Facebook, **740** (2018)
10. Salloum, S.A., Shaalan, K.: Adoption of e-book for university students. In: International Conference on Advanced Intelligent Systems and Informatics, pp. 481–494 (2018)
11. Salloum, S.A., Alhamad, A.Q.M., Al-Emran, M., Monem, A.A., Shaalan, K.: Exploring students' acceptance of e-learning through the development of a comprehensive technology acceptance model. IEEE Access **7**, 128445–128462 (2019)
12. Alhashmi, S.F.S., Salloum, S.A., Abdallah, S.: Critical success factors for implementing artificial intelligence (AI) projects in Dubai Government United Arab Emirates (UAE) health sector: applying the extended technology acceptance model (TAM). In: International Conference on Advanced Intelligent Systems and Informatics, pp. 393–405 (2019)
13. Santoro, G., Vrontis, D., Thrassou, A., Dezi, L.: The Internet of Things: Building a knowledge management system for open innovation and knowledge management capacity. Technol. Forecast. Soc. Change **136**, 347–354 (2018)
14. Al-Emran, M., Mezhuyev, V., Kamaludin, A., Shaalan, K.: The impact of knowledge management processes on information systems: a systematic review. Int. J. Inf. Manage. **43**, 173–187 (2018)
15. Al-Emran, M., Mezhuyev, V., Kamaludin, A.: Towards a conceptual model for examining the impact of knowledge management factors on mobile learning acceptance. Technol. Soc. (2020)
16. Al-Emran, M., Mezhuyev, V., Kamaludin, A.: An innovative approach of applying knowledge management in m-learning application development: a pilot study. Int. J. Inf. Commun. Technol. Educ. **15**(4), 94–112 (2019)
17. Al-Emran, M., Mezhuyev, V.: Examining the effect of knowledge management factors on mobile learning adoption through the use of importance-performance map analysis (IPMA). In: International Conference on Advanced Intelligent Systems and Informatics, pp. 449–458 (2019)
18. Al-Emran, M., Teo, T.: Do knowledge acquisition and knowledge sharing really affect e-learning adoption? An empirical study. Educ. Inf. Technol. (2019)
19. Al-Emran, M., Mezhuyev, V., Kamaludin, A., AlSinani, A.: Development of M-learning application based on knowledge management processes. In: 2018 7th International conference on Software and Computer Applications (ICSCA 2018), pp. 248–253 (2018)
20. Al-Emran, M., Mezhuyev, V., Kamaludin, A.: Students' perceptions towards the integration of knowledge management processes in M-learning systems: a preliminary study. Int. J. Eng. Educ. **34**(2), 371–380 (2018)
21. Dalkir, K.: Knowledge Management in Theory and Practice. Routledge (2013)
22. Mao, H., Liu, S., Zhang, J., Deng, Z.: Information technology resource, knowledge management capability, and competitive advantage: the moderating role of resource commitment. Int. J. Inf. Manage. **36**(6), 1062–1074 (2016)

23. Jimenez-Jimenez, D., Martínez-Costa, M., Sanchez Rodriguez, C.: The mediating role of supply chain collaboration on the relationship between information technology and innovation. J. Knowl. Manag. **23**(3), 548–567 (2019)
24. Altamony, H., Alshurideh, M., Obeidat, B.: Information systems for competitive advantage: Implementation of an organisational strategic management process. In: Proceedings of the 18th IBIMA Conference on Innovation and Sustainable Economic Competitive Advantage: From Regional Development to World Economic, Istanbul, Turkey, 9th-10th May (2012)
25. Shannak, R., Masa'deh, R., Al-Zu'bi, Z., Obeidat, B., Alshurideh, M., Altamony, H.: A theoretical perspective on the relationship between knowledge management systems, customer knowledge management, and firm competitive advantage. Eur. J. Soc. Sci. **32**(4), 520–532 (2012)
26. Geisler, E., Wickramasinghe, N.: Principles of Knowledge Management: Theory, Practice, and Cases: Theory, Practice, and Cases. Routledge (2015)
27. Kasemsap, K.: The roles of information technology and knowledge management in project management metrics. In: Global Business Expansion: Concepts, Methodologies, Tools, and Applications, IGI Global, pp. 1191–1221 (2018)
28. Sedighi, M., Jalalimanesh, A.: Mapping research trends in the field of knowledge management. Malaysian J. Libr. Inf. Sci. **19**(1) (2017)
29. Webb, S.P., Knowledge Management: Linchpin of Change. Routledge (2017)
30. Wang, H., Xu, Z., Fujita, H., Liu, S.: Towards felicitous decision making: an overview on challenges and trends of big data. Inf. Sci. (Ny) **367**, 747–765 (2016)
31. Razmerita, L., Phillips-Wren, G., Jain, L.C.: Advances in knowledge management: an overview. In: Innovations in Knowledge Management. Springer, pp. 3–18 (2016)
32. Alshurideh, M., Alhadid, A., Al kurdi, B.: The effect of internal marketing on organizational citizenship behavior an applicable study on the University of Jordan employees. Int. J. Mark. Stud. **7**(1), 138 (2015)
33. Ammari, G., Al kurdi, B., Alshurideh, M., Alrowwad, A.: Investigating the impact of communication satisfaction on organizational commitment: a practical approach to increase employees' loyalty. Int. J. Mark. Stud. **9**(2), 113–133 (2017)
34. Serenko, A., Dumay, J.: Citation classics published in knowledge management journals. Part I: articles and their characteristics. J. Knowl. Manag. **19**(2), 401–431 (2015)
35. Inkinen, H.T., Kianto, A., Vanhala, M.: Knowledge management practices and innovation performance in Finland. Balt. J. Manag. **10**(4), 432–455 (2015)
36. Alshurideh, M., Salloum, S.A., Al Kurdi, B., Al-Emran, M.: Factors affecting the social networks acceptance: an empirical study using PLS-SEM approach. In: 8th International Conference on Software and Computer Applications, pp. 414–418 (2019)
37. Saqib, M., Udin, Z.M., Baluch, N.: The impact of knowledge management on organizational performance in today's economy. South East J. Contemp. Business, Econ. Law, **12**(3) (2017)
38. Marangunić, N., Granić, A.: Technology acceptance model: a literature review from 1986 to 2013. Univers. Access Inf. Soc. **14**(1), 81–95 (2015)
39. Kitchenham, B., Charters, S.: Guidelines for performing systematic literature reviews in software engineering. Softw. Eng. Group, Sch. Comput. Sci. Math. Keele Univ., 1–57 (2007)
40. Al-Saedi, K., Al-Emran, M., Abusham, E., El_Rahman, S.A.: Mobile payment adoption: a systematic review of the UTAUT model. In: International Conference on Fourth Industrial Revolution (2019)
41. Saa, A.A., Al-Emran, M., Shaalan, K.: Factors affecting students' performance in higher education: a systematic review of predictive data mining techniques. Technol. Knowl. Learn. (2019)
42. Al-Qaysi, N., Mohamad-Nordin, N., Al-Emran, M.: A systematic review of social media acceptance from the perspective of educational and information systems theories and models. J. Educ. Comput. Res. **57**(8), 2085–2109 (2020)
43. Paul, P., Aithal, P.S., Bhuimali, A.: MCA (Information science and management): the next generation interdisciplinary specialization for better social informatics and digital humanities practice. Int. J. Sci. Res. Math. Stat. Sci. **4**(5), 27–32 (2017)

44. Zu'bi, Z., Al-Lozi, M., Dahiyat, S., Alshurideh, M., Al Majali, A.: Examining the effects of quality management practices on product variety. Eur. J. Econ. Financ. Adm. Sci. **51**(1), 123–139 (2012)
45. Ligęza, A., Potempa, T.: Artificial intelligence for knowledge management with bpmn and rules. In: IFIP International Workshop on Artificial Intelligence for Knowledge Management, pp. 19–37 (2012)
46. Ordóñez de Pablos, P., Lytras, M.: Knowledge Management, Innovation and Big Data: Implications for Sustainability Policy Making and Competitiveness. Multidisciplinary Digital Publishing Institute (2018)
47. Sarka, P., Heisig, P., Caldwell, N.H.M., Maier, A. M., Ipsen, C.: Future research on information technology in knowledge management. Knowl. Process Manag. (2019)
48. Antunes, H. de J. G., Pinheiro, P. G.: J. Innovat. Knowl. (2019)
49. Cheng, M.-Y., Roy, A.F.V.: Evolutionary fuzzy decision model for construction management using support vector machine. Expert Syst. Appl. **37**(8), 6061–6069 (2010)
50. Saberi, M., Azadeh, A. Saberi, Z., Pazhoheshfar, P.: A knowledge management system based on artificial intelligence (AI) methods: a flexible fuzzy regression-analysis of variance algorithm for natural gas consumption estimation. In: 2012 International Conference on Information Retrieval & Knowledge Management, pp. 143–147 (2012)
51. Tan, L.P., Wong, K.Y.: A neural network approach for predicting manufacturing performance using knowledge management metrics. Cybern. Syst. **48**(4), 348–364 (2017)
52. Sanzogni, L., Guzman, G., Busch, P.: Artificial intelligence and knowledge management: questioning the tacit dimension. Prometheus **35**(1), 37–56 (2017)
53. Aljaaf, A.J., Al-Jumeily, D., Hussain, A.J., Fergus, P., Al-Jumaily, M., Abdel-Aziz, K.: Toward an optimal use of artificial intelligence techniques within a clinical decision support system. Sci Inf Conf (SAI) **2015**, 548–554 (2015)
54. Nilakanta, S., Miller, L., Peer, A., Bojja, V.M.: Contribution of knowledge and knowledge management capability on business processes among healthcare organizations. In: 2009 42nd Hawaii International Conference on System Sciences, pp. 1–9 (2009)
55. Shahid, N., Rappon, T., Berta, W.: Applications of artificial neural networks in health care organizational decision-making: a scoping review. PLoS ONE **14**(2), e0212356 (2019)
56. Al Dmour, H., Alshurideh, H., Shishan, F.: The influence of mobile application quality and attributes on the continuance intention of mobile shopping. Life Sci. J. **11**(10), 172–181 (2014)
57. Costa, V., Monteiro, S.: Knowledge processes, absorptive capacity and innovation: a mediation analysis. Knowl. Process Manag. **23**(3), 207–218 (2016)

Technology in Education

A Novel Approach for Predicting the Adoption of Smartwatches Using Machine Learning Algorithms

Ibrahim Arpaci, Mostafa Al-Emran, Mohammed A. Al-Sharafi, and Khaled Shaalan

Abstract The essential purpose of this research is to examine the behavioral intention to adopt smartwatches by individuals. A theoretical model was developed by integrating the Technology Acceptance Model (TAM) with the common smartwatch features, including mobility, availability, and perceived enjoyment. A quantitative approach using an online survey was employed to collect data from 491 students enrolled at seven higher educational institutes in Malaysia. This study employs a novel approach for testing the developed research model using six popular machine learning classification algorithms, including a meta classifier (LogitBoost), a logistic regression classifier (SimpleLogistic), a lazy classifier (KStar), a Bayesian classifier (NaïveBayes), a rule learner (OneR), and a decision tree classifier (LMT). The results indicated that SimpleLogistic and LMT performed better than the other classifiers in predicting the behavioral intention with an accuracy of 61%. This study has a novel contribution to the Information Systems (IS) literature since it was the first attempt to apply machine learning techniques in predicting the individuals' adoption of smartwatches. The limitations and future research directions are also discussed.

Keywords Machine learning · Classifiers · Smartwatches · Wearable technologies

I. Arpaci (✉)
Department of Computer Education and Instructional Technology, Tokat Gaziosmanpasa University, Tokat, Turkey
e-mail: ibrahim.arpaci@gop.edu.tr

M. Al-Emran
Department of Information Technology, Al Buraimi University College, Al Buraimi, Oman
e-mail: mustafa.n.alemran@gmail.com

M. A. Al-Sharafi
Faculty of Computer Systems and Software Engineering, Universiti Malaysia Pahang, Gambang, Malaysia
e-mail: alsharafi@ieee.org

K. Shaalan
Faculty of Engineering and IT, The British University in Dubai, Dubai, UAE
e-mail: khaled.shaalan@buid.ac.ae

M. Al-Emran et al. (eds.), *Recent Advances in Intelligent Systems and Smart Applications*, Studies in Systems, Decision and Control 295,
https://doi.org/10.1007/978-3-030-47411-9_10

1 Introduction

Wearable devices are getting prominence step by step soon. These life-style tools have become a fundamental piece of our daily lives [1]. In education, the scope of wearable technologies available to educators has been rapidly increasing [2]. Wearables provide students and educators with "hands-free" access to immersive information that can be viewed as a digital 3D environment or overlaid on smart glasses, which can be accessed through a smartwatch or wristband [2]. Smartwatches represent the most popular type of wearable devices in the early stages of diffusion [3]. Smartwatch apps have many features that go beyond time display. Such tools are one of the latest developments in the growth of IT [3]. Although greater awareness and understanding of the use of these tools in education is required, other factors affecting technological innovation in higher education need to be recognized and addressed [2].

Smartwatch adoption is a particularly important topic in the domain of wearable technologies [4]. Although the importance of the quality interpretation of the interactions of individuals has been commonly used on technology products, little debate has been made about the experiential value of using smartwatches for higher education services. Specifically, the topic of what drives the adoption of smartwatches in higher education stays unanswered. Also, the adoption of IoT in which wearables are a piece of is still rare, particularly in developing nations [5].

Previous research on technology acceptance developed different theoretical models to investigate the adoption of new technologies by individuals, such as the "technology acceptance model (TAM)" developed by Davis [6]. There has been little work, however, to study the adoption of smartwatches through TAM lenses. Therefore, additional smartwatch features like "mobility", "availability", and "enjoyment" will contribute significantly to the evaluation of smartwatches by individuals. Knowing what drives the adoption of smartwatches will help decision-makers to develop highly successful products, and thus, increase their rate of diffusion. This research develops a theoretical model by expanding the TAM with smartwatch features, including "mobility", "availability", and "enjoyment" to understand the students' adoption of smartwatches in a developing country in order to fill the gaps in the literature and find answers to the aforementioned issues. Unlike the related previous studies, this study employs a novel approach to evaluate the developed research model through the use of machine learning classification algorithms.

2 Research Model and Hypotheses

As new advances are always being created and marketed in the present time of the inexorably computerized world, different hypothetical models have been proposed

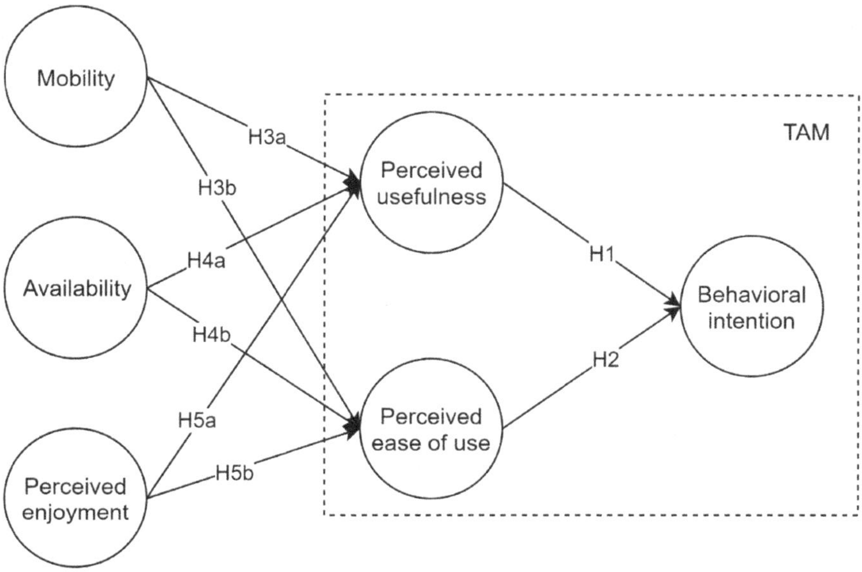

Fig. 1 Research model

to explain the adoption of several technologies. In general, the "technology acceptance model (TAM)" developed by Davis [6] is considered one of the most common theoretical models used to analyze the adoption of ICTs by individuals, and it has been validated across several technology acceptance studies [7–9]. In this way, the present study expands the TAM with the important smartwatch features, including "mobility", "availability", and "enjoyment" to comprehend the students' behavioral intention to utilize smartwatches in instructive purposes. The proposed hypothetical model is outlined in Fig. 1.

2.1 Perceived Usefulness (PU)

PU is defined as "the degree to which a person believes that using a particular system would enhance his or her job performance" [6]. Prior research revealed that PU has a positive association with the behavioral intention to adopt different systems [10–16]. By extending this association to the context of this study, the following hypothesis is suggested:

H1: PU would predict the behavioral intentions to use smartwatches.

2.2 *Perceived Ease of Use (PEOU)*

PEOU is defined as "the degree to which a person believes that using a particular system would be free from effort" [6]. Previous studies pointed out that PEOU has a positive relationship with the behavioral intention to adopt different systems [10–17]. By extending this relationship to the context of this study, the following hypothesis is suggested:

H2: PEOU would predict the behavioral intentions to use smartwatches.

2.3 *Mobility (MOB)*

Mobility refers to the degree to which individuals believe that they can move to various places and use their devices in transit [4, 18]. Mobility was shown to be an essential determinant for the adoption of several systems [19, 20]. This study suggests that the higher the mobility through smartwatch devices, the higher the PU and PEOU of these devices would be. Thus, we hypothesize the following:

H3a. Mobility would predict the PU of smartwatches.
H3b. Mobility would predict the PEOU of smartwatches.

2.4 *Availability (AVA)*

Availability is defined as the degree to which individuals believe that their devices provide real-time access to information and services [4]. Availability was found to be an important determinant for the adoption of smartwatches in the past [4]. This study suggests that the higher the availability through smartwatch devices, the higher the PU and PEOU of these devices would be. Hence, we hypothesize the following:

H4a. Availability would predict the PU of smartwatches.
H4b. Availability would predict the PEOU of smartwatches.

2.5 *Perceived Enjoyment (PE)*

PE refers to the extent to which the activity of using the technology is perceived to be enjoyable in its own right, aside from any performance consequences that might be anticipated [21]. PE was found to be a critical determinant for the adoption of various technologies [22, 23]. This study postulates that the higher the enjoyment perceived through smartwatch devices, the higher the PU and PEOU of these devices would be. Therefore, we hypothesize the following:

H5a. Perceived enjoyment would predict the PU of smartwatches.
H5b. Perceived enjoyment would predict the PEOU of smartwatches.

3 Method

3.1 Procedure and Sample

The participants were selected from seven technical and research-based higher educational institutes in Malaysia. The data were collected through the medium of an online survey. The employment of surveys for data collection helps in effectively testing the research hypotheses [24]. The participants were expected to complete the survey within 10 min. A total of 491 valid responses were collected. The study included 265 females (54%) and 226 (46%) males, whose ages ranged between 18 and 40. The majority of the participants had a bachelor's degree (52.1%), followed by diploma (46%), and master's degree (1.8%), respectively. Moreover, 60.1% of the participants were enrolled under the technology stream. Further, 52.3% of them stated that they use a wearable technology as shown in Table 1.

Table 1 Demographics of the participants

		Frequency	Percent
Gender	Male	226	46.0
	Female	265	54.0
Age	18–22	280	57.0
	23–28	130	26.5
	29–35	46	9.4
	Above 35	35	7.1
Program	Science	27	5.5
	Technology	295	60.1
	Engineering	109	22.2
	Management	60	12.2
Education	Diploma	226	46.0
	Bachelor	256	52.1
	Master	9	1.8
WT Use	Yes	257	52.3
	No	234	47.7

3.2 Research Instrument

The employed online survey consists of two parts. The first part includes questions related to students' demographics, whereas the second one contains questions related to measure the factors in the research model. The items in the second part were measured using a five-point Likert scale, with values ranging from "1 = Strongly disagree" to "5 = Strongly agree". All the items were adapted from the existing literature with further modifications to fit the smartwatch domain. The items used to measure "perceived usefulness", "perceived ease of use", and "behavioral intention" were adapted from the relevant studies conducted in the past [6, 25]. Further, the items used to measure "mobility" were adapted from Huang et al. [19], whereas the items used to measure "availability" were adapted from Kim and Shin [4]. The items used to measure "perceived enjoyment" were adapted from the previous relevant studies [19, 25]. It is worth mentioning that the survey was deployed in English language.

4 Results

4.1 Reliability and Descriptive Statistics

IBM SPSS (ver. 25) was used to test the internal reliability of the scales. Each scale has a Cronbach's alpha value greater than the critical threshold value of 0.70, which suggested an acceptable reliability for each scale. Table 2 shows the internal reliability coefficients of the scales and descriptive statistics for the scale items.

4.2 Hypotheses Testing

This study applied machine learning algorithms, which have different methodologies such as decision trees, bayesian networks, neural networks, and if-then-else rules, to design and test the predictive model. The developed theoretical model was tested by using Weka (ver. 3.8.3) based on the most popular classifiers, including a meta classifier (LogitBoost), a logistic regression classifier (SimpleLogistic), a lazy classifier (KStar), a Bayesian classifier (NaïveBayes), a rule learner (OneR), and a decision tree classifier (LMT) [26–28]. The results indicated that SimpleLogistic and LMT classifiers performed better than other classifiers in predicting the behavioral intention. SimpleLogistic and LMT classifiers predicted the behavioral intention with an accuracy of 61.08% and 61.71% for the percentage-split and 10-fold cross-validation, respectively. These algorithms have a better performance in terms of TP rate (0.617), precision (0.621), and F-measure (0.581) for the 10-fold cross-validation compared to other classifiers, as shown in Table 3. The results suggested that the attributes of

Table 2 Instrument reliability and descriptive statistics

Code	Scale and items	α	Mean	Std. Dev.
Mobility (MOB)		0.82	12.09	1.92
MOB1	"Smartwatch has a good mobility"		3.96	0.741
MOB2	"I feel that I can use smartwatch anywhere"		4.09	0.768
MOB3	"I would like to use smartwatch when I am in transit from one place to another"		4.04	0.731
Availability (AVA)		0.90	14.79	2.55
AVA1	"I can access information and desired contents at any time via smartwatch"		3.72	0.741
AVA2	"I can use smartwatch at any time I want to get desired information and service"		3.68	0.700
AVA3	"Smartwatch offers the sense of real-time connectedness"		3.68	0.746
AVA4	"Smartwatch offers immediate and timely access to information and services"		3.71	0.716
Perceived enjoyment (PE)		0.88	17.78	3.2
PE1	"I experience great pleasure when using smartwatch"		3.55	0.792
PE2	"I feel satisfied when I use smartwatch"		3.56	0.763
PE3	"I feel pleasure when I use smartwatch"		3.64	0.760
PE4	"I find enjoyment when I use smartwatch"		3.55	0.755
PE5	"Overall, using smartwatch is fun"		3.48	0.831
Perceived usefulness (PU)		0.79	15.75	2.86
PU1	"It would be convenient for me to have smartwatch"		3.98	0.734
PU2	"Using smartwatch increases my learning achievement"		3.88	0.759
PU3	"I think smartwatch can help me with many things"		3.82	0.735
PU4	"Overall, smartwatch is useful to me"		4.06	0.690
Perceived ease of use (PEOU)		0.78	15.33	2.42
EOU1	"I find it easy to use smartwatch even when using it for the first time"		3.80	.821
EOU2	"It is easy for me to become skillful in using smartwatch"		3.84	0.750
EOU3	"Smartwatch is flexible to interact with"		3.85	0.755
EOU4	"Overall, smartwatch is easy for me to use"		3.83	0.776
Behavioral intention (BI)		0.84	12.10	2.01
BI1	"I intend to start using smartwatch to interact with colleagues and lecturers"		4.03	0.754
BI2	"I will strongly recommend the use of smartwatch to my peers"		4.01	0.797
BI3	"Overall, I intend to continue using smartwatch in the future"		4.07	0.767

Table 3 Predicting BI by PU and PEOU

Classifier	CCI[a] (%)		TP[b]-rate		Precision		F-Measure	
	10-fold	PS[c]	10-fold	PS	10-fold	PS	10-fold	PS
NaïveBayes	61.30	59.88	0.613	0.643	0.605	0.667	0.599	0.565
SimpleLogistic	61.71	61.08	0.617	0.599	0.621	0.601	0.581	0.569
KStar	59.88	57.49	0.599	0.575	0.602	0.590	0.561	0.499
LogitBoost	60.48	58.08	0.605	0.581	0.600	0.603	0.583	0.538
OneR	61.30	59.88	0.613	0.599	0.606	0.631	0.601	0.529
LMT	61.71	61.08	0.617	0.611	0.621	0.638	0.581	0.569

[a]*CCI* Correctly Classified Instances, [b]*TP* True Positive, [c]*PS* Percentage Split

both PU and PEOU predicted the behavioral intention with an accuracy of 60% in most cases, and thereby, H1 and H2 were supported.

The results also indicated that KStar performed better than the other classifiers in predicting the PU by the attributes of mobility, availability, and perceived enjoyment, as shown in Table 4. KStar predicted the PU with an accuracy of 71.26% and 70.47% for the percentage-split and 10-fold cross-validation, respectively. Accordingly, H3a, H4a, and H5a were supported.

Table 4 Predicting PU by MOB, AVA, and PE

Classifier	CCI (%)		TP-rate		Precision		F-Measure	
	10-fold	PS	10-fold	PS	10-fold	PS	10-fold	PS
NaïveBayes	67.21	65.27	0.672	0.653	0.656	0.648	0.658	0.649
SimpleLogistic	69.86	71.85	0.699	0.719	0.683	0.693	0.659	0.697
KStar	70.47	71.26	0.705	0.713	0.701	0.692	0.672	0.690
LogitBoost	69.45	73.65	0.695	0.737	0.679	0.723	0.663	0.724
OneR	63.34	67.07	0.633	0.671	0.576	0.654	0.590	0.661
LMT	68.84	71.86	0.688	0.719	0.670	0.693	0.659	0.697

Table 5 Predicting PEOU by MOB, AVA, and PE

Classifier	CCI (%)		TP-rate		Precision		F-Measure	
	10-fold	PS	10-fold	PS	10-fold	PS	10-fold	PS
NaïveBayes	61.10	58.68	0.611	0.587	0.606	0.585	0.600	0.568
SimpleLogistic	63.14	61.08	0.631	0.611	0.634	0.608	0.610	0.586
KStar	58.45	56.29	0.585	0.563	0.574	0.550	0.548	0.539
LogitBoost	62.12	58.08	0.621	0.581	0.620	0.603	0.604	0.538
OneR	63.95	62.87	0.640	0.629	0.642	0.625	0.620	0.606
LMT	63.14	61.08	0.631	0.611	0.634	0.608	0.610	0.586

The results shown in Table 5 indicated that OneR performed better than the other classifiers in predicting the PEOU by the attributes of mobility, availability, and perceived enjoyment. OneR predicted the PEOU with an accuracy of 62.87% and 63.95% for the percentage-split and 10-fold cross-validation, respectively. Therefore, H3b, H4b, and H5b were supported.

5 Discussion and Conclusion

5.1 Research Implications

The present study developed a predictive model and tested the proposed model by employing a machine learning approach. Six popular machine learning algorithms were used to investigate whether the individuals' PU and PEOU perceptions would predict the behavioral intention to use smartwatches. Further, the attributes of mobility, availability, and perceived enjoyment would predict the PU and PEOU. The study employed popular classification algorithms having different methodologies such as decision trees, bayesian networks, association rules, neural networks, and if-then-else rules. The results indicated that NaïveBayes (a Bayesian classifier) and LMT (a decision tree) performed better than the other classifiers. They predicted the behavioral intention with an accuracy of 61%. The results suggested that attributes of mobility, availability, and perceived enjoyment predicted the PU with an accuracy of 70% in most cases. Further, the results suggested that the attributes of mobility, availability, and perceived enjoyment predicted the PEOU with an accuracy of 60% in most cases.

This study provides several implications. First, the study added a novel contribution to the IS literature since it was the first attempt to apply machine learning techniques for predicting individuals' adoption of wearable technologies. Second, this study is believed to be the first that tried to examine the behavioral intention to use smartwatches by students in the higher educational institutions in developing countries.

5.2 Limitations and Future Research Directions

Although the study has yielded significant results, a number of caveats need to be considered. First, the developed theoretical model was examined by collecting survey data from students only. What is furtherly required is to examine the developed model by collecting data from educators in order to draw a full picture of the employment of smartwatches in higher educational institutes. Second, it is unfortunate that the developed theoretical model did not consider any moderating variables. It would be interesting to evaluate the effects of gender and smartwatch ownership as moderating variables on students' behavioral intention to adopt smartwatches.

References

1. Çiçek, M.: Wearable technologies and its future applications. Int. J. Electr. Electron. Data Commun. **3**(4), 45–50 (2015)
2. Alvarez, V., Bower, M., De Freitas, S., Gregory, S., de Wit, B.: The use of wearable technologies in Australian universities: examples from environmental science, cognitive and brain sciences and teacher training. In: Mobile Learning Futures—Sustaining Quality Research and Practice in Mobile Learning (2016)
3. Chuah, S.H.W., Rauschnabel, P.A., Krey, N., Nguyen, B., Ramayah, T., Lade, S.: Wearable technologies: the role of usefulness and visibility in smartwatch adoption. Comput. Human Behav. (2016)
4. Kim, K.J., Shin, D.H.: An acceptance model for smart watches: implications for the adoption of future wearable technology. Internet Res. (2015)
5. Al-Emran, M., Malik, S.I., Al-Kabi, M.N.: A survey of internet of things (IoT) in education: opportunities and challenges. In: Toward Social Internet of Things (SIoT): Enabling Technologies, Architectures and Applications, pp. 197–209. Springer (2020)
6. Davis, F.D.: Perceived usefulness, perceived ease of use, and user acceptance of information technology. MIS Q. **13**(3), 319–340 (1989)
7. Al-Emran, M., Arpaci, I., Salloum, S.A.: An empirical examination of continuous intention to use m-learning: An integrated model. Educ. Inf. Technol. (2020)
8. Al-Emran, M., Teo, T.: Do knowledge acquisition and knowledge sharing really affect e-learning adoption? An empirical study. Educ. Inf. Technol. (2019)
9. Arpaci, I., Basol, G.: The impact of preservice teachers' cognitive and technological perceptions on their continuous intention to use flipped classroom. Educ. Inf. Technol. (2020)
10. Arpaci, I.: Antecedents and consequences of cloud computing adoption in education to achieve knowledge management. Comput. Human Behav. **70**, 382–390 (2017)
11. Arpaci, I.: Understanding and predicting students' intention to use mobile cloud storage services. Comput. Human Behav. (2016)
12. Al-Maroof, R.A.S., Al-Emran, M.: Students acceptance of Google classroom: An exploratory study using PLS-SEM approach. Int. J. Emerg. Technol. Learn. **13**(6), 112–123 (2018)
13. Mezhuyev, V., Al-Emran, M., Fatehah, M., Hong, N.C.: Factors affecting the metamodelling acceptance: a case study from software development companies in Malaysia. IEEE Access **6**, 49476–49485 (2018)
14. Mezhuyev, V., Al-Emran, M., Ismail, M.A., Benedicenti, L., Chandran, L.: The acceptance of search-based software engineering techniques: an empirical evaluation using the technology acceptance model. IEEE Access (2019)
15. Salloum, S.A., Alhamad, A.Q.M., Al-Emran, M., Monem, A.A., Shaalan, K.: Exploring students' acceptance of e-learning through the development of a comprehensive technology acceptance model. IEEE Access **7**, 128445–128462 (2019)
16. Alshurideh, M., Salloum, S.A., Al Kurdi, B., Al-Emran, M.: Factors affecting the social networks acceptance: an empirical study using PLS-SEM approach. In: 8th International Conference on Software and Computer Applications, pp. 414–418 (2019)
17. Salloum, S.A., Al-Emran, M.: Factors affecting the adoption of E-payment systems by university students: extending the TAM with trust. Int. J. Electron. Bus. **14**(4), 371–390 (2018)
18. Shin, D.H.: What makes consumers use VoIP over mobile phones? Free riding or consumerization of new service. Telecomm. Policy (2012)
19. Huang, J., Lin, Y., Chuang, S.: Elucidating user behavior of mobile learning: a perspective of the extended technology acceptance model. Electron. Libr. **25**(5), 585–598 (2007)

20. Yen, Y.S., Wu, F.S.: Predicting the adoption of mobile financial services: the impacts of perceived mobility and personal habit. Comput. Human Behav. **65**, 31–42 (2016)
21. Davis, F.D., Bagozzi, R.P., Warshaw, P.R.: Extrinsic and intrinsic motivation to use computers in the workplace. J. Appl. Soc. Psychol. **22**(14), 1111–1132 (1992)
22. Ha, I., Yoon, Y., Choi, M.: Determinants of adoption of mobile games under mobile broadband wireless access environment. Inf. Manag. (2007)
23. Yang, K.: Consumer technology traits in determining mobile shopping adoption: an application of the extended theory of planned behavior. J. Retail. Consum. Serv. (2012)
24. Al-Emran, M., Mezhuyev, V., Kamaludin, A.: PLS-SEM in Information Systems Research: a Comprehensive Methodological Reference. In: 4th International Conference on Advanced Intelligent Systems and Informatics (AISI 2018), pp. 644–653 (2018)
25. Venkatesh, V., Bala, H.: Technology acceptance model 3 and a research agenda on interventions. Decis. Sci. **39**(2), 273–315 (2008)
26. Arpaci, I.: A hybrid modeling approach for predicting the educational use of mobile cloud computing services in higher education. Comput. Human Behav. **90**, 181–187 (2019)
27. Arpaci, I.: What drives students' online self-disclosure behavior on social media? A hybrid SEM and artificial intelligence approach. Int. J. Mob. Commun. (2020)
28. De War, R., Neal, D.L.: WEKA Machine Learning Project: Cow Culling. New Zealand (1994)

Vocabulary Improvement by Using Smart Mobile Application—A Pilot Study

Petra Poláková, Blanka Klímová, and Pavel Pražák

Abstract The aim of this study is to explore the usefulness of smart mobile application in the process of English language learning with special focus on vocabulary acquisition. As vocabulary learning is considered to be a boring, difficult and long process, it is beneficial to transform it into an easy, practical, and pleasurable experience, which can be achieved by using technology. Recently, mobile devices have become an essential part of our everyday life. Regarding this, it is reasonable to incorporate them into the learning process, including language learning. The current pilot study describes an experiment dealing with using a smart vocabulary learning mobile application in the process of vocabulary improvement. The results show the main differences between the experimental group (using mobile devices) and the control group (exposed to the traditional methods of vocabulary learning and given the vocabulary summary through using an electronic notice board).

Keywords Vocabulary · Vocabulary acquisition · Blended learning · Mobile assisted language learning · Mobile application

P. Poláková (✉)
Department of Language Pedagogy and Intercultural Studies, Constantine the Philosopher University in Nitra, Trieda Andreja Hlinku 1, Nitra 949 74, Slovakia
e-mail: petra.polakova@ukf.sk

B. Klímová
Department of Applied Linguistics, Faculty of Informatics and Management, University of Hradec Králové, Rokitanského 62, Hradec Králové 500 03, Czech Republic
e-mail: blanka.klimova@uhk.cz

P. Pražák
Department of Informatics and Quantitative Methods, Faculty of Informatics and Management, University of Hradec Králové, Rokitanského 62, Hradec Králové 500 03, Czech Republic
e-mail: pavel.prazak@uhk.cz

M. Al-Emran et al. (eds.), *Recent Advances in Intelligent Systems and Smart Applications*, Studies in Systems, Decision and Control 295,
https://doi.org/10.1007/978-3-030-47411-9_11

1 The Importance of Vocabulary in Second Language Acquisition

Wilkins [1] is convinced that without grammar very little can be conveyed, but without vocabulary nothing at all can be conveyed. In other words, language learners are required to master vocabulary in order to be able to express themselves.

Vocabulary, or the collection of words, is an essential part of foreign language acquisition. It is of great importance to language teachers and learners since the meaning of a word is normally emphasized in the classroom. Therefore, learners´ vocabulary extension is an important aspect of language development. It is often viewed as a critical tool for second language learners because a limited vocabulary disrupts successful communication [2].

According to Nation [3], the relationship between vocabulary knowledge and language use is complementary as vocabulary knowledge enables language use and, vice versa, language use leads to increasing vocabulary knowledge. For that reason, the acquisition of vocabulary plays an important role in the formation of spoken and written texts. Furthermore, acquisition, of all the language skills, such as listening, speaking, reading, and writing, is influenced by vocabulary knowledge.

Seashore [4] also claims that vocabulary size is related to the development of all language skills. In his opinion, it is the best index for the prediction of achievement in nearly all of the other language skills, i.e. diction, composition, reasoning, general intelligence and even grades in specific subject.

Given the assumption that second language learning depends on the vocabulary size and knowledge, influencing the acquisition of all the other language skills, it is necessary to consider what it means to know words properly.

1.1 Knowing Vocabulary Words

Previously, knowledge of a word was defined as the ability to apply the word correctly in many different situations. It involves the understanding of different meanings of the word. It was also noted that appropriate knowledge of a word required precision of the meaning, flexibility in the word application, understanding and usage of metaphors, the recognition of synonyms, the awareness of the word definition, and the ability to use the word expressively [5].

In accordance with the previous statements, Mahdi [6] writes that knowing a word involves much more than knowing its meaning. He assumes that it is important to consider some elements in the process of vocabulary extension since many learners know the orthography of the words, but they fail in pronunciation, or other learners understand the meaning of the words, but they are unable to use them in different contexts. Regarding this, Nation [3] believes that orthographical knowledge, or word form, is one of the key components in vocabulary acquisition, and language learning in general. According to him, knowing a word form includes knowing three interrelated aspects required to know the proper meaning of the word. These aspects are spoken form, written form, and word parts. Each of them incorporates both the receptive and productive dimensions. Firstly, the spoken aspect includes recognizing the word when it is heard and the ability to express the meaning of the word. Secondly, the written aspect includes the spelling of words, and thirdly, knowing the word parts means to know that a word is made up of affixes. However, it is not sufficient to acquire the vocabulary form in order to understand its meaning. Therefore, it is important to know grammatical functions, the collocations, and the different constraints of word use, too.

In addition to the importance of vocabulary knowledge, one of its significant dimensions is the ability to distinguish between receptive and productive vocabulary. Knowing a word means having both receptive and productive knowledge of the word. From beginners up to advance students, they all have a larger receptive vocabulary than a productive one. It means that their passive vocabulary is bigger than active and they understand more words than they can actually use in their speech or writing.

Receptive vocabulary use includes perception of the word form while listening and reading [7]. On the other hand, productive vocabulary use involves expression of the meaning through speaking or writing [6]. Thus, each foreign language student has receptive oral vocabulary, receptive written vocabulary, productive oral vocabulary, and productive written vocabulary. The relationships among them change over time. However, according to Tesol [7], all four types of vocabulary have the same level of importance in the process of language enrichment.

Based on the statements above, it is to say that vocabulary knowledge is one of the most important components in language learning. Naturally, the size and the correct use of words depend on the second-language proficiency of the language learner. The English language includes a huge number of words, therefore, the goal is to help students to develop an extensive vocabulary, and to help them understand its proper meaning, in order to avoid misunderstanding and master required language skills.

1.2 Ways of Learning Vocabulary Words

In order to increase vocabulary size, both incidental and instructional vocabulary acquisitions are required. Nation [3] believes that there is a distinction between them since vocabulary learnt incidentally refers to all the words that have been learnt from a special context, and, on the other hand, vocabulary that is learnt intentionally is learnt in another way. Robinson [8] writes that intentional vocabulary learning refers to any activity aiming at committing lexical information to memory. Nation [3] considers the difference in the way of the vocabulary acquisition, but Tode [9] states that there is a qualitative difference in what the learner does with the word, i.e., one has to differentiate between what is being done with the word, quality, and how often the word is being met, quantity.

1.2.1 Incidental Vocabulary Learning

Nation [3] believes that vocabulary knowledge and reading comprehension are closely, even if not directly, related to each other, since vocabulary knowledge can help reading and reading can lead to vocabulary growth. Correspondingly, Schmitt [10] found out that ten or more encounters with vocabulary had lead to its sizeable gain since the repetition affects incidental vocabulary learning. According to his findings, the learners who encounter an unknown word more times in informative contexts are able to demonstrate significantly larger gains in vocabulary knowledge than learners who have fewer encounters with an unknown word. On the other hand, Schmitt also points out other research results which show that repetitive exposure to vocabulary leads to its recognition more than to its knowledge.

It can be said that there is an unknown boundary between the word which has been recognized and the word which has been fully understood and learnt in the process of incidental vocabulary learning. Regarding this, even by having a great amount of exposure there is a need of some systematic approach to ensure vocabulary acquisition, which can be reached by intentional learning.

1.2.2 Intentional Vocabulary Learning

Intentional, or direct, vocabulary learning is the way of learning where there are certain predetermined words covered thoroughly and methodically. Vocabulary acquisition is done by rules and rotes [11]. Brown [12] claims that using the right learning strategies can help students to achieve success in vocabulary acquisition, and language learning in general. Strategies such as examining how words are built up; how the root can be found; finding affixes, suffixes, and prefixes; knowing synonym clues, antonym clues, and collocation clues, etc., can be involved there.

Brown [12] also recommends the use of such unfashionable procedures as regular rehearsal of words, rote learning, and training in automatic word recognition as he believes that traditional methods might prove to be more useful in the process of vocabulary acquisition than expected.

As written above, both ways, or methods, of vocabulary acquisition affect the process of gaining new lexical items. Therefore, an exposure to incidental vocabulary acquisition in conjunction with direct, instructional vocabulary learning, is considered eligible and useful.

Moreover, as stated in Mahdi [6], vocabulary acquisition can occur in different environments, too. They are technology-based environment and traditional environment. The technology-based environment refers to the use of new technology in the process of vocabulary learning. On the other hand, traditional methods of vocabulary acquisition include frontal lecturing engaging word lists, word cards, or dictionaries. A combination of educational technology with traditional methods can lead to effective vocabulary acquisition, and it can also help to achieve the balance in the learning process.

2 Blended Learning

Commonly, there are several methods concerning vocabulary acquisition whose main goal is to help students to remember new vocabulary or lexical items. In order to prevent students from forgetting, the vocabulary needs to be learnt, practiced, and revised. It can be done through different methods, however, their employment in the learning process depends on factors, such as content, time availability, and its value for the learners. There is usually a combination of more than one method since it is suggested to provide vocabulary presentation in many different ways [2].

The combination of more than one method in learning environment is known as blended learning. Driscoll [13] defines it as a combination of different web technologies to achieve educational goals. It can be a combination of various pedagogical approaches ensuring the optimal learning results with or without use of smart technologies. It is also stated that blended learning is a mixture of any type of educational technology with frontal lecturing.

Blended learning offers many opportunities in the language learning process. They are as following: the combination of learning online and offline, a combination of individual learning and common learning, a combination of structured and unstructured learning, a combination of general content and content built on the needs of the learner, and a combination of learning, practice and support activities [14].

According to Veselá [15], blended learning is the use of different elements of education in such a way that the learning process has a positive effect on learner achievement. Its philosophy is focused on increasing motivation, personal intuition, knowledge and skills acquirement, and co-responsiveness of the learner to the learning outcomes.

Poláková [16] sees blended learning as an approach to education that combines online interaction with traditional classroom methods. According to her, it is essential to combine face-to-face practises with technology-based activities in order to make learning process more interesting and to help students to increase their motivation.

Recently, ownership of smartphones amongst the students has grown fast. Regarding their availability and constant use among learners, mobile devices have a huge potential to offer a new, powerful learning environment. It is certain that the traditional methods of teaching foreign languages are still useful, they just need to be updated and combined with new trends in education. Therefore, it is suggested to innovate, to technify, and to adapt reasonably. In accordance with this, it is essential to bring the mobile technology into the class and to try to reach the optimal results in the process of language learning by using it [17].

2.1 Mobile Technology in Vocabulary Learning

Learning while moving around or using mobile devices to learn something is defined as mobile assisted language learning (MALL). This method of learning offers advantages such as learning without space constraints, learning in a real context, and sharing information with peers. Taking in consideration the ubiquitous nature of MALL, the classroom is no longer the only learning environment as students become part of the context [18].

According to Kukulska-Hulme [19], mutual combination of learning and mobile technology in such a way that learning materials can be used inside and outside of the classroom is the primary reason for MALL.

Nowadays, many language learners use smart mobile applications (apps) to support their language learning both inside and outside the classroom. As vocabulary is one of the most important aspects in foreign language learning, students use smart vocabulary learning apps, primarily. In formal learning, teachers are supposed to choose the right apps suited to the particular level of learners [20].

MALL was engaged in the current pilot study which was conducted at a secondary vocational school in Slovakia. Traditional pedagogical approaches were supported by using mobile devices in the process of vocabulary acquisition.

3 Research Design

This research aims to prove usefulness of smart mobile application in language learning process with special focus on vocabulary acquisition and show how mobile application can be helpful in learning new words. Since mobile technology is expected to be appropriate and helpful in the language learning process, research hypothesis was as follows: "The students who use the vocabulary learning mobile application achieve significantly higher learning outcomes than the students who do not use the application."

3.1 Participants

One class of 21 vocational high school students was invited to join the study during the period of 3 months, from March to April 2019. They have been learning English for 11 years on average. Their English level was B1 (intermediate level of English) according to the Common European Reference Framework for languages [21]. The participants have attended an English language course for three 45-min hours per week. The students attended the English classes on Thursdays and Fridays. The curriculum was based on a textbook called "Solution" containing 10 units divided into different topics [22]. Only 3 units were chosen to be used in current research.

The participants were divided into two groups; 10 students were enrolled in the non-mobile application group with the possibility to use EduPage in order to revise vocabulary (the control group), and 11 students in the smart mobile application group (the experimental group). There were 18 girls and 3 boys participating in this study. As the students were 16 years old, their parents were asked to sign the informed consent letter in order to express their agreement with the experiment.

3.2 Material

According to a survey conducted by Klímová and Berger [23], the lack of vocabulary is one of the most serious weaknesses in the English learning process. Therefore, the students from the current study had the opportunity to use a method of blended learning, which consisted of the traditional classroom methods, and practicing target vocabulary through the mobile application which is divided into two parts: a mobile platform and a server part. The server part, designed as a web interface for the teachers is responsible for storing information, authenticating users, collecting large data, processing, and distributing messages. The second application part is presented with a mobile application for students.

As Klímová and Berger [23] write, their web application offers a number of features, such as lesson management, registration of the students, distribution of new notifications, and opportunity to respond to students´ comments. Thank to these options, teachers can interact with their students and bring attention to upcoming events. The key element of the web interface is the visualisation of the student results. Based on this, it is possible to evaluate each student separately or to modify the study plan.

3.3 Procedures

During the research, quantitative data were collected. The aim of the quantitative study was to compare the test results of the control group, using traditional learning methods of vocabulary acquisition in combination with vocabulary summary available on EduPage, with the experimental group, using smart mobile application in the learning process. The experimental group test score was expected to be higher than the test score of the control group. In order to support the study, the current research collected quantitative data gathered from vocabulary pre-test and a post-test. The standardized vocabulary test was arranged according to the multiple test approach. The test was divided into eight matching definition, filling gap, and filling gap with given letters exercises. The purpose of the pre-test was to determine whether the vocabulary knowledge of both groups was at the same proficiency level. The post-test aimed at examining the usefulness of using the smart mobile application in the learning process. In this pilot study, mobile application treatment was used. Students

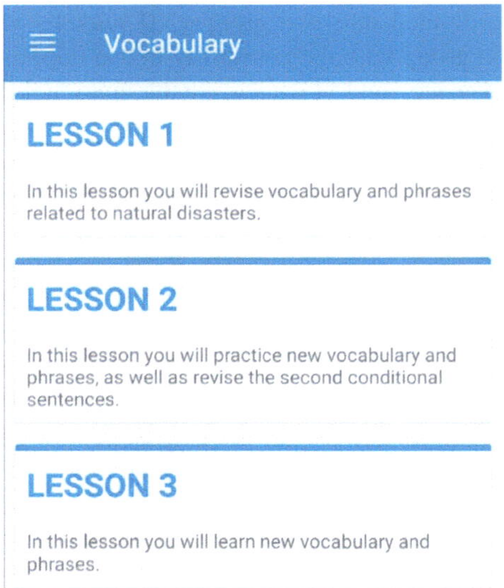

Fig. 1 An overview of available lessons

from the experimental group were enrolled into a specific course. After presenting the new vocabulary in the class, the app offered them the ability to study and test specific vocabulary and phrases, during the period of 3 months. For each phrase or word, they could get a translation, as well as pronunciation. Furthermore, the application enabled communication with teacher, thus the interaction was ensured (Fig. 1).

Meanwhile, the control group adopted the traditional vocabulary learning approach which means that the participants went through the article provided in the textbook, and they were requested to review the target words through finishing the exercises in the textbook and workbook, including answering the questions, matching the definition, and filling the gap. In addition, students from the control group were given the vocabulary summary through the electronic notice board on EduPage which is an academic IT system providing communication among teachers, students, and parents. After treatment, both groups were tested on their vocabulary achievement. The data from the experimental group and the control group were collected and the results are described below.

4 Results

A group of 21 students was divided into two independent groups. The research (experimental) group, whose size was $n1 = 11$, used the prepared mobile application for teaching and learning. The control group, which was $n2 = 10$, did not use the mobile application. Students' results were measured by pre- and post-tests and their values were recorded. The results were calculated using IBM SPSS Statistics 25 software. In order to assess students´ proficiency level, a pre-test was applied in both groups. The results are shown in Table 1, in which the student success rate is given in %.

Due to the small number of students in the current pilot study, it was not appropriate to use parametric t-test to compare the mean values of test scores. Therefore, a non-parametric Mann-Whitney test was used for two independent samples ($U = 52.5$, asymp. $P\text{-value} = 0.860$), according to which the hypothesis of median equality was not rejected, and it was probable that there were no differences in proficiency level of both groups, at the beginning of the study.

A box diagram of the results in the post-test of both groups is shown in Fig. 2. According to this, it could be concluded that the students from the research group achieved higher results. This conclusion is in line with the hypothesis expectation. However, confirmation of this observation was done through calculation.

Descriptive statistics of the post-test results of both groups are shown in Table 2.

Due to the small number of students in both groups, the non-parametric Mann-Whitney test was used for two independent samples ($U = 26.5$, asymp. $P\text{-value} = 0.044$) to compare the results in the post-test. Regarding the fact that there is 5% difference in the results of both groups, the hypothesis can be accepted.

Furthermore, it was considered whether there has been a significant improvement of students from the experimental group or not. The comparison of their results in

Table 1 Descriptive statistics of the experimental (E) and control groups (C) in the pre-test

Group	Mean (%)	Std. Deviation (%)	Median (%)	Minimum (%)	Maximum (%)
E (nl = 11)	60.91	22.92	57.0	35.00	96.00
C (n2 = 10)	61.00	21.83	61.0	18.00	94.00

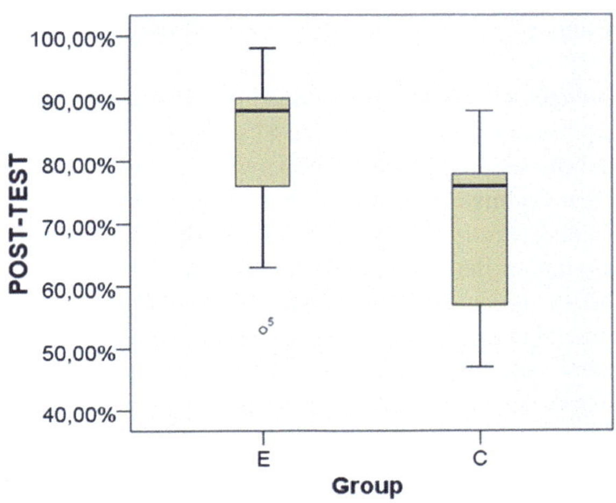

Fig. 2 Box diagram of post-test results for the experimental (E) and control (C) groups

Table 2 Descriptive statistics of the experimental (E) and the control (C) groups in the post-test

Group	Mean (%)	Std. Deviation (%)	Median (%)	Minimum (%)	Maximum (%)
E (nl = 11)	82.00	14.04	88.0	53.00	98.00
C (n2 = 10)	70.40	13.47	70.7	47.00	88.00

pre-test and post-test was done by using the paired variant of Wilcoxon (Z = -2.71, asymp. *P*-value = 0.007). Based on the results of this test, it can be concluded that there is 5% significant difference between the pre-test and post-test results. Since the results achieved in the post-test are higher, it can be stated that the use of the mobile application leads to a significant improvement in language learning. It is required to work with larger groups to confirm these results. This will be done in our future research.

5 Discussion and Conclusion

On the basis of the findings of the current pilot study, it might be concluded that the use of smart vocabulary learning mobile application in English classes is useful in improving student vocabulary knowledge. The positive results of the impact of the use of smart mobile applications on English vocabulary learning in foreign language learning have been also confirmed by other recent research studies [24–26]. For example, Chen et al. [24] conducted a four-week experiment with 20 Taiwanese university students who were divided into two groups: experimental, using a mobile game-based English vocabulary app (MGBL) and control group, using a mobile app without the game effects. The results of their study indicate that the students using MGBL achieved better learning outcomes and were more involved in learning than the students in the control group. Klímová and Pražák [25] in their study compared results of two experiments performed during the year of 2017/18 and the year of 2018/19. Altogether 59 Czech university students participated in both experiments. Students were randomly divided into two independent groups. The study group (an experimental group) included 41 students, using a mobile application in their studies and the second, control group, consisting of 18 students, not using the mobile app. Their findings confirmed that the mobile app was an effective tool in learning new English words and phrases. In addition, Klímová and Pražák [25] in their study emphasize the key features of learning vocabulary, which have been described in the theoretical sections above. They report that by using a mobile app, students can remember and acquire the words better because they have to write the word, they can hear its pronunciation, they can repeat it, and see its correct form [3, 6]. Furthermore, they believe that continuous repetition in small chunks can help students remember the vocabulary more easily and effectively. They also point out that a tailored made mobile app according to students' language needs may stimulate them to learn these words and phrases more often. Mahdi [6] expands that adult students benefit from using mobile devices in vocabulary learning more than young learners.

There are, of course, certain limitations to these experimental studies. Most often, the students in the control groups are disadvantaged because they do not have as many opportunities for exposure to the target language, which means the extra practice time may play a role in achieving better scores by the students in the experimental group. Another limitation of the research studies is a small subject sample. Therefore, further research is needed to verify and validate the findings and reliability of these investigations.

As far as the use of mobile applications in foreign language is concerned, the researchers [26–28] agree that smart mobile applications should be used as an effective and stimulating complementary tool to traditional, face-to-face classes. The reason is that some more demanding language skills such as extensive reading or writing essays are more difficult to acquire via mobile apps. Moreover, students should also learn how to socially interact in foreign language in the sense what paralinguistic features to use.

Generally, the authors conclude that using mobile technologies in the classroom is considered useful since learning through mobile devices can engage the students and allow them to enjoy and learn at the same time. Implementation of mobile learning in the class can help to improve the level of students´ comprehension and to raise their motivation for studying.

Acknowledgements This study is supported by the SPEV project 2020, run at the Faculty of Informatics and Management, University of Hradec Kralove, Czech Republic. The authors thank Ales Berger for his help with the data collection.

References

1. Wilkins, D.: Linguistics in Language Teaching. London, England (1972)
2. Alqahtani, M.: The importance of vocabulary in language learning and how to be Taught. Int. J. Teach. Educ 21–34 (2015)
3. Nation, I.S.P.: Learning Vocabulary in Another Language. Cambridge University Press, Cambridge, England (2001)
4. Seashore, R.H.: The importance of vocabulary in learning language skills. Elementary Engl. 137–152 (1948). Available: https://www.jstor.org/stable/41383521?read-now=1&seq=2#page_scan_tab_contents
5. Graves, M.F.: Vocabulary learning and instruction. Rev. Res. Educ. 49–89 (1986)
6. Mahdi, S.H.: Effectiveness of mobile devices on vocabulary learning: a meta-analysis. J. Educ. Comput. Res. 1–21 (2017)
7. Tesol: Learning Words in a Second Language. Available: https://www.tesol.org/docs/books/bk_teachingvocab_750.pdf
8. Robinson, P.: Cognition and Second Language Instruction. Cambridge, England (2001)
9. Tode, T.: Effects of frequency in classroom second language learning. Peter Lang, Bern, Switzerland (2008)
10. Schmitt, N.: Researching Vocabulary: A Vocabulary Research Manual. Palgrave Macmillan, Great Britain (2010)

11. Björgvinsson, L.A.: Incidental and instructional vocabulary (2012). Available: https://www.academia.edu/3321504/Incidental_and_Instructional_Vocabulary_Acquisition
12. Brown, H.D.: Teaching by Principles, An Interactive Approach to Language Pedagogy. Pearson Education Inc, White Plains (2007)
13. Driscoll, M.: Blended learning: let's get beyond the hype. IBM Glob. Serv. (2003). Available: https://www-07.ibm.com/services/pdf/blended_learning.pdf
14. Khan, B.: Learning features in an open, flexible and distributed environment. AACE J. (2005)
15. Veselá, K.: Učebné pomôcky pre počítačom podporované vyučovanie cudzích jazykov. UKF, Nitra (2009)
16. Poláková, P.: Using mobile technologies in foreign language learning. In: Inovatívne trendy v odborových didaktikách. Prepojenie teórie a praxe výučbových stratégií kritického a tvorivého myslenia. Zborník štúdií z medzinárodnej vedeckej konferencie, pp. 520–524 (2019)
17. Woodock, B., Middleton, A., Nortcliffe, A.: Considering the smartphone learner: an investigation into student interest in the use of personal technology to enhance their learning. Student Engagem. Experience J. (2012)
18. Wang, B.: Designing mobile apps for english vocabulary learning. Int. J. Inf. Educ. Technol. 279–283 (2017)
19. Kukulska-Hulme, A.: Will mobile learning change language learning? ReCall 157–165 (2009)
20. The Open University. User evaluation of language learning mobile applications: a case study with learners of Spanish. (2016) Available: http://www.crtvup.com.cn/zyxz/dj.asp?id=9
21. CERF: The common european framework of reference for languages. Cambridge University Press (2018). Available: https://www.coe.int/en/web/common-european-framework-reference-languages
22. Davies, P.A., Falla, T.: Solutions. Oxford University Press (2012)
23. Klímová, B., Berger, A.: Mobile application for the teaching of english. Adv. Multimedia Ubiquit Eng. (2018)
24. Chen, C.M., Liu, H., Huang, H.B.: Effects of a mobile game-based English vocabulary learning app on learners' perceptions and learning performance: a case study of Taiwanese EFL learners. ReCALL 170–188 (2019)
25. Klímová, B., Pražák, P.: Mobile blended learning and evaluation of its effectiveness on students' learning achievement. In: Cheung, S., Lee, L.K., Simonova, I., Kozel, T., Kwok, L.F. (eds.), Blended Learning: Educational Innovation for Personalized Learning, pp. 216–224 (2019)
26. Avci, H., Adiguzel, T.: A case study on mobile-blended collaborative learning in an English as a foreign language (EFL) context. Int. Rev. Res. Open Distrib. Learn. pp. 1–14 (2017). Available: https://files.eric.ed.gov/fulltext/EJ1163180.pdf
27. Glahn, C., Gruber, M.R., Tartakovski, O.: Beyond delivery modes and apps: a case study on mobile blended learning in higher education. In: Conole, G., Klobučar, T., Rensing, C. (eds.), Design for teaching and learning in a networked world. Lecture Notes in Computer Science, vol. 9307, Springer, Cham (2015)
28. Klimova, B.: Mobile phones and/or smartphones and their apps for teaching English as a foreign language. Educ. Inf. Technol. 1091–1099 (2017)

Examining the Acceptance of WhatsApp Stickers Through Machine Learning Algorithms

Rana A. Al-Maroof, Ibrahim Arpaci, Mostafa Al-Emran, Said A. Salloum, and Khaled Shaalan

Abstract WhatsApp stickers are gaining popularity among university students due to their pervasiveness, specifically in educational WhatsApp groups. However, the acceptance of stickers by university students is still in short supply. Thus, this research aims to empirically examine the determinants affecting the acceptance of WhatsApp stickers through a proposed theoretical model by integrating the technology acceptance model (TAM) with the uses and gratifications theory (U&G). A questionnaire survey was circulated to collect data from 372 university students who have been engaged in a "Group Talk" in WhatsApp. A novel approach was employed to analyze the hypothesized relationships among the constructs in the research model through the use of machine learning algorithms. The results pointed out that IBk and RandomForest classifiers have performed better than the other classifiers in predicting the actual use of stickers with an accuracy of 78.57%. The research findings are believed to provide future directions for stickers developers to better promote stickers in educational activities.

R. A. Al-Maroof (✉)
Department of English Language, Al Buraimi University College, Al Buraimi, Oman
e-mail: rana@buc.edu.om

I. Arpaci
Department of Computer Education and Instructional Technology, Tokat Gaziosmanpasa University, Tokat, Turkey
e-mail: ibrahim.arpaci@gop.edu.tr

M. Al-Emran
Department of Information Technology, Al Buraimi University College, Al Buraimi, Oman
e-mail: mustafa.n.alemran@gmail.com

S. A. Salloum
Research Institute of Sciences and Engineering, University of Sharjah, Sharjah, UAE
e-mail: ssalloum@sharjah.ac.ae

K. Shaalan
Faculty of Engineering and IT, The British University in Dubai, Dubai, UAE
e-mail: khaled.shaalan@buid.ac.ae

M. Al-Emran et al. (eds.), *Recent Advances in Intelligent Systems and Smart Applications*, Studies in Systems, Decision and Control 295,
https://doi.org/10.1007/978-3-030-47411-9_12

Keywords WhatsApp stickers · Technology acceptance model · Uses and gratifications theory · Machine learning algorithms

1 Introduction

Social media technologies have become part of our daily lives [1–5]. Online communication through these technologies enables individuals to interact away from time and space limitations. Online communication is a replacement for writing messages and post them on the chat rooms [6]. Stickers represent one of the up-to-date tools in online communications. Recently, stickers become a way of communication that encourages people all over the world to discuss various social, education, and political issues. One of the benefits of stickers that sets it apart from other mechanisms in online communication is that it encourages users to create their sticker forms that meet their personal needs [7]. The mechanism of creating stickers adds a new experience to the users that connect them emotionally with stickers. It raises the idea of belonging and feeling of possessing [8, 9]. Stickers have their own unique features, such as size, uniqueness, and animation. In this respect, it is argued that stickers are bigger, static, and animated [10]. They can be added or deleted and could be sent separately without insertion in text messages. This paves the way to another important benefit of stickers, which is motivation. Users feel that they are being motivated whenever the question of stickers is raised. Motivation affects the frequency of using stickers and makes them more pervasive. Users do not only send stickers to represent emotions but also use them for strategic or functional purposes [11].

Even though emojis have been investigated by researchers and attracted the attention of academic and non-academic situations [12–14], a few studies have focused on the role of stickers in online communication. More specifically, Ghobadi and Taki [15] examined the role of stickers in Telegram. The study aimed at examining the impact of using Telegram stickers on EFL learners' vocabulary learning. The results have proven that teaching vocabulary through Telegram stickers could assist the learning process and provide learners with a variety of advantages. In addition, Van De Bogart and Wichadee [16] focused on the usage of stickers in LINE networking sites. The results showed that perceived usefulness and attitude toward using stickers by college students had a positive relationship with their intention to use.

Although previous studies have highlighted the importance of emojis in social media [17–22], there is a scarce of knowledge regarding the understanding of the factors affecting the use of stickers in everyday communication in general, and educational activities in particular. To verify the usage of stickers in online communications by university students, this research adopts the technology acceptance model (TAM) and the uses and gratifications theory (U&G) by integrating them together to measure the students' acceptance of stickers. The reason behind the adoption and integration of these two theories stems from several causes. First, previous studies have shown that these two theories have been successfully adopted to pinpoint the acceptance or adoption of several technologies. Second, each theory handles certain features that are critical to understand the acceptance of stickers and their features among university students. More specifically, the TAM takes into account the characteristics of the technology itself, whereas the U&G focuses on the experiences generated by that technology. Third, based on the existing literature, the integration of TAM and U&G has not been experienced to examine the usage of stickers through WhatsApp text messages among students at the university level. In line with the aforementioned reasons, this research aims at integrating the TAM and U&G to understand the factors affecting the students' acceptance of WhatsApp stickers at the university level.

2 Theoretical Framework and Research Model

2.1 Technology Acceptance Model (TAM)

TAM is considered as a model to predict the acceptability of technology as it is governed by two main variables, namely "perceived usefulness (PU)" and "perceived ease of use (PEOU)" [23]. PU refers to "the degree to which a person believes that using a particular system would enhance his or her job performance", whereas PEOU refers to "the degree to which a person believes that using a particular system would be free from effort" [24]. Many researchers have emphasized the importance of these two variables in predicting the acceptance or adoption of technology. In that, these two variables were shown to have a significant positive impact on the adoption of Google classroom [25], e-learning [26–28], m-learning [29], e-payment systems [30], social networks [31], metamodelling [32], and search-based software engineering techniques [33], among many others. Based on that, the TAM is adopted in this research as it focuses on the "ease of use" and "usefulness" variables, which are closely related to the powerful features of stickers, including animation, size, and social or emotional connection with other users.

2.2 Uses and Gratifications Theory (U&G)

The uses and gratification theory (U&G) has been used by many scholars in the domain of media and communication [34]. The theory has a functional approach that is grounded within the concept of "functionalism" [35]. The U&G theory is generated to measure the usage of media by users to gratify their needs. It is believed that users of media have certain needs, motivations, and expectations that can be fulfilled and gratified through the use of media. The U&G focuses on individuals' motivation, which is one of the influential factors in using the media [36]. Gratification can cover various aspects, including hedonic, content, utilitarian, and social gratification.

Within each aspect of gratification, multiple dimensions may be found, such as self-documentation, self-expression, social interaction, and information sharing. The main aim behind using these features is entertainment and spending time [37].

Further, the main essential factors that are embedded inside the U&G theory are the inclusion of both "novelty" and "being there" [17]. Novelty is considered a funny way to create innovation, whereas "being there" refers to the social media users who may click social buttons to feel immersed in a mediated reality. The two factors are crucial to the usage of stickers in communication Apps. Given that stickers are one of the featured additions in WhatsApp, users may use it to fulfill the sense of novelty. Besides, the U&G theory pays attention to the feature of "being there" as it is one of the distinctive features behind using stickers. Stickers can be used in different religious and social situations to reflect the idea of the uniqueness of being there.

Accordingly, the U&G dimensions can serve the basic aim of this study. It has a close relation with motivation and entertainment, which in turn, are considered as influential features that urge users to use stickers. Furthermore, social interaction and information sharing are critical aspects that can measure the importance of using stickers by students.

2.3 Research Model

Due to the narrow applicability of TAM concerning the "usefulness" and "ease of use" variables, the U&G comes as an assistant theory that provides more insights into these variables. The U&G provides an in-depth understanding of learners' perspectives with respect to the extension of TAM [38]. The U&G consists of three core utilities, which were believed to have a significant relationship with the TAM constructs, such as "social utility", "hedonic utility", and "functional utility". Social utility is used to reflect the interpersonal use, which has a close correlation with PU. WhatsApp users can design stickers that indicate their social attitude towards different social events in their life. Similarly, the hedonic utility is related to the PU from the novelty and enjoyment perspectives. Users invented various forms of stickers to arouse the feeling of fun and amusement through WhatsApp communications. When the users of technology use a specific form of stickers to accomplish a specific task, it is then related to the functional utility.

Based on the above discussion, the research model of this study is constructed as depicted in Fig. 1. In that, this research suggests that social utility (including socialization and self-presentation), hedonic utility (including enjoyment and novelty), and functional utility (including unique function) are significantly influenced by both PEOU and subjective norm. It was also postulated that the students' intention to use stickers are affected by PEOU, subjective norm, social utility, hedonic utility, and functional utility. Further, the actual use of stickers is suggested to be influenced by the students' intentions. It is important to report that social, hedonic, and functional utilities are embedded in PU. In line with this, the following hypotheses were formulated:

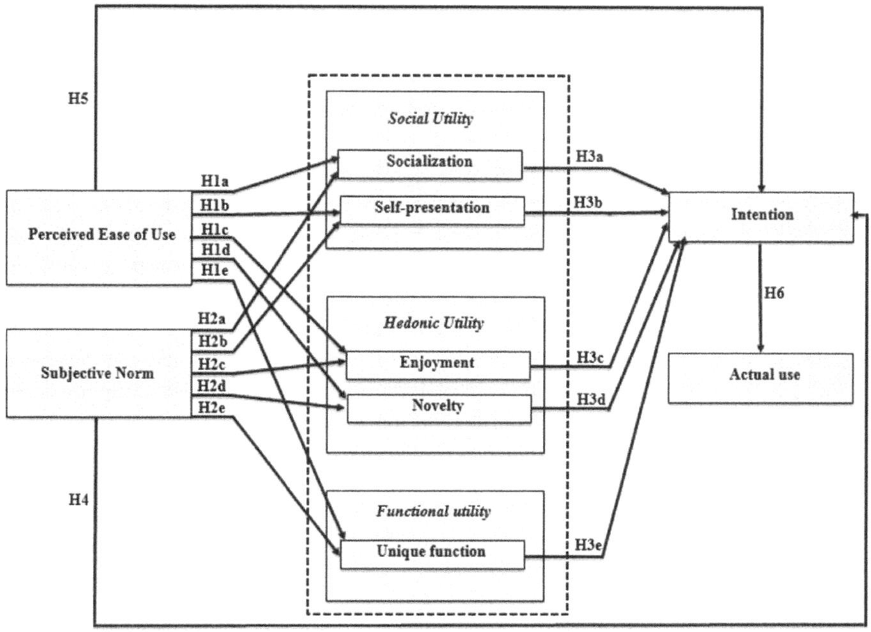

Fig. 1 Research model

H1: PEOU has a significant effect on (a) socialization, (b) self-presentation, (c) enjoyment, (d) novelty, and (e) unique function.

H2: Subjective norm has a significant effect on (a) socialization, (b) self-presentation, (c) enjoyment, (d) novelty, and (e) unique function.

H3: socialization (a), self-presentation (b), enjoyment (c), novelty (d), and unique function (e) have significant effects on students' intention to use stickers.

H4: Subjective norm has a significant effect on students' intention to use stickers.

H5: PEOU has a significant effect on students' intention to use stickers.

H6: Students' intention to use stickers has a significant effect on stickers actual use.

3 Research Methodology

3.1 Procedure and Sample

To understand the students' intention to use stickers, a self-administered survey was circulated to several groups of university students who have been engaged in a "Group Talk" in WhatsApp. The use of questionnaire surveys in such type of research studies is considered the best tool as it can effectively analyze the relationships among the constructs in the proposed research model [39]. The sample was collected from the University of Fujairah located in the United Arab Emirates (UAE). The purpose of selecting this sample is attributed to the reason that most of the selected students have already experienced the use of stickers in their daily university lives. It is imperative to mention that the majority of the discussed issues in the WhatsApp groups are related to educational matters such as courses, assignments, exams, and social activities. The data were collected between February and March 2019. A total of 400 questionnaires surveys were distributed. Among those, 28 responses were discarded due to the large number of missing values, and thus, a total of 372 responses were retained for further analysis.

3.2 Research Instrument

The research instrument consists of two main parts. The first part is dedicated to collect students' demographics, whereas the second part involves the items that measure the constructs in the proposed research model. The second part was measured using a five-point Likert scale ranging from "1 = strongly disagree" to "5 = strongly agree". The items used to measure the actual use, intention, and subjective norm were taken from Shao and Kwon [17]. The items used to measure the PEOU were adopted from Davis [24]. The items used to measure the enjoyment and self-presentation were adopted from Gan [40], whereas the items used to measure novelty were taken from Sundar and Limperos [41]. The items used to measure socialization were adopted from the previous relevant studies [42, 43].

4 Results

4.1 Descriptive Analysis

The participated students were selected from different faculties, enrolled in different majors, and different levels. 53% of the participants were females, and 47% were males. The majority of the respondents were aged between 18 and 29 years old with 66%, whereas the rest were above 29 years old.

4.2 Hypotheses Testing Using Machine Learning Algorithms

The current study employs supervised machine learning classification algorithms through a wide range of methodologies such as decision trees, if-then-else rules, meta classifiers, Bayesian networks, functions, and neural networks in order to examine the relationships among the constructs in the proposed research model. Weka (version 3.8.3) was used to analyze the data by employing the percentage split (66%) test mode based on the following classifiers: a bayesian classifier (NaiveBayes), a logistic-regression classifier (SMO), a lazy-classifier (IBk), a meta-classifier (Bagging), a rule-learner (PART), and a decision-tree (RandomForest) [44–46].

The results shown in Table 1 indicate that RandomForest had a better performance in predicting socialization by the attributes of PEOU and subjective norm. The classifier predicted the socialization with an accuracy of 84.92%. The classifier also had a better performance in TP rate (0.849), ROC area (0.949), and precision (0.846). Thereby, H1a and H2a were supported.

The results shown in Table 2 indicate that both RamdomForest and IBk performed better than the other classifiers in predicting self-presentation by the attributes of PEOU and subjective norm. The classifiers predicted the self-presentation with an accuracy of 84.92%. Further, they had a better performance in TP rate (0.833), ROC area (0.952), and precision (0.856). Thereby, H1b and H2b were supported.

As per the readings in Table 3, the results suggested that RandomForest performed better than the other classifiers in predicting the enjoyment by the attributes of PEOU and subjective norm. The classifier predicted the enjoyment with an accuracy of 76.98%. Further, it had a better performance in TP rate (0.770), ROC area (0.918), and precision (0.784). Thereby, H1c and H2c were supported.

Table 1 Predicting the socialization by PEOU and subjective norm

Classifier	CCI[a] (%)	TP[b] rate	FP[c] rate	Precision	Recall	F-Measure	ROC[d] area
NaiveBayes	69.05	0.690	0.194	0.690	0.690	0.665	0.804
SMO	62.70	0.627	0.243	0.649	0.627	0.659	0.712
IBk	83.33	0.833	0.087	0.828	0.833	0.829	0.956
Bagging	72.22	0.722	0.156	0.708	0.722	0.710	0.885
PART	74.60	0.746	0.141	0.738	0.746	0.734	0.900
RandomForest	84.92	0.849	0.082	0.846	0.849	0.844	0.949

[a]*CCI* Correctly Classified Instances, [b]*TP* True Positive, [c]*FP* False Positive, [d]*ROC* Receiver Operating Characteristic

Table 2 Predicting the self-presentation by PEOU and subjective norm

Classifier	CCI (%)	TP rate	FP rate	Precision	Recall	F-Measure	ROC area
NaiveBayes	64.29	0.643	0.236	0.666	0.643	0.636	0.823
SMO	63.49	0.635	0.251	0.684	0.635	0.620	0.774
IBk	84.92	0.849	0.104	0.856	0.849	0.848	0.952
Bagging	80.95	0.810	0.124	0.814	0.810	0.810	0.909
PART	83.33	0.833	0.117	0.843	0.833	0.833	0.907
RandomForest	84.92	0.849	0.104	0.856	0.849	0.848	0.952

Table 3 Predicting the enjoyment by PEOU and subjective norm

Classifier	CCI (%)	TP rate	FP Rate	Precision	Recall	F-Measure	ROC area
NaiveBayes	51.59	0.516	0.305	0.543	0.516	0.517	0.716
SMO	56.35	0.563	0.289	0.634	0.563	0.559	0.648
IBk	78.57	0.786	0.113	0.829	0.786	0.788	0.923
Bagging	72.22	0.722	0.164	0.794	0.722	0.718	0.873
PART	75.40	0.754	0.137	0.789	0.754	0.757	0.889
RandomForest	76.98	0.770	0.136	0.784	0.770	0.772	0.918

The results shown in Table 4 indicate that both SMO and PART performed better than the other classifiers in predicting novelty by the attributes of PEOU and subjective norm. The classifiers predicted the novelty with an accuracy of 52.38%. Thereby, H1d and H2d were supported.

The results shown in Table 5 reveal that SMO and PART performed better than the other classifiers in predicting unique function by the attributes of PEOU and subjective norm. The classifiers predicted the novelty with an accuracy of 42.07%. Thereby, H1e and H2e were supported.

Table 4 Predicting the novelty by PEOU and subjective norm

Classifier	CCI (%)	TP rate	FP Rate	Precision	Recall	F-Measure	ROC area
NaiveBayes	50.00	0.500	0.479	0.393	0.500	0.416	0.510
SMO	52.38	0.524	0.524	0.524	0.524	0.688	0.500
IBk	38.89	0.389	0.362	0.388	0.389	0.388	0.552
Bagging	50.00	0.500	0.472	0.456	0.500	0.416	0.534
PART	52.38	0.524	0.524	0.524	0.524	0.688	0.500
RandomForest	38.10	0.381	0.362	0.391	0.381	0.385	0.548

Table 5 Predicting the unique function by PEOU and subjective norm

Classifier	CCI (%)	TP rate	FP rate	Precision	Recall	F-Measure	ROC area
NaiveBayes	38.89	0.389	0.420	0.273	0.389	0.282	0.491
SMO	420.07	0.421	0.421	0.421	0.421	0.592	0.500
IBk	37.30	0.373	0.382	0.356	0.373	0.353	0.547
Bagging	40.48	0.405	0.407	0.427	0.405	0.372	0.532
PART	42.07	0.421	0.421	0.421	0.421	0.592	0.500
RandomForest	38.89	0.389	0.398	0.3691	0.389	0.356	0.528

The results shown in Table 6 suggest that RandomForest performed better than the other classifiers in predicting the intention. The RandomForest predicted the intention by the attributes of all dependent variables with an accuracy of 88.89%. The classifier also had a better performance in TP rate (0.889), ROC area (0.981), and precision (0.894). Therefore, H3a, H3b, H3c, H3d, H3e, H4, and H5 were supported.

The results shown in Table 7 indicate that IBk and RandomForest performed better than the other classifiers in predicting the actual use by intention. The classifiers

Table 6 Predicting the intention

Classifier	CCI (%)	TP rate	FP rate	Precision	Recall	F-Measure	ROC area
NaiveBayes	56.35	0.563	0.240	0.573	0.563	0.560	0.728
SMO	55.56	0.556	0.309	0.745	0.556	0.501	0.660
IBk	69.84	0.698	0.170	0.708	0.698	0.698	0.768
Bagging	74.60	0.746	0.141	0.767	0.746	0.744	0.903
PART	81.75	0.817	0.109	0.822	0.817	0.817	0.920
RandomForest	88.89	0.889	0.065	0.894	0.889	0.890	0.981

Table 7 Predicting the actual use by intention

Classifier	CCI (%)	TP rate	FP rate	Precision	Recall	F-Measure	ROC area
NaiveBayes	73.81	0.738	0.178	0.735	0.738	0.736	0.846
SMO	68.25	0.683	0.297	0.761	0.683	0.651	0.754
IBk	78.57	0.786	0.135	0.787	0.786	0.786	0.862
Bagging	75.40	0.754	0.130	0.773	0.754	0.756	0.854
PART	78.57	0.786	0.135	0.787	0.786	0.786	0.840
RandomForest	78.57	0.786	0.135	0.787	0.786	0.786	0.862

predicted the actual use with an accuracy of 78.57%. The classifiers had a better performance in TP rate (0.786), ROC area (0.862), and precision (0.787). Accordingly, H6 was supported.

5 Conclusion

5.1 Research Implications

WhatsApp stickers have recently become widely accepted by students. This research was designed with the aim of understanding the factors affecting the students' acceptance of WhatsApp stickers. To serve this purpose, this research integrates the TAM with the U&G theory to understand the determinants affecting the students' decisions to accept these tools.

Drawing on the research results, this study offers several implications. First, this study is believed to be one of the few that attempts to understand the students' acceptance of WhatsApp stickers in the UAE context. Second, this research integrates the TAM with the U&G theory to understand the determinants affecting the students' acceptance of stickers. Thus, this research validates the applicability of TAM and U&G in predicting the acceptance of stickers by students. Third, this study employs a novel approach in analyzing the hypothesized relationships among the constructs in the proposed theoretical model through the use of machine learning algorithms. This approach was rarely used in the previous literature, and thus, it is believed that this approach would add a significant contribution to the IS literature. Fourth, stickers developers can gain more insights from the research findings of this study by considering the prominent factors and take them into account while developing future stickers specified for educational purposes.

5.2 Limitations and Future Research Directions

Despite the results achieved, this research was limited in some ways. First, this research examined the factors affecting the stickers acceptance by students and did not evaluate the academics' acceptance of these tools. Therefore, what is now needed is another research which could consider the role of academics in using these stickers and understand the determinants affecting their acceptance. Second, the data were collected from only one institution in the UAE. Thus, the generalization of the results to the other UAE institutions should be treated with caution. Third, the data were collected through questionnaire surveys only. Therefore, it would be interesting to use other qualitative methods such as focus groups or interviews to further enlighten the quantitative results.

Acknowledgements This is an extended version of a conference paper published by the International Conference on Advanced Intelligent Systems and Informatics 2019.

References

1. Al-Qaysi, N., Mohamad-Nordin, N., Al-Emran, M., Al-Sharafi, M.A.: Understanding the differences in students' attitudes towards social media use: a case study from Oman. In: 2019 IEEE Student Conference on Research and Development (SCOReD), pp. 176–179 (2019)
2. Al-Qaysi, N., Mohamad-Nordin, N., Al-Emran, M.: What leads to social learning? Students' attitudes towards using social media applications in Omani higher education. Educ. Inf. Technol. (2019)
3. Al-Qaysi, N., Mohamad-Nordin, N., Al-Emran, M.: An empirical investigation of students' attitudes towards the use of social media in omani higher education. In: International Conference on Advanced Intelligent Systems and Informatics, pp. 350–359 (2019)
4. Mhamdi, C., Al-Emran, M., Salloum, S.A.: Text mining and analytics: a case study from news channels posts on Facebook, **40** (2018)
5. Salloum, S.A., Al-Emran, M., Abdallah, S., Shaalan, K.: Analyzing the arab gulf newspapers using text mining techniques. In: International Conference on Advanced Intelligent Systems and Informatics, pp. 396–405 (2017)
6. Lin, T.C., Fang, D., Hsueh, S.Y., Lai, M.C.: Drivers of participation in Facebook long-term care groups: applying the use and gratification theory, social identification theory, and the modulating role of group diversity. Health Inf. J. (2019)
7. Prahalad, C.K., Ramaswamy, V.: Co-creation experiences: the next practice in value creation. J. Interact. Mark. (2004)
8. Jussila, I., Tarkiainen, A., Sarstedt, M., Hair, J.F.: Individual psychological ownership: concepts, evidence, and implications for research in marketing. J. Mark. Theory Pract. (2015)
9. Pierce, J.L., Kostova, T., Dirks, K.T.: Toward a theory of psychological ownership in organizations. Acad. Manag. Rev. (2001)
10. Zhou, R., Hentschel, J., Kumar, N.: Goodbye text, hello emoji: mobile communication on wechat in China. In: Proceedings of the 2017 CHI Conference on Human Factors in Computing Systems, pp. 748–759 (2017)
11. Lee, J.Y., Hong, N., Kim, S., Oh, J., Lee, J.: Smiley face: why we use emoticon stickers in mobile messaging. In: Proceedings of the 18th international conference on human-computer interaction with mobile devices and services adjunct, pp. 760–766 (2016)
12. Ernungtyas, N.F., Sarwono, B., Eriyanto, E., Irwansyah, I.: Mobile applications: integrated user acceptance model. Adv. Sci. Lett. **23**(11), 10573–10576 (2017)
13. Acheampong, P., Zhiwen, L., Boateng, F., Boadu, A.B., Acheampong, A.A.: Determinants of behavioral intentions of 'Generation-Y'adoption and use of computer-mediated communication tools in Ghana. Br. J. Interdiscip. Res. **8**(1), 34–47 (2017)
14. Ozboluk, T., Kurtoglu, R.: Üniversite Öğrencilerinin Emoji Kullanımları ve Emoji Kullanan Markalara Karşı Tutumları Üzerine Bir Araştırma. Bus. Econ. Res. J. (2018)
15. Ghobadi, S., Taki, S.: Effects of telegram stickers on english vocabulary learning: focus on iranian EFL learners. Res. English Lang. Pedagog. (2018)
16. Van De Bogart, W., Wichadee, S.: Exploring students' intention to use LINE for academic purposes based on technology acceptance model. Int. Rev. Res. Open Distance Learn. (2015)
17. Shao, C., Kwon, K.H.: Clicks intended: An integrated model for nuanced social feedback system uses on Facebook. Telemat, Informatics (2019)
18. Sutton, S., Lawson, S.: A provocation for rethinking and democratising emoji design. In: DIS 2017 Companion—Proceedings of the 2017 ACM Conference on Designing Interactive Systems (2017)

19. Feng, Y., Qiu, M., Li, Y., Yang, H.: Cross-culture business communication by Emoji in GMS (2016)
20. Stark, L., Crawford, K.: The conservatism of emoji: work, affect, and communication. Soc. Media Soc. (2015)
21. Ledbuska, L. (214) Emjoi, emoji, what for art thou? Harlot (2014)
22. Zhao, Q., Der Chen, C., Wang, J.L.: The effects of psychological ownership and TAM on social media loyalty: an integrated model. Telemat. Info. **33**(4), 959–972 (2016)
23. Davis, F.D., Bagozzi, R.P., Warshaw, P.R.: User acceptance of computer technology: a comparison of two theoretical models. Manage. Sci. **35**(8), 982–1003 (1989)
24. Davis, F.D.: Perceived usefulness, perceived ease of use, and user acceptance of information technology. MIS Q. **13**(3), 319–340 (1989)
25. Al-Maroof, R.A.S., Al-Emran, M.: Students acceptance of Google classroom: An exploratory study using PLS-SEM approach. Int. J. Emerg. Technol. Learn. **13**(6), 112–123 (2018)
26. Salloum, S.A., Alhamad, A.Q.M., Al-Emran, M., Monem, A.A., Shaalan, K.: Exploring students' acceptance of e-learning through the development of a comprehensive technology acceptance model. IEEE Access **7**, 128445–128462 (2019)
27. Al-Emran, M., Teo, T.: Do knowledge acquisition and knowledge sharing really affect e-learning adoption? An empirical study. Educ. Inf. Technol. (2019)
28. Salloum, S.A., Al-Emran, M., Habes, M., Alghizzawi, M., Ghani, M.A., Shaalan, K.: Understanding the impact of social media practices on e-learning systems acceptance. In: International Conference on Advanced Intelligent Systems and Informatics, pp. 360–369 (2019)
29. Al-Emran, M., Arpaci, I., Salloum, S.A.: An empirical examination of continuous intention to use m-learning: an integrated model. Educ. Inf. Technol. (2020)
30. Salloum, S.A., Al-Emran, M.: Factors affecting the adoption of E-payment systems by university students: extending the TAM with trust. Int. J. Electron. Bus. **14**(4), 371–390 (2018)
31. Alshurideh, M., Salloum, S.A., Al Kurdi, B., Al-Emran, M.: Factors affecting the social networks acceptance: an empirical study using PLS-SEM approach. In: 8th International Conference on Software and Computer Applications, pp. 414–418 (2019)
32. Mezhuyev, V., Al-Emran, M., Fatehah, M., Hong, N.C.: Factors affecting the metamodelling acceptance: a case study from software development companies in Malaysia. IEEE Access **6**, 49476–49485 (2018)
33. Mezhuyev, V., Al-Emran, M., Ismail, M.A., Benedicenti, L., Chandran, D.A.: The acceptance of search-based software engineering techniques: an empirical evaluation using the technology acceptance model. IEEE Access (2019)
34. Al-Qaysi, N., Mohamad-Nordin, N., Al-Emran, M.: A systematic review of social media acceptance from the perspective of educational and information systems theories and models. J. Educ. Comput. Res. **57**(8), 2085–2109 (2020)
35. Samani, M.C., Guri, C.J.: Revisiting uses and gratification theory: a study on visitors to Annah Rais Homestay. J. Komun. Malaysian J. Commun. (2019)
36. Hossain, M.A., Kim, M., Jahan, N.: Can 'liking' behavior lead to usage intention on facebook? Uses and gratification theory perspective. Sustain (2019)
37. Li, X., Chen, W., Popiel, P.: What happens on Facebook stays on Facebook? the implications of Facebook interaction for perceived, receiving, and giving social support. Comput. Human Behav. (2015)
38. Aburub, F., Alnawas, I.: A new integrated model to explore factors that influence adoption of mobile learning in higher education: an empirical investigation. Educ. Inf. Technol. 1–14 (2019)
39. Al-Emran, M., Mezhuyev, V., Kamaludin, A.: PLS-SEM in information systems research: a comprehensive methodological reference. In: 4th International Conference on Advanced Intelligent Systems and Informatics (AISI 2018), pp. 644–653 (2018)
40. Gan, C.: Understanding WeChat users' liking behavior: an empirical study in China. Comput. Human Behav. **68**, 30–39 (2017)
41. Sundar, S.S., Limperos, A.M.: Uses and grats 2.0: New gratifications for new media. J. Broadcast. Electron. Media **57**(4), 504–525 (2013)

42. Ellison, N.B., Steinfield, C., Lampe, C.: The benefits of facebook 'friends:' social capital and college students' use of online social network sites. J. Comput. Commun. **12**(4), 1143–1168 (2007)
43. Lee, S.-Y., Hansen, S.S., Lee, J.K.: What makes us click 'like' on Facebook? Examining psychological, technological, and motivational factors on virtual endorsement. Comput. Commun. **73**, 332–341 (2016)
44. Arpaci, I.: A hybrid modeling approach for predicting the educational use of mobile cloud computing services in higher education. Comput. Human Behav. **90**, 181–187 (2019)
45. Arpaci, I.: What drives students' online self-disclosure behavior on social media? A hybrid SEM and artificial intelligence approach. Int. J. Mob. Commun. (2020)
46. Frank et al.: Weka-A machine learning workbench for data mining. In: Data Mining and Knowledge Discovery Handbook (2009)

Exploring the Effects of Flipped Classroom Model Implementation on EFL Learners' Self-confidence in English Speaking Performance

Mohamad Yahya Abdullah, Supyan Hussin, Zahraa Mukhlif Hammad, and Kemboja Ismail

Abstract Learners of English as a foreign language (EFL) face several challenges which impact negatively on their attempts at achieving English proficiency. This study aimed at exploring the effects of Flipped Classroom Model (FCM) implementation on EFL learners' level of self-confidence in English speaking performance. A mixed methodology design comprising both qualitative and quantitative approaches was used to accomplish the objective of the present study. Data was gathered through analysis of self-confidence in English Speaking Performance Questionnaire (SCESPQ), students' reflective journals, and focus group interviews. The sample was made up of purposively selected 27 undergraduate students who were attending the Advanced Communication Skills course at Buraimi University College (BUC), Sultanate of Oman. The analysis of the quantitative and qualitative data showed a remarkable occurrence of improvement at students' self-confidence in English speaking performance across the time of FCM implementation. The results of ANOVA and MANOVA tests recorded significant differences ($p < 0.001$) in students' self-confidence levels before and during the implementation of FCM. It was found that using FCM contributed significantly to progressively increasing students' self-confidence in English speaking performance over time. For future studies, it is vital to explore the effects of using FCM on other learning factors and language skills.

M. Y. Abdullah (✉) · S. Hussin · K. Ismail
Universiti Kebangsaan Malaysia, Selangor, Malaysia
e-mail: mabdullah@buc.edu.om

S. Hussin
e-mail: supyan@ukm.edu.my

K. Ismail
e-mail: kembojaismail@gmail.com

M. Y. Abdullah
Buraimi University College, Al Buraimi, Sultanate of Oman

Z. M. Hammad
University of Baghdad, College of Languages, Baghdad, Iraq
e-mail: Zahraa82.mukhlif@gmail.com

M. Al-Emran et al. (eds.), *Recent Advances in Intelligent Systems and Smart Applications*, Studies in Systems, Decision and Control 295,
https://doi.org/10.1007/978-3-030-47411-9_13

Keywords Flipped classroom model · Self-confidence · English speaking performance · Google classroom

1 Introduction

Over the past two decades, the English language has spread to all corners of the globe. This spread plays a vital role in evolving most of life's domains such as education, internet, business, science, technology, media, communication, transportation, and others. Its widespread has drawn the attention of many countries towards its essential role to attain the demands of these aspects of life. Thus, these countries became aware of its importance, which represents the bridge towards national progress and modernization [1, 2]. In fact, learning English has become a necessity for most people who seek to create many opportunities in different domains, particularly the global marketplace. More specifically, speaking English fluently and accurately has become an important concern since the ability to speak a language is judged the same as mastering that language [3, 4]. This fact has led to a demand for evolution in efforts to utilize effective and practical methods to improve English speaking performance [5]. Therefore, the mastery of spoken English has received considerable importance and become a priority for most instructors and students.

Nevertheless, many students of English as a Foreign Language (EFL) still encounter different challenges when they are exposed to English speaking in terms of fluency and accuracy. Previous studies underlined the variables of limited linguistic knowledge, lack of self-confidence, anxiety, fear of evaluation, and reluctance to communicate, as the main factors that cause English speaking difficulty [6–10]. According to Al-Hebaish [7], among these factors, self-confidence is the most influential personal factor that plays a supportive role in learners' English speaking performance. In line with this, Gürler [11] contended that the personal factor of self-confidence has an influential role in starting any action in the process of language learning; particularly for speaking whereas it facilitates the process of starting a conversation. Gürler [11] also concluded that self-confidence is one of the keys that unlocks most of the communication challenges and stimulates learner's willingness to communicate the target language context. Accordingly, the researchers of these previous studies recommended refining the language process by adopting more effective teaching methods to promote oral communication skills. In this sense, Al-Issa [12] and Abdullah et al. [13] underlined the value of creating motivational conditions in the classroom to help students boost their self-confidence by building up positive perceptions about themselves and the language learning process. He concluded that teachers should not undermine learners' confidence but instead keep cultivating it.

Despite all the efforts in investigating the problems behind the English-speaking difficulties, very little attention has been paid to find strategies for pedagogical improvement. However, the advanced technology in the twenty-first century brings about new vital opportunities for instructors to adopt new pedagogical approaches in

English classrooms built on the concept of learner-centeredness. This may help teachers to create a classroom environment featured in being more interactive, comfortable, encouraging, and engaging. This learning environment promotes learners to be more independent in their learning process, helps them to build up their self-confidence, and consequently allows them to foster the sense of achievement to produce better English speaking performance [14, 15]. In this regard, the implementation of the Flipped Classroom Model (FCM) as an alternative teaching method may provide authentic opportunities to the EFL learners to consolidate their self-confidence in their English speaking performance.

Across the past ten years, FCM has been viewed by many instructors, educators, and researchers as a very useful and effective instructional method. The application of FCM where the direct instruction in-class time and homework are reversed embraces features that make it suitable for use in different fields and contexts of higher education. This teaching technique allows teachers to make use of the class time for more student-student interaction [16–20]; provides diverse communicative in-class activities [21]; helps students to attain a high level of self-efficacy and achievement [22]; improves students' communicative skills and content knowledge [23]; contributes effectively to boost students' engagement in the learning process [24] and improves students' motivation and English speaking performance [20, 25]. All these benefits are vital indicators that give further insights into how the implementation of FCM may provide EFL learners with plenty of opportunities to practice English speaking and to consolidate their self-confidence. Therefore, the main argument of this study is that FCM may have a positive impact on the level of EFL learners' self-confidence in their English-speaking performance. The primary goal of this research is to explore the effects of FCM implementation on the EFL learners' self-confidence in English speaking performance. Its objective has been investigated by the use of both quantitative and qualitative data This paper is outlined as follows: the following sections present a theoretical discussion for improving the factor of self-confidence in FCM environment. Then a thorough explanation for the methodology utilized in this study is introduced, followed by a presentation, interpretation and discussion of the research results. After that, research conclusion and limitations are discussed.

2 Theoretical Discussion: Improving Self-confidence in FCM Environment

In the context of foreign language learning, the term speaking is generally understood to mean an interactive process that involves a set of psychological factors such as anxiety, motivation, and self-confidence which determine the efficiency of the speaking performance [26–30]. However, there is a general agreement about the supportive and influential role of the factor of self-confidence starting the conversation [7, 11]. Accordingly, it is necessary here for teacher to adopt the notion of leaner-centeredness to create a classroom environment that is characterized in being

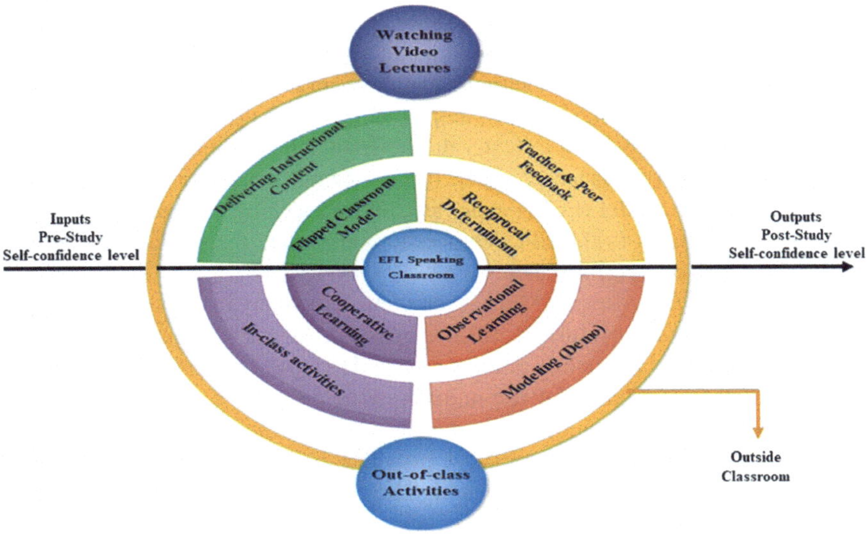

Fig. 1 Improving self-confidence in FCM Environment

more, engaging, interactive, safe, and comfortable. This kind of learning environment promotes student-student and teacher-student interaction and gives learners more responsibility for their process of learning instead of being passive. According to Prawat et al. [31], adopting the notion of learner-centeredness empowers the characteristics of the active learners through cultivating and promoting their roles as contributors and creators of meaning rather than passive receivers of knowledge. This shows a need to frame the classroom environment theoretically based on certain principles that are drawn from different language theories or models to be the basis of this study. Figure 1 illustrates the theoretical framework for the current study.

Based on the theoretical framework of this study, the perspective of the social cognitive theory lies in Bandura's [32] conception of reciprocal determinism and observational learning. The principle of reciprocal determinism interprets the functions of human beings such as the learning process as a result of dynamic and reciprocal interaction of personal, behavioural, and environmental influences. According to Bandura' view, these three fundamentals contribute efficiently in the process of human development. Moreover, it considers cognition as a key element which determines people abilities to construct knowledge, self-regulate, and behavioural performances [33]. In the context of language acquisition, Atkinson [34] postulated that "language and second language acquisition as sociocognitive phenomena." On the other hand, the concept of observational learning in social cognitive theory refers to the people capability to evolve their knowledge and skills throughout the information presented by the impacts of modelling [35]. Bandura [32] elaborated the notion of observational learning that a single model can convey a vast amount of new information through different ways of behaviours to different people and it is achieved through observing people's actions. Stefanone et al. [36]. postulated that modelling in social cognitive perspective is used to "characterize the process through which individuals observe others, interpret their behavior, and adjust their own behavior in response."

Furthermore, the adoption of cooperative learning principle would ensure the achievement of the dynamic and reciprocal interaction inside the classroom. A considerable amount of literature showed that the principle of cooperative learning has a significant impact on language learning achievement. According to Abdullah [37], "cooperative learning is chosen for implementation at the class in order to increase the amount of interaction among students in English and other classes. By interacting with peers, it is expected that students would increase their oral skills, help each other, learn and become less dependent on teachers". It allows enlarging the benefits of group activities by providing opportunities for meaningful negotiation and interaction. Its activities promote students to be more active than passive and provide them with autonomy in learning construction. It helps to provide the concept of scaffolding which was firstly presented by Vygotsky [38]. It enables weak students to learn from their more skillful peers through observing their performance (modelling), peer feedback, and the process of meaningful negotiation and effective interaction. Besides, it allows teachers to take the role of facilitator and guidance [28, 36].

Besides, the noteworthy presence of emerging technology in learning over the last few years has led to a noteworthy contribution to educational settings [13, 27, 39]. This contribution creates richer learning environments to promote more adaptable learning programs. Thus, it can be widely utilized in different pedagogical practices like flipped instruction [40]. The conception of flipped instruction is built to support the notion of a learner-centered approach that makes a great emphasis on active and collaborative learning. Flipped classroom model has been defined as "an alternative model of instruction in which digital technologies are used to shift lectures out of the class hours in the form of screen-cast videos to introduce new concepts" [41]. Thus, students spend class time in more discussions with their teacher and with each other about various subjects in depth. Moreover, it provides plenty of opportunities to involve students in various activities and tasks that promote active interaction and meaningful negotiation.

More interestingly, the characteristics of FCM seem to promote autonomous and cooperative learning, whereas students can express themselves and learn the target language from their experiences in their real world and their autonomous self-regulation. Moreover, the second part of FCM that embodies the face to face interaction will engage the students in more critical thinking processes to promote knowledge construction collaboratively and to develop their language skills [22, 42]. Consequently, FCM learning environment paves the way to involve students in more advanced learning activities during the class time and gives them more responsibilities in their own learning process by creating a larger communication context. Thus, the class time can be better devoted in providing authentic opportunities to EFL learners to consolidate their self-confidence in English speaking performance.

3 Course Design

The teaching method of the Flipped Classroom Model was implemented in "Advanced Communication Skills" (ENGL 154) course at "Buraimi University College" (BUC). This course plays a crucial role in addressing the goals of English department which aim to provide students with realistic opportunities for improving self-confidence, speech delivery skills, extended communication, advanced presentation, and public speaking skills in various life aspects. As concerns, the FCM was implemented via performing a set of procedures. First, students are exposed to watch instructional videos outside the classroom to "learn on their own time and pace before the class sessions; besides, the class time was used for practices, discussions, activities, and interaction" [25]. All lectures, which were supposed to be given in a traditional way at the classroom, were recorded by the teacher (researcher) as instructional video lectures ranging from 5 to 10 min to prepare a bank of short instructional video lectures for Advanced Communication Skills course. Then the researcher edited these clips through, reviewing the content, testing sound quality, dividing them into parts, and giving them titles. The students enrolled in this course attended the class 3 times a week (180 min per week).

Via google classroom, the teacher delivered the instructional video lectures and the instructional guide to provide students with information about the in-class conversational tasks and activities, or the out-of-class activity based on the study plan. This helped students to prepare for the class session in advanced.

3.1 Technology Tools: Google Classroom Application

In this research, a variety of technological tools were used to facilitate the learning process by using FCM such as Google Classroom Application, videos, internet sources, and active whiteboard. However, Google Classroom was used as the main technological tool (platform) that allowed the teacher to deliver the video lectures, instructional guidelines, and any other related materials.

Besides, there were several logical reasons behind using Google Classroom. Firstly, it is the most common social platform in BUC. Most of the lecturers are using it to communicate with their students since communication via any social media tools such as WhatsApp, Twitter, Facebook, or Instagram is not allowed between the teachers and students in accordance with the regulations of the "Ministry of Higher Education" in Oman. Secondly, all teachers and students of BUC were given an institutional email account they can use to log into the Google Classroom. Then they can use it to communicate with their teachers and classmates. Thirdly, Google Classroom has numerous features that promote teachers to employ it in the model of the flipped classroom. It is a functional repository to keep all the instructional video lectures and other materials. Teachers can re-use any of the instructional materials from class to class and at any time. It provides the students with the feature of downloading video lectures or any instructional materials. It has a mobile application which makes the classroom accessible through different digital devices. This option allows students to watch the lecture videos when they need at anytime and anywhere. Besides, it helps teachers and students to communicate outside the classroom as needed. Finally, this platform is considered as a vital opportunity for peer learning. It allows students to post their comments, ask and answer questions, respond to their peers' replies, and share their knowledge and ideas.

4 Methodology

This study was carried out based on a "mixed methodology design" comprising both "qualitative and quantitative approaches" to accomplish its main objective. A Self-confidence in English Speaking Performance Questionnaire (SCESPQ), students' reflective journals, and focus group interviews were used to explore the effects of FCM implementation on the EFL learners' levels of self-confidence in English speaking performance. After data analysis, data triangulation was performed to maintain findings' validity and reliability to accomplish the objective of this research.

4.1 Methods of Data Collection

4.1.1 Participants

Based on the objective of this research, a purposeful sampling technique, specifically the quasi-experimental procedure, was embraced to determine the participants who can provide adequate data for the current research [43]. The samplings for the present study were all the undergraduate EFL students who registered the course of Advanced Communication Skills (ENGL 154) at BUC in the academic year 2017–2018/Semester 2. The sample of the current study included (n = 27) respondents with the age range of 20–24 years old.

4.1.2 Research Instruments

To explore the effects of FCM implementation on the factor of self-confidence in students' English-speaking performance, the researcher used three research instruments comprising a questionnaire, students' reflective journals, focus group interviews.

The first instrument was a Self-confidence in English Speaking Performance Questionnaire (SCESPQ). This 12-item questionnaire was adapted from previous self-confidence questionnaires by Abdallah and Ahmed [44] This adaptation was performed according to the nature of the present study. Furthermore, all required methodological steps have been taken to ensure the validity and reliability of this

Table 1 Four subscales of the SCESPQ

Dimension	Total items (12)	Items no.
Verbal communication	5	1, 2, 3, 4, 5
Non-verbal communication	2	8, 9
Fear of evaluation	2	6, 7
Speaking activities	3	10, 11, 12

scale. The test of "Kaiser-Meyer-Olkin" (KMO) and the "Bartlett test of Sphericity" were performed to measure the sampling adequacy and the overall significance of all the correlations within the correlation matrix. The results reported that the test of KMO scored 0.778 for the scale of self-confidence and the "Bartlett test of sphericity" was $p < 0.001$ which is very significant. This questionnaire was distributed for three times, before, in the mid, and after the implementation of FCM.

SCESPQ consists of 12 items (see Appendix). These items are grouped into four dimensions, as shown in Table 1. The items of the first group identify the level of self-confidence in *verbal communication* like "I feel confident speaking English even if my English is poor" while the items included in the second group examine the construct of self-confidence in *non-verbal communication* such as "I do not feel confident when I need to make eye to eye contact with my audience". The items of the third group measure this variable in regard to *fear of evaluation* in the English speaking performance like "I feel confident when no one pays attention to my mistakes while I am speaking". The items of the final group assess the level of self-confidence associated with *speaking activities* such as "I feel confident to speak English when I have individual presentation".

The second instrument is the students' reflective journals. This research instrument is the first flexible and powerful qualitative method that is recruited to enrich the data via authentic records about participants' natural behaviours, perceptions, and reactions toward the implementation of FCM. It is a vital source of information about progress and development [45]. Therefore, three forms of handwritten reflective journals were designed to be distributed on three phases during the FCM implementation.

The third instrument is the focus group interview. This instrument is the second powerful qualitative method that is employed to obtain reliable and valid data relevant to the phenomenon under investigation [46] cited in [39]. Twelve students were selected and divided into three groups of 4 students (low-performance, mid-performance, and high-performance) based on their marks obtained in the post-oral proficiency test. At the end of the semester, the interview sessions were conducted with the three allocated groups.

4.2 Data Analysis

The findings of overall SCESPQ distributed before, in the mid, and after the implementation of FCM were analyzed using SPSS (version 21). First, the test of "one-way repeated measured ANOVA" was run to assess the effects of FCM implementation on students' level of self-confidence in English speaking performance at three times (before, in the mid, and after the implementation of FCM). This test was carried out to compare the changes in mean scores at three points across the time of FCM implementation. Moreover, "one-way repeated measure Multivariate Analysis of Variance" (MANOVA) was performed to determine whether there were differences among the total mean scores of the subscales within the factor of self-confidence at three times. Second, the qualitative data obtained from the two qualitative research instruments were coded, categorized, and then analyzed thematically. Besides, the quantitative and qualitative data were triangulated to verify and add credibility to the findings.

5 Findings and Discussion

This study aimed at finding out the effects of FCM implementation on the EFL learners' self-confidence in English speaking performance. With the aim of comparing the mean scores of students' self-confidence throughout the implementation of FCM, ANOVA was run to calculate whether there were differences among these mean scores over time with a statistical test at Pre-test (**time 1**, before the implementation of FCM), Post-test 1 (**time 2**, after 6 weeks from utilizing the FCM), and Post-test2 (**time 3**, after 12 weeks i.e. after the implementation). Table 2 presents the means and standard deviations.

Table 2 illustrates that the total mean scores of students' self-confidence ($N = 27$) before the implementation of FCM was ($mean = 1.950$; $SD = 0.511$), then after 6 weeks of the implementation of FCM, an obvious increase occurred at the total mean ($mean = 2.549$; $SD = 0.493$). However, after 12 weeks of using FCM, the total mean score of self-confidence increased noticeably to ($mean = 3.987$; $SD = 0.214$). The results indicate that students' level of self-confidence in English speaking performance has significantly increased across the time of implementing FCM. Figure 2 shows the changes in students' self-confidence levels during the application of FCM.

Table 2 Descriptive "one-way repeated measured ANOVA" for self-confidence

Self-confidence tests	Mean	Std. Deviation	N. of students
Pre-test	1.950	0.511	27
Post-test 1	2.549	0.493	27
Post-test 2	3.987	0.214	27

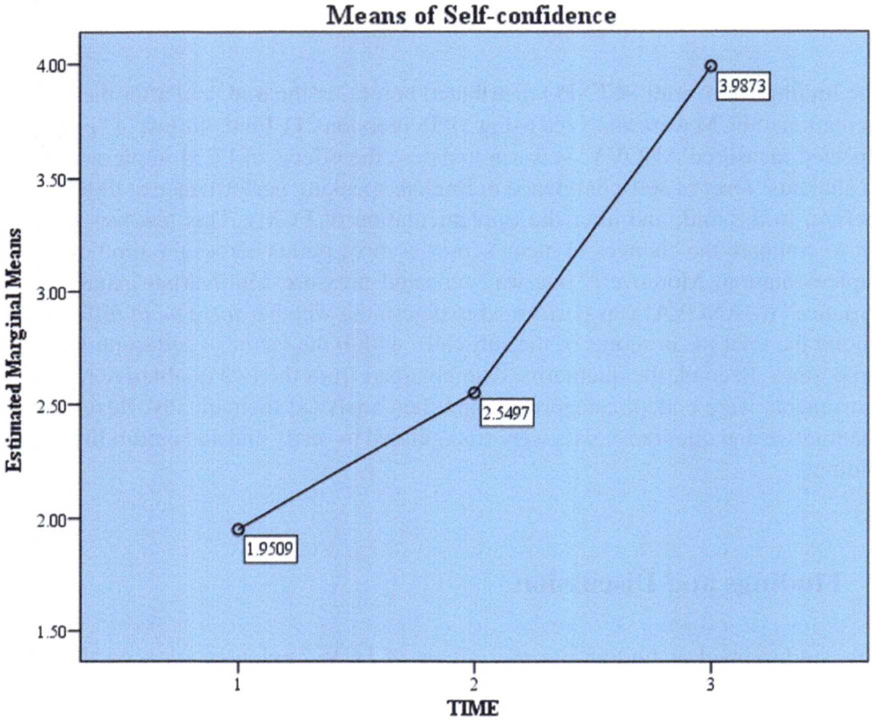

Fig. 2 Self-confidence mean scores over time

Moreover, Fig. 3 illustrates that the percentage of the improvement occurred with students' level of self-confidence across time. It was found that the improvement in the total mean score of self-confidence recorded (30.69%) of increase at the Post-test 1 (after 6 weeks) compared to the total mean score of the Pre-test. Besides, the results of Post-test 2 (after 12 weeks) showed a significant increase (56.38%) in the

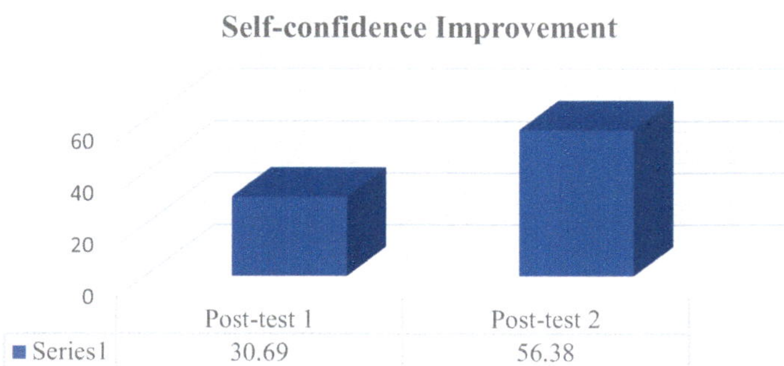

Fig. 3 Percentage of improvement for the self-confidence compared with the pre-test

Table 3 Pairwise comparison of self-confidence over three times

Measure	(I) time	(J) time	Mean Difference (I-J)	SE	*P* value	95% CI	
						LB	UB
Time	1	2	−0.599*	0.110	<0.001	−0.880	−0.318
	1	3	−2.036*	0.122	<0.001	−2.348	−1.725
	2	3	−1.438*	0.108	<0.001	−1.715	−1.160

Note The asterisk (*) means The Mean Difference is significant

percentage of improvement compared to the total mean score of the Pre-test of self-confidence. The continual growth of self-confidence during the FCM implementation can be clearly seen in Fig. 3.

To examine whether this yielded improvement in total mean scores was significant, the inferential statistics were carried out. The result of Mauchly's test indicated that the Sphericity assumption was met, $\chi^2 = 0.565$, $p = 0.754$. Since the "Sphericity" assumption was not violated, there was no need to perform any other correction tests. The findings of "ANOVA" for the variable of self-confidence across the time indicated that the implementation of FCM affected significantly on participants' levels of self-confidence $F_{(2, 52)} = 170.179$, $p < 0.001$, $\eta 2 = 0.867$).

Then, to assess the details for mean differences of the self-confidence variable over time, the "Bonferroni post hoc test" was applied to compare the mean scores. The findings of the "Bonferroni test" (Table 3) pointed out that the difference among all tests was very significant ($p < 0.001$).

For the sake of evaluating the differences in the mean scores of participants' self-confidence in English speaking performance including the subscales of *verbal communication*, *non-verbal communication*, *fear of evaluation*, and *speaking activities* a MANOVA was carried out to examine whether there were differences among these four subscales over the time; as the four subscales were significantly correlated. The findings of "MANOVA" yielded a (Wilks Lambda = 0.038, F = 59.420, $\eta 2 = 0.962$) indicating a significant difference for these four subscales over time. Table 4 presents the means and standard deviations.

Table 4 shows that the subscale of *verbal communication* scored three different mean scores across time including Pre-test (mean = 2.266; SD = 0.460), Post-test 1 (mean = 2.488; SD = 0.789), and Post-test 2 (mean = 3.659; SD = 0.322). Similarly, the subscale of *non-verbal communication* reported three different mean scores over time, Pre-test (mean = 1.907; SD = 0.920), Post-test 1(mean = 3.018; SD = 0.379), and Post-test 2 (mean = 3.981; SD = 0.489). Likewise, the subscale of *fear of evaluation* recorded three different mean scores over the time, Pre-test (mean = 1.814; SD = 0.606), Post-test 1(mean = 2.370; SD = 0.779), and Post-test 2 (mean = 4.148; SD = 0.456). Also, the subscale of *speaking activities* resulted three different mean scores over the time, Pre-test (mean = 1.814; SD = 0.655), Post-test 1(mean = 2.321; SD = 0.526), and Post-test 2 (mean = 4.160; SD = 0.250).

Based on the aforementioned results, all subscales of self-confidence in English speaking performance have increased noticeably across of the FCM implementation.

Table 4 Descriptive one way repeated measured MANOVA for self-confidence

Self-confidence tests	Mean	Std. Deviation	N. of students
Pre. verbal communication	2.266	0.460	27
Post1. verbal communication	2.488	0.789	27
Post2. verbal communication	3.659	0.322	27
Pre. non-verbal communication	1.907	0.920	27
Post1. non-verbal communication	3.018	0.379	27
Post2. non-verbal communication	3.981	0.489	27
Pre. fear of evaluation	1.814	0.606	27
Post1. fear of evaluation	2.370	0.779	27
Post2. fear of evaluation	4.148	0.456	27
Pre. speaking Activities	1.814	0.655	27
Post1. speaking Activities	2.321	0.526	27
Post2. speaking Activities	4.160	0.250	27

Figure 4 illustrates the changes that occurred at the four subscales. What stands out in Fig. 4 is the progressive increase in all subscales of students' self-confidence.

Figure 4 shows the percentage of the improvement that happened to the four subscales of self-confidence over the period of using the FCM. The results of Post-test 1compared to the Pre-test revealed that the improvement of self-confidence subscales increased differently. The highest increase was recorded in the subscale of non-verbal communication (31.9%); the lowest increase was in the subscale of *verbal communication* (9.8%). Besides, the percentage of the improvement of the *fear evaluation* recorded (30.6%) and *speaking activities* reported (27.9%) which were also lower

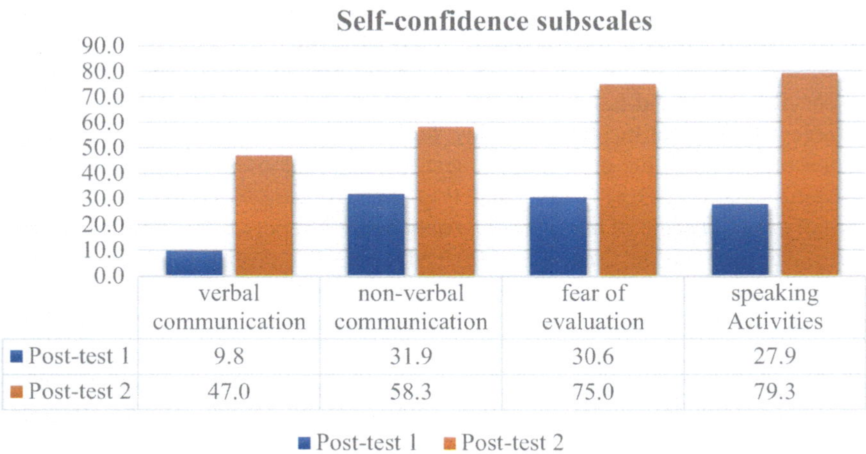

Fig. 4 Percentage of improvement for the self-confidence four subscales compared to the pre-test

than the subscale of *non-verbal communication*. Moreover, the results of the Post-test 2 showed a considerable increase in all self-confidence subscales. It was indicated that the subscale of *non-verbal communication* recorded the biggest improvement in the Post-test1 (31.9%). Besides, the subscale of *speaking activities* also reported the biggest improvement in Post-test 2 (79.3%). Based on the results of Post-test 2, the *verbal communication* subscale recorded the lowest improvement (31.9%) compared to the other three subscales of self-confidence. These results indicate that the participants of the study have experienced positively with all subscales of self-confidence that have been examined in the current study gradually and dissimilarly. In addition, it was found that they perceived their teacher's and peer's feedback positively and became more confident to participate in the speaking activities over the time of the FCM implementation. Then it is essential to look at the inferential statistics to ascertain whether these observed increment (improvements) are significant.

The result of "Mauchly's test" for the four subscales of self-confidence indicated that the "Sphericity" assumption had been met for the subscales of *fear of evaluation* ($\chi^2 = 0.651$, $p = 0.722$) and the subscale of *speaking activities* ($\chi^2 = 1.767$, p $= 0.413$). However, the results of "Mauchly's test" indicated that the Sphericity assumption for the subscales of *verbal communication* ($\chi^2 = 6.570, p = 0.037$) and *non-verbal communication* ($\chi^2 = 10.409, p = 0.005$), has been violated. Accordingly, "Greenhouse-Geisser estimation" was performed to measure the differences among the total mean scores of the questionnaire subscales across time. The findings of "MANOVA" for the four subscales indicated that the difference among the three times (Pre-test, Post-test 1, Post-test 2) were statistically significant whereas the *verbal communication subscale* recorded $F_{(1.6, 42.2)} = 44.685, P < 0.05, \eta 2 = 0.632$), *the nonverbal communication* subscale recorded $F_{(1.4, 52)} = 71.477, P < 0.05, \eta 2 = 0.733$), the subscale of *fear of evaluation* recorded $F_{(2, 52)} = 112.072, P < 0.05, \eta 2 = 0.812$), and the subscale of *speaking activities* scored $F_{(2, 52)} = 173.745, P < 0.05, \eta 2 = 0.870$). Additionally, the findings of the "MANOVA test" revealed that the above subscales were also statistically different across the three times. Besides, to assess the effect of FCM implementation on each subscale a univariate analysis was computed and means of the four subscales were compared using the "Bonferroni post hoc test."

In regard to the mean differences of the four subscales of the self-confidence across time, the "Bonferroni post hoc test" was computed to compare the total mean scores. The findings of the "Bonferroni test" (Table 5) point out that the difference among most of the tests was statistically significant ($p < 0.05$).

In accordance with the results of the Bonferroni test, the mean difference of the four subscales between time 1 to time 2 was not significant (P value more than 0.05). However, the results of "pairwise comparison" also indicated that the mean difference for all these subscales was statistically significant ($P < 0.001$) in the comparison between time 1 and time 3, and between time 2 and time 3. Moreover, these results showed that the four subscales of students' self-confidence in English speaking performance improved significantly during the FCM implementation. Additionally, this increase might record higher scores with time.

Table 5 Pairwise comparison of self-confidence subscales over time

Measure	(I) time	(J) time	Mean difference (I-J)	SE	*P* value	95% CI	
						LB	UB
Verbal communication	1	2	−0.222	0.166	0.576	−0.647	0.202
	1	3	−1.393*	0.117	<0.001	−1.691	−1.094
	2	3	−1.170*	0.185	<0.001	−1.642	−0.698
Non-verbal communication	1	2	−1.111*	0.174	<0.001	−1.555	−0.667
	1	3	−2.074*	0.213	<0.001	−2.619	−1.529
	2	3	−0.963*	0.122	<0.001	−1.275	−0.651
Fear of evaluation	1	2	−0.556*	0.174	0.011	−1.000	−0.111
	1	3	−2.333*	0.151	<0.001	−2.720	−1.947
	2	3	−1.778*	0.163	<0.001	−2.195	−1.360
Speaking activities	1	2	−0.506*	0.137	0.033	−0.857	−0.155
	1	3	−2.346*	0.144	<0.001	−2.714	−1.977
	2	3	−1.840*	0.114	<0.001	−2.132	−1.547

Note The asterisk (*) means The Mean Difference is significant

Due to the data analysis of the reflective journals and interviews' transcripts, it was found that the qualitative results matched with the quantitative ones in regard to the significant impact of FCM on students' self-confidence over the time. Students viewed the implementation of FCM has helped them to be more confident by providing them with authentic opportunities to practice speaking English. For example, RJ2S03 answered the question, ***how do you rate your confidence after five weeks from the implementation of FCM?***, "*Approximately 60%, but after more rehearsal and practice, I think my confidence will be greater. I believe in myself a lot.*" She stated an acceptable range of improvement occurred to her self-confidence. However, she indicated that she became more confident when she was asked to rate her self-confidence at the end of the implementation of FCM, "*I am very much confident, and I am ready to speak without any hesitation*" (*RJ3S03, translated*). Besides, the same student (S03) as one of the interviewees of the second focus group, mentioned that providing plenty of speaking practices during the application of FCM helped her to build her self-confidence "*In fact, for me, I feel that these three factors have improved significantly because of the many opportunities to practice speaking English provided during the use of FCM and this reduced my fears and* ***increased my self-confidence****, motivation, and enthusiasm" (2nd FGS03, translated*).

Another example, S08 was so excited to realize the improvement that occurred on her self-confidence level at the end of the implementation, "*Now, I am feeling very interested because I have more confidence than before*" (*RJ3S08, translated*). However, she was not totally satisfied with the improvement of her self-confidence after five weeks from starting the application of FCM, "*I feel I am confident but I don't have enough self-confidence when I present in some topics and occasions*"

(RJ2S08, translated). This matched the quantitative results which showed progressive improvement in participants' level of self-confidence. There is more evidence from the students' reflections and interviewees' responses which support the gradual increase occurred on students' self-confidence level in English speaking performance over the time of implementing FCM. For example,

Translated extracts from the second and third phases of students' reflective journals:

RJ2S20: *My self-confidence to speak English has increased slightly.*

RJ3S20: *The self-confidence has increased more compared with previous.*

RJ2S11: *I cannot rate my self-confidence, but I can speak English if I am not asked to speak in front, my classmates!*

RJ3S11: *It has increased compared with previous times, but I still need more practice.*

Translated extracts from interviews' transcripts:

3rd FGS23: I've changed a lot and turned out to be a confident *and more responsible person. Also, I used to work all by myself and avoided speaking with others in the classroom because I always feared the idea that others might laugh at me but now, I've changed this idea, and I never cared about being laughed at.*

3rd FGS04: *Honestly, in the past, I used to avoid doing presentations because* ***I wasn't that confident one and I was afraid about being confused and getting embarrassed*** *while presenting or even worse I was afraid that my mind will be numbed the moment I reach the stage and that I might forget what to say… Now, as you've noticed that* ***we've become more interactive and confident***.

Obviously, the match between the qualitative and quantitative data indicated considerable effects of using FCM on students' self-confidence level in English speaking performance. The data also showed that the improvement occurred at students' self-confidence gradually during the implementation of FCM. The improvement of students' self-confidence could be attributed to the significant contribution of collaborative and interactive learning to help students feel more independent in their learning process. They believed that they became more responsible for their learning, and they had authentic opportunities to express and share their thoughts, ideas, and experiences more freely. Additionally, students' increasing desire and enthusiasm to participate in all activities and communicate with their classmates might be attributed to the comfortable, safe, encouraging, and interactive classroom environment. This finding is in concord with the findings of some past studies such as [47] which showed that learners' self-confidence could be built and strengthened by utilizing particular teaching techniques. While Fauzan [48, 49] found that using debate technique and peer assessment as interactive tools to build and enhance learners' self-confidence, Göktürk [47] pointed out that using digital video recordings strengthened participants' self-confidence in oral communication performance. However, neither of these two studies adopted FCM to arrive at the findings.

6 Conclusion

The strong presence of the Flipped Classroom Model as an effective teaching method in the pedagogical process is leading to significant improvements in learning outcomes of different language skills. Besides, the remarkable increase of students' satisfaction towards the implementation of this instructional method can be attributed to the improvements that have been occurred on their achievements. This led to a growing trend towards the use of FCM in a wide range to help students improve their learning skills and overcome their learning difficulties. This study aimed at exploring the effects of FCM implementation on the level of EFL learners' self-confidence in English speaking performance. Taken as a whole, the findings revealed that students' level of self-confidence in English speaking performance has significantly increased across the time of implementing FCM. The findings of the quantitative and qualitative data showed a progressive improvement that occurred considerably at students' self-confidence during the implementation of FCM. The results of ANOVA and MANOVA tests indicated that the differences among the total mean scores of students' levels of self-confidence and the four subscales of self-confidence in English speaking performance were statistically significant ($p < 0.001$). According to students' reflections and interviews' responses, the continual growth of their self-confidence in English speaking performance could be attributed to the embracing of the collaborative and active learning principles along with the support of embedding technology in the learning process. The nature of FCM instruction and the well-designed in-class and out-of-class activities provided the students plenty of opportunities to improve their self-confidence in a creative, engaging, safe, comfortable, and encouraging classroom environment.

Although the implementation of FCM was proven as an effective and useful teaching technique in this study, the findings may not be generalized since the participants were specific to one group of EFL students at BUC. Nevertheless, the findings of this study can be generalized in the context of Oman, wherein this empirical study was conducted. For further study, it is vital to investigate the effects of FCM on other learning factors and language skills. Also, it is crucial to conduct a true experimental design to involve two groups of a bigger number of participants.

Appendix

See Table 6.

Table 6 Self-confidence in English speaking performance

Strongly disagree (SD)	Disagree (D)	Neutral (N)	Agree (A)	Strongly agree (SA)
1	2	3	4	5

No.	Items	SD	D	N	A	SA
1.	I feel confident when I speak in English	1	2	3	4	5
2.	I think I will speak perfect English someday	1	2	3	4	5
3.	I never feel sure of myself when I speak in English	1	2	3	4	5
4.	I can speak English well even if the audience is large	1	2	3	4	5
5.	I feel confident speaking English even if my English is poor	1	2	3	4	5
6.	I become hesitant when my English-speaking teacher corrects my mistakes	1	2	3	4	5
7.	I feel confident when no one pays attention to my mistakes while I am speaking	1	2	3	4	5
8.	I do not feel confident when I need to make eye to eye contact with my audience	1	2	3	4	5
9.	I don't feel comfortable with my posture when I am giving a speech in English	1	2	3	4	5
10.	I like to be the head in any speaking group activities	1	2	3	4	5
11.	I avoid participating in English speaking activities in the classroom with a large number of students	1	2	3	4	5
12.	I feel confident to speak English when I have individual presentation	1	2	3	4	5

References

1. Nunan, D.: Practical English Language Teaching. McGraw-Hill/Contemporary (2003)
2. Ciprianová, E., Vanco, M.: English in the age of globalization: changing ELT models, restructuring relationships. J. Linguist. Intercult. Educ. **3**, 123 (2010)
3. Karahan, F.: Language attitudes of Turkish students towards the English language and its use in Turkish context, pp. 73–87 (2007)
4. Al-mahrooqi, R.: A student perspective on low English proficiency in Oman. Int. Educ. Stud. **5**(6), 263–271 (2012)
5. Harb, J., Bakar, N.A., Krish, P.: Gender differences in attitudes towards learning oral skills using technology. Educ. Inf. Technol. **19**(4), 805–816 (2014)
6. Aydin, S.: An investigation on the language anxiety and fear of negative evaluation among turkish EFL learners. Asian EFL J. 421–444 (2008)
7. Al-hebaish, S.M.: The correlation between general self-confidence and academic achievement in the oral presentation course **2**(1), 60–65 (2012)
8. Al Hosni, S.: Speaking difficulties encountered by young EFL learners. Int. J. Stud. English Lang. Lit. **2**(6), 22–30 (2014)
9. Yalçın, Ö., İnceçay, V.: Foreign language speaking anxiety: the case of spontaneous speaking activities. Procedia - Soc. Behav. Sci. **116**, 2620–2624 (2014)
10. Klimova, B.F.: Detecting the development of language skills in current English language teaching in the Czech Republic. Procedia-Social Behav. Sci. **158**, 85–92 (2014)
11. Gürler, İ.: Correlation between self-confidence and speaking skill of English language teaching and English language and literature preparatory students **1** (2015)

12. Al-Issa, A.S.M.: A critical examination of motivation in the Omani English language education system. J. Lang. Teach. Res. **5**(2), 406–418 (2014)
13. Abdullah, M.Y., Al Ghafri, H.M.H., Al Yahyai, K.S.H.: A qualitative study on the best motivational teaching strategies in the context of Oman: perspectives of EFL teachers. Engl. Lang. Teach., **12**(3), 57 (2019)
14. Kvashnina, O.S., Martynko, E.A.: Analyzing the potential of flipped classroom in ESL teaching. Int. J. Emerg. Technol. Learn. **11**(3), 71–73 (2016)
15. Thi, Q., Tran, T.: Flipped model for improving students' English speaking performance flipped model for improving students' English speaking performance, no. June (2019)
16. Zainuddin, Z., Attaran, M.: Malaysian students' perceptions of flipped classroom: a case study Flipped classroom Background of flipped learning in the University of Malaya. Innov. Educ. Teach. Int. **53**(6), 660–670 (2016)
17. Moore, A.J., Gillett, M.R., Steele, M.D.: The mathematics teacher **107**(6), 420–425 (2015)
18. Al-Mahrooqi, R., Denman, C.: Issues in English education in the Arab world **19**, 350 (2015)
19. Basal, A.: The implementation of a flipped classroom in foreign language teaching. Turkish Online J. Distance Educ. **16**(4), 28–37 (2015)
20. Abdullah, M.Y., Hussin, S., Ismail, K.: Investigating the effects of the flipped classroom model on Omani EFL learners' motivation level in English speaking performance. Educ. Inf. Technol. **24**(5), 2975–2995 (2019)
21. Webb, M., Doman, E., Pusey, K.: Flipping a Chinese university EFL course: what students and teachers think of the model. J. Asiat. **11**(4), 53–87 (2014)
22. Kurt, G.: Implementing the flipped classroom in teacher education: Evidence from Turkey. Educ. Technol. Soc. **20**(1), 211–221 (2017)
23. Sun, Y.C.: Flipping every student? A case study of content-based flipped language classrooms. E-Learning Digit. Media **14**(1–2), 20–37 (2017)
24. Bergmann, J., Sams, A.: Flip your classroom: reach every student in every class every day. Int. Soc. Technol. Educ. (2012)
25. Abdullah, M.Y., Hussin, S., Ismail, K.: Implementation of Flipped Classroom Model and its Effectiveness on English Speaking Performance. Int. J. Emerg. Technol. Learn. **14**(09), 130–147 (2019)
26. Du, X.: The affective filter in second language teaching. Asian Soc. Sci. **5**(8), 162–165 (2009)
27. Hussin, S., Abdullah, M.Y., Ismail, N., Yoke, S.K.: The effects of CMC applications on ESL writing anxiety among postgraduate students. Engl. Lang. Teach. **8**(9), 167 (2015)
28. Abdullah, M.Y., Hussin, S., Shakir, M.: The effect of peers' and teacher's E-feedback on writing anxiety level through CMC applications. Int. J. Emerg. Technol. Learn. **13**(11), 196–207 (2018)
29. Brown, H.D., Lee, H.: Teaching by Principles: An Interactive Approach to Language Pedagogy, vol. 1. Prentice Hall Regents Englewood Cliffs, NJ (1994)
30. Zrekat, Y., Abu Bakar, N., Latif, H.: The level of anxiety among Jordanian EFL undergraduates in oral communication performance. SSRN **3**, 188–202 (2016)
31. Prawat, R.S., American, S., May, N.: Teachers' beliefs about teaching and learning : a constructivist perspective **100**(3), 354–395 (2017)
32. Bandura, A.: Social cognitive theory: An agentic perspective. Annu. Rev. Psychol. **52**(1), 1–26 (2001)
33. Harare, S., Branch, A.A.: Socio-cognitive theory approach in second language acquisition : the state of the art **1**(4), 145–152 (2016)
34. Atkinson, D.: Toward a sociocognitive approach to second language acquisition
35. Bussey, K., Bussey, K., Bandura, A.: Social cognitive theory of gender development and differentiation social cognitive theory of gender development and differentiation, 1999 (2017)
36. Stefanone, M.A., Lackaff, D., Rosen, D.: The relationship between traditional mass media and "social media": reality television as a model for social network site behavior. J. Broadcast. Electron. Media **54**(3), 508–525 (2010)
37. Maria Abdullah, G.J.: Promoting cooperative learning at primary school. Electron. J. Engl. Second Lang. **6**(1), 37–58 (2004)
38. Cole, M., John-steiner, V.: Mind in society the development (1978)

39. Challob, A.I., Bakar, N.A., Latif, H.: Collaborative blended learning writing environment: effects on EFL students' writing apprehension and writing performance. Engl. Lang. Teach. **9**(6), 229 (2016)
40. Grgurovic, M.: Technology-enhanced blended language learning in an ESL class : a description of a model and an application of the diffusion of Innovations theory (2010)
41. Khalil, R., Rania, M., Fahim, S.S.: Assessment as a learning tool in a flipped English language classroom in higher education (2017)
42. Sun, J.C.Y., Wu, Y.T., Lee, W.I.: The effect of the flipped classroom approach to OpenCourseWare instruction on students' self-regulation. Br. J. Educ. Technol. **48**(3), 713–729 (2017)
43. Creswell, J.W., Creswell, J.D.: Research Design: Qualitative, Quantitative, and Mixed Methods Approaches. Sage publications (2017)
44. Abdallah, S.A.A., Ahmed, A.G.A.: The impact of self-confidence on EFL sudanese tertiary level students (2015)
45. Hubbs, D., Brand, C.F.: Learning from the inside out : a method for analyzing reflective journals in the college classroom **33**(1), 56–71 (2010)
46. Louis, C., Lawrence, M., Keith, M.: Research Methods in Education. New York Routledge (2007)
47. Göktürk, N., et al.: Examining the effectiveness of digital video recordings on oral performance of EFL learners. Teach. Engl. Technol. **16**(2), 71–96 (2016)
48. Fauzan, U.: Enhancing speaking ability of EFL students through debate and peer assessment. EFL J. **1**(1), 49–57 (2016)
49. Fauzan, A., Ngabut, M.N.: EFL students' perception on flipped learning in writing class. J. Engl. Foreign Lang. 8(2), 115–129 (2018)

Developing an Educational Framework for Using WhatsApp Based on Social Constructivism Theory

Noor Al-Qaysi, Norhisham Mohamad-Nordin, and Mostafa Al-Emran

Abstract Social media applications are one of the promising tools in facilitating the educational process. Among these applications, WhatsApp was recently suggested as one of the most popular applications used by students. However, there is a scarce of a framework that can fully utilize the WhatsApp for educational purposes. To bridge this gap and to draw upon the social constructivism framework, the aim of this study is to develop a social constructivism framework for WhatsApp (SCFW) that enables both formal (in-class) and informal (off-class) learning environment. The SCFW framework is based on the principle that educators can use WhatsApp closed groups to enable learners to construct and share knowledge with their peers. This has been accomplished through the integration of four aspects, including device aspect, social aspect, learner aspect, and teacher aspect.

Keywords Educational framework · WhatsApp · Social constructivism theory

1 Introduction

Social media has become a part of our daily lives [1, 2]. Undoubtedly, the penetration of social media applications in users' lives made them addicted to their use. In education, these applications allow social knowledge construction in order to change the traditional methods of learning [3]. Higher educational institutions are beginning to

N. Al-Qaysi (✉) · N. Mohamad-Nordin
Faculty of Art, Computing and Creative Industry, Universiti Pendidikan Sultan Idris, Tanjung Malim, Malaysia
e-mail: noor.j.alqaysi@gmail.com

N. Mohamad-Nordin
e-mail: norhisham@fskik.upsi.edu.my

N. Mohamad-Nordin
College of Economics, Management and Information Systems, University of Nizwa, Nizwa, Oman

M. Al-Emran
Department of Information Technology, Al Buraimi University College, Al Buraimi, Oman
e-mail: mustafa.n.alemran@gmail.com

M. Al-Emran et al. (eds.), *Recent Advances in Intelligent Systems and Smart Applications*, Studies in Systems, Decision and Control 295,
https://doi.org/10.1007/978-3-030-47411-9_14

use social media applications to change the traditional learning and teaching methods [4–6]. The integration of different applications of social media could diversify the learning and teaching methods. Recently, WhatsApp was found to be the most common application used for educational purposes by learners [7, 8]. This stems from the fact that WhatsApp enables users to exchange ideas and opinions at anytime anywhere settings. WhatsApp also allows its users to form a closed group of participants to communicate with their peers, teachers, and the community at large. Furthermore, WhatsApp enables the convergence of formal and informal learning by creating a closed group that is a sub-community of students who can interact in a broader information context. Concerning the learning and teaching process, how can WhatsApp be efficiently employed in formal and informal learning? How can a student fully benefit from the WhatsApp application? How can experts design activities and materials appropriate for learning through WhatsApp? Through the lenses of the social constructivism theory, these questions can be fully answered.

Drawing on the "framework for the rational analysis of mobile education (FRAME)" [9], this research develops a social constructivism framework that considers the social and personal learning aspects. It also considers the features of technological devices. Besides, it stresses the active role of employing technology in social and educational processes. The developed framework spots the light on social constructivism theory in which knowledge is constructed through social interaction. The developed framework also emphasizes the concepts of zone of proximal development and mediation. The developed framework, in general, portrays a learning method that enables learners to interact with their peers and instructors, and move among various virtual and physical settings at any time suitable for them.

2 Problem Statement

Koole [9] developed the Framework for the Rational Analysis of Mobile Education (FRAME), which described mobile learning as a process resulting from the integration of three different aspects, including learner's aspect, device aspect, and social aspect. Stemming from the fact that the educational process is a bidirectional interaction between learner and teacher, Koole [9] discarded the role of the teacher in formulating his framework. In line with this gap and in order to provide a comprehensive framework in which the role of both entities (i.e., student and instructor) is considered, the primary purpose of this study is to extend the framework developed by Koole [9] with the teacher role by taking the WhatsApp closed group as a case study.

3 The Developed Framework

This study develops a Social Constructivism Framework for WhatsApp (SCFW) to illustrate both formal (in-class) and informal (off-class) learning within the educational environment. The framework is based on the principle that educators can use WhatsApp closed groups to enable learners to construct and share knowledge with their peers. This kind of interaction makes the educational process more useful and meaningful. The SCFW framework is depicted in Fig. 1. The SCFW framework describes WhatsApp closed group as an experience of combining device aspect, social aspect, learner aspect, and teacher aspect. The framework addresses the knowledge construction concern through the social interaction between students and teachers using the WhatsApp closed groups.

The framework clearly shows that there is an overlap between the four circles; the device (D), learner (L), teacher (T), and social (S) aspects. The intersections that join two circles include features that fit in both aspects. The characteristics of the device usability (DTL) and social technology (DS) designate the technology affordances. The interaction learning (TLS) intersection emphasizes the social constructivism theory. The combination of the four aspects signifies the WhatsApp (DTLS) intersection.

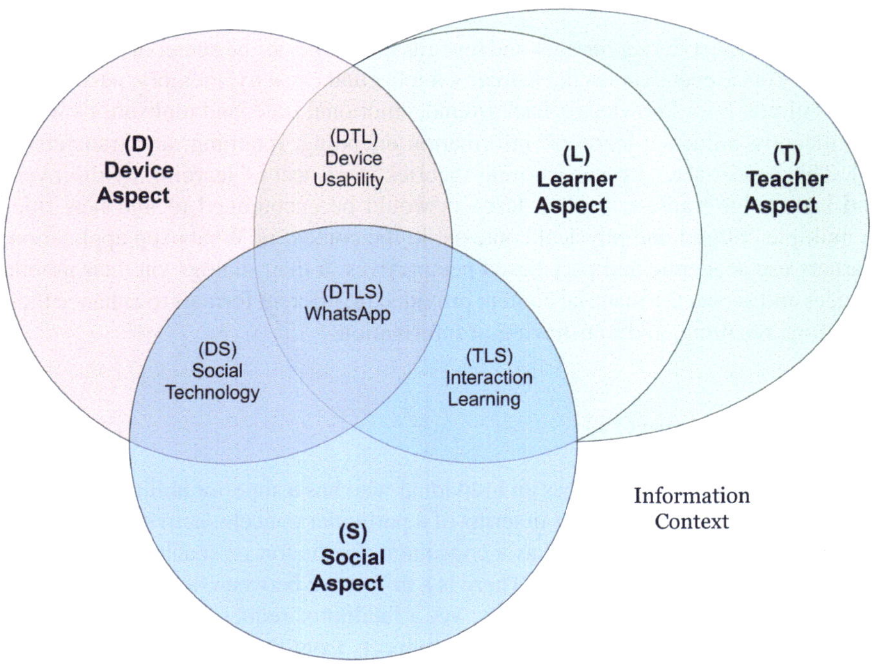

Fig. 1 Social constructivism framework for WhatsApp (SCFW)

4 SCFW Framework Aspects

4.1 *Main Aspects*

4.1.1 Device Aspect (D)

The device aspect (D) is defined as the physical features (e.g., input and output capabilities along with internal processes like storage, power, processor speed, compatibility, and expandability), technical features, and functional features of mobile devices that consistently impact the connection between the student and the learning activities [9]. In the present study, this definition is extended to incorporate the teacher aspect into the learning activities. Such features emerge from the device hardware and software designs to consequently influence the user's levels of psychological and physical comfort. In this sense, mobile devices connect the student and teacher with the technology to maintain high psychological and physical comfort levels. In that, students and teachers are required to own well-designed mobile devices for the purpose of focusing on cognitive tasks themselves rather than mobile devices.

4.1.2 Learner Aspect (L)

The learner aspect (L) encourages and rewards the learner as the center of the learning process. This aspect denotes the learner's intellectual capacity, memory, past experience, culture, prior knowledge, background, emotional state, and motivations which significantly influence learners' information encoding, retaining, and transferring [9]. This aspect also stresses learning theories in respect of learning by discovery and knowledge transfer. In that, learners would be encouraged to find new rules in multiple cultural and physical contexts. In the context of WhatsApp application, learners use academic and peer-based perspectives in their dialogs via their mobile devices and access the material content provided in different formats to enhance their encoding, recalling, and transferring of information.

4.1.3 Teacher Aspect (T)

The teacher aspect (T) describes an individual who has a superior ability and understanding levels than the student in terms of a particular concept, activity, or process. The teacher is being introduced as a consultant, facilitator, or coach who helps the learner to understand the content. There is a difference between being a teacher and a facilitator. The teacher usually tells, yet a facilitator requests. Also, the teacher teaches from the front, yet the facilitator supports from the back. Additionally, the teacher provides answers based on the curriculum, while the facilitator provides guidelines for the students to find answers and conclusions. Moreover, the teacher typically gives a speech, while the facilitator mostly enters into a constructive dialogue with the students [10]. In the context of WhatsApp closed group, the teacher plays the role of a facilitator in assisting students who struggle in using discussion groups. The teacher also assists the students to tackle their problems by offering instructions and hints (i.e., scaffolding) since they work through the zone of proximal development (ZPD) levels. Consequently, the teachers' assistance would be regularly reduced until they become not any longer relied on to enable the struggling students to learn independently.

4.1.4 Social Aspect (S)

Social aspect (S) is considered the core of interaction in a social context [9, 11]. Learners and lecturers are required to follow the interaction and cooperation rules to allow exchanging information, constructing knowledge, and sustaining cultural practices [9]. Through interaction, learners receive feedback to reinforce their cultural, social behaviors, and beliefs [12]. Learner's culture in which interactions occur determines the conversation and cooperation rules in using WhatsApp application for learning and teaching purposes.

4.2 Intersections

4.2.1 Device Usability Intersection (DTL)

The device usability intersection (DTL) refers to the interface between the device and the student [9]. In this study, this intersection covers the attributes belonging to the device (D), teacher (T), and learner (L) aspects. It highlights four significant concepts: device portability, information accessibility, psychological comfort, and satisfaction [9]. First, device portability refers to the device physical features, like weight, size, materials, and peripherals number. The device portability has a direct effect on the device mobility among different environments, as extremely mobile devices should resist shock, dust, and humidity. Second, information accessibility allows the information to travel, so that students can access and retrieve information at anytime anywhere to enjoy the learning experience. The device portability and information availability are significant concepts in device usability. Third, the psychological comfort refers to the student's quick capability to recognize and begin using the mobile device and perform the desired tasks as quickly as possible. The psychological comfort influences both the speed and mental load that enables users to perform tasks. In that sense, reducing mental load requires reducing the actions related to completing a task through the use of memorial devices, metaphors, and chunking information by providing appropriate training and offering simple displays based on the students' prior experiences or socio-cultural knowledge [13]. Thus, designers are required to minimize mental load on users [14]. Consequently, an increasing amount of information can be stored based on the individual's awareness of the information chunks and patterns [14]. Fourth, satisfaction is a highly personal and culturally determined characteristic that is not easy to be predicted. Flexibility allows learners to select functionality and themes, which in turn, allows increasing comfort and satisfaction levels. To conclude, the DTL intersection facilitates the connection between the learners' activities and teachers' needs with device hardware and software characteristics. In that, highly portable, and intuitive devices allow learners to concentrate on the task itself rather than the tools, which consequently supports reducing mental load and increasing task completion levels.

4.2.2 Social Technology Intersection (DS)

Social technology intersection (DS) refers to the ability of mobile devices to support collaboration and communication among different systems and individuals [9]. This intersection highlights several features of mobile tools. First, the device technical services (e.g., telephony, internet access, and short messaging service) have a direct effect on information exchange and interaction among individuals with different needs, aims, and intentions [9]. Second, system connectivity describes devices hardware and software instruments used to connect to multiple systems by different connectivity means [9]. Mobile devices are required to be equipped with mechanisms used for connecting to different systems through different means [9]. In addition, networks necessitate different kinds of wired (e.g., Ethernet cables, and telephone lines) or wireless technology standards (e.g., Bluetooth, infrared, WiFi, CDMA, and GSM) to be used for WhatsApp in the learning process.

Third, the use of collaboration tools (e.g., calendars, project management tools, and authoring tools) allows co-authoring documents, attending or providing lectures, and coordinating tasks to be achieved via different electronic technologies such as workflow tools, shared calendars, project management tools, and automated schedulers [15]. Such collaboration tools enable individuals to take part in various collaboration types. Fourth, wireless networks describe how people exchange relevant information in multiple collaborative situations that could occur at anytime anywhere settings [9]. To conclude, designers and developers are required to consider offering computer-based communication environments or mobile-based media spaces that help learners to interact, although they are separated temporally and physically [15].

4.2.3 Interaction Learning Intersection (TLS)

The interaction learning intersection (TLS) illustrates social learning theories, and basically relies on social constructivism theory [9]. Social constructivism examines that language, knowledge, and social development are developed in accordance with others through social interactions and social communication of ideas [16]. Social constructivism stresses the significant role of culture and context for knowledge construction [17].

Social constructivism theory relies basically on reality, knowledge, and learning [18]. Concerning reality, social constructivism is represented as a product revealed by individuals through human activity [18]. In terms of knowledge, social constructivism believes that knowledge is constructed by individuals socially and culturally [19]. In that, Individuals create meaning through interactions with their peers and the place they live in. With regard to learning, social constructivism believes that learning is the development of psychological processes that occur first socially and then used by individuals internally [20]. Learning in social constructivism is seen as a movement from the existing intellectual level to a superior standard that approaches the learners' potential level [20]. This movement takes place in the ZPD as a result of social interaction.

ZPD is defined as the difference between what a person can learn on his/her own and what that person can learn when learning is supported in collaboration with a more knowledgeable individual [20]. ZPD is an instructional tool that describes the difference between the learner's current ability and the learner's potential with assistance from more knowledgeable individuals to increase growth opportunities and reduce learning ambiguity. ZPD also refers to the area where learning occurs through a process of scaffolding [20]. Scaffolding, in turn, refers to the support provided by more knowledgeable individuals like peers, instructors, or reference sources (e.g., dictionaries), which assist students to perform progressively well [20]. Scaffolding also refers to the learning process in which the instructor plays the role of a knowledge facilitator [21]. In this sense, the instructor provides a structure that supports students to increase their knowledge and control of the involved tasks. In that, instructors are required to design tasks that challenge students to perform beyond their current capacity.

The TLS intersection highlights three criteria, including interaction, situated cognition, and learning communities and cognitive apprenticeships [9]. First, interaction can happen between learner with content, learner with instructor, or learner with other learners [22]. Interaction in learning is a central gateway for mental and physical development along with knowledge acquisition, which in turn, results in a new influential learning form [23]. Second, situated cognition describes the learning task as a situated process within reliable settings to increase learners' motivation [24]. In that, the term reliable settings refer to the learning activities products that are designed for action-oriented learners within actual and larger communities [25]. Third, learning communities and cognitive apprenticeships are both illustrations of social learning approaches that offer different levels of learner control. In terms of learning communities, learners work as a group to achieve mutual goals [26]. Given that, they use technology to communicate and do problem-solving activities with other learners in other places. Concerning cognitive apprenticeship, a learner observes a model functioning in a real situation and then tries the techniques in a comparable situation [9]. In this sense, the learner gradually gains less support from the lecturer until the learner becomes confident enough to work independently [25]. Participating in learning communities and cognitive apprenticeships enable the learners to negotiate meaning in a different situation.

4.2.4 WhatsApp Closed Group (DTLS)

WhatsApp was thus found as a matching application for the social constructivism theory [27]. The WhatsApp application, in this framework, offers an effective mental setting where distance learners can interact with their peers, lecturers, course materials, and computing devices. The WhatsApp application also enables learners to select information, access relevant information, redefine their goals, and reconsider concepts in an information context. WhatsApp use for educational purposes is reshaped and well-defined by the interface among the device (D), learner (L), teacher (T), and social (S) aspects. The interaction nature changes as students interact with peers, tools, contexts, and information [20].

With regard to the access and selection of information, learners are required to deeply identify and assess the accuracy and appropriateness of relevant information as the available variety of information increases [9]. In that, the WhatsApp application enables learners to gain access to information, lecturers, and peers who in turn, support them to determine the significance of information within an information-rich world.

An available growing body of information has been easily accessible and publishable. Given that, lecturers may respond to better flexible knowledge management methods to assist students in recognizing and navigating through knowledge to manipulate, select, and apply the present information in exceptional environments [28]. Learners, in turn, are required to gain the skills and use tools needed to navigate through and filter such information. In that sense, formal and informal educational methods are combined in which teachers become the mentors.

5 Conclusion and Future Research Directions

There is no doubt on how social media applications have become an integral part of our daily lives [29–31]. Among these applications, WhatsApp has changed the landscape in which users can communicate, interact, and share mutual interests among each other. In education, WhatsApp can also be used as a tool for interaction among the students, teachers, and course materials. In line with the literature, there is a scare of knowledge on how WhatsApp can be effectively and efficiently used in higher educational institutions. In other words, there is a need for a framework that can be used to identify the role of the student, teacher, device, and social aspects.

Although Koole [9] developed the FRAME, which described the m-learning as a process resulting from the integration of three different aspects, including learner, device, and social aspect, Koole [9] discarded the role of the teacher in formulating his framework. In line with this gap and in order to provide a comprehensive framework in which the role of both entities (i.e., student and instructor) is considered, the main purpose of this study was to extend the framework developed by Koole [9] with the teacher role by taking the WhatsApp closed group as a case study. The new extended framework (i.e., SCFW) can assist practitioners and scholars in using WhatsApp as a tool for teaching and learning.

Although the SCFW has considered the role of both entities (i.e., students and teachers), which was believed to close the gap of the FRAME [9], there is a need for evaluating this framework by putting it into practice. Thus, this study calls for another research which can evaluate the developed framework in a quantitative and qualitative manner.

References

1. Salloum, S.A., Al-Emran, M., Shaalan, K.: Mining text in news channels: a case study from Facebook. Int. J. Inf. Technol. Lang. Stud. **1**(1), 1–9 (2017)
2. Salloum, S.A., Al-Emran, M., Shaalan, K.: The impact of knowledge sharing on information systems: a review. In: International Conference on Knowledge Management in Organizations, pp. 94–106 (2018)
3. Al-Qaysi, N., Mohamad-Nordin, N., Al-Emran, M.: A systematic review of social media acceptance from the perspective of educational and information systems theories and models. J. Educ. Comput. Res. **57**(8), 2085–2109 (2020)
4. Kong, X.L., Kongkasem, S., Teplitsky, F., Khoo, B.K.S.: Mobilizing learning in higher education: developments in teaching and learning practice. In: Wireless Telecommunications Symposium (2017)
5. Al-Qaysi, N., Mohamad-Nordin, N., Al-Emran, M., Al-Sharafi, M.A.: Understanding the differences in students' attitudes towards social media use: a case study from Oman. In: 2019 IEEE Student Conference on Research and Development (SCOReD), pp. 176–179 (2019)
6. Al-Qaysi, N., Al-Emran, M.: Code-switching usage in social media: a case study from Oman. Int. J. Inf. Technol. Lang. Stud. **1**(1), 25–38 (2017)
7. Al-Qaysi, N., Mohamad-Nordin, N., Al-Emran, M.: An empirical investigation of students' attitudes towards the use of social media in Omani heducation. In: International Conference on Advanced Intelligent Systems and Informatics, pp. 350–359 (2019)
8. Al-Qaysi, N., Mohamad-Nordin, N., Al-Emran, M.: What leads to social learning? Students' attitudes towards using social media applications in Omani higher education. Educ. Inf. Technol. (2019)
9. Koole, M.L.: Mobile learning—A model for framing mobile learning. Mob. Learn. Transform. Deliv. Educ. Train. (2009)
10. Rhodes, L.K., Bellamy, G.T.: Choices and consequences in the renewal of teacher education. J. Teach. Educ. **50**(1), 17–26 (1999)
11. Rambe, P., Bere, A.: Using mobile instant messaging to leverage learner participation and transform pedagogy at a South African University of Technology. Br. J. Educ. Technol. (2013)
12. Kearsley, G.: The nature and value of interaction in distance education. In: Distance Education Symposium 3: Instruction (1995)
13. Shneiderman, B., Plaisant, C.: Designing the user interface : strategies for effective human-computer interaction (5th Ed.).. In: Proceedings 2006 AVI Work. BEyond time errors Nov. Eval. methods Inf. Vis. - BELIV'06 (2006)
14. Bransford, J.D., Brown, A.L., Cocking, R.R.: How People Learn: Brain, Mind, Experience, and School: Expanded Edition (2000)
15. Preece, H.S., Rogers, J.Y.: Interaction Design: Beyond Human-Computer Interaction. Wiley, New York (2002)
16. Mathis, C.: What are supports and barriers to using social constructivism and the reggio emilia approach in state funded preschool programs in California? (2011)

17. Mcmahon, M.: Social constructivism and the World Wide Web—A paradigm for learning. In: What Works And Why : Reflections on Learning with Technology, pp. 1–7, ASCILITE (1997)
18. Kim, B.: Social constructivism. In: Emerging Perspectives on Learning, Teaching, and Technology, pp. 55–61 (2001)
19. Ernest, P.: Social constructivism as a philosophy of mathematics. SUNY Ser. Sci. Technol. Soc. xiv, 315 (1998)
20. Vygotsky, L.S.: Mind in Society: The Development of Higher Psychological Processes. Harvard University Press, Cambridge, MA (1978)
21. Tharp, R.G., Gallimore, R.: Rousing Minds to Life: Teaching, Learning, and Schooling in Social Context. Cambridge University Press (1991)
22. Moore, M.G.: Three types of interaction. Am. J. Distance Educ. (1989)
23. Barker, P.: Designing interactive learning. Des. Prod. Multimed. Simul.-Based Learn. Mater. 1–30 (1994)
24. Smith P.L., Ragan, T.J.: Instructional Design. Wiley (2004)
25. Farmer, J.A., Buckmaster, A., LeGrand, B.: Cognitive apprenticeship: Implications for continuing professional education. New Dir. Adult Contin. Educ. **55**, 41–49 (1992)
26. Reigeluth, C.M., Squire, K.: Emerging work on the new paradigm of instructional theories. Educ. Technol. (1998)
27. Dabbagh, N., Bannan-Ritland, B.: Online Learning: Concepts, Strategies, and Application. Pearson/Merrill/Prentice Hall, Upper Saddle River, NJ (2005)
28. Brown, T.H.: Beyond constructivism : exploring future learning paradigms. Educ. Today **2**(2), 1–11 (2005)
29. Al-Emran, M., Malik, S.I.: The impact of Google Apps at work: higher educational perspective. Int. J. Interact. Mob. Technol. **10**(4), 85–88 (2016)
30. Salloum, S.A., Al-Emran, M., Shaalan, K.: Mining social media text: extracting knowledge from Facebook. Int. J. Comput. Digit. Syst. **6**(2), 73–81 (2017)
31. Salloum, S.A., Mhamdi, C., Al-Emran, M., Shaalan, K.: Analysis and classification of Arabic Newspapers' Facebook pages using text mining techniques. Int. J. Inf. Technol. Lang. Stud. **1**(2), 8–17 (2017)

Research Trends in Flipped Classroom: A Systematic Review

Rana A. Al-Maroof and Mostafa Al-Emran

Abstract Various systematic reviews were conducted to provide a valuable insight into the flipped classroom research trend. However, these reviews ignored the examination of the flipped classroom studies with regard to the information system (IS) theories or models. Based on the previous assumption, the main objective of this study is to review and synthesize the flipped classroom studies related to the IS theories or models based on 15 articles that were published between 2014 and 2018. The significant findings indicated that evaluating the students' and teachers' perceptions towards the usage of the flipped classroom was the main research purpose among all the analyzed studies. Besides, questionnaire surveys were regarded as the dominant method for data collection. Additionally, most of the analyzed flipped classroom articles achieved positive research outcomes. Moreover, the Technology Acceptance Model (TAM) was found as the most common model used for examining the students' and academics' acceptance of the flipped classroom. Furthermore, the majority of the analyzed flipped classroom research articles were conducted in China, followed by Turkey, Taiwan, and Malaysia, respectively. To that end, the results of this study deepen our knowledge and understanding of the flipped classroom studies related to the IS theories or models and enlarge our picture regarding the beneficial trends for future work.

Keywords Flipped classroom · Systematic review · Information systems models · Acceptance · Perceptions

R. A. Al-Maroof (✉)
Department of English Language, Al Buraimi University College, Al Buraimi, Oman
e-mail: rana@buc.edu.om

M. Al-Emran
Department of Information Technology, Al Buraimi University College, Al Buraimi, Oman
e-mail: mustafa.n.alemran@gmail.com

M. Al-Emran et al. (eds.), *Recent Advances in Intelligent Systems and Smart Applications*, Studies in Systems, Decision and Control 295,
https://doi.org/10.1007/978-3-030-47411-9_15

1 Introduction

A growing body of literature has been conducted regarding the use of information technologies in education [1–5]. Building upon these technologies, the notion of the flipped classroom (FC) has been widely spread recently. It has been one of the main concerns of many researchers in varied fields of study [6]. Medical, engineering, English (ELT), and other fields have adopted this technology to enhance the teaching and learning processes. Therefore, learning theories have focused on the importance of this technology as it is a self-directed and independent approach in learning. The flipped classroom is a process where learners are inherently motivated to learn outside the class. The learners will be able to learn in two different circumstances. The first one is individually (off-class by watching videos), and the second one is in a group (in-class group activities) where they got the opportunity for group discussions and collaborative problem-solving in the classroom. Even better, they can critically analyze essential concepts with the help of the teacher in the classroom [7].

The negative or positive perception of this technology has been a vital issue. Both teachers and students have evaluated this technology differently. This fact is clearly stated in some studies, which have indicated some improvements in the flipped model and higher achievement from the perspective of the students. However, students' perceptions may vary. Less satisfaction feedback with the learning experiences provided by the flipped approach has also been noticed. For some researchers, the contrast between improved performance and decreased satisfaction is particularly interesting [8]. Other research studies, on the other hand, have focused on the perception of instructors. It is suggested that proper motivation on the part of the instructor needs to be provided to help students prepared before class in the flipped classroom courses. College instructors should have a clear plan and well-organized instructional goals to promote flipped classroom technology teaching. Besides, they should provide students appropriate, timely, and in-class support [9].

To assess the students' perceptions, researchers have designed different tools such as questionnaire surveys [10–12], open-ended questions, semi-structured interviews, and focus group interviews [13, 14]. However, only a few studies have implemented the information system (IS) models such as the "Technology Acceptance Model (TAM)", "unified theory of acceptance and use of technology (UTAUT)", or "Task-technology fit (TTF)" to examine the students' perceptions towards the use of flipped classroom [15, 16]. Therefore, this study is an attempt to cover this aspect. Accordingly, the current study tries to thoroughly review, analyze, and synthesize the studies that adopted the IS models or theories by classifying the positive and negative feedback collected from the concerned learners.

Recent review studies have focused on the importance of flipped classroom in teaching and learning. Their main criteria for inclusion have focused on either a specific field of study or on the level of students. However, the present study incorporates various fields of study, and all levels are admissible. This study also takes into consideration the IS model or theory used to assess the students' perceptions toward the acceptance of the flipped classroom. In addition, this study attempts to classify the main themes under the flipped classroom and correlate them with the results and place of application. The results, in turn, are classified into positive, negative, and neutral and are connected with the IS theory or model used. To sum up, this study systematically reviews and synthesizes the relevant literature through a meta-analysis approach to provide more comprehensive results as far as the students' perceptions are concerned. Specifically, the current study tries to provide sufficient answers to the following research questions:

RQ1. What are the research aims, adopted methods, and outcomes achieved from the analyzed studies?
RQ2. What are the IS models or theories used to measure the students' perceptions?
RQ3. What are the main disciplines of the analyzed flipped classroom studies?
RQ4. What are the common countries that have adopted the flipped classroom technology?
RQ5. How do the results of this systematic review differ from/or similar to the previous systematic reviews?

2 Literature Review

According to the existing literature, several systematic reviews were carried out in the context of flipped classroom. Although these systematic reviews focused on the importance of the role played by the flipped classroom in teaching and learning, they also have several limitations. Some of them put emphasis on the time-span. Thus, they limited their scope of research to specific years, and others highlight the importance of place. The majority, on the other hand, focuses on one area or field of study.

Previous systematic reviews have attempted to show the effect of using the flipped classroom over a specific period. For example, a study conducted by Uzunboylu and Karagözlü [17] has collected papers from the ScienceDirect database from 2010 to 2015. The study showed that 72% of the analyzed articles took place in the USA, whereas others in other countries such as the Philippine. It also came up with a result concerning the approach used in the application of the flipped classroom. One of them is the apparent differences in the number of studies performed using qualitative, quantitative, and combined methods. The use of qualitative methods was more evident than the quantitative method. Most researchers in recent years did not prefer the use of mixed research methods. Another result of this study is that the flipped classroom approach has been applied to 18 different subject areas between 2010 and 2015. Furthermore, the most preferred research techniques were the experimental study and the case study. Similarly, a study carried out by Zainuddin and Halili [6] has analyzed the trends and contents of flipped classroom research based on 20 publications between the years (2013–2015). It was found that researchers implemented various methodologies with respect to various fields of studies. The flipped classroom has proven to be fruitful for the students academically and motivationally because the students could study at home and come to the class well-prepared. Thus, the learning technology has contributed to the students' learning achievement, motivation, engagement, and interaction.

In the engineering field, two systematic reviews were carried out. The first one stated that the positive gains in problem-solving skills, conceptual understanding, student retention, and satisfaction with flipped classes are widely spread among researchers. Besides, the context and instructional design of the course are vital aspects in reflecting the students' experience and performance. However, most of the revised studies have failed to integrate out the in-class activity [18]. Similarly, the second systematic review showed that most of the researchers in the area of engineering education emphasized the equal-learning opportunities and achievements. It is stated that there is no difference in the students' achievement with or without the use of the flipped classroom because students who used the flipped classroom had learned as much as those who follow the traditional lecture-style format, if not more. Besides, the flipped classroom atmosphere provides many benefits and comes up with some challenges for instructors and students [19].

In the medical field, two systematic reviews were also carried out. The first one provided a strong recommendation for the usage of flipped classroom as the students' satisfaction level with pre-class learning resources was very high. Therefore, they generally express strong satisfaction with the learner-centered approaches in classes over the traditional lecture-based learning. However, technology does not hinder the learning and can develop valuable lifelong learning skills in undergraduate medical students [20]. The second systematic review also provided a positive feedback concerning the use of the flipped classroom in teaching medical courses, assuming that the flipped classroom is a promising teaching approach, specifically when the intent is to enhance the students' motivation, task value, and engagement [21].

In addition, a systematic review conducted by Bergfjord and Heggernes [22] has focused on the area of management education. It is reported that the flipped classroom has improved the students' preparation for lectures in particular and facilitated the process of revising the material after lectures. In addition, the use of watch videos at home makes it easier for weak students to understand the basic and important topics in the newly introduced material. Another systematic review conducted by Choi [23] has focused not only on a specific area (i.e., English classroom) but on a specific place (i.e., Korea) where the flipped classroom has been implemented. The study came up with the result, in which the reviewed studies do not provide tangible evidence for the effects of the flipped classroom in English classrooms. Moreover, another systematic review conducted by Lo and Hew [24] has paid more attention to the students' educational level. The study was limited to K-12 education only. It is argued that the students' perceptions and achievements were very strong. The only negative feedback was related to the instructional approach which has to be improved. The challenges are different. Some of the challenges related to the students and teachers while others related to the operational plan.

Table 1 Inclusion and exclusion criteria

Inclusion criteria	Exclusion criteria
a. Must evaluate students' and/or instructors' perceptions or acceptance b. Must include an IS model or theory c. Must specify the educational level of the participants d. Must specify the field or disciple of study e. Must be a journal article or conference proceeding f. Must be written in English language	a. Dissertation and thesis are not included b. Review papers and book chapters are excluded c. Papers that do not follow an IS model or theory d. Papers that were written in languages other than English

3 Method

A systematic analysis was conducted based on different databases (e.g., IEEE, SAGE, Science Direct, ERIC, Springer, and Google Scholar). The keywords that are used in the search include (("Flipped classroom" OR "Inverted classroom") AND ("perception" OR "acceptance")). The search for these keywords was carried out in April 2018. The current review study follows the Kitchenham and Charters's rules and strategies for carrying out a systematic review [25], and other relevant systematic reviews [26–29]. Based on this assumption, the review process was undertaken in four distinct stages, as illustrated in the following subsequent sections.

3.1 Inclusion/Exclusion Criteria

The steps for the inclusion and exclusion criteria are given in Table 1.

3.2 Articles Selection Process

In this review study, the search and refinement stages were conducted based on the Preferred Reporting Items for Systematic Reviews and Meta-Analysis (PRISMA) [30]. Figure 1 shows the PRISMA flowchart. In that, the total collected articles, the identification, screening, and eligibility of the final papers are all described in numbers.

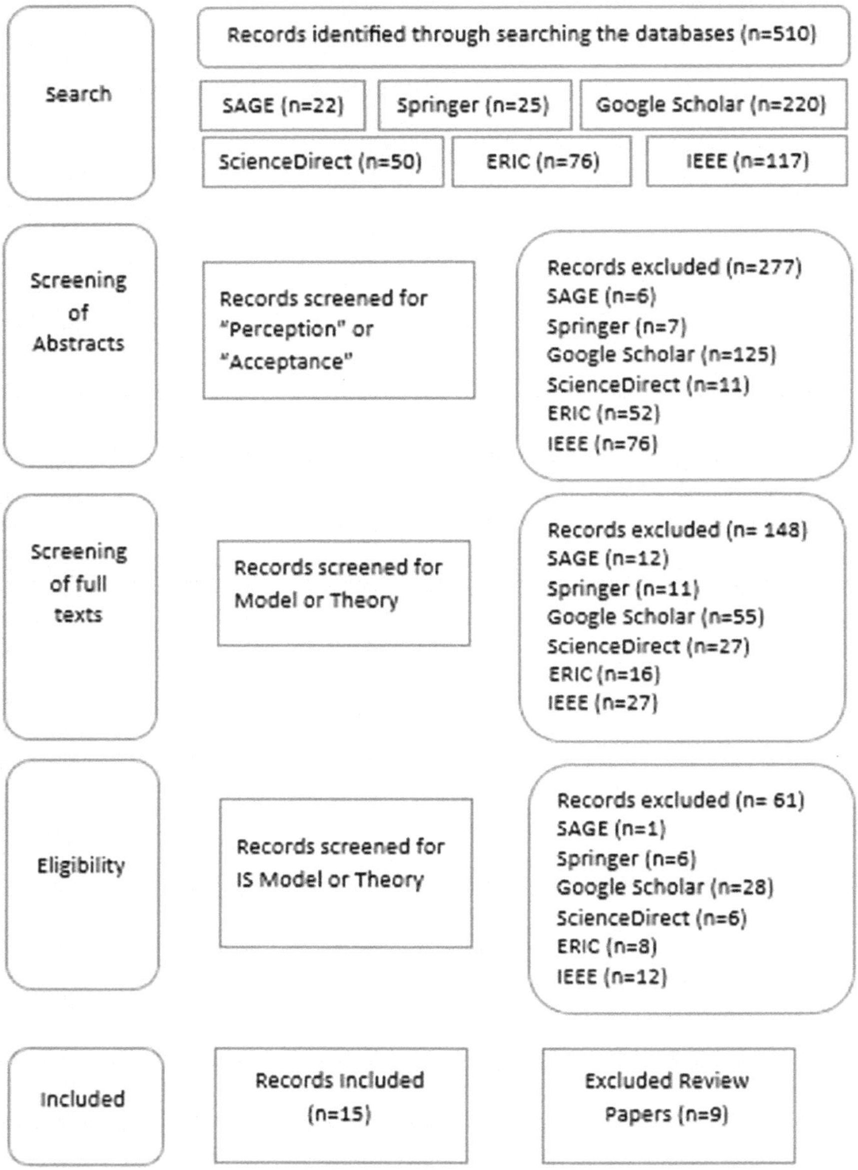

Fig. 1 PRISMA flowchart for the selected studies

3.3 Data Coding and Analysis

Seven features related to the quality of study research methodology were coded including: (a) research purpose, (b) method (e.g., survey, interview, etc.), (c) participants demographic (e.g., school students, college students, trained teachers, etc.), (d) IS model or theory (e.g., TAM, TTF, UTAUT, etc.), (e) discipline (e.g., humanities, social sciences, natural sciences, etc.), (f) research outcome (i.e., positive, negative, and neutral), and (g) country. During the preliminary data analysis stage, studies that do not adopt a specific theory or model were excluded.

3.4 Quality Assessment

The quality assessment of the reviewed papers is a critical factor that can be used along with the inclusion and exclusion criteria [29]. The main advantage of this assessment is to measure the weighting of the studies individually. It also enhances the interpretation of the findings. The instruments that are used in the quality assessment are a group of checklists of factors that need to be evaluated. The quality items are given numerical scales. The checklist is prepared based on the factors of design, conduct, analysis, and conclusions [31]. Based on this assumption, a group of questions under each of the previously mentioned criterion is selected from a checklist given by Budgen and Brereton [31]. Table 2 shows the questions that can help in evaluating the quality of the studies that were retained for further analysis (N = 15). Each question was scored based on the three-point scale, in which "Yes being worth 1 point", "No being worth 0 point", and "Partially being worth 0.5 point". Table 3

Table 2 Quality assessment questions

Design:
1. Are the study aims clearly stated?
2. Is the technology defined clearly?
Conduct:
3. Are the data collection methods adequately described?
4. Are the participants' numbers and qualifications mentioned clearly?
Analysis:
5. Are negative findings presented?
6. Are the quantitative and/or qualitative analyses adequately described?
7. Are the statistical (reliability and validation) or descriptive (content) methods described?
8. Do the numbers add up across different tables?
Conclusion:
9. Does the study deepen your understanding of the topic under analysis?
10. Are the results significantly comprehensive?

Table 3 Quality assessment results

RP No.	Q1	Q2	Q3	Q4	Q5	Q6	Q7	Q8	Q9	Q10	Total
RP-1	1	1	1	1	0	1	1	1	1	1	9
RP-2	1	1	1	1	0	1	1	1	1	1	9
RP-3	1	1	1	1	0.5	1	1	1	1	1	9.5
RP-4	1	1	0.5	0.5	0	1	1	1	1	1	8
RP-5	1	1	1	1	0	1	1	1	1	1	9
RP-6	1	1	1	1	0.5	1	1	1	0.5	0.5	8.5
RP-7	1	1	1	1	1	1	0.5	1	1	1	9.5
RP-8	1	1	1	1	0	1	1	1	1	1	9
RP-9	1	1	1	1	0.5	1	1	0.5	1	0.5	8.5
RP-10	1	1	1	1	1	1	1	1	1	1	10
RP-11	1	1	1	1	0	1	1	0	1	1	8
RP-12	1	1	1	1	0	1	1	1	1	1	9
RP-13	1	1	1	1	0	1	1	1	1	1	9
RP-14	1	1	1	1	0	1	0.5	0	1	0.5	7
RP-15	1	1	0.5	0.5	0	1	0.5	1	1	1	7.5

illustrates the outcomes of the quality assessment. It seems that all the studies have been successfully accepted as they match the quality assessment. In short, all the studies are competent to be used for further analysis.

4 Results

Out of the total number (N = 510) of collected studies on the flipped classroom that involve the students' perception or acceptance, it seems that only a few number of studies (N = 15) have adopted the IS theory or model and these studies are listed in the Appendix.

4.1 *RQ1: What are the Research Aims, Adopted Methods, and Outcomes Achieved from the Analyzed Studies?*

The purpose of each study in the retained papers has been classified under three main themes, including: (1) evaluating the students' and teachers' perceptions towards the usage of flipped classroom with a theoretical model (11 studies), (2) integrating the flipped classroom with other technologies such as LINE and Web 2.0 (2 studies), and (3) assessing the relationship between the flipped classroom and achievement or motivation based on a theoretical model (2 studies). Figure 2 illustrates the distribution of the analyzed studies with regard to research aims.

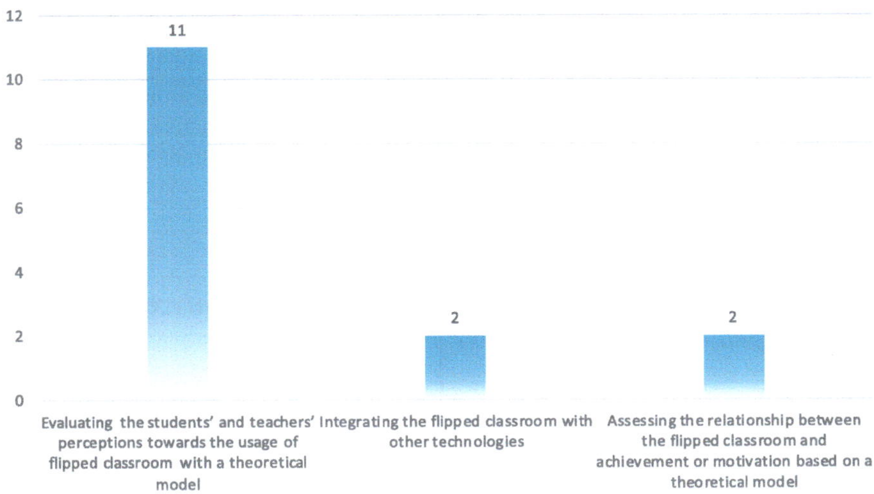

Fig. 2 Distribution of studies by research purpose

The distribution of research methods is classified in accordance with the number of methods used. This includes: (1) Model-evaluation based paper, and (2) Flipped classroom-evaluation method based. The first investigates the integration of a new model or theory in a flipped classroom environment, whereas the second type focuses on the impact, advantages, achievement, and motivation of students towards the flipped classroom. In accordance with the previously suggested evaluation, the reviewed papers were classified into two groups. The first group is one-method (11 studies), and the second is the multi-method studies (4 studies). Some papers have made use of only one method in their assessment, such as a survey, while others intended to use a survey along with interviews and focus groups discussion.

The research outcomes were classified into three different categories, including positive outcomes, neutral outcomes, and negative outcomes. The positive results were the highest. Most of the studies end up with recommendations that were in support of using the flipped classroom (13 studies). Only two studies come up with either negative or neutral results. The negative results stem from the fact that technology becomes a burden on the part of the students. The neutral results have both positive and negative results due to the use of technology.

4.2 RQ2: What are the IS Models or Theories used to Measure the Students' Perceptions?

Based on the results obtained from the analyzed articles, the studies were classified either as the TAM-based model (13 studies) or the non-TAM based-models (2 studies). These two types of classification refer to the fact that TAM was the most widely used model for evaluating the students' acceptance in the analyzed flipped classroom studies. The methods used in the TAM-based studies include surveys, interviews, and focus groups. The non-TAM based model was adopted whenever the UTAUT model or other models were used. The main purpose of these studies was either to integrate two different models or to investigate a relation between two factors, such as perception and motivation. The standard tool was the survey to evaluate the non-TAM based model.

4.3 RQ3: What are the Main Disciplines of the Analyzed Flipped Classroom Studies?

The disciplines of the analyzed studies reflect various academic branches where the flipped classroom has been applied. Based on the Appendix, the application of flipped classroom can be found in different disciplines like humanities, formal sciences, and applied sciences. This is an indicator that the flipped classroom technology has become very popular across various academic disciplines.

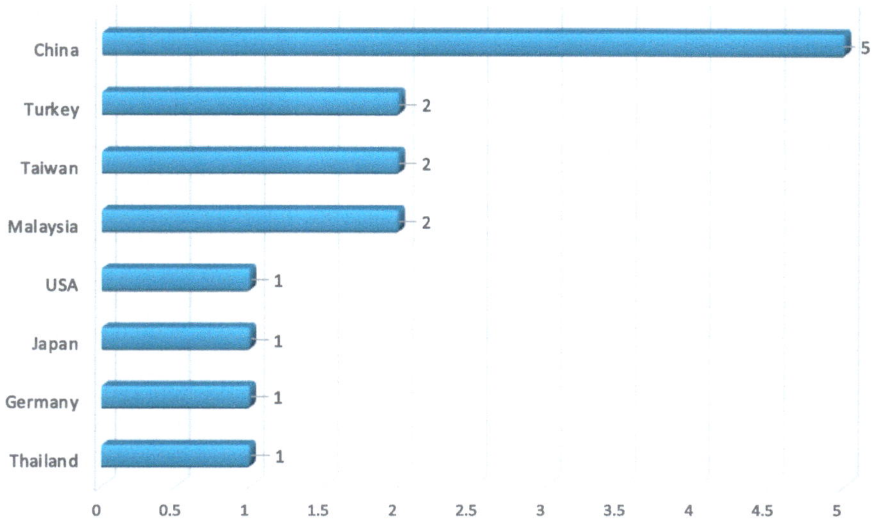

Fig. 3 Distribution of studies by country

4.4 *RQ4: What are the Common Countries that have Adopted the Flipped Classroom Technology?*

Figure 3 depicts the distribution of the analyzed studies with respect to the countries in which those studies were undertaken. It is evident that the majority of the examined flipped classroom studies were conducted in China (N = 5), followed by Turkey, Taiwan, and Malaysia with 2 studies each, and USA, Japan, Germany, and Thailand with 1 study each, respectively. These results contradict with the results observed in a previous study [17], in which most of the analyzed studies were undertaken in the USA and the Philippine.

4.5 *RQ5: How do the Results of this Systematic Review Differ from/or Similar to the Previous Systematic Reviews?*

The difference between this systematic review and other reviews lies in the fact that it does not limit its scope to a particular period, place, or field of study in terms of articles collection. It incorporates different periods, places, and fields of study. A review study conducted by Bergfjord and Heggernes [22] focused on the management field only. Other systematic reviews focused on the medical domain only [20, 21]. However, this systematic review has incorporated different fields of study, including English, built environment, programming, Marxist principles, and computer and instructional technologies. The critical factor that this study emphasizes is the use of a specific

IS model or theory, such as TAM or UTAUT, to assess the students' acceptance or perceptions.

This review study comes up with the result that the positive effects of the flipped classroom are higher than the negative effects. It also emphasized that though there are certain challenges, the flipped classroom is still considered to be an effective tool. On the contrary, a systematic review conducted by Karabulut-Ilgu et al. [19] presented different results, in which, most of the researchers in the area of engineering education emphasized the equal-learning opportunities and achievements. No difference is evident between students' achievement with or without the use of the flipped classroom. Thus, the flipped classroom has many advantages, but it comes up with some challenges for instructors and students. In the medical domain, it was argued that the flipped classroom had enhanced the students' satisfaction level with pre-class learning resources, and their satisfaction was extremely high [20, 21].

The similarities between the reviews lie in the approach used in the application of the flipped classroom. This study and previous studies emphasized the use of various research methods to achieve the results, such as using qualitative, quantitative, and mixed methods. Similarly, it seems that this review and previous reviews arrived at the conclusion that researchers implement various methodologies, fields of studies, and technology tools. In addition, the high percentage of acceptance is evident in this review and other reviews. The flipped classroom has proven to be fruitful for the students academically and motivationally as the students could study at home and come to the class well-prepared. Thus, this learning technology has contributed to the enhancement of students' learning achievement, motivation, engagement, and interaction.

5 Discussion

The main objective of this study is to go through various research articles that have used the flipped classroom technology in their teaching and learning environment. Then, the next step is to review and analyze those articles in a systematic process taking into consideration the students' acceptance and perceptions towards the flipped classroom usage in general, and to focus on the IS theories and models used as a complementary aim. This review study focuses on analyzing the studies with respect to various perspectives like research purposes, research methods, research outcomes, IS theories/models, disciplines, countries, and participants. Concerning the research purpose, three main themes were determined based on the analyzed studies, including: (1) evaluating the students' and teachers' perceptions towards the usage of flipped classroom with a theoretical model, (2) integrating the flipped classroom with other technologies such as LINE and Web 2.0, and (3) assessing the relationship between the flipped classroom and achievement or motivation based on a theoretical model. Based on these results, it has been observed that there is a lack of studying the impact of flipped classroom on students' academic performance. In addition, the integration of flipped classrooms with other technologies is still in its early stage and needs further investigation. Thus, future work should consider these issues while conducting further studies in the flipped classroom context.

With regard to the adopted methods, it seems that the use of "questionnaire surveys" is clearly evident as a basic method to collect the required data. This stems from the fact that these tools are competent enough to reflect the relationships among the factors in the conceptual models [32]. Moreover, it has been observed that interviews and focus groups were less considered while conducting the analyzed studies. Thus, further research should take this point into consideration. With respect to research outcomes, the majority of the analyzed studies achieved positive research outcomes. This is an indicator that the flipped classroom is a successful technique in the learning-teaching process. With regard to the IS theory or model, most of the analyzed studies were relied on developing and extending the TAM to examine the students' acceptance and perceptions towards the usage of flipped classroom. The main factors that are combined under this model enhance the usefulness and the effectiveness of the flipped classroom. The dominant usage of TAM refers to its simplicity, soundness, and applicability to work with different technologies [33, 34]. Further research should consider these results and attempt to study the applicability of other IS theories or models with the flipped classroom.

In terms of discipline, the application of flipped classrooms can be found in different disciplines, including humanities, formal sciences, and applied sciences. This is an indicator that the flipped classroom technology has become very popular across various academic disciplines. Further research could focus on the disciplines that were missed from the current analysis. Other areas of interest in this study are the distribution of the analyzed studies among different countries all over the world. The current study shows that China is the most common country where the flipped classroom has been the main area of interest to the majority of research articles. The other countries that follow China are Turkey, Taiwan, and Malaysia, respectively. This finding supports the fact that the flipped classroom is highly appreciated as a tool in the teaching and learning environments in China, and the researchers are highly attracted to newly adopted strategies of teaching in different disciplines. Concerning the participants who took part in the analyzed studies, it has been noticed that the majority of those participants were university students. This indicates that the flipped classroom is very common in higher educational institutions. In contrast, a few number of studies were undertaken in schools. Therefore, future studies should tackle the identified gap and start examining the effectiveness of flipped classrooms in school environments.

6 Conclusion and Future Work

Although there have been various studies that reviewed the use of flipped classrooms in different disciplines and try to give a comprehensive picture about the importance of this tool in each field, they fail to examine the relation between the flipped classroom as a learning tool and the IS theories or models. Therefore, the main objective of this study is to review and synthesize the flipped classroom studies related to IS theories or models. To put it more specifically, this review study sheds light on the research purposes, research methods, research outcomes, IS theories/models, disciplines, countries, and participants of the analyzed studies. The difference between this systematic review and the previous reviews lies in the fact that it doesn't limit its scope to a particular period, place, or field of study in terms of articles' collection.

This review study revealed seven new findings. First, evaluating the students' and teachers' perceptions towards the usage of flipped classrooms with a theoretical model was the main research purpose among all the analyzed studies. Second, questionnaire surveys were regarded as the dominant method for data collection. Third, the majority of the analyzed flipped classroom studies achieved positive research outcomes. Fourth, TAM is the most common model used for examining the students' and academics' acceptance and perceptions towards the usage of flipped classrooms. Fifth, the application of flipped classrooms was found in different disciplines, including humanities, formal sciences, and applied sciences. Sixth, China, Turkey, Taiwan, and Malaysia were found as the most common countries in which the flipped classroom plays an influential role in the teaching and learning environments. Seventh, university students were regarded as the most frequent participants who took part in the analyzed studies.

Although TAM was found as the most frequent model that has been used by various flipped classroom studies, there is still an open door for other IS theories or models to be examined with respect to the flipped classroom acceptance. To that end, future work should consider this gap and attempt to use other IS theories or models to study the students' and academics' acceptance and perceptions towards the usage of flipped classrooms. In short, the results of this systematic review afford a better understanding of the flipped classroom studies related to IS theories or models and form a constructive reference for future studies in the topic.

Like any other study, this study was limited in two ways. First, the study focused on collecting articles related to students' and academics' acceptance and perceptions only. Further research is encouraged to extend this study by eliminating this restriction in an attempt to include more studies. Second, this review study has considered some databases in terms of articles collection like IEEE, Springer, SAGE, ERIC, Science Direct, and Google Scholar. Future studies might focus on the analysis of flipped classroom studies in other databases like Web of Science and Scopus, among many others.

Appendix

Analysis of flipped classroom studies related to various IS theories or models

No.	Study	Year	Country	Discipline	Sample	Model	Method	Purpose of the study	Outcomes
RP-1	[16]	2017	Malaysia	Built environment	Undergraduate students	TAM + TTF	Survey	To examine students' usage of FC with the embedded TAM and TTF model	Positive results. Encourage to use FC
RP-2	[35]	2018	China	IT	Secondary School Students	TAM	Survey	To investigate the impact of the flipped classroom on students' acceptance of programming	Positive results. Encourage to use FC
RP-3	[36]	2017	China	Marxist principles	Teachers	UTAUT and Connected Classroom Climate (CCC)	Survey	To examine a model that integrates the (UTAUT) and (CCC) in FC instructional model	Positive results. Encourage to use FC

(continued)

(continued)

No.	Study	Year	Country	Discipline	Sample	Model	Method	Purpose of the study	Outcomes
RP-4	[37]	2017	Turkey	Not specified	University Students	TD	Survey	To investigate the structural relationships among the community of inquiry, transactional distance perception, and motivation	Positive results with certain exceptions
RP-5	[38]	2017	China	Statistical learning	University Students	TAM	Survey, individual interviews, and focus groups	To investigate the students' acceptance of the statistics learning platform	Positive results. Encourage to use FC
RP-6	[39]	2017	Taiwan	English language	University Students	TAM	Survey, semi-structured interview, and focus groups	To explore the benefits of the flipped classroom model for learners of English as a Foreign Language	Positive results. Encourage to use FC

(continued)

(continued)

No.	Study	Year	Country	Discipline	Sample	Model	Method	Purpose of the study	Outcomes
RP-7	[40]	2018	Germany	Project management	University Students	TAM	Survey	To present an evaluation instrument to assess the learning quality of flipped classroom courses	Neutral results. Attention needs to be paid to improve the model
RP-8	[41]	2017	Bangkok	Management information system	University Students	TAM	Survey	To examine the antecedents of engagement intention after using the flipped classroom	Positive results. All groups accepted the technology
RP-9	[42]	2017	China	Not specified	Secondary School students	TAM	Survey	To examine the students' technology acceptance, learning motivation, and self-efficacy in FC	All groups accepted the new learning system

(continued)

(continued)

No.	Study	Year	Country	Discipline	Sample	Model	Method	Purpose of the study	Outcomes
RP-10	[15]	2017	Taiwan	Not specified	University Students	TAM	Not specified	To examine the effectiveness of a flipped oral training instructional design incorporating LINE, based on the evaluation of how mobile-assisted flipped learning affected EFL learners' attitudes in light of TAM	Neutral, negative, and positive results have been gained when using different technologies
RP-11	[9]	2017	USA	Water and civilization	University Students	TAM	Interview	To examine the instructors' and students' attitudes towards their teaching and learning experiences in the FC sections	Positive results. Improving students' skills in problem solving and collaboration

(continued)

(continued)

No.	Study	Year	Country	Discipline	Sample	Model	Method	Purpose of the study	Outcomes
RP-12	[43]	2014	Malaysia	English language	University Students	TAM	Survey	To examine the students' perceptions towards the use of LMS at the Defence University	Negative results. LMS may become a burden to the students in their learning process.
RP-13	[44]	2016	Japanese	Educational studies	University Students	TAM	Survey	To identify the contents and structure of learners' perceived usefulness of "flipped learning"	Positive results. It enhances the effectiveness of the FC

(continued)

(continued)

No.	Study	Year	Country	Discipline	Sample	Model	Method	Purpose of the study	Outcomes
RP-14	[45]	2018	Turkey	IT	University Students	TAM	Open-ended Questions	To encourage the integration of Web 2.0-supported flipped learning in the LMS course and to provide the views of students regarding the model	Positive results. Findings indicated that flipped learning helps the students in sharing ideas and information with peers. It generates an active teaching-learning environment
RP-15	[46]	2014	China	Educational studies	University Students	TAM	Survey	To explore the students' achievement emotion in the flipped classroom setting	The students' achievement tends to be positive

References

1. Al-Emran, M., Arpaci, I., Salloum, S.A.: An empirical examination of continuous intention to use m-learning: an integrated model. Educ. Inf. Technol. (2020)
2. Al-Emran, M., Mezhuyev, V.: Examining the effect of knowledge management factors on mobile learning adoption through the use of importance-performance map analysis (IPMA). In: International Conference on Advanced Intelligent Systems and Informatics, pp. 449–458 (2019)
3. Al-Emran, M., Mezhuyev, V., Kamaludin, A.: An innovative approach of applying knowledge management in M-Learning application development: a pilot study. Int. J. Inf. Commun. Technol. Educ. **15**(4), 94–112 (2019)
4. Salloum, S.A., Alhamad, A.Q.M., Al-Emran, M., Monem, A.A., Shaalan, K.: Exploring students' acceptance of e-learning through the development of a comprehensive technology acceptance model. IEEE Access **7**, 128445–128462 (2019)
5. Al-Emran, M., Alkhoudary, Y.A., Mezhuyev, V., Al-Emran, M.: Students and educators attitudes towards the use of M-learning: gender and smartphone ownership differences. Int. J. Interact. Mob. Technol. **13**(1), 127–135 (2019)
6. Zainuddin, Z., Halili, S.H.: Flipped classroom research and trends from different fields of study. Int. Rev. Res. Open Distance Learn. (2016)
7. Bradford, M., Muntean, C., Pathak, P.: An analysis of flip-classroom pedagogy in first year undergraduate mathematics for computing. In: Proceedings—Frontiers in Education Conference, FIE (2015)
8. Wasserman, N.H., Quint, C., Norris, S.A., Carr, T.: Exploring flipped classroom instruction in calculus III. Int. J. Sci. Math. Educ. (2017)
9. Long, T., Cummins, J., Waugh, M.: Use of the flipped classroom instructional model in higher education: instructors' perspectives. J. Comput. High. Educ. (2017)
10. Gilboy, M.B., Heinerichs, S., Pazzaglia, G.: Enhancing student engagement using the flipped classroom. J. Nutr. Educ. Behav. (2015)
11. Simpson, V., Richards, E.: Flipping the classroom to teach population health: increasing the relevance. Nurse Educ. Pract. (2015)
12. Liebert, C.A., Mazer, L., Bereknyei Merrell, S., Lin, D.T., Lau, J.N.: Student perceptions of a simulation-based flipped classroom for the surgery clerkship: a mixed-methods study. In: Surgery (United States) (2016)
13. Wanner, T., Palmer, E.: Personalising learning: exploring student and teacher perceptions about flexible learning and assessment in a flipped university course. Comput. Educ. (2015)
14. Kong, S.C.: Developing information literacy and critical thinking skills through domain knowledge learning in digital classrooms: an experience of practicing flipped classroom strategy. Comput. Educ. (2014)
15. Chen Hsieh, J.S., Huang, Y.M., Wu, W.C.V.: Technological acceptance of LINE in flipped EFL oral training. Comput. Human Behav. (2017)
16. Ajayi, I.H., Iahad, N.A., Ahmad, N., Yusof, A.F.: A conceptual model for flipped classroom: influence on continuance use intention. In: International Conference on Research and Innovation in Information Systems, ICRIIS (2017)
17. Uzunboylu, H., Karagözlü, D.: The emerging trend of the flipped classroom: a content analysis of published articles between 2010 and 2015. RED. Rev. Educ. a Distancia, pp. 1–13 (2017)
18. Kerr, B.: The flipped classroom in engineering education: a survey of the research. In: 2015 International Conference on Interactive Collaborative Learning (ICL) (2015)
19. Karabulut-Ilgu, A., Cherrez, N.J., Jahren, C.T.: A systematic review of research on the flipped learning method in engineering education. Br. J. Educ. Technol. (2017)
20. Ramnanan, C.J., Pound, L.D.: Advances in medical education and practice: student perceptions of the flipped classroom. Adv. Med. Educ. Pract. (2017)
21. Chen, F., Lui, A.M., Martinelli, S.M.: A systematic review of the effectiveness of flipped classrooms in medical education. Med. Educ. (2017)

22. Bergfjord, O.J., Heggernes, T.: Evaluation of a 'flipped classroom' approach in management education. J. Univ. Teach. Learn. Pract. (2016)
23. Choi, J.: Critical review of empirical studies on the flipped classroom in english education in Korea. English Lang. Lit. Teach. **22**(4), 1–19 (2016)
24. Lo, C.K., Hew, K.F.: A critical review of flipped classroom challenges in K-12 education: possible solutions and recommendations for future research. Res. Pract. Technol. Enhanc. Learn. (2017)
25. Kitchenham, B., Charters, S.: Guidelines for performing systematic literature reviews in software engineering. Softw. Eng. Group, Sch. Comput. Sci. Math. Keele Univ. 1–57 (2007)
26. Al-Saedi, K., Al-Emran, M., Abusham, E., El_Rahman, S.A.: Mobile payment adoption: a systematic review of the UTAUT model. In: International Conference on Fourth Industrial Revolution (2019)
27. Saa, A.A., Al-Emran, M., Shaalan, K.: Factors affecting students' performance in higher education: a systematic review of predictive data mining techniques. Technol. Knowl. Learn. (2019)
28. Al-Qaysi, N., Mohamad-Nordin, N., Al-Emran, M.: A systematic review of social media acceptance from the perspective of educational and information systems theories and models. J. Educ. Comput. Res. **57**(8), 2085–2109 (2020)
29. Al-Emran, M., Mezhuyev, V., Kamaludin, A., Shaalan, K.: The impact of knowledge management processes on information systems: a systematic review. Int. J. Inf. Manage. **43**, 173–187 (2018)
30. Moher, D., et al.: Preferred reporting items for systematic reviews and meta-analyses: the PRISMA statement. PLoS Med. (2009)
31. Budgen, D., Brereton, P.: Performing systematic literature reviews in software engineering. In: Proceeding of the 28th international conference on software engineering—ICSE '06 (2006)
32. Al-Emran, M., Mezhuyev, V., Kamaludin, A.: PLS-SEM in information systems research: a comprehensive methodological reference. In: 4th International Conference on Advanced Intelligent Systems and Informatics (AISI 2018), pp. 644–653 (2018)
33. King, W.R., He, J.: A meta-analysis of the technology acceptance model. Inf. Manag. **43**(6), 740–755 (2006)
34. Al-Emran, M., Mezhuyev, V., Kamaludin, A.: Technology acceptance model in m-learning context: a systematic review. Comput. Educ. **125**, 389–412 (2018)
35. Yan, O.S., Cheng, G.: Exploring the impact of flipped classroom on students' acceptance of programming in secondary education. In: Proceedings of 2017 IEEE International Conference on Teaching, Assessment and Learning for Engineering, TALE 2017 (2018)
36. Yang, H.H., Feng, L., MacLeod, J.: Understanding college students' acceptance of cloud classrooms in flipped instruction: integrating UTAUT and connected classroom climate. J. Educ. Comput. Res. (2017)
37. Karaoglan Yilmaz, F.G.: Predictors of community of inquiry in a flipped classroom model. J. Educ. Technol. Syst. (2017)
38. Song, Y., Kong, S.-C.: Investigating students' acceptance of a statistics learning platform using technology acceptance model. J. Educ. Comput. Res. (2017)
39. Chen Hsieh, J.S., Wu, W.C.V., Marek, M.W.: Using the flipped classroom to enhance EFL learning. Comput. Assist. Lang. Learn. (2017)
40. Vogelsang, K., Hoppe, U.: Development of an evaluation for flipped classroom courses, pp. 821–832 (2018)
41. Thongmak, M.: Flipping MIS classroom by Peers: gateway to student's engagement intention. In: Proceedings of the 26th International Conference on World Wide Web Companion, pp. 387–396 (2017)
42. Zheng, J., Li, S., Zheng, Y.: Students' technology acceptance, motivation and self-efficacy towards the eSchoolbag: an exploratory study. Int. J. Infonomics (2017)
43. Juhary, J.: Perceived usefulness and ease of use of the learning management system as a learning tool. Int. Educ. Stud. (2014)

44. Yoshida, H.: Perceived usefulness of 'flipped learning' on instructional design for elementary and secondary education: with focus on pre-service teacher education. Int. J. Inf. Educ. Technol. (2016)
45. Tosun, N.: Implementation of web 2.0-supported flipped learning in the learning management systems course: an experience from Turkey. Malays Online J. Educ. Manag. (2018)
46. Luo, J.T., Sun, M., Wu, B., Gu, X.Q.: Exploring the effectiveness of a flipped classroom based on control-value theory: a case study. In: Workshop Proceedings of the 22nd International Conference on Computers in Education, ICCE 2014 (2014)

Perceptions and Barriers to the Adoption of Blended Learning at a Research-Based University in the United Arab Emirates

Rawy Thabet, Christopher Hill, and Eman Gaad

Abstract This study investigates the academic staff's perception of blended learning adoption in higher education (tertiary level) as well as the barriers to this adoption. To this end, 6 academic staff, joining the faculty of education, in one of the research-based universities in the UAE, completed a self-administered survey. A mixed methods approach was used to collect both quantitative and qualitative data. The results suggest that the academic staff are fully aware of the importance of integrating blended learning. They are also ready to adopt any innovation that would help students advance in their study provided that the adoption and provision of these technologies are accompanied by professional training. The results also show that the only barrier that inhibits the adoption of technology is the lack of a clear policy on this integration and also the lack of blended learning-enabling software. This research paves the way for more research on the integration of blended learning at the tertiary level. It also raises some recommendation to the management of the university, where the study was conducted, to adopt more advanced features of Blackboard such as Blackboard collaborate.

Keywords Blended learning · Adoption · Barriers · Tertiary education · UAE

1 Introduction

The integration of Information and Communications Technology (ICT) into instruction has become more critical than before because of the desired changes that this new method has added to the modes of delivery [1]. ICT policy has been widely implemented by higher education institutions all over the world, for example, as per

R. Thabet (✉) · C. Hill · E. Gaad
The British University in Dubai, Dubai, UAE
e-mail: Rawy.sabet@buid.ac.ae

C. Hill
e-mail: Christopher.hill@buid.ac.ae

E. Gaad
e-mail: Eman.gaad@buid.ac.ae

M. Al-Emran et al. (eds.), *Recent Advances in Intelligent Systems and Smart Applications*, Studies in Systems, Decision and Control 295,
https://doi.org/10.1007/978-3-030-47411-9_16

a survey conducted by National Postsecondary Education Cooperative (NPEC), 61% of tertiary institutions offer some kind of online modules [2]. However, in some parts of the world, the advantages of integrating ICT and online learning are still underestimated. This underestimation or lack of adoption is usually attributed to some institutional, economic or attitudinal factors [3].

It is also important to note that education has gone through many different phases of transformation. It started with very basic tool and is currently witnessing many advances such as technology integration. Among the latest trends in innovative teaching is the use of blended learning (BL) which means using both face-to-face and online (or virtual) teaching [4, 5].

This study focuses on a postgraduate research university in the United Arab Emirates (UAE). The university is accredited by the Ministry of Higher Education in the UAE and it has some affiliation with international universities. The education system in the university is based on offering taught modules for each programme. Students are required to achieve 70% of the classes offered. All course materials for all programmes are offered through learning management system (LMS) called Blackboard. This platform incorporates numerous functions and capabilities which would allow any institution to run its courses online without any physical presence of students [6]. This study intends to investigate the available opportunities for technology integration into the education system at this university. Furthermore, the study intends to find what would be the barriers that prevent the academic staff from using Blackboard to its full capacity.

The current study plans to fill a gap in literature as there is a scarcity of research in this area. Many studies were carried to find the effect of technology integration on education, but the majority of these studies focused on pre-tertiary level in the United Arab Emirates, apart from few studies such as [7, 8]. This current study also generates some recommendations for the university management on how to better integrate technology.

In addition to identifying the hurdles that prevent the implementation of ICT into higher education, the study also intends to find the perspective of the faculty members towards the implantation of ICT. To this end, the following questions have been developed.

1. Is there any form of blended learning happening at the university? And if so, what is it?
2. What are the barriers to technology integration into instruction at the university?

The argument in this paper proceeds as follows. In the first part, blended learning and its value is explored from the perspective of some academic staff working at a higher education institution. In the second part, barriers of integrating blended learning are identified. Finally, some recommendations for improving services and removing barriers are raised to the university management.

2 Literature Review

Integrating ICT in education is one of the current trends and there is a good number of studies from literature that discussed this topic in details. However, there is an apparent need to run this study at the university which is used as a case study. The urgency of this need arises from the lack of knowledge about this topic among teaching staff and the potentials that this new method has for improving the quality of delivery at the university. Studies from literature are discussed as follows. First, blended learning is defined then the researcher explores the challenges that are available in other institutions and how they are different or similar to the current case. After that, technology status in the UAE in general and in the university, in particular, is investigated. Then, some of the theoretical underpinnings for technology integration are discussed. Finally, the attitude of academics towards this trend is explored.

Blended learning (BL) is variedly defined. The variability in defining BL is attributed to the different methods through which the instructional content is designed and delivered. According to Valiathan [9], BL is defined as a mixture of different instructional methods such as Electronic Performance Support Systems (EPSS) and collaboration software. When defined BL, Boelens et al. [10], Mitchell and Honore [11] explained that it is a combination of activities that take place inside classroom and the learning that happens online (e-learning) as well. Mitchell and Honore believe that learning is so complicated that it cannot be limited to classroom and this was their reason for including the e-learning feature. Munro and Rice-Munro [12] concluded that many of us have been exposed to this type of e-learning which has many advantages such as low cost; the accessibility of material all the time; the ability of employees to simultaneously get trained and carry out their normal activities at work. Dalmolin et al. [13] explained that BL has an advantage of combining the best of traditional teaching in addition to the integration of communication and information technology. Despite the fact that there is variability in the definitions of BL, given by different scholars, there are some common factors among these definitions. They all call for combing different instructional modalities (medium) which usually include five categories i.e. in person; blended; online asynchronous; online synchronous; distant learning. They all also call for combining different instructional methods [14]. Finally, they all recommend mixing online and offline instruction [15].

2.1 Barriers to Blended Learning Integration

Teaching using technological methods is one of the way to achieve accessibility to education for all deprived students who are usually from different parts of the world. Additionally, being educated through this innovative way of teaching changes students into lifelong learners. Moreover, in some countries where there are cultural restrictions which prevent men to mingle with women in the educational environments, this innovative way of learning would remove these restrictions and allow both genders to equally participate in the same learning environment [16, 17].

Although this innovative method of teaching has many advantages, there are some challenges that hinder its implementation on a wider spectrum [18]. However, integrating technology is not a 'quick fix' technique as it should be done in a gradual way over a long period of time. The gradual implementation of technology was highly considered by the Indonesian Ministry of Education when developed the Indonesian Qualification Framework (IQF). This IQF included nine levels and at each level, there are specific tasks to be carried out by teachers who are at these levels, for example, at levels six to eight, it is expected from the teachers with bachelor or master degree to implement technology in their teaching while for teachers who are at lower level, it is not expected from them to adopt technology [19].

Barriers to the integration have always been a frequent topic in literature. Some of the main barriers identified by literature are the low competence of teaching staff, insufficient support (whether financial or technical) from the workplace management, and the lack of time [20–22]. Researchers have differently categorised the barriers into different taxonomies. However, with deeper investigation, it can be easily recognised that those researchers titled the barriers using different names but in essence, all these different titles referred to the same thing. As an evidence, Ertmer [20] divided the integration barriers into first-order and second-order. By first-order, he meant the incremental and institutional barriers. When he defined second order barriers, he meant the fundamental and personal barriers. Coleman and Mtshazi [23] referred to the first order as internal and to the second order as external factors. However, the work of Coleman and Mtshazi has an advantage over Ertmer as the former gave more details about both the internal and external factors. By external factors they meant "insufficient technical support, lack of hardware, software and no recognition for using online teaching" (p. 32).

2.2 *The Status of Technology Worldwide and Locally*

The World Conference held in Dakar stressed the importance of education as a means to the development and prosperity for the different populations. One of the core values this conference adopted is the right of education for all (EFL). To get the most benefit from the education systems, new technologies should be harnessed. One of the ways of harnessing technology is to integrate it into education [24]. In the same context, the Global Competitiveness Report, which is issued by the World Economic Forum (WEF), compares the different countries based on twelve pillars. One of these pillars is the technological readiness. According to this report, in 2010, the UAE was taking the 14th place worldwide; however, in 2012, the country position dropped to 32nd worldwide (see Table 1). This drop could be attributed to the fact that the development rhythm or pace of the other countries was faster than the UAE [25]. It could also be argued that the 12 pillars are interrelated which means if a

Table 1 UAE readiness for technology

	2012	2011	2010	2009	2008	2007	2006
Technological readiness	32	30	14	17	28	33	35

country scored low in one of these pillars, this low score affects the scores of the other pillars [26, 27]. However, according to the latest WEF [28]'s report, the UAE is now taking the 2nd place globally after South Korea in adopting the information and communication technology.

According to the WEF [27]'s report, the UAE had a plan to spend about 3.3 billion dollars on health and education sectors. The budget allocated for education was to be spent on purchasing ICT hardware that could be used by schools. The report also showed that the high income of a country does not positively correlate with having advanced technologies as there are some countries with high per capita income such as Kuwait and Kingdom of Saudi Arabia while they occupy low positions compared to other countries with lower income. Ahmed and Al-Roubaie [29], in justifying the low rank of some countries, claimed that technology integration or usage is affected by the culture of this country.

This case study focuses on a non-profit university. Its main source of income is confined to the education fees collected from students and some financial subsidiaries donated by some sponsors. The university offers an on-campus instruction with very limited opportunities for blended or e-learning. Student, joining the university, have to physically attend classes to pass modules. The platform used by academics to communicate with students is Blackboard. Although this platform has many capabilities, it is not used to its full capacity. According to Carnevale [30], in 2005, Blackboard was occupying 45–50% of the market of learning management systems. In 2006, this percentage of Blackboard usage in the surveyed United States universities increased to 70%. However, in the subsequent years, the percentage dropped to 57 due to the competition from other software developer such as 'Desire2Learn' and 'Moodle' [31]. The literature on the usage of Blackboard in the UAE lacks these important statistics.

2.3 *Theoretical Underpinnings*

Resta [32] developed a framework to identify the purposes of integrating ICT into teacher education. He devised four ICT competencies. Example of these competencies are "content and pedagogy; technical issues…" (p. 41). These four areas of competencies are supported by four themes. These are examples of these themes; "leadership and vision; lifelong learning…" (p. 41) (see Fig. 1).

The cultural theme in this framework is very important as it recommends paying due attention to the culture when integrating technology into the curriculum of teacher education. In other words, before integrating technology, the appropriateness of this

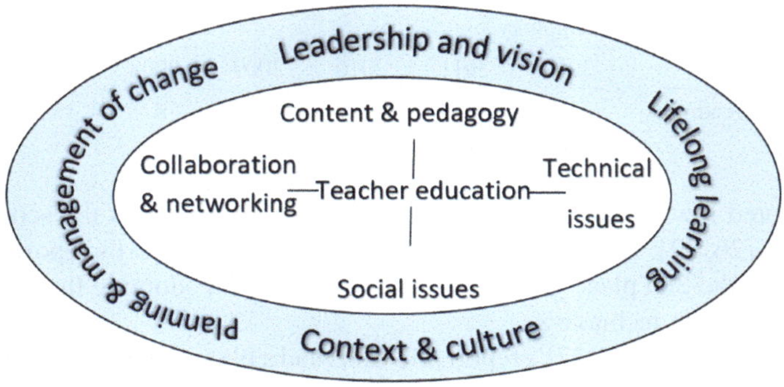

Fig. 1 ICT integration framework: [32, p. 41]

technology to the culture should be considered. The lifelong learning theme is also equally important and this idea of continuous learning should be instilled in the minds of teachers.

In the same vein, the advent of technology and its advanced devices made it necessary for scholars to develop some theoretical frameworks in order to understand ICT and its diffusion and implementation [33]. According to Al-Senaidi et al. [8], the most influential theories were 'Diffusion of Innovation'; 'Technology Acceptance Model (TAM)'; and the third one was the 'five-step model for infusing technology into instruction'. The TAM postulates that there are four aspects that impact the person's usage of technology. The first aspect is 'behavioural intention (BI)' which means the likelihood of a person whether s/he use the technology or not. The second aspect is the 'perceived usefulness'. The third is the 'perceived ease of use'. The fourth is 'attitude'. There are casual links among these five factors/aspects [34].

2.4 Perception Towards ICT Integration

Of course, the instructors' attitude towards this integration is an important factor and sometimes represents a key challenge [4, 35].

In Oh and Park's [4] study, most of the participants had a positive standpoint towards the integration of blended learning as they believe that this innovative method improved the quality of their teaching. The word 'most' is somehow vague, and it would have been better if the authors mentioned the exact number of participants who had positive attitude towards blended learning. In the same study, it was also reported that the universities, to which the participants belong, provided them with all possible resources such as instructional designers and technical support. However, in another study conducted by Al-Senaidi et al. [8] in one of the governmental university in Oman, concluded that academic staff had negative attitude towards the integration of technology and they attributed this negativity to the lack of resources and support. Therefore, it can be easily concluded that some external factors such as lack of support plays an important role in forming the attitude of teaching staff towards the integration of technology.

In their endeavour to bridge the gap of having little research written on institutional adoption of technology, Porter and Graham [36], developed a framework that would help the different institution to adopt technology. There are two main features of this framework. The first is the gradualness of technology adoption as, according to this frame, the implementation process starts with spreading awareness about this technology: then an initial implementation is achieved: finally, the mature implementation takes place. Porter and Graham's framework also identified the strategies (e.g. policies and purposes of implantation), structure (e.g. administrative framework) and support (e.g. technical support) that the institutions tackle at each stage of the three stages mentioned above. In addition to their proposed framework, Porter and Graham also use Roger's model i.e. Diffusion of Innovation to decide the extent to which the institution's strategy, structure and support help facilitate or hinder the adoption of BL among higher education faculty. One of the main principles that Roger's model is based on is the categorisation of innovation adopters into five groups which are the 'innovators, early adopters, early majority, late majority, laggards'. It can be easily discerned that these five categories are in descending order, for example, people who fall under the 'innovators' category are considered the first to use any new innovation while people in 'laggards' category are considered that last to adopt technology and they adopt it when they are forced to do so. Both models of Porter and Graham and Roger depend on the gradualism of technology adoption. Porter and Graham found that the availability of technological support and adequate infrastructure significantly correlate with the adoption of technology by faculty.

It seems that the idea of categorising technology adopters into different groups appealed to many researchers. Bensona et al. [37], for example, interviewed 16 faculty members and from these interview, they could identify three approaches towards blended learning. These approaches are 'bolts on', 'technology is all' and 'purely pedagogic'.

'Technology is all' approach includes a group of a small number of academic staff who are eager to use any technology available for them regardless of its importance for students. The 'bold-on' group seems more relatively reasonable in their use of technology as they consider it as an added advantage which can be used to improve teaching but not to replace teachers nor to be given preference to pedagogy. However, this group sometimes show reluctance to learn how to use or adopt a new technology. To overcome this kind of reluctance, Powell et al. [38] suggest giving those reluctant teachers the opportunity to express their opinions and perceptions about these new technologies. Those reluctant people could also be encouraged to suggest modification. The third group, 'purely pedagogic', tends to keep a balance between the importance of pedagogy and technology. They do not show any reluctance to adopt technology provided that it can help students learn unlike the 'Technology is all' whose main priority is to use technology regardless whether it would help students or not.

The lack of training and professional development are two main barriers identified by many studies, for example, Aldosemani et al. [39], identified similar barriers such as the need for continuous training and also the need for workshops which should cover technology-related topics such as e-learning and learning management systems.

3 Methodology

The approach adopted by this study is the mixed methods as the study includes quantitative data collected from surveys with close-ended responses and qualitative data collected by interviewing some of the participants. The mixed method is an ideal research method as it combines both the advantage of both methods namely the qualitative and quantitative [40]. The Statistical Package for the Social Sciences (SPSS- version 23) was used to analyse quantitative data while the qualitative data collected from interviews were analysed thematically.

A self-reported survey has been used. This tool is adopted from some surveys found in literature such as Rizvi et al. [41], Yilirim et al. [42]. The adopted survey was uploaded to Google Documents and a link was emailed to the targeted sample.

The survey content has been validated using face-validity which means the degree to which the experts judge whether the components of the survey suit the objectives of this survey [43]. To achieve the face-validity of the survey, it has been presented to one of the authorities on technology integration to judge whether the survey really measures what it intends to measure [44].

As for the sample, it consisted of six participants. These participants were academics, joining the faculty of education. The targeted population was the whole academic at the university (31 in total). However, as per Slovin's formula [45], this small sample is big enough to represent the whole population of the university. As per this formula, 'n' stands for the size of the sample, 'N' stands for the targeted population and 'e' is the margin of errors which is usually considered as either 0.05 or 0.001. This method of calculating the sample size was scrutinized by some scholars due to its simplicity [45]. Slovin's formula is calculated below. According to this formula, 4 participants are representative of the 31 academics.

$$
\begin{aligned}
n &= N/\left(1 + Ne^2\right) \\
n &= 31/\left(1 + 31 \times 0.5^2\right) \\
n &= 31/(1 + 7.75) \\
n &= 31/(8.75) \\
n &= 3.5428 \text{ sampling}
\end{aligned}
$$

The survey had three sections. The first section included some demographic data. The second section tried to find out about the participants' perspective towards the

Table 2 Demographic data of participants

Variable	Values	Frequency	Percent
Gender of participants	Male	6	100.0
Age of participants	30–44 years old	3	50.0
	45 or above	3	50.0
	Total	6	100.0
Academic rank of participants	Assistant professor	2	33.3
	Associate professor	2	33.3
	Professor	2	33.3
	Total	6	100.0

integration of technology into instruction. The third included some of the expected obstacles/barriers for integrating blended learning. In this section, the participants had to rate their level of agreement on a scale of five points (viz strongly agree (= 1) and strongly disagree (= 5)).

3.1 *The Study Participants*

The surveyed participants are academics at the university under review. The response rate was high (6 out of 7) as all of the targeted participants completed the survey except one. As for the gender, all of the respondents were male. When asked about their age group, 50% reported that they are between 30 and 44 and the other half were 45 or above. They differ in their academic ranks as 33% (2) were professors; 33% were associate professors; and 33% were assistant professors (see Table 2).

4 Findings

4.1 *Quantitative Findings*

As mentioned earlier, the qualitative data were retrieved from the interview and analysed using thematic analysis [46] while the quantitative data were collected from the survey close-ended questions and analysed using SPSS.

The first question in the second section of the survey asked participants whether they know the meaning of BL. To this question, all of them reported that they know what BL means. The second question asked about the frequency of using BL. The participants were asked to choose one from five adverbs of frequency with 'always' as

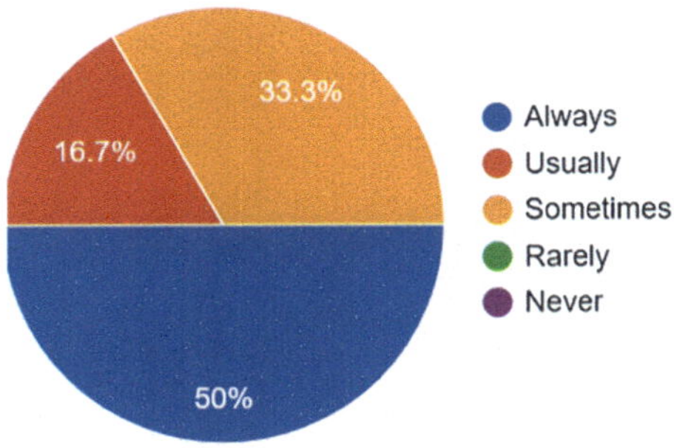

Fig. 2 Frequency of BL use

the first option and denotes highest frequency while 'Never' the last and denotes the least frequency. Fifty percent of the participants confidently admitted they 'always' used BL to deliver their courses. Thirty-three percent of participants reported that they 'sometimes' use BL and the remaining seventeen percent admitted that 'usually' use BL (see Fig. 2).

The third question in this section asked participants about their view of the value of blended learning, to which one hundred percent agreed that BL positively improved their ability to teach. The fourth question in this section asked participants to tick the Blackboard feature that they use and also to say how frequently they use it. The most two agreed features were 'Turnitin direct assignment' and 'Discussion Board' as five out of the six participants (83%) agreed that they 'always' and 'usually' use these two features respectively. The most disagreed feature was 'grade centre' as each one of them selected different adverb of frequency. As shown by Fig. 3 below, when participants rated 'grade centre' each one of the adverb of frequency was used.

When rated their frequency of use of 'wikis', two of them (40%) admitted that they never used this feature. The responses of the remaining three participants were 'usually' (20%), 'sometimes' (20%) and 'rarely' (20%). The last feature that the participants responded to is 'Journals'. Three of them (60%) agreed that they 'always' use 'Journals'. The remaining 40% of the participants responded negatively as they admitted that they 'rarely' and 'never' used this feature.

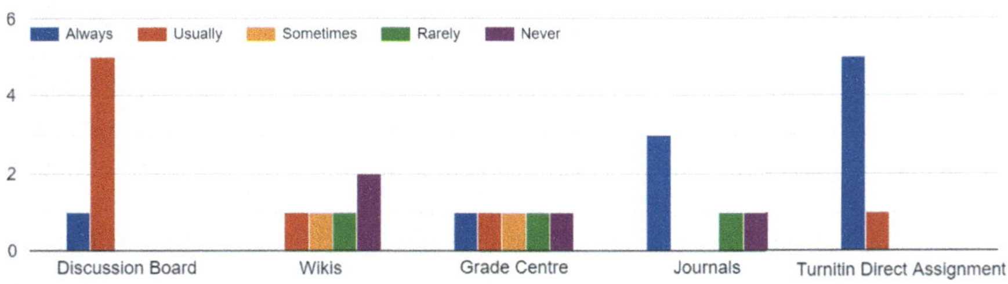

Fig. 3 Blackboard features and their frequency of use

Table 3 Basic knowledge

I do not have the basic knowledge		
	Frequency	Percent
Strongly agree	1	16.7
Disagree	2	33.3
Strongly disagree	3	50.0
Total	6	100.0

Table 4 Unreliability and enough time

Technology is unreliable			I do not have enough time to adopt technology		
	Frequency	Percent		Frequency	Percent
Neutral	1	16.7	Neutral	1	16.7
Disagree	2	33.3	Disagree	2	33.3
Strongly disagree	3	50.0	Strongly disagree	3	50.0
Total	6	100.0	Total	6	100.0

The third section of the survey was about the barriers that the academics face when they try to implement BL. For each statement, the participants had to rate their agreement on a scale of five points where '1' means 'strongly agree' and '5' means 'strongly disagree'. To increase readability and reduce the number of figures in the paper, close responses such agree and strongly agree are added up [47].

The first statement in this section was read as 'I do not have the basic knowledge required to integrate technology…'. Surprisingly one participant strongly agreed with this statement. Fifty percent and thirty-three percent of the participants strongly disagree and disagree respectively (see Table 3).

When responded to the statement which says 'technology is unreliable', the majority of them (83.3% = 50% strongly disagreed + 33.3% disagreed) disagreed. Only 16.7% were indecisive as they selected 'neutral' option (see Table 4).

The statement about having enough time to adopt technology received the same degree of disagreement and indecisiveness that the previous statement received.

When asked to judge the inappropriateness of time during which the university offered training, 50% were neutral and 33% (= 16.7% + 16.7%) disagreed to the statement. Only 16.7 agreed that the time is not convenient (see Table 5).

The responses to the three 'lacks' that the survey included varied widely, for example, while fifty percent agreed that there is a lack of administrative support, only around seventeen agreed that there is a technical support lack (see Table 6). When the participants responded to the statement about the financial support lack, 50% agreed (= 16.7 strongly agree + 33.3% agreed) agree to this statement. The degree of indecisiveness about the three lacks was almost the same. When they responded to the lack administrative support, 16.7% (1) were neutral. However, when responded to the other two lacks namely technical support and financial support, the level of

Table 5 Convenience of training time

The university does not provide convenient time for training		
	Frequency	Percent
Agree	1	16.7
Neutral	3	50.0
Disagree	1	16.7
Strongly disagree	1	16.7
Total	6	100.0

Table 6 Admin and technical support

Lack of administrative support			Lack of technical support for technology		
	Frequency	Percent		Frequency	Percent
Agree	3	50.0	Agree	1	16.7
Neutral	1	16.7	Neutral	2	33.3
Disagree	1	16.7	Disagree	2	33.3
Strongly disagree	1	16.7	Strongly disagree	1	16.7
Total	6	100.0	Total	6	100.0
Lack of adequate financial support					
	Frequency	Percent			
Strongly agree	1	16.7			
Agree	2	33.3			
Neutral	2	33.3			
Strongly disagree	1	16.7			
Total	6	100.0			

indecisiveness increased almost to the double (33.3% of them were neutral). The level of disagreement to the lack of administrative support, technical support and financial support varied widely; these levels of disagreement were 33.4%, 50% and 16.7% respectively.

When responded to the statement that says 'Technology does not fit well…', 16.7% were neutral and 83.3% disagreed with the statement. However, when responded to the statement which says 'I am not interested in using..', all of them disagreed (Table 7).

The survey items were divided into two main factors which are the 'knowledge and perspective' and 'perceived barriers for technology adoption'. SPSS was used to confirm that the items under each factor can load together to serve the same construct. The items under each factor were also computed together as one variable, namely, all items related to knowledge were added up and the same step was taken with the items under 'barriers'. Each one of the new variables i.e. 'knowledge' and 'barriers' was correlated to the other independent variables such as age and academic rank. Table 8

Table 7 Suitability for the courses and the degree of interest

Technology does not fit well for the courses taught			I am not interested in using technology in teaching and learning		
	Frequency	Percent		Frequency	Percent
Neutral	1	16.7	Disagree	1	16.7
Disagree	3	50.0	Strongly disagree	5	83.3
Strongly disagree	2	33.3	Total	6	100.0
Total	6	100.0			

Table 8 Correlations

		Academic rank of participants	Knowledge2
Academic rank of participants	Pearson correlation	1	−0.552
	Sig. (2-tailed)		0.335
	N	6	5
Knowledge2	Pearson correlation	−0.552	1
	Sig. (2-tailed)	0.335	
	N	5	5

below shows an example of how academic rank correlated with knowledge. It is very clear that there is a negative correlation between these two variables. However, it is not a significant one as $p = 0.335$ which is greater than 0.05. The small size of the sample could be one of the reasons for having this insignificant correlation.

4.2 *Qualitative Findings*

When the participants were interviewed, they were asked one main question which was 'what would encourage you to do more blended learning?'. Their answers came as follows.
Theme 1: Adding new software

> Adding the necessary features to Black Board, such as the "Collaborate" feature
>
> Enough resources to support blended learning; make software available to all students and academics e.g. NVIVO; SPSS; Adobe connect to hold online lessons
>
> Greater off campus access
>
> less teaching load, more supplement resources including software, online adequate library,….etc.

Theme 2: Students' perspective

> We need students' perspectives about this. If it's something they want, perhaps more faculty will adopt it

5 Discussion of the Results

This study tried to find how knowledgeable faculty members are about technology integration. It also tried to identify some of the barriers for the implementation of BL. The positive responses of the faculty members show that they have positive attitude towards integrating technology into instruction. They admitted that they have the basic knowledge which is adequate enough to let them run blended instruction. However, the results showed that the main hurdles to implement technology are the insufficient administrative and financial support. These findings are in line with Al Khawaldi [48] and Al Musawi [49].

Looking back at the results, it can be easily seen that there are three survey items which received a 100% level of agreement. These items were as follows. All of the participants know the meaning of blended learning; all of them agree that blended learning has positively impacted their teaching; all of them are interested in using technology. These 100% results suggest that the participants have positive standpoint towards the integration of technology into instruction. This finding can also be corroborated by the high level of disagreement to items no. 9 (technology is unreliable) and 13 (technology does not fit the courses taught). Each one these two items received 83.3% of disagreement. However, the other findings from survey especially the ones about barriers, show that there are some hurdles that prevent them from fully implementing technology integration. The lack of financial support and administrative support reached the same level of agreement as both of these two items received 50% of agreement. However, 50% of the same participants acknowledged that there is no problem with the technical support provided by the university. These findings are somehow contradicting and questionable. The researcher believes that the three 'lacks' i.e. lack of admin support, lack of technical support and lack financial support got confused by the faculty members. This can be supported by the low frequency of use of some Blackboard features such as 'wikis' for which 40% of the participants admitted that they never used. If there was an adequate amount of technical support, some Blackboard features such as 'grade centre' would have been used more frequently than what was reported in this study. The most frequently used features were 'discussion board' and 'Turitin assignment directory'. The latter is one of the requirements for assignment submission as each student has to submit his/her final assignment to 'Turnitin' for plagiarism detection. Therefore, each faculty has to be aware of this feature. There is no specific training that the new faculty receives on how to use this feature or the other features. What happens is that the head of programme inducts the new member on some of these features and in the future it's the responsibility of this new member to solve any upcoming technical issues. They can also seek help from the IT department.

As mentioned previously, the case study is of a research-based university which offers face-to-face instruction. However, with a deeper look at the current platform i.e. Blackboard, it can be easily discerned that there is some kind of unrecognised blended learning happening at the university. Each faculty member has the right to conduct two virtual classes in which students do not have to be physically available on campus. This information came to the knowledge of the researcher verbally through another faculty member. When the researcher tried to find an evidence for this, he could not find any document/policy mentioning this piece of information. The university works hard to incorporate blended learning. For this purpose, there is a committee at the university whose main objectives are to develop blended learning framework.

The eagerness to implement blended learning was also very clear from the interviews' content given by the participants. One of them recommended adding 'Blackboard Collaborate' which is considered the main tool on Blackboard used by lecturers to conduct synchronous classes. Unfortunately, this feature is not included in the current version of Blackboard used by faculty members at the university. While having an interview with one of faculty members, who recommended 'Blackboard collaborate', it was found that he was not aware that the 'wikis' feature is active within this version of Blackboard.

In the same context, another faculty member recommended getting some software such 'NVivo' and 'Adobe connect' which can help them conduct online classes. Another faculty member recommended taking the view of the students about blended learning and if the university found that the students like to have blended learning, the faculty members would be happy to adopt it. Another one approached the topic from different angle which is the load of work that the faculty members have. This member recommended having less workload and providing the members with more tools and '*adequate library*' so that they could run blended learning.

The sample size of this study is too small to generalise its findings. The survey itself did not include some important barriers such as computer/technology anxiety and student perspective towards the integration of technology. Finally, all the participants were only from the faculty of education and the other faculties such as engineering and business were not included.

6 Conclusion

This research paper shows that the teaching staff at the faculty of education are willing to adopt technology provided that they are supplied with the required tools to run such classes. This finding is inconsistent with Al-Senaidi et al. [8]'s which showed that the participating staff are dissatisfied with technology adoption. Although the participants in both of these two studies (i.e. current study and Al-Senaidi, Lin and Poirot's study) complained about the lack of technical support, the participant of this current study were fully satisfied and ready to adopt technology. Additionally, all of the faculty members know the meaning of blended learning, all of them agree that blended learning has positively impacted their teaching, and finally, all of them are interested in using technology.

The finding of this research paves the way for more research in this area on a larger scope. It also raises some recommendations to the management at the university such as providing more professional training and purchasing more advanced software that enable the teaching staff to run virtual classes such as Blackboard Ultra which combines many added features such as Blackboard collaborate and Ally. This last feature targets people with disability as it allows them to access the same content uploaded to Blackboard in different formats such as audio and electronic Braille. Based on the recommendations of this research, the university started exploring the possibility of upgrading the current version of Blackboard Learn 9.1 to Blackboard Ultraa. Further research on the impact of using this new version of Blackboard and integration of technology in the university is needed.

Acknowledgements This research paper was a part of a project submitted to the British University in Dubai during the Ph.D. study of the first author.

References

1. Pajo, K., Wallace, C.: Barriers to the uptake of web-based technology by university teachers. J. Distance Educ. **16**(1), 70–84 (2001)
2. Parsad, B., Lewis, L., Tice,, P.: Distance education at degree-granting post- secondary institutions: 2006–2007. National Postsecondary Educ. Coop. (2008)
3. Straub, D., Keil, M., Brenner, W.: Testing the technology acceptance model across cultures: A three country study. Inf. Manag. **33**(1), 1–11 (1997)
4. Oh, E., Park, S.: How are universities involved in blended instruction? Educ. Technol. Soc. **12**(3), 327–342 (2009)
5. Picciano, A.G.: Blending with purpose: the multimodal model. J. Asynchronous Learn. Netw. **13**(1), 7–18 (2009)
6. Kabassi, K., Dragonas, I., Ntouzevits, A., Pomonis, T., Papastathopoulos, G., Vozaitis, Y.: Evaluating a learning management system for blended learning in Greek higher education. Springerplus **5**(1), 1–12 (2016)
7. Vrazalic, L., MacGregor, R., Behl, D., Fitzgerald, J.: E-learning barriers in the United Arab Emirates: preliminary results from an empirical investigation. Innov. Knowl. Manag. Twin Track Econ. Challenges Solut. - Proc. 11th Int. Bus. Inf. Manag. Assoc. Conf. IBIMA 2009, **1–3**, 1782–1787 (2009)
8. Al-Senaidi, S., Lin, L., Poirot, J.: Barriers to adopting technology for teaching and learning in Oman. Comput. Educ. **53**(3), 575–590 (2009)
9. Valiathan, P.: Designing a blended learning solution. Learn. Circuits (2002). [Online]. Available: http://www.learningcircuits.com/2002/aug2002/valiathan.html. Accessed 28 May 2019
10. Boelens, R., Van Laer, S., De Wever, B., Elen, J.: Blended learning in adult education: towards a definition of blended learning (2015)
11. Mitchell, A., Honore, S.: Criteria for successful blended learning. Ind. Commer. Train. **39**(3), 143–149 (2007)

12. Munro, R.A., Rice-Munro, E.J.: Learning styles, teaching approaches, and technology. J. Qual. Particip. **27**(1), 26–32 (2004)
13. Dalmolin, A., Mackeivicz, M.T., Pochapski, G.A.O., Pilatti, G.L., Santos, F.A.: Learning styles preferences and e-learning experience of undergraduate dental students. Rev. Odontol. da UNESP **47**(3), 175–182 (2018)
14. Rossett, A.: The ASTD E-learning Handbook. McGraw-Hill, New York (2002)
15. Rooney, J.: Blending learning opportunities to enhance educational programming and meetings. Assoc. Manage. **55**(5), 26–32 (2003)
16. Aldosemani, T., Shepherd, C., Thomson, J., Raddaoui, A.: Second life as third place for English language learners' crosscultural interaction. Q. Rev. Distance Educ. **17**(2), 29–40 (2016)
17. Al-Hassan, S., Shukri, N.: The effect of blended learning in enhancing female students' satisfaction in the Saudi context. English Lang. Teach. **10**(6), 190–203 (2017)
18. Zhao, F.: Enhancing the quality of online higher education through measurement. Qual. Assur. Educ. **11**(4), 214–221 (2003)
19. Mali, Y.C.G.: Beyond words Vol. 4. No. 1. May 2016 graduate school, Widya Mandala Catholic University Surabaya, Indonesia. Beyound Words **4**(1), 17–26 (2016)
20. Ertmer, P.A.: Addressing first- and second-order barriers to change: strategies for technology integration. Educ. Technol. Res. Dev. **47**(4), 47–61 (1999)
21. Snoeyink, R., Ertmer, P.A.: Thrust into technology: how veteran teachers respond. J. Educ. Technol. Syst. **30**(1), 85–111 (2005)
22. Johnson, A.M., Jacovina, M.E., Russell, D.E., Soto, C.M.: Challenges and solutions when using technologies in the classroom. In: Crossley, S.A., McNamara, D.S. (eds.) Adaptive Educational Technologies for Literacy Instruction, pp. 13–29. Taylor & Francis, New York (2016)
23. Coleman, E., Mtshazi, S.: Factors affecting the use and non-use of Learning Management Systems (LMS) by academic staff. S. Afr. Comput. J. **29**(3), 31–63 (2017)
24. UNESCO: The dakar framework for action. Education for all: meeting our collective commitments (2000). [Online]. Available: http://unesdoc.unesco.org/images/0012/001211/121147e.pdf. Accessed 30 Apr 2019
25. Abdalla Alfaki, I.M., Ahmed, A.: Technological readiness in the United Arab Emirates towards global competitiveness. World J. Entrep. Manag. Sustain. Dev. **9**(1), 4–13 (2013)
26. Bachellerie, I.J.: Knowledge creation and diffusion: the role of UAE universities', Gulf Research Center (GRC), Report. In: The WAITRO 20th Biennial Congress, leadership for innovation (2010)
27. World Econmic Forum.: The global competitiveness report 2015–2016 (2011)
28. World Econmic Forum: The global competitiveness report 2019 (2019)
29. Ahmed, A., Al-Roubaie, A.: Building a knowledge-based economy in the Muslim world. World J. Sci. Technol. Sustain. Dev. **9**(2), 76–98 (2012)
30. Carnevale, D.: Government's request for data may delay Blackboard's purchase of WebCT. Chron. High. Educ. **52**(16), A29 (2005)
31. Green, K.C.: Campus computing, 2010: The campus computing project (2010). [Online]. Available: http://www.campuscomputing.net/sites/www.campuscomputing.net/files/Green-CampusComputing2010.pdf. Accessed 30 Apr 2019
32. Resta, P.: Information and Communication Technologies in Teacher Education: A Planning Guide. UNESCO, Paris (2002)
33. Scurry, D., Ensminger, D., Haab, M.: A model for integrating instructional technology into higher education. Br. J. Educ. Technol. **36**(2), 327–329 (2005)
34. Ajzen, I., Fishbein, M.: Understanding Attitudes and Predicting Social Behaviors. Prentice-Hall, Englewood Cliffs, NJ (1980)
35. Kim, C.M., Baylor, A.L.: A virtual change agent: motivating pre-service teachers to integrate technology in their future classrooms. Educ. Technol. Soc. **11**(2), 309–321 (2008)
36. Porter, W.W., Graham, C.R.: Institutional drivers and barriers to faculty adoption of blended learning in higher education. Br. J. Educ. Technol. **47**(4), 748–762 (2016)
37. Bensona, V., Andersona, D., Ooms, A.: Educators' perceptions, attitudes and practices: blended learning in business and management education. ALT-J. Res. Learn. Technol. **19**(2), 143–154 (2011)

38. Powell, A., et al.: Blending learning: the evolution of online and face-to-face education from 2008–2015 (2008)
39. Aldosemani, T., Shepherd, C.E., Bolliger, D.U.: Perceptions of instructors teaching in Saudi blended learning environments. TechTrends **63**(3), 341–352 (2019)
40. Creswell, J.: Research Design: Qualitative, Quantitative & Mixed Methods Approaches, 5th edn. Sage, Los Angeles (2018)
41. Rizvi, N.F., Gulzar, S., Nicholas, W., Nkoroi, B.: Barriers in adopting blended learning in a private university of Pakistan and East Africa: faculty members' perspective. mHealth **3**, 18–18, (2017)
42. Yilirim, G.Y., Yilirim, S., Yildirim, Z.: Main barriers and possible of ICT integration into pre-service teacher education porgrams. Educ. Technol. Soc. **12**(1), 193–204 (2009)
43. Hardesty, D.M., Bearden, W.O.: The use of expert judges in scale development. Implications for improving face validity of measures of unobservable constructs. J. Bus. Res. **57**(2), 98–107 (2004)
44. Sangoseni, O., Hellman, M., Hill, C.: Development and Validation of a questionnaire to assess the effect of online learning on behaviors, attitudes, and clinical practices of physical therapists in the United States regarding evidenced-based clinical practice. Internet J. Allied Heal. Sci. Pract. **11**(2), 1–12 (2013)
45. Tejada, J., Punzalan, J.R.: On the misuse of Slovin's formula Jeffry. Appl. Opt. **50**(19), 3187 (2011)
46. Alkahtani, K.: Teachers ' knowledge and use of assistive technology for students with special educational needs. Apr 2013 (2017)
47. Gaad, E., Thabet, R.: Behaviour support training for parents of children with down syndrome: perceptions, feedback, and attitudes. Int. J. Divers. Identities **17**(3), 19–34 (2017)
48. Al Khawaldi, H.: Faculty perceptions towards ET status at Omani colleges of education. Yarmouk Univ. (2000)
49. Al-Musawi, A.: Existing formats and functions of media units in the Omani higher education. J. Educ. Psychol. Sci. **3**(2), 33–51 (2002)

Applying a Flipped Approach Via Moodle: New Perspectives for the Teaching of English Literature

Emira Derbel

Abstract Technological innovations created alternative pathways in the educational field, which has revolutionized the learning of English and has yielded a motivating teaching environment. Due to the latest incorporation of online management systems into the learning process, teachers and students have been inspired by new teaching approaches and effective methods of learning. This paper intends to shed light on the efficiency of the flipped approach as a pedagogical teaching method mediated via Moodle that creates shared learning communities and champions a learner-centred experience. By adopting both qualitative and quantitative methods, this paper focuses on a pre-online activity guided by study questions and a post-class group discussion based on the flipped approach within the context of A Survey of American literature course. By engaging 35 students in a pre-class activity and in-class flipped lecture, the study seeks to examine the efficiency of the flipped approach and its influence on students' performance as opposed to traditional or classical learning process. The study concludes by dwelling on the usefulness of applying a flipped approach via Moodle. Such blended learning experience has proved to champion a motivating learning experience and to forge the portrait of an autonomous student.

Keywords Moodle · Flipped learning · Pedagogy · Motivation · Blended learning · Learning management system

1 Introduction

The fusion between learning and technology has marked the educational field yielding multiple and varied teaching pedagogies and learning modes. Due to an increasingly developing digital world and language management systems, a smarter and faster learning environment has come to impose its efficiency against traditional

E. Derbel (✉)
The University of Manouba, Tunis, Tunisia
e-mail: emira@buc.edu.om

Al Buraimi University College, Al Buraimi, Oman

M. Al-Emran et al. (eds.), *Recent Advances in Intelligent Systems and Smart Applications*, Studies in Systems, Decision and Control 295,
https://doi.org/10.1007/978-3-030-47411-9_17

learning strategies. As a learning strategy that complements rather than ignores classical methods of teaching, flipped learning has gained attention among teachers as it boosts students' self-reliance and technological awareness during the out-of-class learning activities, while fuelling their motivation during the in-class follow-ups.

The flipped method has been introduced as one type of blended learning that inverts the role of homework and classroom learning practice while gaining technological proficiency. Blended learning is a learning strategy which has the "potential to transform higher education institutions to supplement the traditional classroom setting and increase the quality of student learning" [1]. Flipped learning, therefore, is an active version of learning whereby "the information–transmission component of a traditional lecture is moved out class time and replaced by a range of interactive activities designed to entice active learning through the use of computer technology and the internet" [2]. Such innovative and creative move is essential mainly when it comes to the learning of English literature as it requires deeper levels of reflection, critical thinking and analytical skills. The utilization of Moodle in the implementation of a flipped learning approach accentuates students' engagement as taking active, independent part in the learning process and students' motivation during the post-class activities accumulated through the process of building or acquiring knowledge reaped during the pre-class learning process [3].

The flipped method, therefore, redefines the role of teachers and students and relocates both parts of the educational process into two different areas. The passive or traditional transmission of knowledge turns to a smart involvement of students during the pre-class guiding questions and creates interactive and engaging in-class learning. On the other hand, students become cognitively, emotionally and behaviorally engaged as they grow trained to self-independent reflection and analysis that leads to collective sharing of ideas in classroom [4]. Such process will not only develop students' skills, but it will also improve their self-confidence.

The use of an ICT-based learning management system as that of Moodle has simplified the implementation of the flipped approach for it builds hyper bridges of communication among teachers and students through the storage of videos, homework, discussion and exams. The integration of technology in the study of English language and literature attributes to the course a new shape "to increase student engagement and to restructure and replace traditional class hours" [5]. In this study Moodle has been used as the main platform to upload a YouTube comic video of Edgar Allen Poe's "The Fall of the House of Usher" along with 10 guiding questions for students to reflect on as part of the flipped method before attending the classroom. The originality of the study lies at the level of grouping two different groups composed of 10 students each, while the first group is already engaged in the flipped model, the second one will take part in a traditional lecture. It is in this context that this paper dwells on the extent to which the flipped methodology as one type of blended learning is a distinctively useful pedagogy in the context of literature course. The study also sheds light on the psychological aftermath of the implementation of the flipped method via Moodle on students/participants and non-participants via a survey distributed to them at the end of the lecture. The current study aims to answer the following research questions:

1. How does ICT-based learning management system as that of Moodle smooth the way towards the implementation of the flipped pedagogy and play the role of motivator and facilitator towards students-centred learning?
2. To what extent could the implementation of the flipped classroom method enhance students' acquisition of literary knowledge and boost their capacity for critical thinking and critical analysis?

2 Literature Review

According to Abeysekera and Dawson [2], 'the classroom flip' concept occurred in Baker [6] while the act of inverting or 'inverted classroom' method has been credited to Ref. [7] and 'just-in-time learning is used by Novak [8]. All previously stated terminologies are used interchangeably in many studies as they alter the process of a traditional lecture by replacing what is typically done in the classroom by outside-classroom activities through watching videos, visiting course-related websites or preparing key concepts through homework guiding question to be discussed during the in-class activity. As an emerging pedagogical tool, the flipped model is essentially about the act of relocating a traditional lecture outside the classroom via technology and assigned as homework, while in-class activities take the shape of collaborative learning [7, 9]. Hence, the flipped method is considered as "an instructional approach that educators use to turn the traditional classroom lecture model into a more active learning classroom" [10].

As one type of blended learning, the flipped method gains its practicality and efficiency from its dependence on technology. As a medium for the implementation of the flipped classroom model, Moodle allows teachers to allocate in-class time for extra group activities, better/deeper involvement in the act of learning since it offers students the opportunity of independent learning prior to class by affording them with useful online videos, lectures and reflection on a given subject. Abeysekera and Dawson [2] dismiss the capacity of the flipped classroom to enhance students' engagement, autonomy and centeredness, they simply define the flipped method as "putting the focus on moving tasks in space and time" (p. 2). The latter fail short to recognize that by directing students to accomplish a certain task prior to official classroom timing via Moodle, a student does not only create a separate space for learning and for research, but he/she embarks in a process of reflection, analysis, critical thinking which results into an intrinsic status of self-reliance and autonomy.

The flipped approach is grounded in the constructivist theory of learning that does not only fuel the method with the capacity of turning students to the main bearers of the learning activity, but it also enriches the course with a variety of learning activities. Sjoberg [11] affirms that "in constructivism, knowledge is actively constructed by the learner, not passively received from the outside" (p. 3). The creativity inherited from the flipped pedagogy is transmitted to the classroom whereby students engage in a wide range of activities that strengthens various aspects of collaboration and engagement. In the flipped approach, students have the opportunity to learn according to their needs, preferences and pace [12]. Such distinctive quality is highly needful in a literature course during which language and theoretical barriers menace students' understanding and appreciation of the subject.

In the case of a literary subject, the flipped approach further strengthens such exchange of responsibilities between teachers and students for the sake of alleviating the burden of a literature class, which consumes teachers and students' timing. As opposed to a flipped classroom, a traditional lecture of American literature demands more explanation from the instructor and in-depth exploration of topics from students. By following a flipped method, students undertake out-of-class work to explore literary concepts, acquire background knowledge about a literary trend or figure and to analyze a pre-assigned literary passage guided with visual video from YouTube or by a list of guiding questions. In this way, students in a flipped classroom create their own learning domains and develop engagement since they are "actively involved in a variety of educational activities that are likely to lead to high quality learning" and that is likely to boost their motivation and satisfaction [13].

By creating a "**f**lexible environment", teachers instil on students a "**l**earning culture" where "**i**ntentional content" develops the cognitive understanding of students and upgrades teachers to the level of "**p**rofessional educators" [14]. The previously mentioned four elements are the main constituents of flipped classrooms from which such pedagogical approach gained a high degree of flexibility because it "is adaptable to teacher's style, methods, and circumstances: teacher can personalize his/her version of flipped learning for their students and play to their individual strengths as educators" [9]. Engaging students on a pre-online activity allows for a more efficient and creatively engaging in—class learning and optimizes students zeal to adapt the flipped approach as "a sustained learning path that continues even after the class ends" [15]. The amalgamation of online system as that of Moodle and teaching approaches as the flipped one, learning communities have been created.

The usage of a learning management system as Moodle attributes flexibility to time and space. Students are no longer enticed to a face-to-face lecture during a particular time frame. They can, instead, move freely choose any kind of setting convenient to their social or cultural status from which they can get access to the online lecture, homework, video or lesson previously downloaded by their teacher. Such process of 'active learning' functions as a remedy from the passive role students play during a traditional lecture [16]. A flipped classroom approach launches students into active learning because it targets their "conscientiousness, concentration" and incline them into a deep approach to learning [17]. Nevertheless, procrastination stays lingering on the background of the flipped approach and remains one of the main negative factors that may hinder the main process.

The flipped approach is a learning method essentially based on the ardent role students' play during the pre-classroom preparation stage. It, therefore, consists of two parts: interactive learning activities during lesson and individual digging and research via computer based system out of lesson. This hyper-transfer of knowledge that takes place before the lesson is essential because only by moving the traditional components of an ordinary lecture out of class that students can benefit from the flipped class method and gain enough background information on their own. This will allow them to play an active role in class, to take advantage from collaborative in-class work and to participate in class discussions.

Kurt [12] argues that the flipped approach is a student-centred instructional model grounded in the constructivist theory of learning (p. 2). As opposed to traditional lecture, the flipped classroom teaching technique has a positive outcome as it stresses an active and interactive learning context. It further supports students' independency and boosts their understanding. The contribution of this paper is the implementation of the flipped method on a literature course as previous studies about the flipped method have mainly dealt with scientific courses or have deployed the method on general English courses as speaking or writing. In addition, the flipped approach is mainly dealt with via students' interaction with a comic adaptation of Edgar Allen Poe's "The Fall of the House of Usher" as part of the pre-class activity. As Derbel [18] puts it "the growing number of graphic adaptations of classic texts and their usage in teaching today proves to be a novelty or a revolution in the teaching of classic works of literature"(p. 56). The fusion between verbal and non-verbal texts offers a unique interplay [19].

Following this line of thought, this paper sheds light on the efficiency of the flipped approach as compared to a traditional lecture of American literature course. This work is primarily concerned with the application of the flipped approach in the Survey of American literature course. By contrasting the positive aftermath of the flipped classroom implementation on students to the passive outcome of traditional lecture, the study illustrates the capacity of such pedagogy to construct an interactive learning environment and to foreground its efficiency in helping students overcome traditional barriers of language and literary concepts while studying literature courses.

3 Methodology

This research uses a mixed-method approach combining statistical data from questionnaire with two main qualitative means of measurement composed of observations and group interviews. Both methods have served the purpose of measuring students' degree of satisfaction, motivation, autonomy and involvement into a flipped learning environment channelled by Moodle.

3.1 Research Context and Participants

This study is conducted on the Survey of American literature course (ENGL 475) at the English department at Al Buraimi University College (BUC) during the 2019/2020 academic year. My choice of BUC as the main setting of this study stems from various reasons. First, BUC is the researcher's working institution, a reason that has smoothed the research process in terms of application and of data collection.

Being acquainted with the main tenets of the university has allowed me to explain the teaching approach adapted in this study to the selected classes and to communicate with students throughout the whole process. The application of the flipped method on this course comes to support and to strengthen the department's goals that insist on the ardent gain of critical thinking and the knowledge of technology by the graduates of English language and literature. The participants of this research are 35 students of English literature from both morning and evening sections. The flipped method was applied in both sections throughout several lectures in order to reach valuable and concrete results. Students from both sections were engaged in pre-class activities followed by in-class discussion, elaboration of guiding questions forwarded to them in advance and group work.

3.2 Procedure

The first stage of this process has as its main constituents a YouTube video of Edgar Allen Poe's "The Fall of the house of Usher" as illustrated in Fig. 1, supported with a number of guiding questions posted in the form of an announcement on the electronic platform of the Survey of American Literature course of both. The pre-class guiding questions target various literary components of the graphic adaptation of the short story such as setting, characterization and thematic dimension. Follow-ups and reminders were constantly sent to students in order to guarantee the smooth running of the learning experience.

Such amalgamation between learning strategies and technology as the integration of a content management system as Moodle enhances students learning, improves

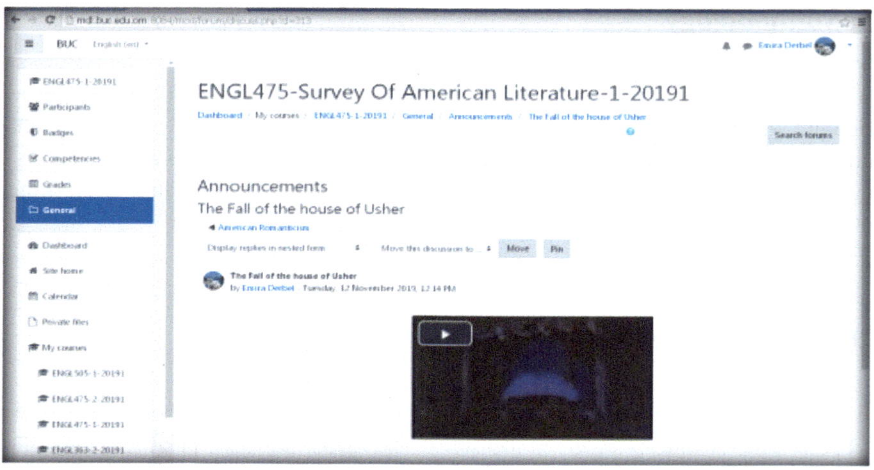

Fig. 1 Video announcement during the pre-class stage

quality teaching and reduces time loss. In fact, one of the main advantages of Moodle that it allows students to have access to the online pre-activity class and to answer questions without regards to neither time nor place. Added to that, the fusion between literary content and image channels an innovative experience of understanding and analysing graphic literature which have become part of the flipped approach and an innovative classroom teaching method. Both have offered new teaching perspective and construct better learning environment [20]. In the second phase of the research process, students were able to answer all six guiding questions which mainly target the setting of Edgar Allen Poe's "The Fall of the House of Usher", its gothic elements, the thematic dimension with a deep reference to characters analysis and narrative point of view. It is worth noting that even the process of writing about previously mentioned literary elements will not constitute an obstacle as opposed to traditional in class writing process because students bear in mind that they themselves are going to constitute the audience among which answer are going to be discussed, corrected or altered [20]. Teachers will, instead guide them through the whole process.

3.3 Instruments

The researcher has deployed an online survey, observation and focus group interviews in order to shed light on the ability of Moodle to facilitate the implementation of a flipped learning approach and to highlight the extent to which a flipped approach can boost students' comprehension of literature in contrast to a traditional lecture model.

3.3.1 Online Questionnaire via Google Forms

The study utilizes an online survey entitled "Implementation of Flipped Classroom Model via Moodle: A Quantitative Study" which was sent to 35 students involved in the flipped approach. According to Kannan et al. [21], the speed and efficiency of the internet, such quantitative tool proved to be useful and innovative. The survey is composed of four paragraph form questions and two 5-point Likert scale sections where participants rated their level of interest from level one to level five where level 1 stands for "strongly disagree" and level 5 stands for "strongly agree". The first part of the Likert rating scale in designed to shed light on the participant's degree of agreement regarding the motivating, helpful and enlightening aspect of the pre-class or homework activity accomplished via Moodle as part of the flipped approach. The second part is dedicated to the evaluation of the flipped classroom usefulness as a pedagogical method in opposition to the traditional classroom lecture.

3.3.2 Observation

Class observation provides valuable opportunities to the researcher to have a close eye on participants interaction, dialogues and emotional behaviour. The tool helps the researcher draw a "thick description" of what might have been left unsaid by other instruments such as online surveys or focus group interviews [22]. As a qualitative instrument, class activities observations has allowed the researcher to explore in depth areas mentioned in the online survey and to scrutinize students' psychological side and spontaneous reactions during the in-class activities, throughout their conversations and non-verbal gestures or facial expressions within a flipped learning environment. Hence, the research turns to a 'guest' performing a 'covert observation study' whereby participants are not aware of it [23]. Furthermore, the notes collected following the observation were used to triangulate data with other research tools. Despite the 'naturalistic' aspect by which observation has been always suspected as an analytical tool, the triangulation of this qualitative tool has served the purpose of resisting such doubts [24].

3.3.3 Focus Group Interview

As a qualitative technique of data collection, the focus group interview serves the purpose of gathering a small number of students "to explore [their] attitudes and perceptions, feelings and ideas about a topic" [25]. It is designed to "investigate the quality of relationships, activities, situations or materials" [26]. I deployed a stratified sampling technique to select the interviewees of this study. I selected 20 students, which I equally sub-divided into 4 groups of 5 students. As Denscombe [25] puts it, a "focus group consists of a small group of people, usually between six and nine in number, who are brought together by a trained moderator (the researcher) to explore attitudes and perceptions, feelings and ideas about a topic" (p. 115) I sought to amalgamate high achievers with average and low achievers in each group for the purpose of giving voice to all levels and to address the topic from different angles. The focus group interview was conducted throughout multiple flipped class sessions.

4 Findings and Discussions

The findings resulting from both quantitative and qualitative methods are discussed in accordance to the research tools deployed in this study.

4.1 Google Form Survey

The online-survey questions were divided to two categories: Paragraph questions and five-point Likert scale (1 = strongly disagree, 5 = strongly agree). The first aim behind such variation is to give room to students to express themselves emotionally and critically in relation to the practicality of using Moodle as the main learning management system to perform their pre-class activity as part of the flipped method. The second goal of affording students with various types of survey questions is to obtain fruitful feedback on the utility of using a flipped learning pedagogy in a literature course and to observe their inclination towards a possible implementation of this inverted learning method in the rest of subjects.

4.1.1 Students' Impression While Dealing with Moodle

The first question of the online survey targets the efficiency of utilizing Moodle as the main learning management system to accomplish the pre-class activity. Generally, students' responses exhibit a positive stand from adopting Moodle as an online instructional medium. A closer perception to their comments, reveals the extent to which students conceive Moodle as "useful, practical and helpful" management system since it allows them to "work independently, easily accessible either from their computers or smart phones and plays a key role in instructor/teacher communication. Moodle, students add, allows us to accumulate prior-knowledge about the short story, it facilitates learning and therefore, it becomes a practical means to understand the in-class lesson. Few students however, two out of twenty-seven responses to this question, have pointed to the need of training on the management system for they are more familiar with Google classroom.

The utility of implementing a flipped method via Moodle is also explored through the first five Likert-scale items, as showed below, which delineates or illustrates the responses of 35 students at the level of their degree of motivation, learning gains and engagement during the performance of the pre-class activity.

As Table 1 illustrates, students' testimonies in regards to the practical aspect of Moodle is positive. As the main management learning system utilized at Al Buraimi University College by all academic departments, students insist with a percentage of 85.71% that Moodle facilitates communication between students and teachers as it is considered the main online platform via which a smooth exchange of homework, pre-class activities or assignments is fulfilled. This is mainly due to students' ability to access the online learning platform regardless of time or place, a crucial characteristic of Moodle upon which 81.14% of participants highly agree. Such practical aspect of Moodle does not only alleviate students frustration in front of any new learning task, but it also favors an autonomous, independent and motivating learning experience to which the participants in this survey have attributed 84%. Further, 78.29% of students show their readiness to make use of Moodle in the rest of courses since they have grown familiar with the different options of the management system, which have

Table 1 Pre-class activity via Moodle

Items	Percentage (%)	Mean
Exchanging pre-class activities via Moodle facilitate my understanding of the short-story and guarantees smooth communication between students and teachers	85.71	4.29
I can access the homework via Moodle at anytime from anywhere	81.14	4.06
Pre-class activities familiarized me with Moodle as a system and with all its options	74.86	3.74
Pre-class activities made me able to work independent	81.71	4.09
Enabled me to have deeper levels of learning and understanding	84.00	4.20
Increased my critical-thinking abilities	80.00	4.00
I intend to use Moodle in the rest of courses and lectures	78.29	3.91
Pre-class activities are helpful, practical and motivating	84.00	4.20

facilitated their engagement in the flipped method. It is worth noting that 84.00% of participants agreed upon the utility of the pre-class activity that has offered them deeper levels of learning and better comprehension skills. By taking sufficient self-learning time to reflect on the guiding questions and to visually analyze Edgar Allen Poe's comic adaptation of the "Fall of the House of Usher, 80.00% of the participants have affirmed the capacity of such form of blended learning to nourish their critical thinking skills. Such extremely positive stand prepared the ground for the researcher to assess students' views vis a vis traditional lecture by proposing a second five Likert scale in the survey comparing and contrasting classical teaching lecture, to a flipped classroom.

Table 2 foregrounds students' new perception of the traditional lecture as opposed to a flipped one.

Table 2 Traditional lecture versus flipped classroom

Items	Percentage	Mean
Traditional lecture embodies language barriers	76.57	3.83
Pre-class activity facilitates my understanding during in-class time	82.29	4.11
Flipped learning increases self-trust, critical thinking and engagement in class	83.43	4.17
Flipped learning allows independent research as opposed to traditional lecture	78.86	3.94
Flipped learning increases my motivation and autonomy as opposed to my passive role in a traditional lecture	81.71	4.09
Traditional lecture makes me a passive learner	71.43	3.57
In-class group work increases my admiration of literature courses	80.00	4.00
Traditional lecture nourished my frustration because of my limited knowledge	72.57	3.63

In the light of the above results, one can notice that participants have developed a positive perception towards the flipped classroom in contrast to the recognition of the limited efficiency of a traditional lecture. In literature classes, literary language, depth of analysis and students' limited literary background stand as barriers in front of students' engagement or participation. These data attest to the fact that 76.57% of participants find literary jargon as a real obstacle in front of their understanding of a literary work. Likewise, feelings of detachment from the group, isolation within a traditional learning environment and even under self-estimation may hinder the process of learning and decrease the level of students at a later stage. What can be noticed, therefore, is that 82.29% of participants postulate that pre-class activities offer them deeper levels of understanding during in-class timing substituting, in this way, the portrait of the passive learner inherited in traditional lectures by an independent, autonomous and motivated (81.71%) one.

As the table above illustrates, students present the traditional lecture as part of a transmissive experience since it constructs them as passive learners and eliminates any kind of competence or autonomy. Participants have channelled such anxiety vis a vis the traditional learning process with a percentage of 72.57% of students who consider that traditional lecture nourishes their frustration and diminished their capacities by remaining simply followers instead of active members in the classroom. In fact, the in-class reading experience of a narrative short story is viewed as a form of intellectual challenge rendered an enjoyable experience to weak and passive students [27, 28]. Students subjected to controlling learning environments have been found to acquire knowledge less effectively, especially when learning is complex or requires conceptual, creative processing [2]. The flipped classroom model, however, is positively accepted by participants as it increases their self-trust, critical thinking and engagement in class (83.43%) and renders them active participants in the pre-class (78.86%), and the in-class activities (80%). The flipped classroom approach generates in students intrinsic motivation for it turns them to autonomous learners in contrast to traditional lectures that tend to create passive learners (71.43%).

The second question of the online questionnaire explores the main characteristics of the flipped approach appreciated by students:

4.1.2 Valuable Aspects of the Flipped Approach

Students answers seem to be fully consistent with the collected data previously examined and recollected from the five-point Likert scale. The collected comments focus on the capacity of the flipped approach to afford students with independent learning period during the pre-class activity, which prompts learner's autonomy and creates time for interaction between teacher and student during the in-class period. A flipped classroom approach supports team working and boosts cooperation among students. The general feedback generated from the students answers also foreground the efficiency of the flipped classroom methods as it fosters their motivation and dissolves the psychological barrier engendered by the quasi-impenetrable literary meaning. Students also point to the usefulness of Moodle as a flexible learning platform responsible for the accomplishment of the flipped method. The third question by which the survey concludes assesses the readiness of the 35 participants to adopt a flipped pedagogy in the rest of courses. Students show a high intent and wish to follow the inverted classroom method instead of the traditional one.

4.2 Focus Group Interview

The focus group interview was oral and it was composed of open-ended questions directed to four groups including five students each. The interviewees were keen to channel their experience and their impression in regards to the implementation of the flipped method on the course of Survey of American literature via Moodle. The questions serve to gain an in-depth understanding of the interviewee's perceptions of the efficacy of a flipped classroom method, while the investigator was engaged in note taking. Three major categories/areas emerged from the analysis of the data: their perceptions towards a learning process via a flipped classroom model, a move from passive to individualized learning, the applicability of the method in literature courses.

The analysis of qualitative date obtained from participants' interviews revealed that a high number of students appreciated the efficiency of pre-class preparation via Moodle because it created better learning opportunities and facilitated their comprehension of American literature. Flipped learning method erased obscure levels of literary concepts since the levels of difficulties were individually transcended before attending the lecture. As one participant puts it "*The application was easy and practical. I understood the short story's elements better because I answered the guiding questions shared* via *Moodle in advance*". Watching the YouTube video of the comic adaptation of the short story at home allowed the students to visualize all events, characters and setting which encouraged participation in the classroom and allocated much more time for group interaction.

Interviewees also mentioned the positivity inherited in the classroom environment during the in-class activities. All four interviewed groups agreed upon the potency of the flipped method to create a student-centred learning and a more energetic and motivating learning context in contrast to traditional lecture. Students kept reiterating key terms such *as* "*independent*", "*positive*", "*helpful*", "*we participate*", "*we understand*". Furthermore, participants articulated the importance of using Moodle each at his own pace without being hurried up by time as in classical lectures. They, for instance, added, "*the application is very good because we can watch the short story video at anytime*". Such feeling of confidence and ease has spread among all participants and has further encouraged them to make use of the flipped method in the rest of English language and literature courses because it championed collaboration and interaction among participants. Learning becomes based on the principle of sharing.

4.3 *Class Observation*

This qualitative measuring tool was mainly led in a silent way while all students were engaged in the class activities. The researcher comments on students' unconscious reactions, participation, feelings, motivation and facial expressions. Not only focus group interviews, observation notes also revealed that students were eagerly responsive by insisting on taking a creative part in the classroom activities. What was highly noticeable is that in addition to the prior knowledge that they have developed from pre-class activities, the graphic genre itself has motivated them because it has offered them freer platforms for interpretations as op[pose to traditional narrative genre [18]. Furthermore, their reactions and comments unveil their motivation and energetic spirit. As part of the observation also, the researcher noticed the students' easiness in dealing with Moodle. It was highly noticeable that they have gained enough practice and have developed new skills of using the application as an integral part of the learning process.

The qualitative data obtained from the focus group interviews and class observations reflect the same line of impressions, collected from participants. In depth observation of students' reaction, impressions or feelings clearly echoed the collected data in the focus group interview and confirm them. Students gained a clear aspect of self-confidence, a positive perception of the flipped method and an awareness of the practicality of using Moodle as an online learning platform.

5 Conclusion and Future Work

As an innovative pedagogical method, flipped learning associates the practical deployment of technology by using Moodle to a motivating learning practice in and outside the confines of the classroom. This study explores the integration of the flipped classroom model in the teaching of literary course because it constitutes an innovative area of research worth of study/discovery for its potential to create alternative learning and teaching paths. This paper makes use of qualitative and quantitative methods to measure the efficiency of amalgamating flipped learning with technology through the integration of Moodle as the main system management responsible for students/teacher communication in order to scrutinize through the obtained results the impact of such mix on the learning process and on students' progress.

Based on the collected data, the study proves the efficiency of the flipped classroom method to stimulate students' self-confidence, interaction and engagement during the pre-class and in-class stages. During this first stage, students developed an independent learning strategy by engaging themselves in a process of research and analysis to tailor appropriate answers to a number of guiding questions announced in advance by the teacher on Moodle. The free access to the short story video supported by guiding questions via Moodle has created a feeling of comfort and ease among students who are not merely limited to a physical presence in the teaching milieu of the classroom to learn about literature, but they can freely access the online educational platform independently from time or place. The main factor behind effectively engaging students at this first video stage of the flipped method is simply by creating the appropriate setting for such learning context [29]. The second part of the flipped method has embodied students' engagement in group-work during the in-class sessions through which shy and reluctant students showed a psychological improvement by overcoming their fear from uttering wrong responses. In short, the flipped approach has proved its potential to overcome cultural barriers by rising students' awareness and motivating them during the pre-class activities [30]. They, instead, reflected a high sense of engagement and creativity that boosted their motivation.

The flipped approach adapted for the teaching of the American Survey course through Moodle has redefined the role of students and teacher. As opposed to the traditional or classical perception of the teacher as the main source of knowledge, the flipped method casts the instructor into the role of the supervisor, the facilitator or guide while creating a student-centred learning environment whereby students assume the biggest share of the learning process. The efficiency of the flipped method is also detected at the level of students' performance in the mid-term exam results that were generally satisfactory in both sections following the implementation of the flipped approach. As illustrated through the collected data from the online survey, students have asserted that it would be motivating and creative enough if the flipped method becomes one of the teaching pedagogies adopted in the rest of courses.

At the end of this paper, it would be worthy to mention that the implementation of a flipped classroom method in a literature course is not without its thorny aspects. One main limitation of the flipped approach is the risk of reluctant participants who would not take part in the flipped process for no apparent reason. Though only a minority showed an inclination towards carelessness by not engaging in the two main stages of the flipped method, their remarkable loss during the in-class activity contrasted to their peers creativity put them in an embarrassing situation that drove them to learn from their classmates through discussing the various elements of the guiding questions. One of the gains students can reap from the flipped approach is independent learning and autonomy, but teachers should be wary not to turn all classes to flipped lectures otherwise autonomy would become burdening for students who may feel that they were left on their own in the process of learning without concrete guidance. Hence, the importance of keeping up with online communication between students and teacher is salient in order not to jeopardize the flipped process. Both literary studies and learning through the deployment of technology evolve continuously which offers to the researcher variable and valuable opportunities to explore new areas for research. One among the well-advanced literary fields in English studies is the development of cyber graphic literature that can interestingly be shared and taught via various technological platforms or applications in a way that would connect fun to learning.

References

1. Garrison, D.R., Kanuka, H.: Blended learning: uncovering its transformative potential in higher education. Internet High. Educ. (2004). https://doi.org/10.1016/j.iheduc.2004.02.001
2. Abeysekera, L., Dawson, P.: Motivation and cognitive load in the flipped classroom: definition, rationale and a call for research. High. Educ. Res. Dev. (2015). https://doi.org/10.1080/07294360.2014.934336
3. Guo, P.J., Kim, J., Rubin, R.: How video production affects student engagement: An empirical study of MOOC videos. In: L@S 2014—Proceedings of the 1st ACM Conference on Learning at Scale (2014). https://doi.org/10.1145/2556325.2566239
4. Fredricks, A., Blumenfeld, J., Paris, P.: School engagement: potential of the concept: state of the evidence. Rev. Educ. Res. **74**, 59–119 (2004)
5. Garrison, D.R., Vaughan, N.D.: Blended Learning in Higher Education: Framework, Principles, and Guidelines (2012)
6. Baker, W.J.: The 'classroom flip': using web course management tools to become the guide by the side, pp. 9–17. Cedarville University: Communication Faculty Publication (2000)
7. Lage, M.J., Platt, G.J., Treglia, M.: Inverting the classroom: a gateway to creating an inclusive learning environment. J. Econ. Educ. (2000). https://doi.org/10.2307/1183338
8. Novak, G.M.: Just-in-time teaching. New Dir. Teach. Learn. (2011). https://doi.org/10.1002/tl.469
9. Bergmann, A., Sams, J.: Flipped Learning for Elementary Instruction (2014)
10. Keengwe, J., Onchwari, G., Oigara, J.N.: Promoting active learning through the flipped classroom model (2014)
11. Sjøberg, S.: Constructivism and learning. In: International Encyclopedia of Education (2010)
12. Kurt, G.: Implementing the flipped classroom in teacher education: evidence from Turkey. Educ. Technol. Soc. **20**(1), 211–221 (2017)
13. Coates, H.: The value of student engagement for higher education quality assurance. Qual. High. Educ. **11**(1), 25–36 (2005)
14. Ozdamli, F., Asiksoy, G.: Flipped classroom Approach. World J. Educ. Technol. (2016). https://doi.org/10.18844/wjet.v8i2.640
15. Stone, B.B.: Flip your classroom to increase active learning and student engagement. In: Proceedings from 28th Annual Conference on Distance Teaching & Learning (2012)
16. Andrews, T.M., Leonard, M.J., Colgrove, C.A. and Kalinowski, S.T.: Active learning not associated with student learning in a random sample of college biology courses. Life Sci. Educ. **10**(4), 394–405 (2011)
17. Tucker, B.: The flipped classroom. Educ. Next **12**(1), 82–83 (2012)
18. Derbel, E.: Feminist graphic narratives: the ongoing game of eluding censorship. Mediterr. J. Soc. Sci. (2019). https://doi.org/10.2478/mjss-2019-0005
19. Mhamdi, C.A.: The use of political cartoons during popular protests: the case of the 2011 Tunisian uprising. J. English Stud. (2017). https://doi.org/10.18172/jes.3063
20. Al-Mohammadi, S.A., Al-Mohammadi, E.D.: To whom do we write? Audience in EFL composition classes. In: Methodologies for Effective Writing Instruction in EFL and ESL Classrooms (2014)
21. Kannan, P.K., Chang, A.M., Whinston, A.B.: Marketing information on the I-Way: data junkyard or information gold mine? Commun. ACM **41**(3), 35–43 (1998)
22. Clifford, G.: Thick description: toward an interpretive theory of culture. In: Geertz Clifford (1973) The Interpretation of Cultures, pp. 3–30, New York
23. Cohen, L., Manion, L., Morrison, K.: Research Methods in Education (2007)
24. Lincon, Y.S., Guba, E.G.: Naturalistic Inquiry. Sage, Beverly Hills, Califórnia (1985)
25. Denscombe, M.: The good research guide for small-scale social research projects (2007)
26. Fraenkel, J.R., Wallen, N.E.: How to Design and Evaluate Research in Education. McGraw-Hill Inc, Boston (2003)
27. Al-Mohammadi, S.: Integrating reading and writing in ELT. In: Focus on EFL Reading Theory and Practice, pp. 260–274 (2014)

28. Al-Mohammadi, S.: Mixed ability classes in English language teaching: challenges and opportunities. In: Issues in English Education in the Arab World, pp. 110–124 (2015)
29. Mhamdi, C.: What Can Video Add to the Learning Experience? Challenges and Opportunities (2017)
30. Derbel, E., Al-Mohammadi, S.: Integration of language skills and culture in English Language teaching: rationale and implications for practice. In: Issues in English Education in the Arab World, pp. 216–232 (2015)

An Integrated Model of Continuous Intention to Use of Google Classroom

Rana Saeed Al-Maroof and Said A. Salloum

Abstract Google Classroom (henceforth GC) has emerged as a tool in e-learning. This paper has adopted an integrated model to investigate the continuous intention to use GC as a facilitator in the learning and teaching environment in higher education. GC provides the students with opportunities to cope up with their teaching environment outside class timing. The two features of on-line and real-time learning communication increase the number of users in addition to the fact that the application is user-friendly. Accordingly, most universities in the Gulf Area tend to use GC due to a large number of students. Limited research has been carried on the construction of the continued usage intention of GC by students in higher education institutions. Thus, the objective of this paper is to propose a model for the continuous use intention of GC. To achieve this goal, a model that integrates three theoretical theories has been used which are Flow Theory, Information System Expectation–Confirmation Model and the DeLone and McLean Information Systems Success Model. The hypothesized model is validated empirically using the responses received from an online survey of 515 students were analyzed using structural equation modeling (SEM-PLS). The results show that concentration, perceived enjoyment, perceived usefulness, perceived ease of use, and satisfaction significantly affected the intention to use Google Classroom by college students. Moreover, perceived usefulness, perceived ease of use, confirmation, Information quality and system quality had a significant influence on satisfaction.

Keywords Google classroom · Flow theory · Information system expectation–confirmation model · McLean information systems success model

R. S. Al-Maroof
Department of English Language and Linguistics, Al Buraimi University College, Alburaimi, Oman

S. A. Salloum (✉)
Faculty of Engineering & IT, The British University in Dubai, Dubai, UAE
e-mail: ssalloum@sharjah.ac.ae

Research Institute of Sciences &Engineering, University of Sharjah, Sharjah, UAE

M. Al-Emran et al. (eds.), *Recent Advances in Intelligent Systems and Smart Applications*, Studies in Systems, Decision and Control 295,
https://doi.org/10.1007/978-3-030-47411-9_18

1 Introduction

An increasing number of research articles was conducted to examine the role of technology in education [1, 2]. Among these technologies, Google classroom (GC) is a web-based platform that can be used for free which makes it so popular in higher education. Its higher importance lies in the fact that it saves time and it can be used easily in creating classes, assigning assignments or quizzes, communicating immediate messages. Therefore, teachers use it often when they want to send quick and urgent messages as well as to share different teaching material and resources with their students [3]. The acceptance of GC has been dealt with in different research papers recently [3–5]. The acceptance of technology has an effective role in measuring the degree of success or failure in using it [6, 7]. Due to the fact that GC as e-learning system has been highly used by students, no studies have focused on its continuous usage. Many researchers tackle the factors which focus on users' acceptance of technology such as Technology Acceptance Model (TAM) but not on the continuous intention to use the technology.

It is well-known that acceptance in using technology is important, but it is considered as the first step in achieving the successful adoption or continued usage. There are many papers which propose that continued usage is a clear indication of students achievement level [8]. In addition, it helps in understanding the elements that increase the opportunities for continuous use intention [9]. This, in turn, will help the teachers, educators and students to develop strategies to make use of this technology [10]. Continuous usage plays an influential role in deciding the high success rate in practically using a technology emphasizing the factors that influence satisfaction and desire to continue using GC.

To be able to practically assess the factors that affect the continuous intention to use GC, Flow Theory is integrated with the other two theories in one unified model. Flow theory puts emphasis on the element of concentration, involvement and enjoyment. In other words, it helps to measure the psychological factors that support the continuous intention to use GC. The high level of concentration is considered a curial element that supports the intention to use where the users do not pay attention to time and place [11]. A similar environment is evident when the level of involvement is high. It reflects that the perceived enjoyment is high. Accordingly, the flow of experience that is derived from concentration, involvement and perceived enjoyment is vital to the continuous intention to use the technology. Most interestingly, the flow theory can affect the students' experience, hence, create enjoyable environment that develops a direct relationship with the intended e-learning tool [12].

Though the psychological factors help in measuring the continuous intention to use the technology, still there are other factors that affect the success in continually using it. Accordingly, ISECM is used to measure the importance of factors such as confirmation, perceived usefulness, perceived used, and satisfaction in the continuous use of GC. These factors can support the fact that success in using the technology has a close relation with the continuous use as it was developed by [13]. All the previously mentioned factors affect directly the users' psychological and practical reasons to continue use the technology. Nevertheless, a third dimension is needed to focus on the effectiveness of the technology itself from the users' perspective which is the system quality and information quality. In this respect [14] state that enjoyment, trust and attitude have been proved to be successful tools that can be used to measure dependent variables such as system quality, intention to use, user satisfaction and individual impact.

Finally, the study intends to focus on the continuous intention to use technology rather than acceptance or post-adoption since continuous intention is an obvious indicator for the success of using a technology. It focuses on long term use which form a crucial factor in judging the users' satisfaction. The lack of such studies may lead some users to stop using the technology, hence, it threatens the long-term success of the technology [15].

2 Literature Review in CUI

CUI has been the focus of intention by many researchers. One of the outstanding studies is by [10] where he proposed various factors related to continuous intention to use technology. He focused on ECM, TAM and TPB models to examine the impact of factors such as satisfaction, perceived behaviour and usefulness, attitude on e-learning. Another comprehensive study is carried out by [16] who propose a model for the continuous use intention of IOLEs. The model combines the Flow Theory, Information System Expectation–Confirmation Model and the DeLone and McLean Information Systems Success Model. The most important results showed that satisfaction, perceived usefulness, perceived enjoyment, perceived ease of use and focused immersion are factors that affect the continuous intention to use. Furthermore, satisfaction as an influential element has its positive influence on information quality, perceived usefulness and perceived enjoyment.

A study by [17] proposed factors that affect the continuous use of intention which are system quality, information quality, service quality, internet self-efficacy, perceived usefulness, intrinsic, user satisfaction, and continuous intention to use. They added the variables including gender, age, and level of education to examine the effect of continuous intention to use by different users. In line with the previous studies, [18] adopted Unified Theory of Acceptance and Use of Technology with some modification to tackle new components that focus on subjective task value and its effect on continuous intention to use.

A more simplified model is used by [19] when they adopted the TAM theory to determine the continuous intention to use government websites. The final results focused on the role of self-efficacy that can positively affect users' continuous intention to use. Besides, perceived ease of use enhances the continuous intention through the perceived usefulness. Furthermore, a study by [20] focused on the integration of subjective norm and expectation—confirmation model to propose a theoretical model of user continuous behaviour. A similar study that focuses on electronic government service by [21] adopted two influential factors which are expectation and satisfaction. They integrate the Social Cognitive Theory and Expectation-Confirmation Theory as well as Social Cognitive Theory and Expectation-Confirmation Theory to study the influence of satisfaction and continuity of use of e-government services. In agreement with the previous studies is a study by [22] which focus on the expectation-confirmation model. The study concluded that perceived service quality and perceived usability have a great influence on continuous intention to use MIM. On the other hand, the study places no importance on the factors of satisfaction and security. [23] tackles the importance of perceived usefulness and perceived quality on the continued use of intention in correlation with users' perceived usage characteristics, quality characteristics and satisfaction. In line with the previous studies, [24] examine the effect of information quality, system equality and service quality on trust and privacy risk that in turn affect the continued usage intention. The study found out that the quality of information, system and service have a positive effect on the perceived trust and consequently on the intention to continuously use the technology. The study showed that perceived privacy risk does not affect continuous use.

On the other hand, some studies tackle the factors of the intention to continuous use from a different perspective. A study by [25] focused on augmented reality (AR) technology and its relation with motivation. Besides, information quality has the most influential role in influencing AR continuous intention to use technology. The perceived enjoyment seems not to affect the continuous intention to use the technology.

3 Research Model and Hypothesis

Based on the available literature concerning the factors affecting the CUI, it seems that most studies focused on the importance of the quality of information, system and service and its effect on the satisfaction and confirmation factors on one hand [17, 21, 22, 24, 25] and the perceived enjoyment perceived ease of use and perceived usefulness on the other hand [18, 19]. This section is devoted to discussing the three models that are going to be used to measure the level of success and achievement of GC in higher education. It should be emphasized here that that some other studies tend to adopt more complicated and comprehensive model that incorporates all the previously stated factors (satisfaction, quality enjoyment, usefulness, ease of use and confirmation) in a unified model that investigate the effect of these factors. A study by Venter and Swart combine flow theory that has a great effect on the psychological aspects of concentration, motivation and enjoyment, the IS success that has effect on the quality factors, the ECM that focuses on satisfaction, confirmation (2018) A similar attempts was proposed by [10] who integrated TAM, TPB and ECM). This study intends to adopt a comprehensive model that integrates various influential factors and infuse them in a unified model to pinpoint the factors that influence the continuous intention to use GC.

It is worthy to mention that the proposed theory incorporates different aspects that are related to regularly used theories in previous studies. Besides, the integrated theories have focused on crucial factors in CI. Concentration factor, for instance, is tackled within the Flow theory. Similarly, the satisfaction factor can be measured through the integration of TAM theory (which focuses on the perceived usefulness and ease of use) and IS theory (which focuses on the information quality and system quality). The proposed model, however, has ignored other factors such as perceived enjoyment (which is part of TAM) and temporal dissociation (which is part of Flow Theory). Thus, future studies can incorporate these factors to have a widely comprehensive model.

3.1 A/Flow Theory

Flow theory is based on a psychological conception which focuses on positive users experience [26]. The flow theory tries to provide an answer to the questions where users are highly interested in activity in the sense that other individual behaviour seems to be meaningless/useless. The user within this theory seems to be overwhelmed with an intense level of concentration and he/she feels that they are controlled by what they are involved in [11]. The high level of involvement and concentration leads to the total ignorance of consciousness and time. Therefore, flow is a vital element to understand the interaction of users in technology as well as their attitudes [27].

The flow theory can be used to examine the holistic experience which implies that this theory touches upon the total involvement by using a valid metric [28] and [29]. This theory has shown that the flow of experience may result in a higher level of communication and engagement as well as satisfaction and acceptance of the technology. Moreover, some studies have linked the flow of experience with learning [30, 31]. Flow experience affects the learning outcomes, thus, it is connected to attitude towards learning both directly and indirectly [28]. The other two important factors within flow theory are motivation and entertainment. The flow of experience is connected with the degree of motivation and entertainment that technology has [10]. The hypotheses that can be derived based on Flow theory and its two factors which are concentrations and perceived enjoyment are:

H1: Concentration based on the flow of experience will positively influence the CUI when using GC.
H2: Perceived enjoyment based on the flow of experience will positively influence CUI when using GC.

3.2 Information System Expectation–Confirmation Model (ISECM)

According to [16], adoption expectations have an immense effect on the perceived usefulness and influence the satisfaction level of the user. Accordingly, it adds more power to the users' continued use of a technology [14]. Many researchers have used the IS-ECM model to predict the CUI and they have expanded the original IS-ECM with various constructs. They, for example, connected the perceived use perceived enjoyment to the IS-ECM to enhance the predictive power of the model [32]. Thus, the success of using new technology is eventually related to the continuous use of the technology that usually follows the initial acceptance. Therefore, the study incorporates the confirmation, perceived usefulness, satisfaction constructs to create the IS continuance intention following the model that was proposed by [13].

Furthermore, TAM theory has witnessed a tremendous change throughout the time, and it has been validated in a number of educational technology studies [33–36]. TAM has evolved to TAM2 and TAM3 where originally TAM has been extended to TAM2. TAM3 has proposed an integrated technology acceptance model which was developed by [37] tackling other crucial constructs such as perceived usefulness, perceived ease of use, intention to use, usage behavior, voluntariness, experience, subjective norm, image, job relevance, output quality, result demonstrability, anchor and adjustment [38]. Though they have integrated new constructs to measure users' satisfaction in TAM2 and TAM3, still certain limitations have raised that leads to the development of UTAUT model (the Unified Theory of Technology Adoption and Use of Technology) to overcome such limitations which includes constructs such as hedonic motivation, price value, and experience and habit to study acceptance and use of technology in a consumer context. These models have formed the baselines for many studies [39, 40].

This implies that ISECM theory has influential factors which are perceived ease of use, perceived usefulness which affect both satisfaction and confirmation. These four factors will have an immense effect on creating continuous intention to use GC by the users. The hypotheses include the following:

H3: Perceived Usefulness will positively affect satisfaction.
H4: Perceived Usefulness will positively influence CUI in using GC.
H5: Perceived ease of use will positively affect satisfaction.
H6: Perceived ease of use will positively influence CUI in using GC.
H7: Confirmation will positively affect satisfaction.

3.3 Information System Success Model (ISSM)

It Information system success embraces crucial factors that include the task, the user, the technology and structure. However, it seems that the user has an influential impact on system success. Attitudes, self-efficacy, technology experience, enjoyment, trust, and user expectations are considered to have the heaviest impact on users to achieve IS success. Enjoyment, trust and attitude have been proved to be successful tools that can be used to measure dependent variables such as system quality, intention to use, user satisfaction and individual impact [14]. Accordingly, the satisfaction construct has a close connection with the prediction of CUI. To be able to measure the satisfaction construct, the system quality and information quality constructs are added to the model and are in turn connected with CUI. The service quality construct is excluded as the study intends to measure the service of a particular system which is irrelevant in this study. The hypotheses that can be formed are:

H8: Information quality will positively influence satisfaction in using GC.
H9: System quality will positively influence satisfaction in using GC.
H10: Satisfaction that is gained due to the high system and information quality will have a high impact on ICU in using GC.

3.4 The Integrated Model

This study intends to adopt a comprehensive model that integrates various influential factors and infuse them in a unified model to pinpoint the degree of these factors on the continuous intention to use GC. The investigation of the literature review has shown that researchers tend to use a simplified model to investigate the factors affect the continuous intention to use technology. Other researchers have tended to use a unified model that integrates different theoretical theories to investigate the importance of a group of factors on the continuous intention to use technology. Table 1 summarized the models that have adopted in investigating the continuous intention to use educational technologies recently between 2015 and 2018.

The analysis of the continuance intention to use technology has indicated that studies tend to use different models based on the type of technology they wanted to investigate, the selected samples, and the relationship between constructs. The table has shown that satisfaction has an impact on developing a regular continuous intention to use technology, especially in the e-learning environment. It is clearly shown that some studies tend to integrate different models whenever the technology is widely-spread and part of the educational technologies. Accordingly, this study has adopted a more comprehensive model following studies that integrate two or three models in one unified model such as [10, 16, 44]. The model comprises the three models of Flow theory, the ISECM and the IS success. Figure 1 below represents the adopted model.

Table 1 The models adopted in investigating the continuous intention to use educational technologies

Authors	Models adopted	Educational technologies	Samples
[41]	An extended model based on Information Systems Continuance Theory (ISCT) and agency theory (PAT)	e-learning	College teachers
[42]	The information systems continuance and expectation-confirmation	MOOCs	MOOC's users
[43]	The extension of Bhattacherjee's expectation–confirmation model by adding two moderating effects (perceived voluntariness and habit) on the intention–usage relationship	Teaching blogs	Teachers
[44]	TAM and IS success	e-learning	Undergraduate students
[45]	The expectation–confirmation model	e-learning	Employees
[46]	Social perspective model	Mobile social apps	Undergraduate Students
[47]	The integration among perceived ease of use, perceived usefulness, expectation-confirmation, satisfaction	Mobile learning management system	Undergraduate students
[48]	Integration of two models which are expectation-confirmation model and technology acceptance model	Mobile book-reading apps	Undergraduate students
[49]	Bhattacherjee's (2001) expectation-confirmation model.	Digital textbook	Secondary school students
[50]	Satisfaction factors	e-learning	Employees
[16]	Integration of three models IS success, ISECM, and Flow theory	Microsoft office	Undergraduate students

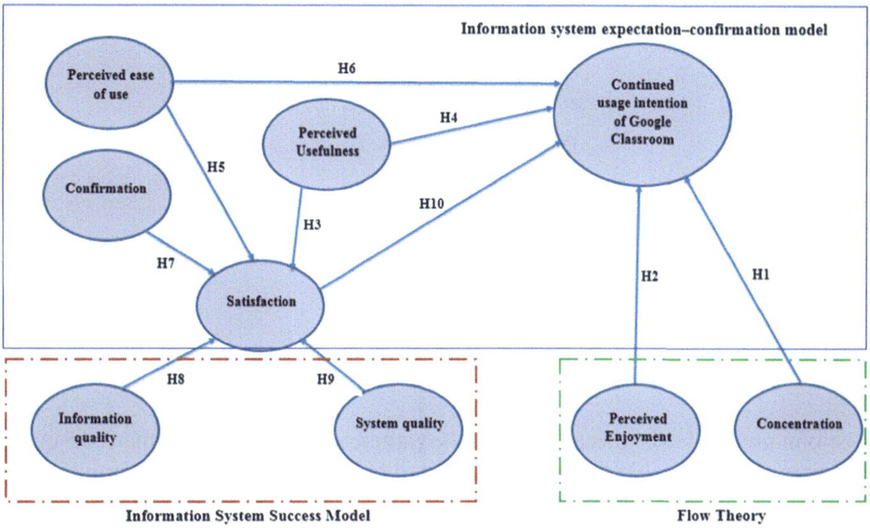

Fig. 1 The proposed model

4 Research Methodology

4.1 Pre-test of the Questionnaire

For the pilot study, the researcher used 10% of the total sample size of the research survey i.e. 600 respondents. The sample size was chosen in light of the standard research practices. In addition to this, all the questions involved in the questionnaire were pre-tested where 60 students were used who were chosen on random basis. Moreover, as per [51], the reliability of the study was found through Cronbach's Alpha whose results showed that the alpha values of all variables exceed 0.7, which has been demonstrated in Table 2. The results indicate the questionnaire used for the study is very reliable. It was ensured that all the respondents understood the final questionnaire as an attempt to further increase the overall reliability and survey quality.

4.2 Data Collection

Data collection process took place during the fall semester 2019/2020 from 01.10.2019 to 20.11.2019. To achieve the objectives of the study, a group of university students were chosen as the target population. The total number of students who provided answers to the questionnaire is 600. Eighty-five incomplete responses

Table 2 Cronbach's Alpha values for the pilot study (Cronbach's Alpha ≥ 0.70)

Construct	Cronbach's alpha
Confirmation	0.823
Continued usage intention of GC	0.836
Perceived ease of use	0.799
Concentration	0.851
Information quality	0.822
Perceived enjoyment	0.890
Perceived usefulness	0.887
Satisfaction	0.815
System quality	0.869

questionnaires had to be discarded from the gathered responses. After that, 515 questionnaires were available that were completed by the respondents which indicate a response rate of 86%. On the rule given by [52], the total number of estimated sampling size is 278 out of a population of 1000 students. Accordingly, 515 participants out of 1000 students are considered as sufficient in comparison with 278. To be able to approve the hypotheses, the structural equation modeling was used for testing with a sufficient group of participants. Furthermore, the five-point Likert Scale was implemented to measure students' intention to use GC. The focus on students in the current study stems from the fact that students tend to use GC regularly. They are considered to have a high professional usage of this e-learning tool. Their input may enrich the study because of their high educational level. The scale levels are distributed from strongly agree (5), agree (4), neutral (3), disagree (3) and strongly disagree (1). The survey is composed of 29 survey items that were given to the participants based on the Likert Scale. The details of the data collection from the survey are presented in Table 3. To be able to test the hypotheses, the survey instrument that was explained in the research. The 29 items of the survey are used for measuring the constructs presented in the questionnaire. The scales and items also represented in Appendix A. Five hundred fifteen students from a famous university in UAE, namely, The British University in Dubai (BUiD) and University of Fujairah (UOF) were chosen as the sample to fill the questionnaire.

Table 3 Participants' details

University	No. of students
The British University in Dubai (BUiD)	265
University of Fujairah (UOF)	250
Total	515

4.3 Students' Personal Information/Demographic Data

The online survey was distributed to five hundred fifty students. Accordingly, five hundred fifty responses were collected. The total number of responses that were assessed for completeness is 480 and 35 responses were ignored for insufficiency reasons. Table 4 shows that the rate of responses was 94% which referred to the sample demographic details. It is also shown that more than half of the respondents (57%) were male whereas (43%) of the total number of participants female which is (68%). The age of the participants is between 18–29 years old. Some of the participants around 20% have no experience in using Google Classroom in their academic life. Table 4 shows the respondents demographic information.

Table 4 Summary of students' demographic characteristic

Variables	No. of respondents	Percent %
Gender		
Male	293	57
Female	222	43
Age		
18–29 years	350	68
30–39 years	88	17
40–49 years	70	14
50–59 years	7	1
College		
College of Business	230	45
College of IT	105	20
College of Education	135	26
College of Humanities	45	9
Level of education		
Bachelor	276	54
Master	150	29
Ph.D.	89	17
Categorization of respondents		
Experienced users	385	75
Non-experienced users	130	25

5 Findings and Discussion

5.1 Data Analysis

This study follows the general guidelines for employing the PLS-SEM in IS research [53]. The latent factors and variable experimental factors of each construct are associated with each other which is depicted by measurement model. Prior to the confirmation of this association in structural model, it is important to attain certain reliability and validity in the measurement model [54, 55].

5.1.1 Convergent Validity

Considering [56], convergent validity is obtained comparative to the values of the indicators such as factor loadings, variance extracted and reliability (consisting of Cronbach's Alpha, composite reliability, and Dijkstra-Henseler's). Various values obtained for constructs depicted internal consistency. This was signified by values of reliability coefficient and composite reliability (CR) obtained for each construct which were higher than 0.7 [57]. Table 5 represents the Cronbach's alpha (CA) values which were higher than normal value of 0.7 and the value of composite reliabilities of each construct which were between 0.739 and 0.892. Since Cronbach's alpha (CA) tend to underestimate and composite reliability (CR) tend to overestimate the actual reliability of construct scores, the authors recommend to refrain from using CA and CR when estimating models with Smart PLS [58]. Instead, researchers should evaluate and report construct reliability by means of the Dijkstra-Henseler's rho (pA) reliability coefficient [58]. Similar to CA and CR, the reliability coefficient ρA should indicate values of 0.70 or higher in exploratory research and values above 0.80 or 0.90 for more advanced stages of research [59–61]. As shown in Table 5, the reliability coefficient ρA of each measurement construct is above 0.70. Furthermore, it was found that AVE values met the standards of elucidation of a minimum of 50% of variance extracted from the essential components of the construct since the values obtained for average variance extracted (AVE) were between 0.596 and 0.810. Consequently, convergent validity was accomplished by the scales involved in the measurement of constructs.

5.1.2 Discriminant Validity

Measurement model demonstrated that each value of AVE was found to exceed the value of squared correlation among constructs [54, 55, 57] which clearly implies that the criteria for accomplishment of discriminant validity are also met. See Tables 6 and 7.

Table 5 Convergent validity results which assures acceptable values (Factor loading, Cronbach's Alpha, composite reliability, Dijkstra-Henseler's rho ≥ 0.70 and AVE > 0.5)

Constructs	Items	Factor loading	Cronbach's alpha	CR	PA	AVE
Confirmation	CON1	0.790	0.782	0.765	0.791	0.660
	CON2	0.789				
	CON3	0.766				
Continued usage intention of GC	CUI1	0.817	0.799	0.739	0.746	0.706
	CUI2	0.893				
	CUI3	0.863				
Perceived ease of use	EOU1	0.824	0.877	0.779	0.784	0.596
	EOU2	0.929				
	EOU3	0.888				
Concentration	CONS1	0.892	0.714	0.880	0.893	0.810
	CONS2	0.836				
	CONS3	0.848				
Information quality	IQ1	0.760	0.720	0.761	0.775	0.753
	IQ2	0.882				
	IQ3	0.850				
	IQ4	0.808				
Perceived enjoyment	PE1	0.779	0.786	0.833	0.845	0.671
	PE2	0.875				
	PE3	0.770				
Perceived usefulness	PU1	0.701	0.801	0.819	0.865	0.777
	PU2	0.866				
	PU3	0.932				
Satisfaction	SAT1	0.800	0.841	0.808	0.848	0.659
	SAT2	0.810				
	SAT3	0.813				
System quality	SQ1	0.809	0.816	0.892	0.895	0.654
	SQ2	0.889				
	SQ3	0.852				
	SQ4	0.849				

5.1.3 Model Fit

SmartPLS offers the following fit measures: The standard root mean square residual (SRMR), exact fit criteria, d_ULS, d_G, Chi-Square, NFI, and RMS_theta indicate the model fit in PLS-SEM. The SRMR implies the difference between observed correlations and model implied correlation matrix [62] and values less than 0.08 are considered as good model fit measures [63]. The NFI values higher than 0.90

Table 6 Fornell-larcker scale

	CON	CUI	EOU	CONS	IQ	PE	PU	SAT	SQ
CON	**0.903**								
CUI	0.254	**0.889**							
EOU	0.260	0.318	**0.787**						
CONS	0.333	0.310	0.499	**0.899**					
IQ	0.450	0.517	0.228	0.644	**0.887**				
PE	0.522	0.229	0.025	0.355	0.518	**0.879**			
PU	0.487	0.369	0.238	0.330	0.478	0.158	**0.921**		
SAT	0.540	0.100	0.111	0.201	0.455	0.410	0.221	**0.820**	
SQ	0.369	0.288	0.366	0.300	0.470	0.244	0.118	0.229	**0.813**

The bold significance is the requirements are confirmed by the Fornell-Larker condition, since all AVEs along with their square roots are greater than its correlation with other constructs

Table 7 Heterotrait-monotrait ratio (HTMT)

	CON	CUI	EOU	FI	IQ	PE	PU	SAT	SQ
CON									
CUI	0.287								
EOU	0.639	0.702							
FI	0.310	0.287	0.400						
IQ	0.147	0.524	0.236	0.300					
PE	0.632	0.488	0.177	0.170	0.123				
PU	0.456	0.227	0.296	0.560	0.286	0.608			
SAT	0.214	0.147	0.471	0.700	0.245	0.588	0.520		
SQ	0.358	0.102	0.240	0.631	0.200	0.218	0.476	0.150	

Table 8 Model fit indicators

	Complete model	
	Saturated model	Estimated mod
SRMR	0.051	0.072
d_ULS	0.988	2.307
d_G	0.664	0.735
Chi-Square	489.936	491.568
NFI	0.826	0.827
Rms Theta	0.089	

indicate good model fit [64]. The NFI is a ratio of Chi2 value of the proposed model to the null model or benchmark model [65]. The larger the parameters, the higher the NFI, and therefore NPI is not recommended as a model fit indicator [62]. The squared Eucledian distance, d_ULS, and the geodesic distance d_G are two metrics that provide discrepancy between empirical covariance matrix and covariance matrix implied by composite factor model [58, 62]. RMS theta is only applicable for the reflective models, and evaluates the degree of outer model residuals correlation [65]. The closer the RMS theta value to the zero, the better the PLS-SEM model, and their values less than 0.12 are considered good fit, and anything else suggest lack of fit [66]. Hair et al. [62] suggested that saturated model evaluates correlation between all constructs, whereas, estimated model takes total effects and model structure into account.

Table 8 shows that RMS_theta value was found to be 0.089 which implies that the concerned goodness-of-fit for PLS-SEM model was large enough to demonstrate global PLS model validity.

5.1.4 Hypotheses Testing—Path Coefficient

The testing of the proposed hypotheses involved evaluation of association between the theoretical constructs related to the structural model [67–84]. This process was conducted with the help of structural equation model. Moreover, the model utilized Smart PLS having the features of maximum likelihood estimation. As appeared in Table 9 and Fig. 2, the model had a high predictive power [85], that's, the percentage of the variance within satisfaction, and continued usage intention of GC are nearly 75% and 81%, separately. Particularly, concentration ($\beta = 0.598$, $P < 0.001$), perceived enjoyment ($\beta = 0.823$, $P < 0.001$), perceived usefulness ($\beta = 0.169$, $P < 0.05$), perceived ease of use ($\beta = 0.387$, $P < 0.001$), and satisfaction ($\beta = 0.455$, $P <$

Table 9 R^2 of the endogenous latent variables

Constructs	R^2	Results
Continued usage intention of GC	0.812	High
Satisfaction	0.752	High

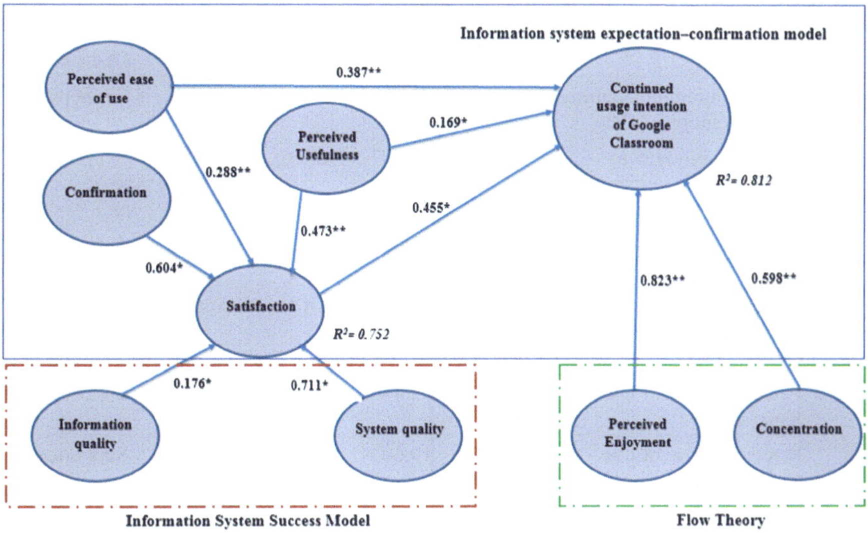

Fig. 2 Path coefficient of the model (significant at $p^{**} <= 0.01$, $p^{*} < 0.05$)

0.05) have positive significant effect on continued usage intention of GC. As the result, hypotheses H1, H2 H4, H6, and H10 are supported. The results showed that satisfaction significantly influenced perceived usefulness ($\beta = 0.473$, $P < 0.001$), perceived ease of use ($\beta = 0.228$, $P < 0.001$), confirmation ($\beta = 0.604$, $P < 0.05$), Information quality ($\beta = 0.176$, $P < 0.05$), and system quality ($\beta = 0.711$, $P < 0.05$) supporting hypothesis H3, H5, H7, H8 and H9 respectively. A summary of the hypotheses testing results is shown in Table 10.

6 Conclusion and Future Work

The current study has contributed to the study of continuous intention to use technology by proposing an integrated model that comprises three models. The findings are in line with previous studies [11, 17, 86] which indicate that there is a direct intention between certain influential variables and the continuous intention to use. For instance, the current study supports most of the previous TAM findings where perceived usefulness and perceived ease of use affect the usage of technology.

Similarly, other factors such as satisfaction and concentration are considered the most influential factors on the continuous intention to use technology. The results of the research have proved that most of the respondents are satisfied with the use of GC and that satisfaction factor plays an important role in the continuous use of the GC. Satisfaction factor is in turn connected with the information quality and system quality. The research result is supported by previous literature as they put a correlation

Table 10 Hypotheses-testing of the research model

H	Relationship	Path	*t*-value	*p*-value	Direction	Decision
H1	Concentration → Continued usage intention of GC	0.598	23.178	0.000	Positive	Supported**
H2	Perceived Enjoyment → Continued usage intention of GC	0.823	21.107	0.000	Positive	Supported**
H3	Perceived Usefulness → Satisfaction	0.473	13.880	0.000	Positive	Supported**
H4	Perceived Usefulness → Continued usage intention of GC	0.169	2.231	0.018	Positive	Supported*
H5	Perceived Ease of Use → Satisfaction	0.228	9.446	0.003	Positive	Supported**
H6	Perceived Ease of Use → Continued usage intention of GC	0.387	19.603	0.001	Positive	Supported**
H7	Confirmation → Satisfaction	0.604	2.322	0.042	Positive	Supported*
H8	Information quality → Satisfaction	0.176	3.320	0.036	Positive	Supported*
H9	System quality → Satisfaction	0.711	3.487	0.029	Positive	Supported*
H10	Satisfaction → Continued usage intention of GC	0.455	4.088	0.016	Positive	Supported*

Significant at $p^{**} <= 0.01$, $p^{*} < 0.05$

between satisfaction with system quality and information quality [87, 88]. The results pinpoint another important issue which is the role of perceived enjoyment on the continuous intention to use the technology. It has been approved by the respondents' responses that perceived continuous highly affects their continuous intention to use GC Thus, in order to make sure that continuous intention to use technology is valid, the university has to ensure that the technology has a level of enjoyment that attracts users directly to use it.

6.1 Future Implications

This study was intended to develop an integrated model that integrates three different theories. Based on the results obtained it seems that instructor and students should pay more attention to technologies such as Google Classroom. This technology is useful as it is easy to use, and it facilitates the communication between instructors and students outside classroom environments. The other reason is that it facilitates the question-making and marking whenever the nature of the material given to the students can be assessed based on multiple choices or short answers.

6.1.1 Theoretical Implications

The study has theoretical implications for literature to be taken into consideration. First, this study is an attempt to analyze the main factors that may affect the continuous intention to use the technology and give reasonable justifications behind the long-term usage of the technology. The study is considered as an example of integrating three different models to study continuous intention at high educational level. Second, the proposed extended model intended to have a strong predictive power in providing reasons behind the continuous intention to use technology and paved the way to the role of GC in teaching environment. Hence, future research may use the adopted model as a theoretical basis to investigate the continuous intention of using other educational e-learning tools. Third, the findings shed light on the fact that users' satisfaction within the framework of long-term usage may increase the chances of using the technology to fulfill their educational needs.

6.1.2 Managerial and Practical Implications

The current study has certain managerial and practical implications. Users' satisfaction is considered a crucial factor in measuring the continuous use of technology. Thus, such studies should provide a decisive classification for the most important factor that is more preferred by the users. Furthermore, they should focus on the benefits behind using this technology for long-term to ensure that the technology meets the updated users' needs. In other words, providing significant reasons behind the continuous use of technology may bring higher satisfaction to the user. Taking into consideration factors such as concentration, usefulness, ease of use, high information quality can add a solid understanding of the reasons behind the continuous use of the technology.

6.2 Limitations and Further Research

This study has several limitations. The first one is that the analysis focuses on the continuous intention to use GC without focusing on the variations between males and females, that is to say, gender variations. Males' and Females' reasons behind continuous intention to use technology may vary. The second limitation is that this study is not longitudinal where the emphasis on changes of willingness to continue using the technology over the item is not part of the study objectives. This study takes into consideration the immediate response within the current year in a short snap-shot. The final limitation is related to the model itself, the model has excluded the use of TPB theory even though it can work in accordance with both TAM and Flow theory [10]. Therefore, the model that makes use of TPB theory can be used to measure GC continuous intention to use by students at universities and colleges.

Appendix A

Constructs and Items

Confirmation

- I have gained better experience than I expected when I used GC.
- The quality of the service from GC was above my expectations.
- Almost all my expectations were confirmed when using GC.

Continued usage intention of GC

- I have the intention to use GC in the future regularly.
- I have the intention to use GC frequently in the future.
- I will advise my colleagues to use GC.

Concentration

- I was absorbed intensely in GC activity.
- My attention was focused on the GC activity.
- I was deeply engrossed in the activity.

Perceived Ease of Use

- GC as a learning tool is user-friendly.
- GC is a technology that I can be skillful at easily.
- Almost all tools in GC are easy to control.

Information Quality

- GC provides users with accurate information.
- GC provides information on timely-basis.
- GC provides us with relevant information to the users.
- GC provides us with information that can be easily understood.

Perceived Enjoyment

- I spend interesting time when I use GC.
- I feel a high level of enjoyment when using GC.
- I have a lot of fun when I use GC.

Perceived Usefulness

- GC helps me to improve my learning performance.
- GC increases my learning effectiveness.

- GC has been a useful tool in learning.

Satisfaction

- All users feel satisfy as far as GC performance is concerned.
- Users feel pleased with the experience they go when they used GC.
- I was willing to use GC based on my wise decision.

System Quality

- One of the advantages of GC is it gives users gquick response in the rush hours.
- GC follows a logic sequence to finalize the task.
- Whenever I use GC, I get the predictable/assumed results.
- GC has displayed organization of information on the screens clearly.

References

1. Al-Emran, M., Mezhuyev, V., Kamaludin, A.: An innovative approach of applying knowledge management in M-learning application development: a pilot study. Int. J. Inf. Commun. Technol. Educ. **15**(4), 94–112 (2019)
2. Al-Emran, M., Alkhoudary, Y.A., Mezhuyev, V., Al-Emran, M.: Students and Educators Attitudes towards the use of M-Learning: Gender and Smartphone ownership differences. Int. J. Interact. Mob. Technol. **13**(1), 127–135 (2019)
3. Jakkaew, P., Hemrungrote, S.: The use of UTAUT2 model for understanding student perceptions using google classroom : a case study of introduction to information technology course. In: Digital Arts, Media and Technology (ICDAMT), International Conference, pp. 205–209 (2017)
4. Berque, D.A., Newman, J.T.: GlassClass: exploring the design, implementation, and acceptance of google glass in the classroom. In: International Conference on Virtual, Augmented and Mixed Reality, pp. 243–250 (2015)
5. Al-Maroof, R.A.S., Al-Emran, M.: Students acceptance of Google classroom: an exploratory study using PLS-SEM approach. Int. J. Emerg. Technol. Learn. **13**(6), 112–123 (2018)
6. Davis, F.D.: User acceptance of information technology: system characteristics, user perceptions and behavioral impacts. Int. J. Man Mach. Stud. **38**(3), 475–487 (1993)
7. Alhashmi, S.F.S., Salloum, S.A., Mhamdi, C.: Implementing artificial intelligence in the United Arab Emirates healthcare sector: an extended technology acceptance model. Int. J. Inf. Technol. Lang. Stud. **3**(3) (2019)
8. Dağhan, G., Akkoyunlu, B.: Modeling the continuance usage intention of online learning environments. Comput. Human Behav. **60**, 198–211 (2016)
9. Al-Emran, M., Arpaci, I., Salloum, S.A.: An empirical examination of continuous intention to use m-learning: an integrated model. Educ. Inf. Technol. (2020)
10. Lee, M.C.: Explaining and predicting users' continuance intention toward e-learning: an extension of the expectation–confirmation model. Comput. Educ. **54**(2), 506–516 (2010)
11. Liu, S.H., Liao, H.L., Pratt, J.A.: Impact of media richness and flow on e-learning technology acceptance. Comput. Educ. **52**(3), 599–607 (2009)
12. Benvenuti, M., Błachnio, A., Przepiorka, A.M., Daskalova, V.M., Mazzoni, E.: Factors related to phone snubbing behavior in emerging adults: the phubbing phenomenon. In: The Psychology and Dynamics Behind Social Media Interactions, IGI Global, pp. 164–187 (2020)
13. Bhattacherjee, A.: Understanding information systems continuance: an expectation-confirmation model. MIS Q. 351–370 (2001)

14. Petter, S., DeLone, W., McLean, E.R.: Information systems success: the quest for the independent variables. J. Manag. Inf. Syst. **29**(4), 7–62 (2013)
15. Zhang, Y., Xiao, S., Zhou, G.: User continuance of a green behavior mobile application in China: an empirical study of ant forest. J. Clean. Prod. **242**, 118497 (2020)
16. Venter, M., Swart, A.J.: An integrated model for the continuous use intention of Microsoft Office simulation software. In: 2018 IEEE Global Engineering Education Conference (EDUCON), pp. 320–329 (2018)
17. Almahamid, S., Rub, F.A.: Factors that determine continuance intention to use e-learning system: an empirical investigation. In: International Conference on Telecommunication Technology and Applications Proceedings of CSIT, vol. 5, pp. 242–246 (2011)
18. Chiu, C.-M., Wang, E.T.G.: Understanding Web-based learning continuance intention: The role of subjective task value. Inf. Manag. **45**(3), 194–201 (2008)
19. Wangpipatwong, S., Chutimaskul, W., Papasratorn, B.: Understanding Citizen's continuance intention to use e-government website: a composite view of technology acceptance model and computer self-efficacy. Electron. J. e-Gov. **6**(1) 2008
20. Kim, B.: Understanding antecedents of continuance intention in social-networking services. Cyberpsychology Behav. Soc. Netw. **14**(4), 199–205 (2011)
21. Alruwaie, M., El-Haddadeh, R., Weerakkody, V.: A framework for evaluating citizens' expectations and satisfaction toward continued intention to use e-government services. In: International Conference on Electronic Government, pp. 273–286 (2012)
22. Oghuma, A.P., Libaque-Saenz, C.F., Wong, S.F., Chang, Y.: An expectation-confirmation model of continuance intention to use mobile instant messaging. Telemat. Informatics **33**(1), 34–47 (2016)
23. Tseng, S.-M.: Exploring the intention to continue using web-based self-service. J. Retail. Consum. Serv. **24**, 85–93 (2015)
24. Wang, E.S.-T., Lin, R.-L.: Perceived quality factors of location-based apps on trust, perceived privacy risk, and continuous usage intention. Behav. Inf. Technol. **36**(1), 2–10 (2017)
25. Kim, K., Hwang, J., Zo, H., Lee, H.: Understanding users' continuance intention toward smartphone augmented reality applications. Inf. Dev. **32**(2), 161–174 (2016)
26. Spreng, R.A., Chiou, J.: A cross-cultural assessment of the satisfaction formation process. Eur. J. Mark. **36**(7/8), 829–839 (2002)
27. Saadé, R., Bahli, B.: The impact of cognitive absorption on perceived usefulness and perceived ease of use in on-line learning: an extension of the technology acceptance model. Inf. Manag. **42**(2), 317–327 (2005)
28. Choi, D.H., Kim, J., Kim, S.H.: ERP training with a web-based electronic learning system: The flow theory perspective. Int. J. Hum Comput Stud. **65**(3), 223–243 (2007)
29. Lee, B.-C., Yoon, J.-O., Lee, I.: Learners' acceptance of e-learning in South Korea: theories and results. Comput. Educ. **53**(4), 1320–1329 (2009)
30. Csikszentmihalyi, M.: Finding Flow: The Psychology of Engagement with Everyday Life. Basic Books (1997)
31. Csikszentmihalyi, M., Abuhamdeh, S., Nakamura, J.: Flow. In: Flow and the Foundations of Positive Psychology, Springer, pp. 227–238 (2014)
32. Thong, J.Y.L., Hong, S.-J., Tam, K.Y.: The effects of post-adoption beliefs on the expectation-confirmation model for information technology continuance. Int. J. Hum Comput Stud. **64**(9), 799–810 (2006)
33. Al-Emran, M., Mezhuyev, V., Kamaludin, A.: Towards a conceptual model for examining the impact of knowledge management factors on mobile learning acceptance. Technol. Soc. (2020)
34. Al-Emran, M., Teo, T.: Do knowledge acquisition and knowledge sharing really affect e-learning adoption? An empirical study. Educ. Inf. Technol. (2019)
35. Al-Qaysi, N., Mohamad-Nordin, N., Al-Emran, M.: A systematic review of social media acceptance from the perspective of educational and information systems theories and models. J. Educ. Comput. Res. **57**(8), 2085–2109 (2020)
36. Al-Emran, M., Mezhuyev, V., Kamaludin, A.: Technology acceptance model in M-learning context: a systematic review. Comput. Educ. **125**, 389–412 (2018)

37. Venkatesh, V., Bala, H.: Technology acceptance model 3 and a research agenda on interventions. Decis. Sci. **39**(2), 273–315 (2008)
38. Yu, Z.: Visualizing Co-citations of technology acceptance models in education. J. Inf. Technol. Res. **13**(1), 77–95 (2020)
39. Cabrera-Sánchez, J.-P., Villarejo-Ramos, Á.F.: Acceptance and use of big data techniques in services companies. J. Retail. Consum. Serv. **52**, 101888 (2020)
40. Ngugi, D.G., Sawe, E., Nguyen, D., Bertsch, A.: Modelling and measuring acceptance and use of internet banking: the systematic development of an instrument. Rev. Integr. Bus. Econ. Res. **9**(2), 24–45 (2020)
41. Bøe, T., Gulbrandsen, B., Sørebø, Ø.: How to stimulate the continued use of ICT in higher education: integrating information systems continuance theory and agency theory. Comput. Human Behav. **50**, 375–384 (2015)
42. Alraimi, K.M., Zo, H., Ciganek, A.P.: Understanding the MOOCs continuance: the role of openness and reputation. Comput. Educ. **80**, 28–38 (2015)
43. Chen, C.P., Lai, H.M., Ho, C.Y.: Why do teachers continue to use teaching blogs? the roles of perceived voluntariness and habit. Comput. Educ. **82**, 236–249 (2015)
44. Mohammadi, H.: Factors affecting the e-learning outcomes: an integration of TAM and IS success model. Telemat. Informatics **32**(4), 701–719 (2015)
45. de Melo Pereira, F.A., Ramos, A.S.M., Gouvêa, M.A., da Costa, M.F.: Satisfaction and continuous use intention of e-learning service in Brazilian public organizations. Comput. Hum. Behav. **46**, 139–148 (2015)
46. Hsiao, C.-H., Chang, J.-J., Tang, K.-Y.: Exploring the influential factors in continuance usage of mobile social apps: satisfaction, habit, and customer value perspectives. Telemat. Informatics **33**(2), 342–355 (2016)
47. Joo, Y.J., Kim, N., Kim, N.H.: Factors predicting online university students' use of a mobile learning management system (m-LMS). Educ. Technol. Res. Dev. **64**(4), 611–630 (2016)
48. Hu, J., Zhang, Y.: Understanding Chinese undergraduates' continuance intention to use mobile book-reading apps: an integrated model and empirical study. Libri **66**(2), 85–99 (2016)
49. Joo, Y.J., Park, S., Shin, E.K.: Students' expectation, satisfaction, and continuance intention to use digital textbooks. Comput. Human Behav. **69**, 83–90 (2017)
50. Fleming, J., Becker, K., Newton, C.: Factors for successful e-learning: does age matter? Educ. Train. **59**(1), 76–89 (2017)
51. Hair, J.F., Black, W.C., Babin, B.J., Anderson, R.E.: Multivariate Data Analysis. New Jersey (2010)
52. Krejcie, R.V., Morgan, D.W.: Determining sample size for research activities. Educ. Psychol. Meas. **30**(3), 607–610 (1970)
53. Al-Emran, M., Mezhuyev, V., Kamaludin, A.: PLS-SEM in information systems research: a comprehensive methodological reference. In: 4th International Conference on Advanced Intelligent Systems and Informatics (AISI 2018), pp. 644–653 (2018)
54. Fornell, C., Larcker, D.F.: Evaluating structural equation models with unobservable variables and measurement error. J. Mark. Res. **18**(1), 39–50 (1981)
55. Ifinedo, P.: Acceptance and continuance intention of web-based learning technologies (WLT) use among university students in a Baltic Country. Electron. J. Inf. Syst. Dev. Ctries. **23**(1), 1–20 (2006)
56. Hair, J.F., Black, B., Babin, B., Anderson, R.E., Tatham, R.L.: Multivariate Data Analysis. Macmillan, New York (1987)
57. Hair, B., Black, W.C., Babin, B., Anderson, R.E.: Tatham, Multivariate data analysis. Pearson Education, New Jersey (2006)
58. Dijkstra, T.K., Henseler, J.: Consistent and asymptotically normal PLS estimators for linear structural equations. Comput. Stat. Data Anal. **81**, 10–23 (2015)
59. Hair, J.F., Ringle, C.M., Sarstedt, M.: PLS-SEM: indeed a silver bullet. J. Mark. theory Pract. **19**(2), 139–152 (2011)
60. Henseler, J., Ringle, C.M., Sinkovics, R.R.: The use of partial least squares path modeling in international marketing. In: New Challenges to International Marketing, pp. 277–319, Emerald Group Publishing Limited (2009)

61. Nunnally, J.C., Bernstein, I.H.: Psychometric Theory (1994)
62. Hair Jr, J.F., Hult, G.T.M., Ringle, C., Sarstedt, M.: A Primer on Partial Least Squares Structural Equation Modeling (PLS-SEM). Sage Publications (2016)
63. Hu, L., Bentler, P.M.: Fit indices in covariance structure modeling: Sensitivity to underparameterized model misspecification. Psychol. Methods **3**(4), 424 (1998)
64. Bentler, P.M., Bonett, D.G.: Significance tests and goodness of fit in the analysis of covariance structures. Psychol. Bull. **88**(3), 588 (1980)
65. Lohmöller, J.B.: Latent variable path modeling with partial least squares. Physica-Verlag, Heidelberg, Germany (1989)
66. Henseler, J., et al.: Common beliefs and reality about PLS: comments on Rönkkö and Evermann (2013). Organ. Res. methods **17**(2), 182–209 (2014)
67. Alshurideh, M.: Pharmaceutical promotion tools effect on physician's adoption of medicine prescribing: evidence from jordan. Mod. Appl. Sci. **12**(11) (2018)
68. Alomari, K.M., AlHamad, A.Q., Salloum, S.: Prediction of the digital game rating systems based on the ESRB
69. Alshurideh, M.T., Salloum, S.A., Al Kurdi, B., Monem, A.A., Shaalan, K.: Understanding the quality determinants that influence the intention to use the mobile learning platforms: a practical study. Int. J. Interact. Mob. Technol. **13**(11), 157–183 (2019)
70. Alghizzawi, M., Habes, M., Salloum, S.A., Ghani, M.A., Mhamdi, C., Shaalan, K.: The effect of social media usage on students'e-learning acceptance in higher education: a case study from the United Arab Emirates. Int. J. Inf. Technol. Lang. Stud. **3**(3) (2019)
71. Habes, M., Salloum, S.A., Alghizzawi, M., Alshibly, M.S.: The role of modern media technology in improving collaborative learning of students in Jordanian universities. Int. J. Inf. Technol. Lang. Stud. **2**(3), 71–82 (2018)
72. Salloum, S.A., Al-Emran, M., Khalaf, R., Habes, M., Shaalan, K.: An innovative study of E-Payment systems adoption in higher education: theoretical constructs and empirical analysis. Int. J. Interact. Mob. Technol. **13**(6) (2019)
73. Salloum, S.A., Shaalan, K.: Adoption of e-book for university students. In: International Conference on Advanced Intelligent Systems and Informatics, pp. 481–494 (2018)
74. Salloum, S.A.S., Shaalan, K.: Investigating students' acceptance of E-learning system in higher educational environments in the UAE: applying the extended technology acceptance model (TAM). The British University in Dubai (2018)
75. Alshurideh, M., Salloum, S.A., Al Kurdi, B., Al-Emran, M.: Factors affecting the social networks acceptance: an empirical study using PLS-SEM approach. In: 8th International Conference on Software and Computer Applications, pp. 414–418 (2019)
76. Salloum, S.A., Shaalan, K.: Factors affecting students' acceptance of e-learning system in higher education using UTAUT and structural equation modeling approaches. In: International Conference on Advanced Intelligent Systems and Informatics, pp. 469–480 (2018)
77. Salloum, S.A., Alhamad, A.Q.M., Al-Emran, M., Monem, A.A., Shaalan, K.: Exploring students' acceptance of E-learning through the development of a comprehensive technology acceptance model. IEEE Access **7**, 128445–128462 (2019)
78. Al-Maroof, R.S., Salloum, S.A., AlHamadand, A.Q.M., Shaalan, K.: A unified model for the use and acceptance of stickers in social media messaging. In: International Conference on Advanced Intelligent Systems and Informatics, pp. 370–381 (2019)
79. Al-Shibly, M.S., Alghizzawi, M., Habes, M., Salloum, S.A.: The impact of De-marketing in reducing jordanian youth consumption of energy drinks. In: International Conference on Advanced Intelligent Systems and Informatics, pp. 427–437 (2019)
80. Alhashmi, S.F.S., Salloum, S.A., Abdallah, S.: Critical success factors for implementing artificial intelligence (AI) projects in Dubai Government United Arab Emirates (UAE) health sector: applying the extended technology acceptance model (TAM), In: International Conference on Advanced Intelligent Systems and Informatics, pp. 393–405 (2019)
81. Alshurideh, M., Al Kurdi, B., Salloum, S.A.: Examining the main mobile learning system drivers' effects: a mix empirical examination of both the expectation-confirmation model (ECM) and the technology acceptance model (TAM). In: International Conference on Advanced Intelligent Systems and Informatics, pp. 406–417 (2019)

82. Habes, M., Salloum, S.A., Alghizzawi, M., Mhamdi, C.: The relation between social media and students' academic performance in Jordan: youtube perspective. In: International Conference on Advanced Intelligent Systems and Informatics, pp. 382–392 (2019)
83. Alghizzawi, M., Habes, M., Salloum, S.A.: The relationship between digital media and marketing medical tourism destinations in Jordan: Facebook perspective. In: International Conference on Advanced Intelligent Systems and Informatics, pp. 438–448 (2019)
84. Salloum, S.A., Al-Emran, M., Habes, M., Alghizzawi, M., Ghani, M.A., Shaalan, K.: Understanding the impact of social media practices on E-Learning systems acceptance. In: International Conference on Advanced Intelligent Systems and Informatics, pp. 360–369 (2019)
85. Chin, W.W.: The partial least squares approach to structural equation modeling. Mod. methods Bus. Res. **295**(2), 295–336 (1998)
86. Hsu, C.-L., Lu, H.-P.: Why do people play on-line games? An extended TAM with social influences and flow experience. Inf. Manag. **41**(7), 853–868 (2004)
87. Lin, T.-C., Chen, C.-J.: Validating the satisfaction and continuance intention of e-learning systems: combining TAM and IS success models. Int. J. Distance Educ. Technol. **10**(1), 44–54 (2012)
88. Lee, J.-K., Lee, W.-K.: The relationship of e-Learner's self-regulatory efficacy and perception of e-Learning environmental quality. Comput. Hum. Behav. **24**(1), 32–47 (2008)

A Game-Based Learning for Teaching Arabic Letters to Dyslexic and Deaf Children

Sahar A. EL_Rahman

Abstract In Arab and Muslim countries, the Arabic language is important because it is the Quran's language. Thus, teaching it to the children is an important issue. However, some of these children might suffer from a learning difficulty or disability such as dyslexia and deaf children. Unfortunately, it has been hardly to find any educational application in Arabic to help them and make their study life easier. Therefore, the aim of this work is to provide a good quality, interesting and encouraging application to teach both the Arabic alphabet and the Arabic sign language for a child who suffers from either a reading difficulty (Dyslexia) or a hearing disability (deaf children). The system teaches the child the Arabic letters systematically in a sequential manner starting for the first letter and ending up with the last one to help the child to learn the letters, step by step and do not move from the first letter to the second one until passing the first letter. It uses an interesting interface and interactive games to provide an encouraging learning environment and achieve a learning and studying life easier for the children with Dyslexia and deaf children.

Keywords Education · E-learning · Mobile application · Learning difficulties · Hearing disability · Deaf children · Arabic alphabets · Reading difficulties · Dyslexia · Sign language

1 Introduction

Education is one of the basic needs for a human being. It is a process by which our mind develops. In other words, a mental and intellectual training helps us to meet challenges and overcome obstacles to progress. The necessity of education is immense. It is a significant factor in the development of children, communities,

S. A. EL_Rahman (✉)
Computer Science Department, College of Computer and Information Sciences, Princess Nourah Bint Abdulrahman University, Riyadh, Saudi Arabia
e-mail: sahr_ar@yahoo.com

Electrical Engineering Department, Faculty of Engineering-Shoubra, Benha University, Cairo, Egypt

M. Al-Emran et al. (eds.), *Recent Advances in Intelligent Systems and Smart Applications*, Studies in Systems, Decision and Control 295,
https://doi.org/10.1007/978-3-030-47411-9_19

and countries [1]. The primary goal of education is to prepare the individual to move forward on the path of learning. However, there are many reasons impede the learning process, and although varies from one person to another. Therefore, they need creative teaching methods to both facilitate their learning process and use the modern technology to help them [2].

Technology is everywhere in education. The effective use of technology in education has changed the face of education and it has created more educational opportunities. Both teachers and students have benefited from various educational technologies. Teachers have learned how to integrate technology in their classrooms and students are getting more interested in learning with technology [2]. Recent advancements in educational technologies have yielded positive results in the education sector. This new educational technology is supporting both teaching and learning processes to increase the student engagement and motivation towards learning [2], especially game-based learning and mobile learning applications. Game-based learning can improve the learning performance and learning motivation of students [3]. Also, mobile learning applications are found improving the students' intent to utilize such applications [4].

Nowadays, many children suffer from learning difficulties. Learning difficulties are neurologically based processing problems. These processing problems can interfere with learning basic skills such as reading, writing and/or math. They can also interfere with higher-level skills such as organization, time planning, abstract reasoning, long or short-term memory and attention. It is important to realize that learning difficulties can affect an individual's life beyond academics and can affect relationships with family, friends and in the workplace. There are so many types of learning difficulties: Math difficulties (Dyscalculia), Writing difficulties (Dysgraphia), Reading difficulties (Dyslexia), Auditory Processing Disorder (APD), Language Processing Disorder, Nonverbal Learning difficulties, Visual Perceptual/Visual Motor Deficit [5, 6].

Moreover, there are many types of disabilities that might impede the learning process such as deaf and hard of hearing, vision loss, and physical disabilities. However, it has been found that the deaf children did not receive enough interest and support in the basic educational learning process especially in the Arab world and there are a few educational applications that are not helpful for them [7].

Furthermore, in Arab and Muslim countries, the Arabic language is important because it is the Quran's language, which leads to the importance of teaching the Arabic language to the children [8]. Therefore, this work focuses on the reading difficulty, which is known as dyslexia, because it is the most common and important one because reading is an essential skill that learners cannot move forward in their educational life without it [9]. In addition, it has been hardly to find any educational application in Arabic to help them and make their study life easier [6].

To conclude, the proposed system is intended to both Arabic dyslexic children and deaf children. It developed an interactive game mobile application to teach the dyslexic and deaf child the Arabic Alphabet in a sequential order followed by short quizzes using interesting interfaces by adding some cartoon characters to motivate the child for learning.

The rest of this paper is organized as follows. The literature review will be introduced in the next section while the proposed solution along with a comparison between the proposed system and existing ones will presented in sections three and four respectively.

2 Background Information

2.1 Reading Difficulty (Dyslexia)

Learning disabilities are neurologically based processing problems. These processing problems can interfere with learning basic skills such as reading, writing and/or math. They can also interfere with higher-level skills such as organization, time planning, abstract reasoning, long or short term memory and attention [9].

Dyslexia is one of the most common learning difficulties and the most attracting of scientists and educators' attention [10]. It is primarily associated with trouble reading. Some doctors, specialists and educators may refer to it as a "reading disorder" or a "reading disability." Nevertheless, it can also affect writing, spelling and even speaking. People with dyslexia can still understand complex ideas. Sometimes they just need more time to work through the information. They may also need a different way to process the information, such as listening to an audiobook, instead of reading it [11].

Early success in reading skills usually leads to later success in reading, while failing to read before the third or fourth year of schooling may be indicative of lifelong reading problems. Thus, early detection is best done in early childhood or during the first year of school where the gap that separates the students at risk of reading failure and the students that are likely to be successful readers is small. Early detection alone, however, will not improve literacy levels unless the student receives appropriate early intervention before reading problems become entrenched [12].

Reading is the process of constructing meaning through the dynamic interaction among the reader, the text and the context of the reading situation. Reading process involves visual motor skills and perception of the symbols by the brain. It is generally broken down into two components: Reading the words, or decoding and understanding what is read, or comprehension. This study intended to define reading the lines. Reading the lines is the process of decoding the words in order (*reading accurately and fluently*) [13].

In the comprehension component, often children with learning difficulties suffer from problems in understanding what is read (*such as absorbing stories in the core classes in spite of these problems may appear as a result of the weakness of skills decoders but it may appear due to poor definition of key story elements and text content strategies*) [14]. However, dyslexia suffers from a variety of difficulties such as [10]:

- Spelling difficulty.
- Inability to pronounce the Arabic letters.
- Delayed speaking or speaking unclear or mixing characters such as:
  - Deleting some letters or adding others.
  - Switching letters position forward or backward.
  - Switching some words with its synonyms.
  - Weakness in distinguishing vowels.
- Difficulty in repeating what is heard.
- Weaknesses in focusing when listening to stories.

2.2 *Hearing Disability*

Hearing disability refers to those who are hard of hearing and use the Sign Language to communicate [9]. In addition, eye gaze and attention are cognitive skills that develop during the growth of a deaf child. These skills are aspects of executive function, the cognitive system that controls and manages other cognitive processes such as planning, working memory, attention, etc. Eye gaze and attention skills are crucial despite the tendency to wait to teach fingerspelling until after a child knows the alphabet, Baker explained in 2010 that deaf children with early exposure to ASL (American sign language) learn to fingerspell well before they develop the alphabetic principle. They see fingerspelling not as individual letters, but as whole units within a movement envelope. As a result, the finger spelt word is comprehended as a sign instead of as letters that make up a word. Later, after they learn the alphabet, children realize that the signs they learned were actually made of letters [15].

As children learn to use fingerspelling in more complex ways, they begin to lexicalize it. The process of lexicalization takes letters and forms them into a sign, and this process has been found to increase the vocabularies of deaf children. At approximately the third grade, fingerspelling, reading, writing, and spelling skills converge. Studies have shown that children who are good finger spellers tend to be good readers and that children who are good readers are usually good finger spellers [15].

Finally, deaf children suffer from challenges in the learning process. Research showed that this category did not receive enough interest and support in the basic educational learning process in the Arab world unlike the western world. It is rare to find educational programs and applications containing strategies suitable for them. It is worth to mention that the Arabic language is Acoustic language while the sign language that is used by deaf individuals is a visual descriptive language that cannot be read or written which is completely far from the nature of the Arabic language [16].

3 Related Work

In this work, an interactive game mobile application was developed to teach the dyslexic and deaf child the Arabic. Therefore, the existing systems related to the proposed solution has been explored and detailed below:

3.1 Reading Difficulty (Dyslexia)

No educational applications to support teaching Arabic Alphabets intended for children who suffer from dyslexia. However, some of the applications that support the teaching of the Arabic Alphabet for normal children, which may be useful for children who suffer from dyslexia, have been explored in the following sub-sections.

3.1.1 The Alphabets Train Application

It is an interactive application that helps the child to pronounce the Arabic letters correctly and to identify the forms of the letters in a fun and attractive way by linking them to the names of the favorite animals for them (see Fig. 1) [17].

The application works on developing the child's skills by attracting his/her attention and using different senses. The alphabet train is colorful and beautiful dragging his/her successor twenty-eight vehicle. In each vehicle, there is a letter of the Arabic language and an animal that begins with that letter [17].

If the figure at the bottom of the screen clicked, then the character will be altered enabling the child to listen to the character, record his/her voice and listen to it again. The application is very easy to use, therefore the child alone can use it and he/she can move between the screens easily [17].

Fig. 1 Alphabets train system [17]

Fig. 2 Abjad application [18]

3.1.2 Abjad Application

This application provides an easy way to teach the Arabic alphabets for children. It is designed to be entertaining and colorful with bright colors to attract the child while teaching the basics of reading and writing letters (see Fig. 2). It contains the following activities [18]:

- If a character is clicked, then a short song for a word that begins with that letter will be played.
- Running a simple test by displayed an image and the child can click on the appropriate letter that corresponds to that displayed image.
- Running a game that links letters with their images.
- Connecting the dots to draw letters.
- Dragging letters to form words.
- Running a memory game, which stimulate the child's memory and keep the focus on the pronunciation of the letters.
- Helping the child to learn the different forms of the character (beginning, middle, end and isolated).

3.1.3 Learn Reading and Writing with Maher Application

This interactive application enables the students to learn reading and writing skills of all Arabic language characters, through a range of educational games, to find him/herself playing, enjoying and learning (see Fig. 3) [19].

3.2 Hearing Disability

There are educational applications for the deaf people but not specialized on deaf children. Each one has been detailed in the following sub sections.

Fig. 3 Maher system [19]

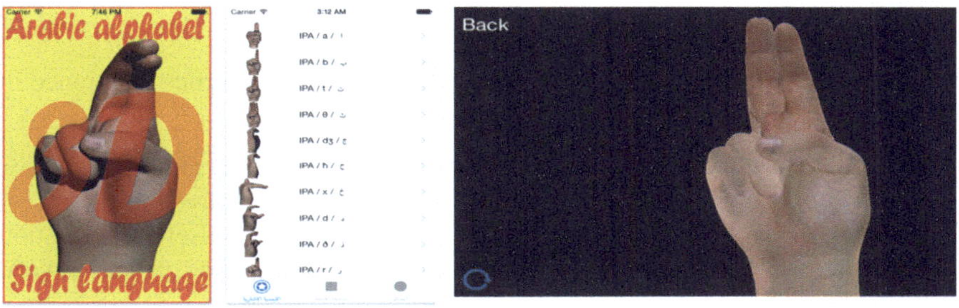

Fig. 4 3D Arab alphabet sign language application [20]

3.2.1 3D Arab Alphabet Sign Language

This application provides an introduction to the Arabic Signs Language (ASL) for deaf people, a quiz with over then 100 questions along with a 3D navigation for each sign (see Fig. 4) [20].

3.2.2 Palestinian Sign Language Application

This application represents both the Arabic alphabet and the numbers and in the Palestinian sign language as indicated in Fig. 5 [21].

3.3 A Comparison Between the Proposed System and Other Systems

The comparison between the existing applications and the proposed application is shown in Tables 1 and 2.

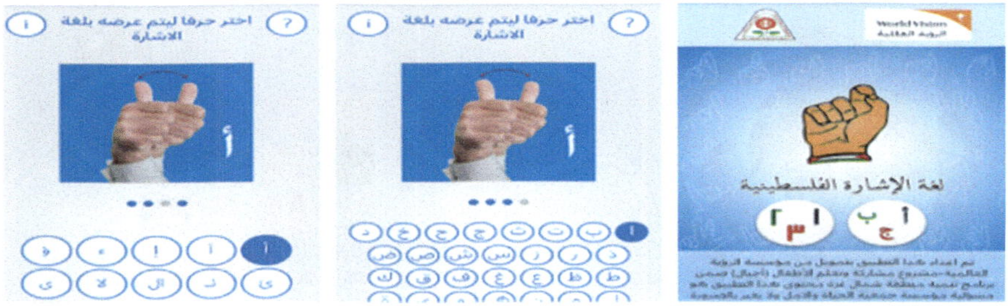

Fig. 5 Palestinian sign language system [21]

4 Methodology

In order to design and develop a learning system with the properties mentioned in the previous section, the steps below have been followed:

1. The research related to this system has been searched and explored. Precisely, research related to dyslexia and hearing disability has been investigated.
2. An interview with a special education teacher has been done. Useful information has been extracted from this interview. As a result, a combination of data between the theory and practice has been gained.
3. Learning systems similar to this proposed one have been explored. A comparison between these systems with the proposed one has been done in terms of problem solved, domain and users, design methods, etc.
4. The proposed system has been:
   (a) Designed and developed to meet the requirements mentioned in the previous section.
   (b) Tested on a number of dyslexic children in a special education center.
5. Some notes and comments on the system have been taken from a specialist.

5 Proposed Solution

It has been mentioned earlier the importance of learning especially learning how to read for either dyslexic children or deaf children and due to the society unawareness and ignorance of the problems and the lack of educational programs these children suffer; this work has aimed to design an entertainment educational application that helps these children.

The application is an android mobile application that teaches the Arabic alphabets. It is intended to be used by children in the age between 6 and 8 years' old who suffer from either reading difficulty (Dyslexia) or hearing disability.

Table 1 Comparison of proposed and similar applications for dyslexia

	أتعلم القراءة والكتابة مع ماهر[19]	أبجد[18]	قطار الحروف[17]	Proposed system
Problem solved	Children boredom and not being attracted to traditional teaching methods	Children boredom and not being attracted to traditional teaching methods	Children boredom and not being attracted to traditional teaching methods	Build a mobile application to Teach reading Arabic Alphabet for dyslexia children
Domain & users	This application for teaching Arabic alphabets, it is intended to be used by children	This application for teaching Arabic alphabets, it is intended to be used by children	This application for teaching Arabic alphabets, it is intended to be used by children	This application for teaching Arabic alphabets, it is intended to be used by children who suffer from reading difficulties (Dyslexia) and deaf children
Design methods	Specially designed for children using the GUI as a fun toolkit with 2D image, icons, graphics, cartoons, sounds and short song matched with each letter for supporting the needs of normal kids	Specially designed for children using the GUI as a fun toolkit with 2D image, icons, graphics, cartoons, sounds and short song matched with each letter for supporting the needs of normal kids	Specially designed for children using the GUI as a fun toolkit with 2D image, icons, graphics, cartoons and sounds with each letter for supporting the needs of normal kids	Specially designed for children suffering from reading difficulties, the GUI as a fun toolkit with 2D image icons, graphics, cartoons and sounds
Output	The target is to help children to learn Arabic alphabet in the easy and attractive way	The target is to help children to learn Arabic alphabet in easy way	The target is to help children to learn Arabic alphabet in easy way	The target is to help dyslexic children overcome any difficulties they face as well as provide them with useful skills

(continued)

Table 1 (continued)

	أتعلم القراءة والكتابة مع ماهر [19]	أبجد [18]	قطار الحروف [17]	Proposed system
Features	• Learning by playing. There are 15 games for each letter which the child must pass to get the certificate for learning this letter • Learning different forms of the letter and how it looks like in each case (beginning, middle, end and isolated) letter • Going forward is done step by step in such a motivating way for the child to learn and develop even more• Generally, children with learning disabilities need encouraging words vehemently	• Learning by Playing. There are 5 different games such as matching words, beginning with that letter with the appropriate picture, and ordering letters to form a word • Letters writing training • Learning letters by singing them	• Helping your child to pronounce the Arabic letters correctly and helping the child to know the forms of letters in a fun and attractive way • The application works on developing child's skills by attracting his/her attention and making use of different senses • If the figure at the bottom of the screen clicked, the character will alter • After that, the child listens to the character, can record his voice and listen to it again	• Containing activities and voices. • Teaching reading letters one by one with repetition • Learning the alphabets by pronouncing Arabic letters • Explaining what the child read by vocal repeating and showing images • Comparing the letters with real things • Learning different forms of the letter and the way it looks like in each case as (first, middle, end and isolated) in order to remember it easily • Special exercises customized for this category

(continued)

Table 1 (continued)

	أتعلم القراءة والكتابة مع ماهر[19]	أبجد[18]	قطار الحروف[17]	Proposed system
Limitations	• The target age group are children from 6 to 7 years old • This includes teaching the entire 28 Arabic alphabets	• The target age group are children from 4 to 7 years old • This includes teaching the entire 28 Arabic alphabets and writing each letter	• The target age group are children from 4 to 7 years old • This includes teaching the entire 28 Arabic alphabets and way of pronouncing each letter	• The target age group are children from 6 to 8 years old • This includes teaching the entire 28 Arabic alphabets and way of pronouncing each letter. The way of writing each letter depending on its position in the word. And teaching Arabic Diacritics (كسرة,ضمة,فتحة)
Disadvantages	• Not suitable for children with reading disabilities as it was not aimed for them • It does not take the amount of learnt material into consideration	• Not suitable for children with reading disabilities as it was not aimed for them • It does not take the amount of learnt material into consideration	• Some letters are not pronounced correctly • The games corner needs to be developed • It is usable for android operating system only • Not suitable for children with reading disabilities as it was not aimed for them • It does not take the amount of learnt material into consideration	• It is suitable for Smartphone with android operating system only

Table 2 Comparison of proposed and similar applications for deaf children

	لغة الإشارة الفلسطينية[21]	3D Arab alphabet sign language [20]	Proposed system
Problem solved	Rare application for deaf people and these people really need this type of applications	Deaf people need to learn the Arabic Alphabetic because it is important for them to communicate with the others who have no idea about sign language	We could not find an application to help reading Arabic alphabetic sign language children with deaf disability
Domain & users	This application for teaching Arabic alphabets, it is intended to be used by deaf people	This application for teaching Arabic alphabets, it is intended to be used by deaf people	This application for teaching Arabic alphabets, it is intended to be used by children who suffer from reading difficulties (Dyslexia) and deaf children
Design method	Specially designed for deaf people using the GUI as with 2D image, icons, graphics, for supporting the needs of deaf people	Specially designed for deaf people using the GUI as with 3D imagery, icons, graphics, for supporting the needs of deaf people	Specially designed for deaf children using the GUI as a fun toolkit with 2D image, icons, graphics and video for supporting the needs of deaf kids
SW & HW	Smartphone with android operating system	Smartphone with IOS operating system	Smartphone with android operating system
Output	The target is to help deaf people to learn Arabic alphabet in easy way	The target is to help deaf people to learn Arabic alphabet in easy wa	The target is to help deaf children overcome any difficulties they face as well as provide them with useful skills
Features	• Learn the Arabic Alphabet and numbers you have to click on the letter button, then you will see the letter in Arabic with the sign picture • There are pictures to show the letter sign for Arabic Alphabetic and numbers	Introduction to Arabic Language Signs (ARSL) for deaf people by alphabet, quiz over 100 questions and a 3D navigation for each sign	• By adding the sign language and showing images in the application to help deaf children to receive information perfectly • Learning different forms of the letter and the way it looks like in each case as (first, middle, end and isolated) in order to remember it easily

(continued)

Table 2 (continued)

	لغة الإشارة الفلسطينية[21]	3D Arab alphabet sign language [20]	Proposed system
Limitations	• The target age group is greater than 10 years old • This includes teaching the entire 28 Arabic alphabets by using the Arabic sign language	• The target age group is greater than 15 years old • This includes teaching the entire 28 Arabic alphabets by using the Arabic sign language	• The target age group are children from 6 to 8 years old • This includes teaching the entire 28 Arabic alphabets by using the Arabic sign language and the way of writing each letter depending on its position in the word
Disadvantages	• The application is not intended for children, therefor the colors are unattractive and there are no carton pictures to attract the child. The application is intended for adults, so there is no sequential plan to teach the characters	• The application is not intended for children, so the colors are unattractive, there are no carton pictures to attract the child, and there are no images to bring the form of the letter. In addition, the application is for adults, so there is no sequential plan to teach the characters	• It is suitable for Smartphone with android operating system only

For dyslexic children, the proposed system teaches the entire 28 Arabic alphabets, their pronunciation and the way of writing each letter depending on its position in the word. In addition, it teaches them the Arabic Diacritics (***Fat'ha***, ***Dammmah***, ***Kasrah***). For deaf children, on the other hand, the proposed system teaches the entire 28 Arabic alphabets by using the Arabic sign language and the way of writing each letter depending on its position in the word. The application is also helpful for normal children who do not suffer from either reading difficulty (Dyslexia) or hearing disability.

Therefore, the proposed system provides general properties for both dyslexic and deaf children:

1. Teaching Arabic alphabets in simple and easy ways.
2. Explaining the ideas and information clearly and avoiding interference and complexity.
3. Encouraging and stimulating child when trying or succeeding in passing quiz/game.
4. Teaching the different forms of the letters (***beginning***, ***middle***, ***end and isolated***).

5. Teaching by playing: where the child can play different games to make sure he/she had mastered the letter.
6. Teaching the Arabic Diacritics.

In addition, there are more properties special for dyslexic children and others for deaf ones. Each one will be explained in the following sub sections.

5.1 Reading Difficulties

The proposed system provides additional properties for only dyslexic children:

1. Providing activities and sounds.
2. Teaching letters' pronunciations one by one with repetition.
3. Explaining what the child is reading by vocal repeating and showing images.
4. Comparing the letters with real things in order to remember it easily.

5.2 Deaf Children

The only property that was added for deaf children is the sign language by displaying images of it to help deaf children to receive information perfectly.

5.3 Proposed System Architecture

First of all, to enable users to use the application services, the child must be registered in the application. The child needs to new register in the application enter your user name and e-mail address and password and determine the type (Reading Difficulties or Deaf Children). Home page contains a series of stages contain teach letters and many of quizzes. All of services are stored in the database so each time, when the user login, the user can access to its data and stages, as shown in Fig. 6.

5.3.1 Application Architecture

In the proposed application architecture, there are three layers and each layer have specific function, as shown in Fig. 7.

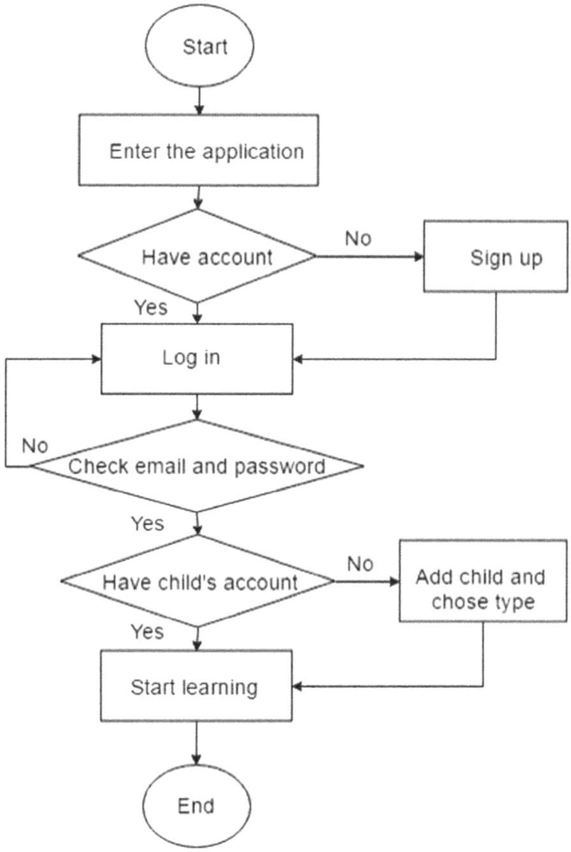

Fig. 6 Process of logging the system

5.3.2 Description of Design Components

The description of each component in the application architecture of mobile application as follows:

1. ***Presentation layer***: the top-most layer of the application which contains user interface. The main function of this layer is to translate interaction, so the user can understand it.
2. ***Business layer***: the middle layer of the application which coordinates the application, processes command, make logical decisions. It also moves and process data between presentation and data access layers.
3. ***Data access layer***: the bottom layer which store and retrieve information from the database. The information passes through the business layer then to the user. SQLite database is used that is local database.

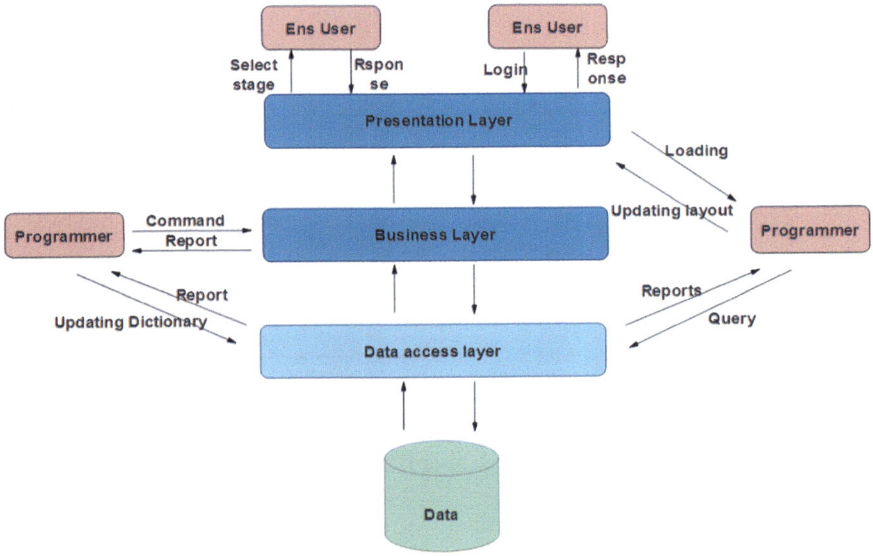

Fig. 7 Application architecture

5.4 *Data Design*

The database design Contain Database schema, as shown in Fig. 8.

5.5 *User Interface Design*

This section represents the User Interface Design in details. First, User Interface Design Overview: shows an overview of the way used to design the interface. Second, User Interface Navigation Hierarchy: explains the way to navigate between the screens. Finally, User Function Interfaces: show all of the application screens and explain the function and benefits of each one, sample of main screens are indicated in Fig. 9.

5.5.1 User Interface Design Overview

The main and important point is to make sure the design is attractive, fun and easy to use by the child. Simply after the user register and login (certainly by help of somebody who helped the child in downloading the application for example his\her parent) the home screen shows four sections, the parent can open up to four profiles for his\her children under one account, only by click on "Add new user" and enter

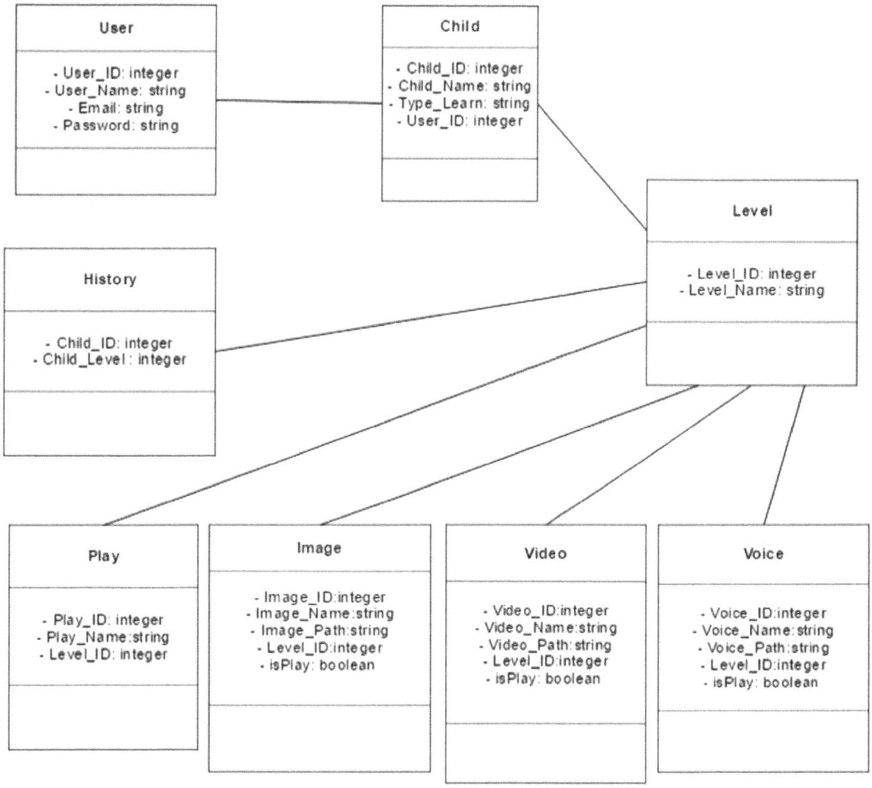

Fig. 8 Database schema

the name and the problem the child suffer from (Dyslexia\Hard of hearing (. The main screen also shows three icons about, information and logout. When the user clicks on his\her name his\her profile will open and show two buttons "Let's play" and "Record", also the home icon which return the user to the home screen.

When the user clicks on "Record" button all the letters the child has done will show with a star on the top. when the user clicks on "Let's play" button the stage, screen shows all 12 stage the user cannot move to the next stage before finish the previous one. Each stage contains different number of letters, the letters divided in the stages depending on its shape. When the user clicks on the letter, then he/she will go through a sequential learning plan containing pictures, voices, activities and games.

5.5.2 User Interface Navigation Hierarchy

The navigation hierarchy in detail is indicated in Fig. 10.

(a) Start Screen

(b) Login Screen

(c) Home Screen

Fig. 9 The main screen in the proposed application

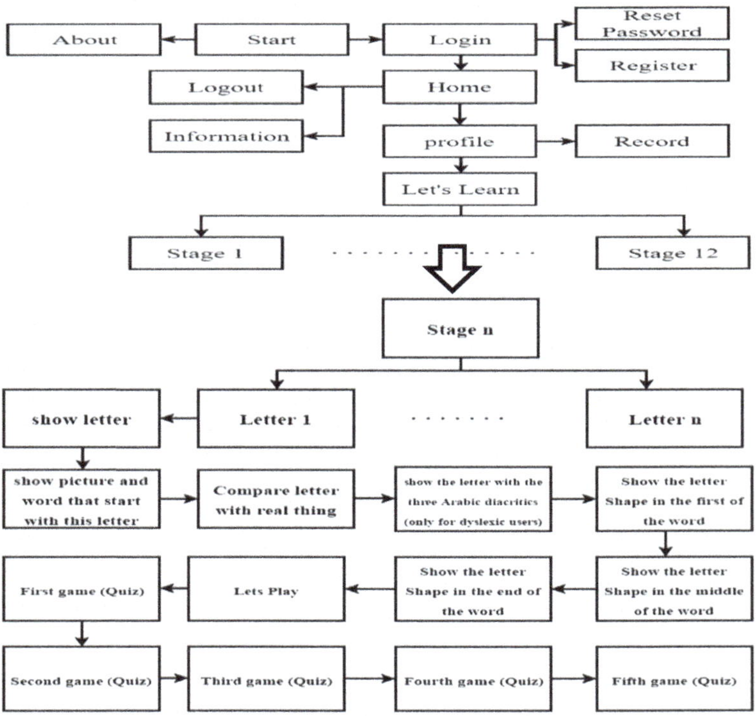

Fig. 10 Navigation hierarchy

6 Results

After following the previous methodology, the system has been developed and a comparison between it and other related systems have been applied. The following sub sections introduces the developed system along with the comparison.

6.1 The Proposed System (Let Us Learn Reading)

The proposed system has been developed as an android mobile application to teach the Arabic alphabet for dyslexic children and the Arabic Sign Language for deaf ones. The android studio has been used to design and develop the application while the firebase technology has been used to save and retrieve data from a cloud storage.

The system contains from two modules: parent module and child module. The parent module is to enable the parent to register up to four children in the system (see Fig. 11). The other module which is the child module is divided into two modes: dyslexia and deaf (see Fig. 12).

In the dyslexia mode, the dyslexic child will be able to learn the Arabic alphabet in a sequential order by pronouncing them. During learning each letter, a group of games is provided to the child, which he/she cannot escape them without answering each one of them.

In the deaf mode, the deaf child will be able to learn the Arabic Sign Language in a sequential order by using photos and videos. During learning each letter, a group of

(a) Dyslexic Users.

(b) Deaf Users.

Fig. 11 Parent module screens

(a) Dyslexic Users.

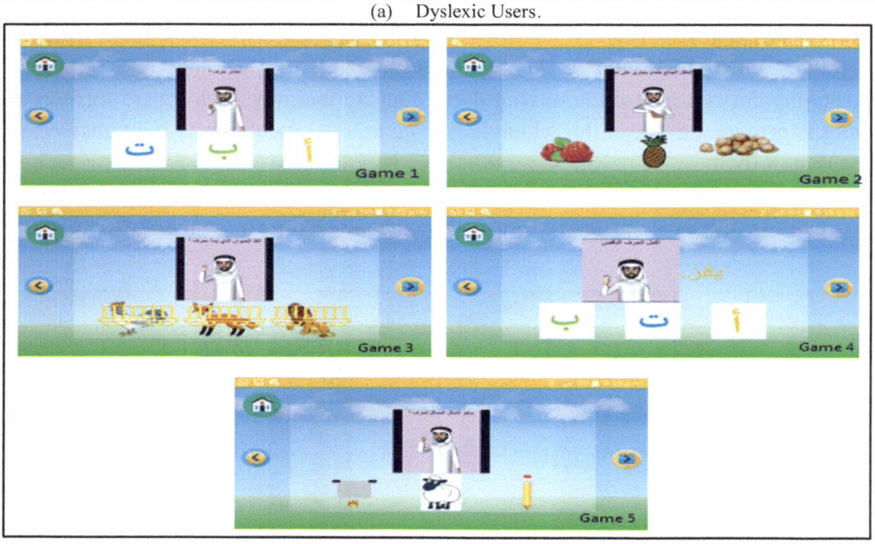

(b) Deaf Users.

Fig. 12 Profile screen

games is provided to the child, which he/she cannot escape them without answering each one of them.

In both modes, the learning system contains twelve stages with each stage contains levels that represent the Arabic alphabets (see Figs. 13 and 14). In addition, a record for each child is kept informing both the teacher and the parents which the Arabic

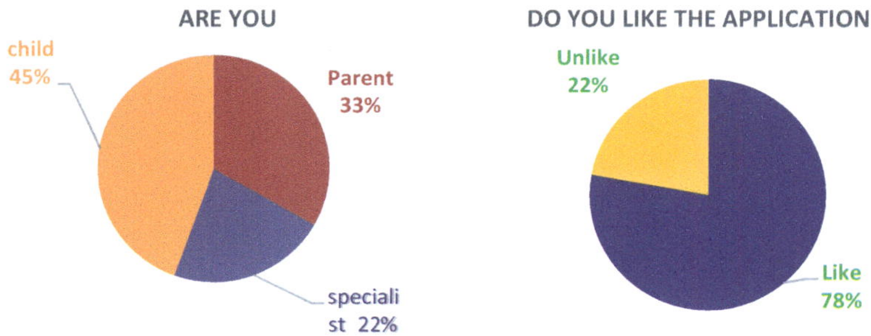

Fig. 13 Stages screen

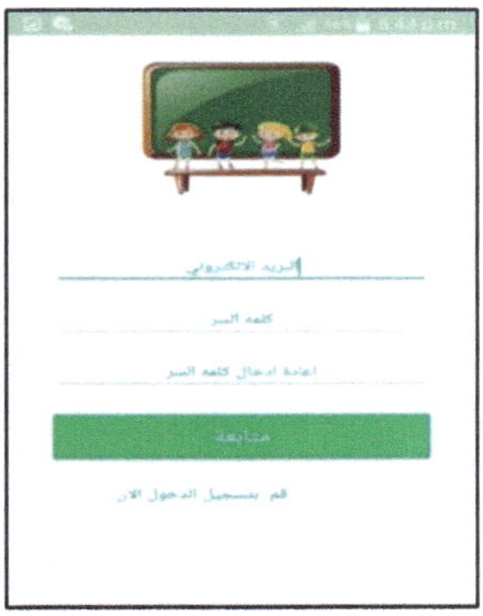

(a) Registration

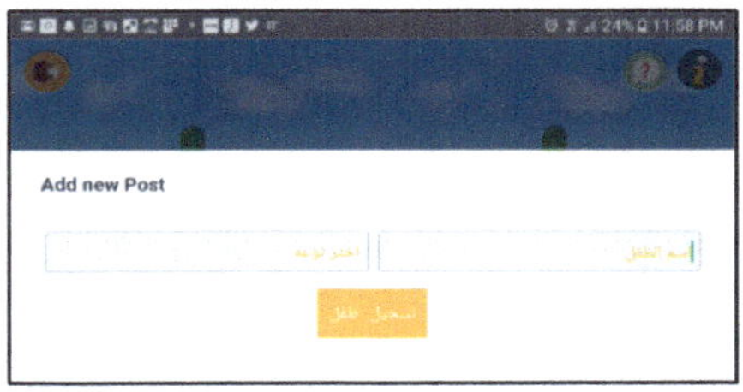

(b) Adding a new child

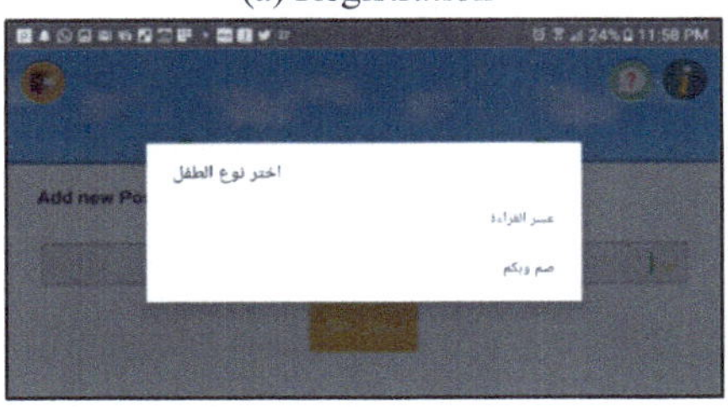

(c) Identifying the child's problem

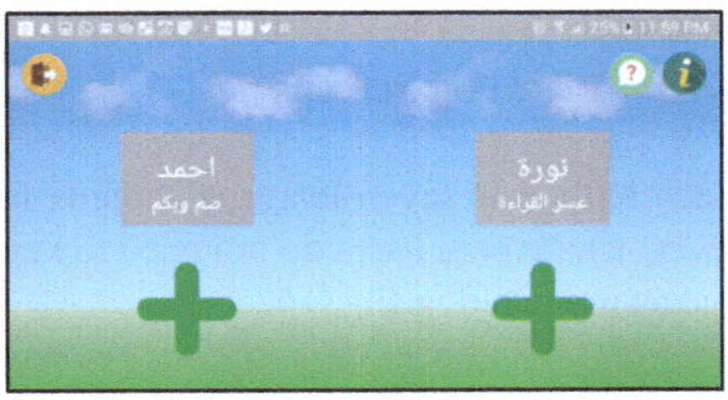

(d) Result of addition

Fig. 14 Levels Screen

letter has been reached. A star (see Fig. 15) signals the Arabic letters that the child succeeded in.

Fig. 15 Child's record

Fig. 16 Learning screens

6.1.1 Learning Screens

Each letter has seven learning screens for dyslexic users and six screens for deaf users, where deaf users do not need to know about the Arabic Diacritics in this early age as indicated in Fig. 16.

6.1.2 Let's Play Screens

After the child finishes the learning about specific letter the game screens will appear that includes five games. The game explanations are done by the voice for dyslexic children and by the video in sign language for deaf children (see Fig. 17).

Fig. 17 Games screens

6.2 *Proposed System Evaluation*

The system has been tested in terms of its functionality and acceptance. The functionality test is for testing the functions of the system to ensure its freeness from bugs. On the other hand, the acceptance testing is for investigating the acceptance level of the application by a four of dyslexic and deaf children, three parents and two specialists (see Fig. 18). It has been found the following:

1. 78% of the participants liked the application (see Fig. 18).
2. Based on the specialist's feedback, it is more appropriate to unforce the child to start learning the Arabic alphabet from the beginning and in a sequential manner. In other words, the sequential order of teaching the Arabic alphabet should be broken down by unlocking all the letters to enable the specialist to choose any letter he/she wants the child to learn.
3. The specialists recommended that adding a video of the human's mouth to show the way of pronouncing the letter.

Fig. 18 Real evaluation of the proposed application

7 Conclusion and Future Work

The Arabic language is an important language especially for both Arab countries because their first language and Muslim countries because it is the language of the Holy Quran. Thus, teaching the Arabic language to the children is an important issue. However, some of these children might suffer from a learning difficulty or disability such as dyslexia and deaf children. Unfortunately, it has been hardly to find an educational application in the Arabic language to help them and make their study life easier. Thus, this work aimed to serve the children who suffer from either a reading difficulty or a hearing disability to enable them to learn the principles of reading. In addition, it teaches them writing the Arabic alphabets taking into consideration the position of the letters within the word. To do so, the system contained multiple stages ranging from completing exercises for each Arabic letters to providing multiple games to test their understanding of the previous stage. The application has been evaluated by a four of dyslexic and deaf children, three parents and two specialists. The findings are 78% of the participants liked the application. Based on the specialist's feedback, the sequential order of teaching the Arabic alphabet should be broken down by unlocking all the letters to enable the specialist to choose any letter he/she wants the child to learn. Also, the specialists recommended that adding a video of the human's mouth to show the way of pronouncing the letter.

Acknowledgements The authors would like to thank all the participants involved into this work at College of Computer and Information Sciences, Princess Nourah bint Abdulrahman University especially to Hadeel AlAteeq, Bedor Al-Ghmlas, Ashwaq Al-Anazi, Wejdan Al-Zahrani, Noura Al-hosini, Buthainh AL-Qhtani, and Al Jawhara Al-Sayari.

References

1. Why is education important in life? [Online] Available: https://www.reference.com/education/education-important-life-2c3b80038e953b9f. Accessed: 4 Nov 2016
2. Ramey, K.: The use of technology—in education and teaching process. [Online] Available: http://www.useoftechnology.com/the-use-of-technology-in-education/. Accessed 4 Nov 2016
3. Sung, H.Y., Hwang, G.J.: A collaborative game-based learning approach to improving students' learning performance in science courses. Comput. Educ. **63**, 43–51 (2013)
4. Al-Emran, M., Mezhuyev, V., Kamaludin, A.: An innovative approach of applying knowledge management in M-learning application development: a pilot study. Int. J. Inf. Commun. Technol. Educ. (IJICTE) **15**(4), 94–112 (2019)
5. Lerner, J., Johns, B.: Learning Disabilities and Related Disabilities. Nelson Education, Stamford, USA (2014)
6. UNICEF: Children and young people with disabilities fact sheet. [Online] Available: https://www.unicef.org/disabilities/files/Factsheet_A5__Web_NEW.pdf. Accessed 2 May 2013
7. Jeannette, W., Jenna, L.: Supporting deaf and hard of hearing students tools, technology and key resources to cultivate academic success. [Online] Available: http://www.accreditedschoolsonline.org/resources/students-with-hearing-impairments. Accessed: 11 Nov 2016
8. Murad, M.: أهمية تعلم اللغة العربية. [Online] Available: http://mawdoo3.com/%D8%A3%D9%87%D9%85%D9%8A%D8%A9_%D8%AA%D8%B9%D9%84%D9%85_%

D8%A7%D9%84%D9%84%D8%BA%D8%A9_%D8%A7%D9%84%D8%B9%D8%B1%D8%A8%D9%8A%D8%A9
9. Sink, C.: Mental Health Interventions for School Counselors. Cengage Learning, Belmont, CA (2011)
10. Jim, R.: Identifying and teaching children and young people with dyslexia and literacy difficulties. An independent report (2009). [Online] Available: http://webarchive.nationalarchives.gov.uk/20130401151715/http://www.education.gov.uk/publications/eOrderingDownload/00659-2009DOM-EN.pdf
11. Emily, L.: Understanding dyslexia. [Online] Available: https://www.understood.org/en/learning-attention-issues/child-learning-disabilities/dyslexia/understanding-dyslexia. Accessed: 07 Nov 2016
12. Kelly, K., Phillips, S.: Teaching Literacy to Learners with Dyslexia: A Multi-sensory Approach. Sage, Chennai, India (2016)
13. Karanja, W.: Effects of reading difficulties on academic performance among form three Students in public secondary schools, Kiambu County, Kenya. Ph.D. dissertations, Kenyatta University, 2015
14. Aljawhara, A.: الحاجات التعليمية لأطفال صعوبات التعلم واستراتيجيات معالجتها. [Online] Available: http://www.dr-khalidh2.com/wp-content/uploads/2013/04. Accessed: 6 Nov 2016
15. Gertz, G., Boudreault, P.: The Sage deaf studies encyclopedia. Sage Publications, Washington D.C., USA (2015)
16. Jamiel, S.: The problem of education of the deaf (1). Almanal Magazine, Societal Issues (2014)
17. قطار الحروف. [Online] Available: https://play.google.com/store/apps/details?id=com.bee.studio.letterstrain. Accessed: 6 Nov 2016
18. Kadri, O.: Abjad on the App Store. [Online] Available: https://itunes.apple.com/us/app/abjad/id570214181?mt=8. Accessed: 20 Oct 2016
19. أتعلم القراءة والكتابة مع ماهر. [Online] Available: https://play.google.com/store/apps/details?id=air.com.horof.maher.ali&hl=ar. Accessed: 6 Nov 2016
20. Brahim, E.: 3D Arab alphabet sign language on the App Store. Internet: https://itunes.apple.com/il/app/3d-arab-alphabet-sign-language/id879930785?mt=8. Accessed: 28 Oct 2016
21. لغة الإشارة الفلسطينية. [Online] Available: https://play.google.com/store/apps/details?id=com.adhamenaya.esharat&hl=ar. Accessed: 6 Nov 2016

Information Communication Technology Infrastructure in Sudanese Governmental Universities

Abdalla Eldow, Rawan A. Alsharida, Maytham Hammood, Mohanaad Shakir, Sohail Iqbal Malik, Ahmed Kh. Muttar, and Kais A. Kadhim

Abstract Information Communication Technology (ICT) infrastructure is the most important aspect of educational organizations and their management systems. The administration of Sudanese Governmental Universities (SGUs) takes into account (ICT) infrastructures when making plans, policies, visions, and missions for achieving their objectives effectively. This study employed the interview technique to explore and evaluate the reality of the (ICT) infrastructures as well as obtaining useful results to guide SGUs for better application of ICT. The study is based on six elements that are considered as fundamentals of the ICT infrastructures. These elements include computers, networks, ICT technology centers, websites, ICT and Computers Centers (CC) manager's qualifications, in addition to the vision, mission, and policy of ICT. These elements are considered in the ICT infrastructure standards decided by the Ministry of Higher education in Sudan. The study found that most of

A. Eldow · S. I. Malik
Department of Information Technology, Al Buraimi University College, Al Buraimi, Oman
e-mail: abdalla@buc.edu.om

S. I. Malik
e-mail: sohail@buc.edu.om

R. A. Alsharida · M. Hammood
Computer Science Department, Tikrit University, Tikrit, Iraq
e-mail: rawan_adel@tu.edu.iq

M. Hammood
e-mail: maythamhammood@tu.edu.iq

M. Shakir (✉)
Department of Business Administration and Accounting, Al Buraimi University College, Al Buraimi, Oman
e-mail: mohanaad@buc.edu.om

A. Kh. Muttar
College of Administrative Sciences, Applied Science University, Eker, Bahrain
e-mail: ahmed.almuhamadi@asu.edu.bh

K. A. Kadhim
Faculty of Language Studies, Sohar University, Sohar, Oman
e-mail: kalwan@su.edu.om

M. Al-Emran et al. (eds.), *Recent Advances in Intelligent Systems and Smart Applications*, Studies in Systems, Decision and Control 295,
https://doi.org/10.1007/978-3-030-47411-9_20

SGUs have LANS and use the fiber optics as transmission media, different network speeds, web sites, and sub web sites designed effectively. Though most of SGUs have ICT centers, only two universities have E-Learning Centers. Additionally, there is no lack in the design of web sites, construction of technology centers, qualifications of the ICT and CC managers, and the documentation of the ICT vision, mission, and policies. These elements are found to be compatible with the ICT standards for higher education institutions in Sudan. However, there are a few numbers of computers as the ratio of users per computer is (15:1). Adding to that, there are few LANs because most of the SGUs have campuses and faculties more than LANs, and some of these campuses are very wide.

Keywords ICT · E-learning · CAL · LANs · Education tools

1 Introduction

The Sudanese Governmental Universities (SGUs) take into account (ICT) infrastructures when making plans, policies, visions, and missions for achieving their objectives effectively. Consequently, the Ministry of Higher Education in Sudan decided to specify the ICT standards to guide the universities to the optimum implementations of ICT which in turn, will improve the higher education field and the Sudanese social community. Currently, education is experiencing a rapid spread of (ICT) in primary and secondary schools as well as universities. With establishing ICT in the national curriculum, students can gain invaluable experience of ICT in the classroom before they engage in a career. Spending on ICT facilitates in educational establishments has increased dramatically over the last decade, and the majority of schools and colleges have assigned staff and management to (ICT) [1]. ICT offers new and innovative modes of learning for all students at all levels. It generates new methods of learning when used appropriately by teachers [2]. ICT centers are important for teachers and learners; therefore, ICT must be employed in the teaching and learning process. Also, it is a must to develop e-learning materials for on-line learning [3]. Hence, all universities have developed centers to drive the initiatives and to provide the necessary manpower support [4]. Accordingly, this study aims at exploring and evaluating the reality of ICT infrastructure as well as obtaining useful results that help in guiding SGUs for better application of ICT. In addition, the study attempts to answer questions related to the ICT standards in the Ministry of Higher Education and Scientific Research in Sudan. These questions are as follows [5]:

1. How many network points are in SGUs?
2. Are they sufficient and suitable as compared with the size of the organization?
3. What is the speed (per m/s)?
4. What are the bands capacities?
5. Are the speed and band capacity satisfactory as compared with the number of users connected to the network?

2 Literature Review

Due to growth in scientific and technological, the sheer amount of knowledge and information is increasing. Chavula and Chekol have discussed the strategic planning steps of the government, which are divided into three parts: "setting up of a structure, developing the documents and designing implementation plans" [6]. Kannappanavar and Vijayakumar [7] presented a survey about using the software and hardware facilities in two libraries in two Agricultural science University in Karnataka. The results show that none of the University libraries at the time of the study were having databases and full implementation of IT applications in their libraries. Hannatu tested the role of ICT in education and found that it is important in teaching and learning processes such as problem-solving and enhancing communication skills [8].

The scholars [9] conducted profundity semi-structured interviews for 23 faculty members in three Australian universities. These interviews revealed their experiments and trends with ICT in education, such as information-oriented approaches and student-oriented, practice-oriented feedback: consisting of activities directed towards activity and industry teaching the router. Many technologies used in learning like data show, videos, computers, etc. Reference [10] assessed the student's behaviors and observation towards utilized videos in learning at the University of Ghana. By conducting semi-structured interviews with 20 students, they discovered that using video as a medium was a successful step in education. In New South Wales Australia, different ICTs & e-learning techniques were used as online tools for projects of students' research. Authors of [11] used personal websites, dialogue boards, and cooperative web pages called wikis to help students for sharing information between them and found that these e-learning tools were a perfect idea to give students more learning than others. E-participation was another aspect of ICTs that was used in human society, especially in Chinese governance universities. Therefore, Ref. [12] used this technique to improve the administration of their university by interacted students' democratic opinions through the internet with the university. Furthermore, Ref. [13] conducted a questioner on smart learning, which means not just used smart devices but also used smart technologies in education on some of the South Korean primary teachers. They revealed that smart learning contributes to fix beliefs of students towards using ICT. Also, they discovered that teachers who have more experience in technologies provide support in using ICT in schools. Thus, it improved smart learning in schools. On the other hand, Ref. [14] explored the perceptions of business education teachers towards the advantage of ICT in Nigerian classrooms and discovered that there was a weakness in the educational process. Also, Ref. [15] in Turkey applied the Internet as an ICT tool used for searching about information instead of sharing information. In Taiwan, another kind of ICT support learning tool was a Speech-Driven PowerPoint (SDPPT) system. It was created in a university classroom. Reference [16] detected that students get more interactivity and enjoyment with information when using such a system.

3 Methodology

In this study, the researchers used the interview technique to collect data for achieving the study objectives. The interviews were conducted with ICT and computer centers (CC) managers in seven SGUs in Khartoum city. The study selected six elements as fundamentals of the ICT infrastructures, including computers, networks, ICT technology centers, websites, ICT and CC manager's qualifications, in addition to the ICT vision, mission, and policy. These elements were selected because they are included in the ICT standards as a measurement specified by the Ministry of Higher Education in Sudan as a guide for all Sudanese higher institutions. The authors follow the research in [17–21]. Also, the researchers utilized the MS-Excel application to analyze the data, as clarified in the following sections.

4 Interviews with ICT and Computers Centers (CC) Managers in SGUs

In this section, the researchers presented and analyzed the data obtained from the interviews with ICT and CC managers concerning the six elements of the infrastructure defined in the previous section. These elements are explained in the following subsections:

4.1 Computers in SGUs

After interviewing the ICT and CC managers, the researchers organized the collected data concerning the number of computers in SGUs, as shown in Table 1.

Based on Table 1, the total number of computers in SGUs in Khartoum city is (13,317) computers. It is clear that UK and AU have the biggest number, which is

Table 1 The number of computers in SGUs

University name	No. of computers	Percentages (%)
University of Khartoum (UK)	4000	30
Al-Neelain University (AU)	4000	30
Sudan University of S. and Technology (SUST)	3514	26.4
National Ribat University (NRU)	650	5
Omdurman Islamic University (OIU)	500	3.6
Holy Quran University (HQU)	453	3.5
Al-Ziem Al_Azhari University (AAU)	200	1.5
Total	13,317	100

(4000) computers for each one of them. In contrast, AAU has the least number represented by (200) computers. The three universities UK, AU, and SUST form 86.4% of the total number of computers, whereas the other four universities form only 13.6%. Fig. 1 presents the distribution of computers in SGUs. Then, the researchers included the number of staff, students, and employees for comparing them, as described in Table 2. Next, the researchers calculated the total number of users per university as well as the ratio of users per computer, as shown in Table 3 and Fig. 2.

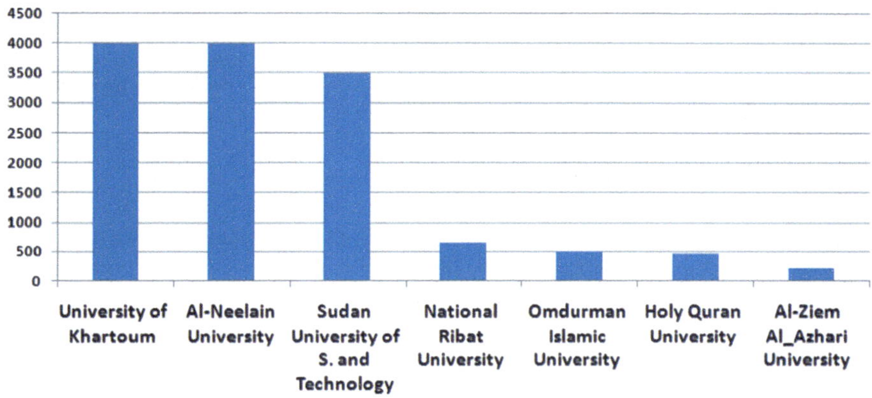

Fig. 1 The distribution of computers in SGUs

Table 2 The number of staff, students, employees, and computers in SGUs

	U of K	SUST	OIU	AU	HQIU	AAU	NRU	Total
Staff	1240	1010	950	900	400	450	361	5311
Students	24,440	24,722	48,658	48,965	12,302	9977	12,204	181,268
Employees	5500	1259	3000	617	408	500	Null	11,284
Computers	4000	3514	500	4000	453	200	650	13,317

Table 3 THE ratio of users per computer in SGUs

University name	No. of computers	No. of users	User/computer
Omdurman Islamic University	500	52,608	105
Al-Ziem Al_Azhari University	200	10,927	55
Holy Quran University	453	13,110	29
National Ribat University	650	12,565	19
Al-Neelain University	4000	50,482	13
University of Khartoum	4000	31,180	8
Sudan University of S. & T.	3514	26,991	8
Total	13,317	197,863	(15) user/computer

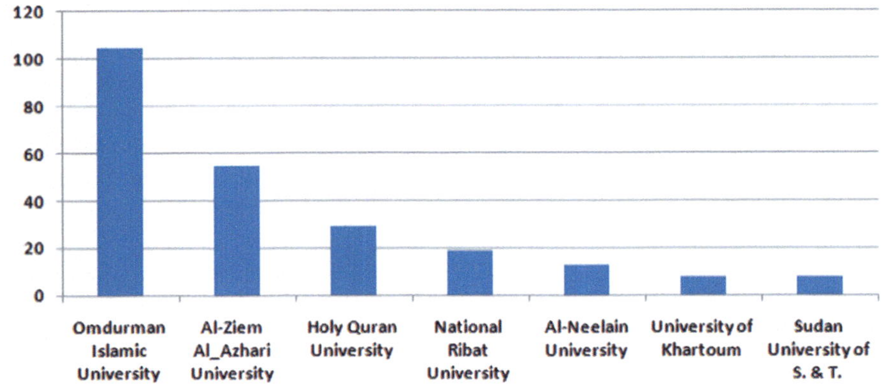

Fig. 2 The ratio of users per computer in SGUs

According to Table 3 and Fig. 2, OIU consists of (52,608) users; however, it has only (500) computers. This result indicates that the ratio of users per computer is (105:1), which reveals the few numbers of computers in it. The ratio of users per computer in Al_Azhari University is (55), forming nearly half of the ratio in (OIU); and in Holy Quran University is (29). The ratio in National Ribat and Al-Neelain Universities is (19) and (13), respectively. However, UK and SUST have the least ratio represented by (8) for each one of them. This result indicates that there is a suitable number of users per computer, denoting that UK, SUST, and AU have the best integration of computers. Table 3 presents the total number of users per computer, which is (15:1), and this result indicates a deficiency in the number of computers in some SGUs.

4.2 *Networks in SGUs*

The researchers organized the data in terms of the number of networks, network points, bandwidths, costs, and types of links in SGUs (see Table 4). All SGUs have LANs. UK and SUST have the biggest and similar numbers of LANs as (18) for each. In contrast, AU and AAU have only one LAN. The number of LANs in UK and SUST is compatible with the number of their computers. Figure 3 shows the gradient level of LANs in SGUs.

SGUs have LANs and different types of transmission media. The fiber optics transmission media is used in most of them. They have different bandwidth. Hence, OIU, UK and SUST have the same and largest bandwidth, which is (54 and 50 Mbs), while (HQIU) has the smallest bandwidth, which is only (5 Mbs). In addition, they have several network points. UK and SUST have the same and the biggest number of network points as (4000), whereas HQIU has the smallest number of network points as (72). In terms of costs per year, UK and SUST have the same and highest cost per

Table 4 The network properties in SGUs

University	UK	SUST	OIU	AU	HQIU	AAU	NRU	Total
No. of LANs	18	18	12	1	9	1	16	75
No. of points	4000	4000	1000	2700	72	Null	230	12,002
Size of band width (Mbs)	50	50	54	10	5	Null	15	50–54
Type of link	Fiber optics	Fiber optics	Hybrid	Fiber optics	Hybrid	Null	Hybrid	Hybrid + fiber
Cost per year	600,000 S.P	600,000 S.P	36,000 S.P	120,000 S.P	8000 S.P	Null	15,000 S.P	1,379,000 S.P

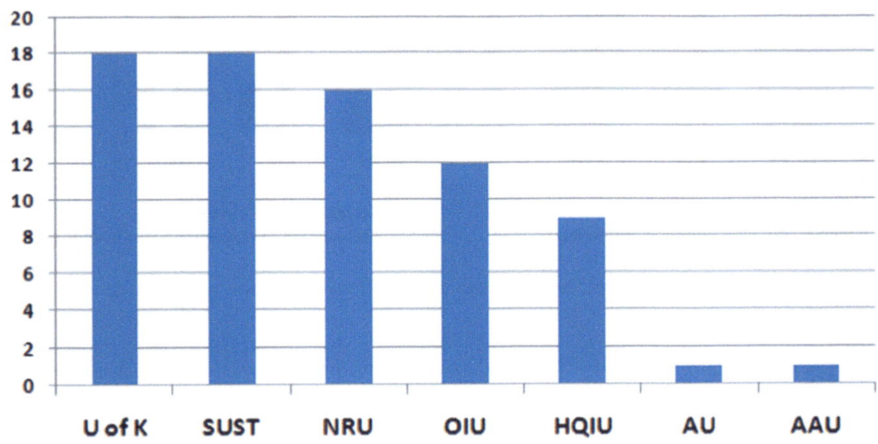

Fig. 3 The distribution of LANs in SGUs

year, which is (600,000 S.P), while HQIU has the smallest cost, which is (8000 S.P). The number of colleges and campuses in each university was considered to compare among LANs of universities with their faculties, centers, and campuses, as shown in Table 5 and Fig. 4.

Data described in Table 5 revealed that the number of faculties is greater than that of LANs in most SGUs. In addition, most universities have various centers, institutes, and campuses.

When comparing the number of LANs with the number of faculties, institutes, and centers, the results indicated the existence of (75) LANs and (183) (faculties, institutes, and centers). So, the ratio of LANs to the number of faculties, institutes, and centers is (1:5). In addition, some of these campuses and faculties have wide areas distributed throughout Khartoum city. This result indicates that there is a lack

Table 5 The number of LANs, faculties, and campuses in SGUs

University	UK	SUST	OIU	AU	HQIU	AAU	NRU	Total
No. of LANs	18	18	12	1	9	1	16	75
No. of faculties	20	21	18	16	9	14	16	114
Centers, institutes	28	9	10	4	18	0	0	69
No. of campuses in Khartoum City	15	10	8	12	5	5	5	60

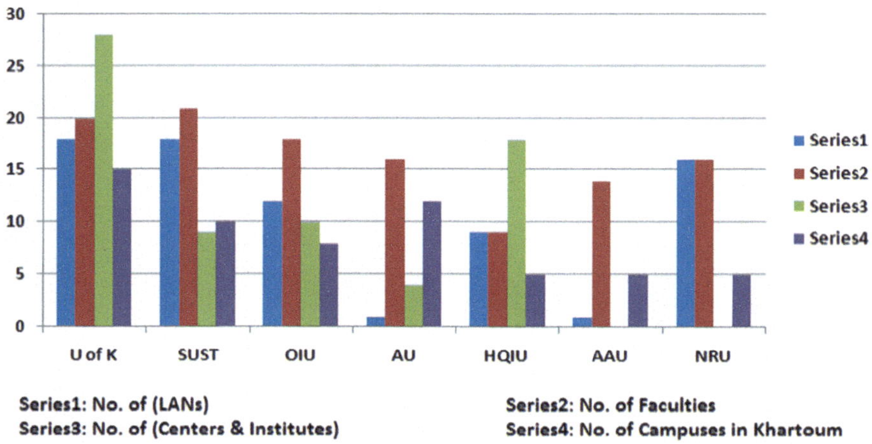

Fig. 4 The distribution of LANs, faculties, centers, and campuses in SGUs

of network infrastructures in SGUs, especially with some colleges that contain a large number of students. There are (18,749) students in the college of business and social studies in AU (7).

4.3 The Technology Centers in SGUs

The researchers organized the data in terms of the number of networks, networks points, bandwidths, costs, and types of links in SGUs (see Table 4). In Table 6 and Fig. 5, the researchers organized the collected data from the ICT managers on technology centers existing in universities. Table 7 illustrates the number of technology centers in each university. Consistent with Table 6 and Fig. 5, the highest percentage was for computer centers and E-Libraries as all SGUs have them by (100%), followed by ICT centers by (71%). In spite of this fact, most SGUs do not have IT and E-learning centers, which are represented by (43%) and (29%), respectively. Concerning the numbers, all SGUs have one computer center except OIU and AAU, which have (4) and (2) centers, respectively. The total number of computer centers in

Table 6 The percentage of technology centers existing in SGUs

University	UK	SUST	OIU	AU	HQIU	AAU	NRU	Percentage (%)
Computer centers	Yes	Yes	Yes	Yes	Yes	Yes	Yes	100
E-library	Yes	Yes	Yes	Yes	Yes	Yes	Yes	100
ICT centers	Yes	No	Yes	Yes	No	Yes	Yes	71
IT centers	Yes	No	No	Yes	No	Yes	No	43
E-learning centers	No	No	No	Yes	No	Yes	No	29

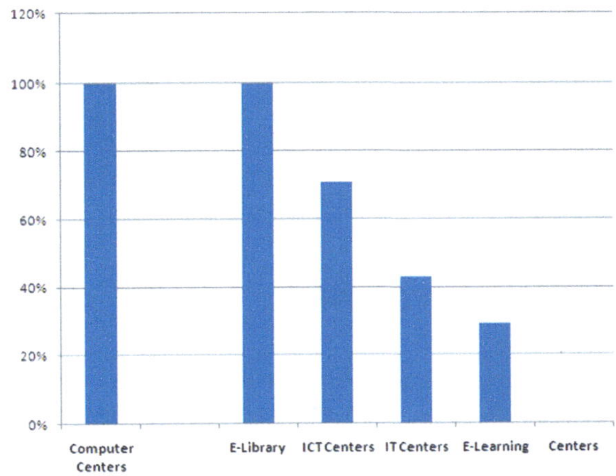

Fig. 5 The percentages of technology centers in SGUs

Table 7 The number of technology centers in SGUs

University	UK	SUST	OIU	AU	HQIU	AAU	NRU	Total
Computer centers	1	1	4	1	1	2	1	11
IT centers	1	0	0	1	0	1	0	3
ICT centers	1	0	2	1	0	1	1	6
E-learning centers	0	0	0	1	0	1	0	2
E-library	1	1	1	1	1	1	1	7
Total	4	2	7	5	2	6	3	29

SGUs is (11), followed by (7) for E-libraries, (6) for ICT centers, and then (3) and (2) for IT and E-learning centers, respectively. This result denotes that the construction of technology centers in SGUs is sufficient. This result is compatible with the ICT standards for higher education in Sudan.

Table 8 The number of web sites in SGUs

University name	UK	SUST	OIU	AU	HQIU	AAU	NRU
Web sites	1	1	1	1	1	1	1
Sub web sites	Many	12	0	0	0	0	0

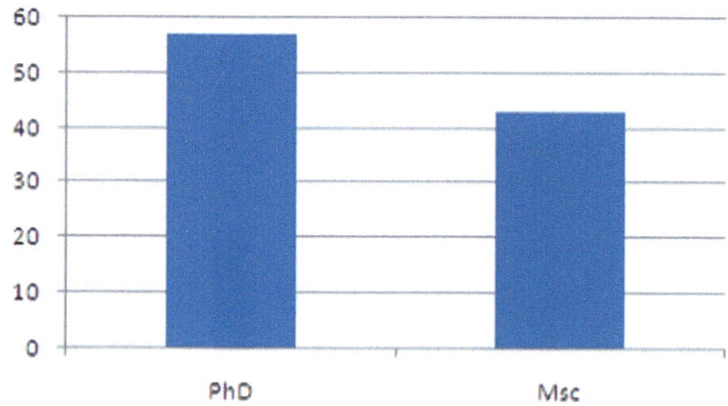

Fig. 6 The percentages of technology centers in SGUs

4.4 Web Sites in SGUs

Table 8 shows the number of web sites in SGUs. All SGUs have one web site, but only UK and SUST have many sub-websites. This result indicates that the creation of web sites in SGUs in Khartoum city is satisfactory. In addition, this result is compatible with the ICT standards of the Ministry of higher education in Sudan [20] (Fig. 6).

4.5 The ICT and CC Managers' Qualifications in SGUs

The qualifications of ICT managers in SGUs present in Table 9, where most of ICT or CC managers have Ph.D. and the others have M.Sc. This result indicates that all ICT and CC managers in SGUs have higher qualifications. This result indicates that the qualifications of the ICT and CC managers in SGUs are satisfactory, which is compatible with the ICT standards of the Ministry of higher education in Sudan.

Table 9 ICT managers' qualifications in SGUs

University name	UK	SUST	OIU	AU	HQIU	AAU	NRU
Qualification of (ICT) managers	Ph.D.	Ph.D.	M.Sc.	Ph.D.	M.Sc.	Ph.D.	M.sc.

Table 10 ICT vision, mission, and policy in SGUs

University	U of K	SUST	OIU	AU	HQIU	AAU	NRU	Percentage (%)
(ICT) Vision and mission	Yes	Yes	Yes	Yes	Yes	Yes	Yes	100
ICT policy	Yes	Yes	Yes	Yes	Yes	Yes	Yes	100

4.6 *ICT Vision, Mission, and Policy in SGU*

Table 10 shows the ICT vision, mission, and policy in SGUs. According to Table 10, all SGUs have ICT vision, mission, and policy, indicating that the ICT vision, mission, and policy are formulated and documented in SGUs by (100%). This shows sufficient documentation of the ICT vision, mission, and policy in SGUs. This is compatible with the ICT standards of the Ministry of higher education in Sudan.

5 Results and Discussion

Based on the analysis of the six elements that are considered as fundamentals of the ICT infrastructures in SGUs in Khartoum city, the study found that SGUs have ICT infrastructures, but some SGUs suffer a lack in some elements. For example, the number of computers in SGUs is insufficient because the ratio of users per computer is (15:1). This result requires from SGUs to pay more attention to the integration of computers, especially in OIU, whose ratio is (105:1), which can be attributed to neglecting the integration of computers in its policies and plans. However, UK and SUST have a satisfactory integration of computers as there are eight users per computer in each of the two universities.

All SGUs have different numbers of LANs. UK and SUST have the biggest number, which is (18) (LANs) for each one. Also, SGUs have different types of transmission media, and most of them use fiber optics as transmission media. They have different bandwidth. OIU, UK, and SUST have the same and largest bandwidth, which is (54 and 50 Mbs). On the other hand, HQIU has the smallest band width, which is only (5 Mbs). Likewise, SGUs have different networks points. Hence, UK and SUST have the same and biggest number of networks points, which is 4000 points, whereas HQIU has the smallest number represented by 72 points. In terms of costs per year, UK and SUST have the same and biggest costs per year represented by (600,000 S.P), while HQIU has the smallest cost, which is (8000 S.P). These facts about LANs properties in SGUs including number of LANs, type of transmission media, network points, band width size, and cost per year denote that UK and SUST share the same and high level of LANs properties, but HQIU has the lowest level (LANs) properties. Finally, the study found that there is a lack in the number of LANs because most of SGUs have centers and faculties greater than the number of LANs, forming the ratio of (1:5). These results require from SGUs to pay more attention to the effective use of LANs to cover all faculties, centers, institutes, and campuses.

All SGUs have one computer center; thus, there are (11) computer centers. Additionally, all SGUs have E-libraries represented by (7). Most of them have ICT centers, forming (6) centers. However, only (3) universities have IT centers, and only (2) universities have EL centers. Findings on technology centers in SGUs reflect that the computer centers and E-libraries exist by (100%) in SGUs. Moreover, IT and ICT centers exist by (71%) and (43%), respectively, whereas E-learning centers exist by 29%. These results lead to the inference that there is no lack in the construction of technology centers in SGUs, and it is compatible with the ICT standards of the Ministry of higher education in Sudan.

Furthermore, all SGUs have one website, indicating that there is no lack in the creation of the websites, and it is compatible with the ICT standards of the Ministry of higher education in Sudan. Adding to that, all of ICT and CC managers in SGUs have Ph.D. and M.Sc. degrees, indicating sufficiency in relation to the qualifications of the ICT and CC managers. Also, this is compatible with the ICT standards of the Ministry of higher education in Sudan. Concerning ICT vision, mission, and policy, all SGUs have them, referring to the adequacy in the documentation of the ICT vision, mission, and policy in SGUs. Also, this is compatible with the ICT standards of the Ministry of higher education in Sudan. The authors concluded that ICT would increase the efficiency and effectiveness of the education resources, thus improve the output, whether in the students or teachers. ICT makes learning easy to understand and enjoyable. There is a lot of not coverage about ICT. In the future, it possible to improve policymaking, enhanced its context of ICT action and further significates development outcomes.

References

1. Heathcote, P.M.: 'A' Level ICT. Payne Gallway (2003)
2. Lee, K.-T.: Using ICT as a subject, tool for curriculum and co-curricular resource (2002)
3. Shakir, M.T.: 3D multimedia model for Educational design engineering. World Acad. Sci. Eng. Technol. Int. J. Educ. Pedagog. Sci. **2**(5) (2015)
4. Saud, M.S., Shuaibu, B., Mohrsquo, L., Buntat, Y., Yarsquo, A., et al.: Applicability of information and communication technologies (ICTs) in the administration of technical and vocational education and training (TVET) in a knowledge-based society. Int. J. Phys. Sci. **6**(28), 6513–6520 (2011)
5. Ministry of Higher Education and Scientific Research in Sudan: The National Standards Guide to guaranty the quality of Education in Sudan, p. 33 (2012)
6. Chavula, H.K., Chekol, A.: ICT policy development process in Africa. In: Adomi, E.E. (ed.) Handbook of Research on Information Communication Technology Policy: Trends, Issues and Advancement. Information Science Reference, vol. 1, no. 3 (2011)
7. Kannappanavar, B.U., Vijayakumar, M.: Use of IT in university of agricultural science libraries of Karnataka: a comparative study. DESIDOC Bull. Inf. Technol. **21**(1), 21–26 (2001)
8. Hannatu, A.: The role of ICT in teaching science education in schools. Int. Lett. Soc. Hum. Sci. **19**, 217–223 (2014)

9. Khan, Md.S.H., Markauskaite, L.: Approaches to ICT-enhanced teaching in technical and vocational education: a phenomenographic perspective. Hig. Educ. **73**(5), 691–707 (2017)
10. Boateng, R., et al.: Videos in learning in higher education: assessing perceptions and attitudes of students at the University of Ghana. Smart Learn. Environ. **3**(1), 1–13 (2016)
11. Gregory, Sue, Bannister-Tyrrell, Michelle: Digital learner presence and online teaching tools: higher cognitive requirements of online learners for effective learning. Res. Pract. Technol. Enhanced Learn. **12**(1), 18 (2017)
12. Li, X., Zhao, G.: Democratic involvement in higher education: a study of Chinese student E-participation in University Governance. High. Educ. Policy, 1–23 (2018)
13. Isiyaku, D.D., Ayub, M.A.F., AbdulKadir, S.: Antecedents to teachers' perceptions of the usefulness of ICTs for business education classroom instructions in Nigerian tertiary institutions. Asia Pac. Educ. Rev. **19**(3), 337–352 (2018)
14. Ha, C., Lee, S.-Y.: Elementary teachers' beliefs and perspectives related to smart learning in South Korea. Smart Learn. Environ. **6**(1), 3 (2019)
15. Ilgaz, H., Mazman, S.G., Altun, A.: A cross cultural perspective on information communication technologies learning survey. Technol. Knowl. Learn. **20**(2), 159–168 (2015)
16. Chen, C.-L.D., et al.: Incorporating a smart classroom 2.0 Speech-Driven PowerPoint System (SDPPT) into university teaching. Smart Learn. Environ. **2**(1), 7 (2015)
17. Al-hashimi, M., Mohanaad, S., Maytham, H., Abdalla, E.: Address the challenges of implementing electronic document system in Iraq e-government-Tikrit city as a case study. J. Theor. Appl. Inf. Technol. **95**(15), 3672–3683 (2017)
18. Tawafak, R.M., Mohammed, M.N., Arshah, R.B.A., Shakir, M., Mezhuyev, V.: Technology enhancement learning reflection on improving Students' Satisfaction in Omani Universities. Adv. Sci. Lett. **24**(10), 7751–7757 (2018)
19. Al-Hashimi, M., Al-Nidawi, W.J., Othman, M., Shakir, M., Sulaiman, H.: Evaluate information security Governance frameworks in cloud computing environment using main and sub criteria. J. Comput. Theor. Nanosci. **16**(3), 996–1006 (2019)
20. Eldow, A., Shakir, M., Yahya Abdu, M., Iqbal, S., Muttar, A.K.H.: Exploring the extent of ICT role as a teaching tool in the public universities of Khartoum State. J. Eng. Appl. Sci. **13**(15) (2018)
21. Hammood, M.M.: Testing the agreement among three free accessibility evaluation tools: a case study of Iraqi public universities. Tikrit J. Pure Sci. **24**(3), 133–140 (2019). ISSN: 18131662

Emerging Technologies

Internet of Things and Cyber Physical Systems: An Insight

Charu Virmani and Anuradha Pillai

Abstract Technology has immense power to transform the life of every individual whom breaths on the blue-planet provided its efficient utilization. From hyperloop to smart homes, it is a future that one can only see to believe. With the ever-expanding population rates, almost exponentially, in developing countries like India, almost 70% of the world population would be living in the cities by 2025 and technology would have to play a rescuer for accommodating a large number of people in the cities with basic amenities available to one and all along with fighting various challenges such as traffic congestion, environmental pollution, so on and so forth. The age-old notions of impossible have long been challenged and bypassed with the ever-evolving technical solutions and rendered us an innovation that is rapidly taking over the world, i.e., IoT (Internet of Things). With its aid, our century can easily be termed as the age of speaking machines that could serve as a solution for our long pending problems for development. This chapter focuses on the introduction to IoT and cyber physical systems with a detailed discussion of potential threats to IoT technology.

Keywords IoT · Wearable devices · CPS · Threats · Countermeasures · Activity trackers and sensors

1 Introduction

IoT—Internet of Things is a network of physical objects containing embedded technology to communicate and sense or interact with their internal states or the external environment [1, 2]. It can simply be called a cloud of interconnected devices wherein microprocessors and micro-controllers, once connected to the internet, can

C. Virmani (✉)
Computer Science and Engineering, Faculty of Engineering and Technology, Manav Rachna International Institute of Research and Studies, Faridabad, India
e-mail: charu.fet@mriu.edu.in

A. Pillai
Computer Engineering, JC Bose University of Science and Technology, YMCA, Faridabad, India
e-mail: anuangra@yahoo.com

M. Al-Emran et al. (eds.), *Recent Advances in Intelligent Systems and Smart Applications*, Studies in Systems, Decision and Control 295,
https://doi.org/10.1007/978-3-030-47411-9_21

communicate with other devices, thus forming an IoT infrastructure. The imperative interconnecting component in IoT is the Internet that ensures the connectivity of various devices and thus enabling them to function on a real-time basis. The origins of the technology could be traced back to the Radio Frequency Identification (RFID) Systems wherein the term IoT was first coined by Ashton in MIT Auto-ID Labs in 1999, that referred to wireless communication abilities integrated with sensors and computing devices, that enabled uniquely identifiable things to share information across the Internet with very little or no human interaction at all [3]. IoT could be understood from two main perspectives, "*internet-oriented*" and "*things oriented*" with the former based on the expansion of internet applications and the latter meaning several things having identities to form a network based on standard communication protocols [4]. The architecture of IoT can largely be divided into four layers as depicted in Fig. 1:

- The Sensing Layer senses and captures the real-time information of resources, devices, and is also responsible for sharing it among the identifiable units through a well-laid wireless network via sensors.
- The Networking Layer is concerned with connecting all things to form the physical network of the manufacturing system and allow devices to share the information with other connected devices.
- The Middleware Layer is supposed to manage, control, and transmit information on a real-time basis through a cost-efficient platform integrated by hardware and software functions.
- The Application Layer's work is to integrate the various methodologies and functions of the system to achieve an IoT-enabled industrial output.

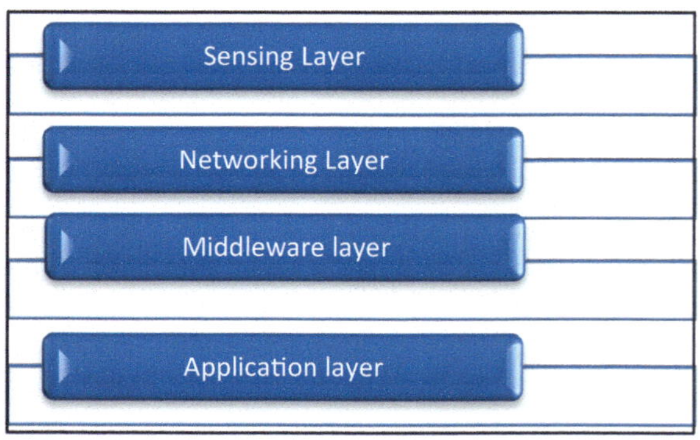

Fig. 1 Layers of IoT architecture

1.1 Key Features of IoT

The features of IoT address the "4Cs" [5], i.e., Connection, Communication, Computing and Control and thus the IoT devices are deployed across various domains to achieve the connection of objects as shown in Fig. 2.

- *Intelligence*: The various algorithms and the interplay of hardware and software make any product behave "smart". The main feature that sets apart IoT is its intelligence.
- *Connectivity*: The connectivity feature enables network accessibility and compatibility. Accessibility allows one to get on the network while compatibility provides the common ability to consume and produce data.
- *Sensing*: The sensing technology provides various means to create experiences that reflect the true physical world with people in it.
- *Expressing*: It enables interactivity with the physical world along with the people on a real-time basis which explains the fame of IoT today.
- *Safety*: This includes the safety of both personal data and also physical well-being.

Due to the aforementioned features, IoT finds varied applications across many different fields and contribute majorly to the business and environmental factors. The next section illustrates significant applications of IoT in technology and industry. The remainder of the chapter is organized as follows. IoT devices and device management is presented in Sect. 2 followed by trenching on wearable devices in Sect. 3. The key enabling advancements of cyber physical system is presented in Sect. 4 followed by its key features in Sect. 4.1 and applications in Sect. 4.2. Section 5 introduces activity trackers and sensors. Section 6 presents a detailed caveat to IOT.

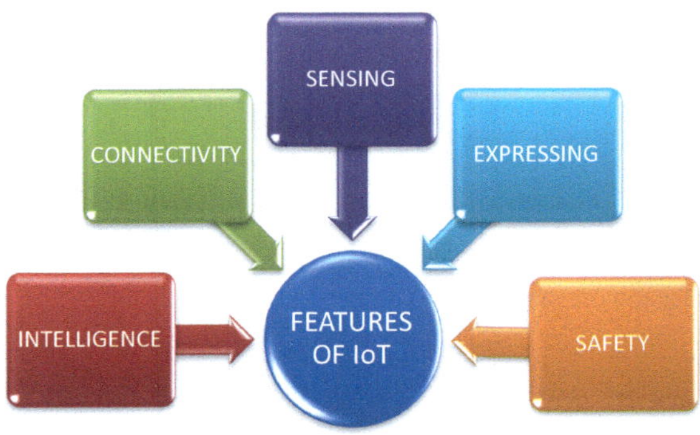

Fig. 2 Features of IoT

1.2 Applications of IoT

Internet of Things finds varying applications in the day-to-day life and also in big organizations to co-ordinate the activities on a real-time basis with technological aid [6–13]. Some of them are illustrated in Fig. 3.

- **Smart Homes**: This involves the connectivity inside homes through smart devices. Some examples include smart home locks, smart thermostats, smart entertainment devices and so on.
- **Smart Farming**: The deployment of information and communication technology in farm activities is included under it. Internet of technology and cloud computing would be the tools for the initiation of smart technological solutions in farm activities.
- **Smart Grid**: This involves the use of information about various stakeholders such as electricity consumers and suppliers to improve upon the performance, reliance, efficiency and also the improvement in economic metrics.
- **Smart City**: The main objective of a smart city is to improve the lives of citizens by solving issues of noise pollution, traffic congestion, etc. through smart technological solutions.

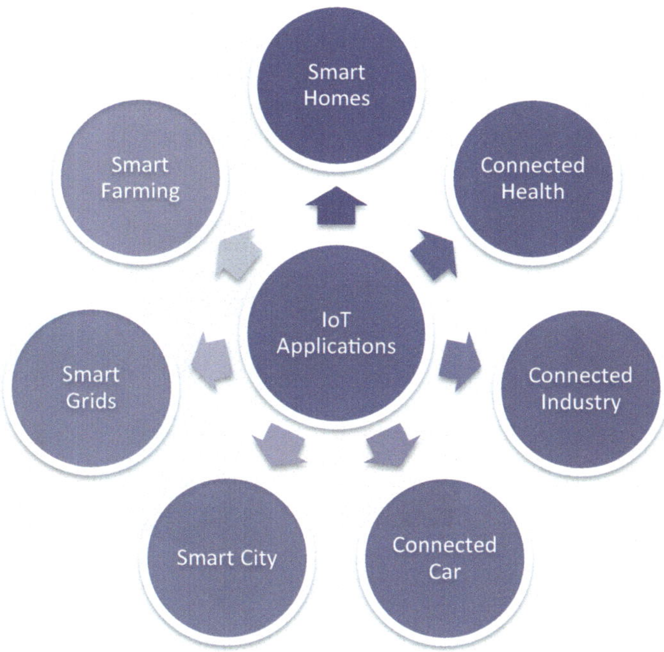

Fig. 3 IoT applications

- **Connected Cars**: This involves the use of technology to enhance driver comfort via traffic control technologies, connectivity with other cars and also mapping services. Remote monitoring could also be an important aspect in future cars.
- **Connected Industry**: In the field of a market researcher, IoT is being seen as the field of utmost potential. Connected industrial equipment and smart industries are some examples of IoT applications in the industrial field.
- **Connected health**: Through the connection of health-systems and smart medical services, the health of citizens could be majorly improved by ensuring last-mile connectivity to people and ensuring reach of services thereof.

IoT Devices are carefully chosen based on its applications. The next section reflects IoT devices and device management in technology, industry and services.

2 IoT Devices

IoT Devices connect wirelessly to a network and capable of transmitting data. Every device talks to the other related devices in an environment to automate home and industry tasks, and to communicate sensor data to users, businesses and other interested parties [14]. The innovative IoT technology could find various applications in major domains such as healthcare, transportation, smart environments and so on. Thus, to utilize the technology various companies such as Amazon, CISCO, MS Azure, Salesforce and Oracle have rolled out devices in major domains mentioned above. IoT devices could be majorly categorized into three types depending upon their area of application as depicted in Fig. 4.

- *Consumer connected devices* are smart home devices such as smart TV's, smart toys, smart coffee makers, smart speakers, smart air conditioning, smart lighting, smart alarm clocks, smart thermostats, etc. that facilitate in the day to day tasks of a common man, making them more interactive and user-friendly.
- *Industrial connected devices* are commonly referred to as IoT and include devices such as intelligent logistics, smart warehousing, smart robotics, connected robotics, smart automotive manufacturing, etc.

Fig. 4 Types of IoT devices

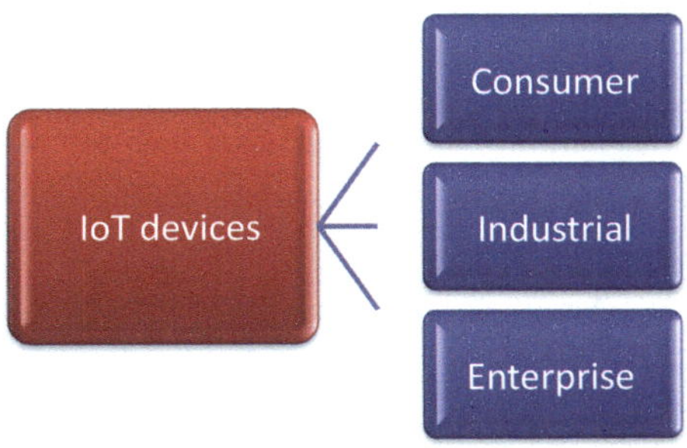

- *Enterprise connected devices* smart city technologies—such as those used to monitor traffic and weather conditions. The devices include smart air quality monitoring devices, smart light panels, smart water systems, voice-activated printers, etc.

The Table 1 summarizes some of the IoT devices found in the market today.

A host of challenges can hinder the successful and efficient deployment of an IoT system and its connected devices, including the security, interoperability, processing capabilities, scalability, and availability. Many of these can be addressed with IoT device management either by adopting standard protocols or using services offered by a vendor. The next section unfolds the features of device management for IoT applications.

2.1 IoT Device Management

Device management helps the companies to integrate, organize, monitor and effectively manage IoT-enabled devices at scale, offering features critical to maintaining the health, connectivity, and security of the IoT devices along their entire lifecycles [15, 16]. Following are the features of IoT Device Management:

- *Device registration*: The IoT devices need to be registered onto a network following the instructions posted at the "Connecting a device that does not have a web browser". This step basically deals with the representation of a device in the registry. The registration information is used to configure the IoT device to connect to a particular platform.
- *Device Authorization*: It is an important part of basic security processes and essentially needed for IoT devices. The authorization is done to identify the device registered. It could be in the form of tokens called, token authorization wherein the generated token uniquely identifies the device registered.
- *Device Configuration*: It is essentially a user-defined data that controls the IoT device. With modification in configuration, a device can be easily controlled. Often the configuration updates are limited to one update per second per device but to enhance security measures, these have been extended to an update per ten seconds.
- *Device Provisioning*: It involves the description of various resources required for an IoT device. To provision a device, there is a need for the creation and registration of a few entities such as:
  - *The certificate*
  - *Policy underlining the certificate*
  - *A unique identifier for the IoT device.*
- *Device Monitoring*: As the networking of devices becomes more and more complex and the generated data grows at an exponential rate, the monitoring of devices

Table 1 IoT devices

Device name	Function	Type
BenjiLock	• Replacement for traditional padlock • Uses biometric technology to unlock via fingerprint	Enterprise
Amazon Echo	• Makes things easy and more accessible • Voice activated	Consumer
Amazon FreeRTOS	• An operating system for Microcontrollers • Makes small, low-power edge devices easy to program, deploy, secure, connect, and manage	Industrial
FootBot Air Quality Monitor	• Measures CO_2, temperature, humidity in a room • Connects to a Smartphone App	Enterprise
Ellipse by Lattis	• Automatic keyless entry using Bluetooth • Backup: Key in a personal code of up to eight characters to lock or unlock • Built-in rechargeable battery using solar	
Tile	• Tile's Bluetooth range is up to 100 ft. Bluetooth is most effective at the 30 ft range, depending on the environment • Loudly ring a lost item from the app and find it by sound • Find your phone. Press Tile to make your phone ring—even if it's on silent	Consumer
August Smart Lock Pro 3rd Generation	• Check your doors status when you are away from home • Provide key codes to friends and family and use auto lock feature when you leave the house	Consumer
Smart Irrigation Sprinkler Controllers	• Simplify home lawn and garden care, and reduce water consumption • Instead of fixed timers, these systems draw on data from sensors, weather forecasts and plant-care databases to determine watering needs and deliver just enough moisture at just the right time • Lower your water bill, and check up on your plants from anywhere in the world using mobile and web apps	Industrial
Schneider Electric—Wonder Ware	• Provides energy-management systems and industrial automation solutions	Industrial

(continued)

Table 1 (continued)

Device name	Function	Type
Green IQ Controller	• Integrations with PlanLink, Parrot, Koubachi, Netatmo weather and TORO® Precision soil moisture sensor • 6 irrigation zones Supports 24VAC valves • Integrates with Echo, IFTTT, Nest, Cresteon, Muzzley	Industrial
Lono	• Supports up to 20 zones • WiFi & Bluetooth Connectivity • Automatically adjusts your watering schedule based on current and upcoming weather	Industrial

becomes indispensable. Various dimensions may be invariable involves such as infrastructure monitoring or environmental monitoring.

- *Device Troubleshooting*: This involves the analysis and correction of the various faults in the IoT devices that may be connected to its business reputation also.

IoT device management ensures efficiency and has a host of advantages as discussed below:

- *Fast Device Onboarding*: It allows users to securely add device attributes such as device name, device type, the year of manufacturing, certificates, and access policies to the IoT registry; also to put a large number of connected devices into service quickly.
- *Easy remote device management*: It can be helpful in remotely updating the software running on devices after being deployed and also to ensure that the latest software is running on the devices. Additionally, other functions such as rebooting, factory resets and security patches can be done remotely.
- *Locate connected devices quickly*: Helps to quickly search and find any IoT device across the fleet of devices on a real-time basis. Devices can be found on the basis of various device attributes such as device ID, device state, and type.
- *Simple IoT device organization*: This facilitates tracing, operating and managing devices according to business and security requirements such as deploying firmware, updating all devices, defining the way of communication of the devices, etc.

  IoT device management allows in connecting a big fleet of devices and also facilitating their monitoring and deployment across industrial sites along with controlling the deployment speed and categorize the devices based on function and security applications.

IoT devices that can be worn as accessories or attached to clothes is the core of IoT and are popular as wearable devices. It can sense, analyze, store, transmit and utilize the data based on application. The next section reflects the capabilities of wearable devices and IoT.

3 Wearable Devices and IoT

Wearable devices have become a hot topic today and are at the very heart of IoT related discussions. Presently, IoT devices function as mere peripherals to Smartphones and the curiosity of whether they could play an even bigger role with the Internet of Things remains to be seen. With the aid of IoT, intelligence can be deployed in about everything in the environment but does the same hold true for wearable devices remains a mystery unsolved [17–19]. The main features of wearable devices are shown in Fig. 5.

The primary function of wearable devices is related to identification and security which is useful within the work environment. Some wearable devices include biometric capabilities such as fingerprint activation, face recognition, etc. Some other capabilities such as location sensing have played important roles in emergency situations and for ensuring the safety and security of women. Health and fitness bands widely available in the markets today offer heart rates, perspiration levels and even

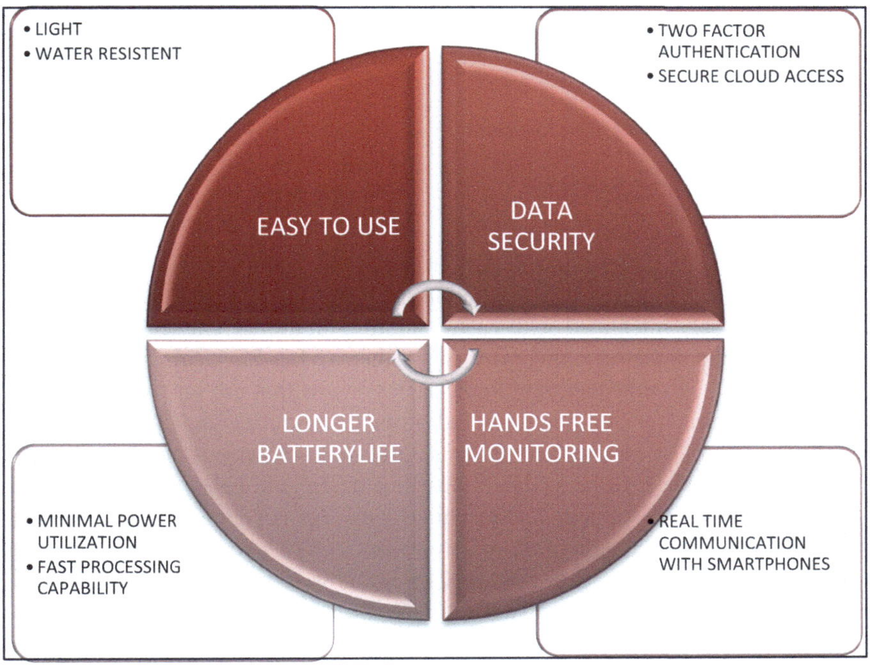

Fig. 5 Features of wearable devices

Table 2 Wearable devices

Wearable device	Usage
Toyota mobility band	• Improves the mobility of blind people • Can be worn around shoulders
Medtronic guardian glucose monitoring	• Helpful for diabetic patients to measure glucose levels • Monitoring device detects and notifies about the levels
Navigation without screen with connected insoles	• Help in navigation with the vibration of insoles

oxygen levels in the bloodstream. More recently they can be used for measuring the alcohol levels. Till now the Smartphones have been used as central points for the operation of these wearable devices but the use of IoT into wearables may completely eliminate this necessity. With the aid of Smartphones, the strategy for device operation could be set and once the strategy is in place, the devices could communicate with each other based on the set guidelines, which could be one possible solution of integrating IoT into wearable devices. Various wearable devices have been made that make the use of IoT and are available easily. Some of them are listed in Table 2.

IoT and Cyber Physical Systems (CPS) refers to smart systems which have physical as well as cyber aspect where former reflects on connectivity and later emphasizes on embedded part. The next section presents cyber physical system with its key features and applications.

4 Cyber Physical Systems

A CPS is a structure that is controlled and monitored by computer-based algorithms and works in close integration with the internet. The software and hardware components are closely integrated with each operating on a different scale, displaying distinct behaviors and interacting with each other. Cyber-Physical Systems (CPS) are the epitomes of embedded intelligent ICT systems that are independent, interconnected, collaborative, autonomous and provide computing and communication, monitoring/control of physical components/processes in various applications [20, 21]. Future CPS needs to be scalable, distributed, decentralized allowing interaction with humans, environment, and machines while being connected to the Internet or to other networks. Adaptability, reactivity, optimality, and security are features to be embedded in such systems, as the CPS are now forming an invisible 'neural network' of the society. Smart grids, Autonomous automobile systems, Robotic systems, Automatic pilot avionics are some examples of CPS applications [22, 23]. The design architecture of CPS is as shown in Fig. 6.

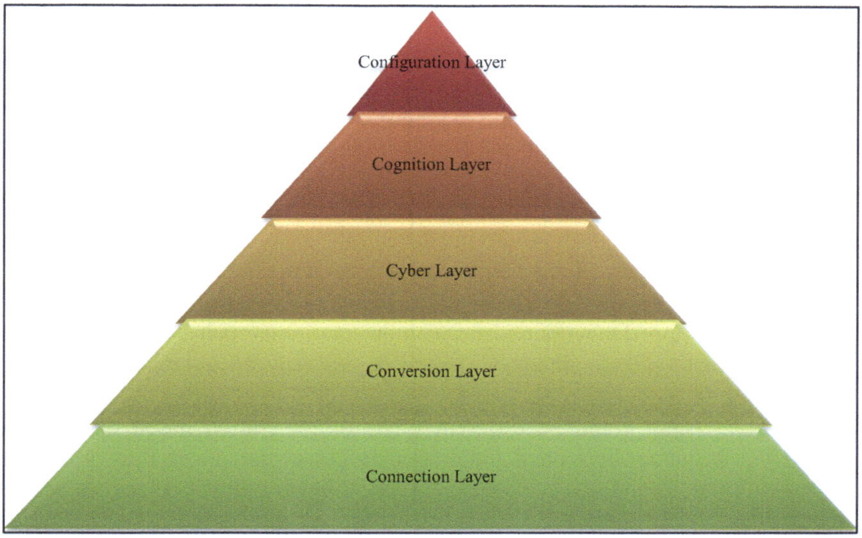

Fig. 6 CPS architecture

Commonly referred to as the 5C architecture based upon which the designing and deployment of cyber-production systems is done [23, 24].

- In the connection level, the connection is established with the devices which can also be a self-connection via sensors.
- In the conversion level, the data from self-connected devices and sensors is converted into information by contemplating features of critical issues.
- In the Cyber layer, each machine creates its twin by using the instrumented features and also categorizes the health pattern of the machine based on the "time-machine" methodology. The twin in the cyber environment can perform self-compare for peer-to-peer performance.
- In the Cognition layer, the outcomes of self-assessment and evaluation are made available to the user based on an infographic.
- In the configuration layer, the machine or the systems can be reconfigured based on priority and risk criteria to achieve pliable performance.

4.1 Key Features of CPS

The traditional systems were closed systems with limited adaptability and autonomy and hence could not play an important role in the evolving technological world. CPS, however due to its features has emerged with scope to contribute majorly in the industrial revolution 4.0. Some features of CPS are as follows:

- *Operational Autonomy*: Due to the high autonomy of cyber-physical systems, they can operate without much human intervention which essentially relieves the load of workers and can prove to be a boon in especially trying or inhuman working conditions. The beauty of automation comes from the fact that it can be extended to any domain and may also be embedded with other technologies and work in compliance.
- *Distributed Control*: Due to connectivity across various systems, they form a distributed system control thus promoting decentralized control that ensures security and prevents autocracy of any form in any decision making process.
- *Adaptability*: As opposed to traditional systems, the cyber physical systems are much more equipped to work under uncertain conditions such as load failure, etc. Due to this feature, the maintenance cost is greatly reduced and renders increased availability.
- *Integration*: There could be different types of integration based on connectivity. For instance, horizontal integration refers to the integration of similar types of functions and services whereas vertical integration works across system hierarchies.
- *Open System*: In contrast to the traditional systems, cyber physical systems are open, i.e., they are connected to various other computing systems that also contribute to a distributed system. The diagnosis and upgrading processes get simplified in an open system.

4.2 Applications of CPS

CPS can be networked at multiple scales and are highly automated. Their dependable operation assures reliability, safety, security, and enhanced usability. Due to the various features, as mentioned above, cyber physical systems find great applications in various domains as discussed below:

- *Transportation*: CPS can ensure faster and safer transportation modes mostly aircraft. It could also bring out more efficient and revolutionary smart cars. The traffic fatalities would be reduced to a great extent due to sensors thus saving human life and also medical expenses with decreased harm to infrastructure.
- *Environment and Sustainability*: It could enable forecasting and mitigating ecological changes. This prior information could be of extreme help to prevent, mitigate or respond urgently to such unprecedented situations and to recover from disasters via human sensors, public sensors or environmental sensors rendering human society more prepared to face even impending ecological challenges.
- *Energy and Automation*: This includes homes and offices that are much energy efficient and operate at cheaper costs. Energy grids are also part of it that allows the intelligent use of energy along with its management.

- *Healthcare*: Increased use of effective in-home care along with more capable devices for diagnosis and new internal and external prosthetics can contribute majorly to this field. Smart healthcare includes various processes such as evaluation, sensing, identification, assessment and lastly intervention.
- *Infrastructure*: A smart critical infrastructure includes a more reliable power grid and also highways that can manage dense traffic with improved safety for all. This would make the static infrastructure much adaptable and secure.

With growing technological innovations in our century, people have bent towards a lethargic way of life and in turn, have fallen victims or are vulnerable to innumerable diseases such as obesity, diabetes, blood pressures, cardiovascular diseases to name a few. Various devices have been manufacturing to raise more awareness about the need for physical well-being and constantly remind of repercussions on heath due to various activities. Activity trackers are meant for this purpose only [25] and are elaborated in next section.

5 Activity Trackers and Sensors

Activity trackers are devices or applications meant for monitoring and tracking fitness-related metrics such as distance walked/run, calories burnt heart rate, blood pressure, quality and quantity of sleep. These are wearable IoT devices that come equipped with a 3-axis accelerometer to track the activity in every direction. The gyroscope measures the orientation and rotation. The altimeter measures the altitude and the sensors provide information about frequency, intensity, duration, acceleration and the various patterns of movement.

Wearable sensors are available in the markets today and widely used in security, sports and medical sciences [26, 27]. They mainly detect unforeseen situations on work on monitoring various symptoms. Without the need to hospitalize a person, the sensors provide many efficient remedies, that too, on a real-time basis. Brain activity, muscle motion, body temperature are a few of the parameters that are considered by sensors. In the sports domain, they are majorly used during training to track down performance and get a better analysis.

Cyber awareness has rendered technologies vulnerable globally. IoT is considered as the pioneer of future technological solutions is the most at risk today. The vulnerabilities are especially dominant because of the involvement of the internet which is, in fact, an enabler or probably an efficient platform for a host of threats to unfold and make an impact beyond imagination at times. Another could be the role-play of connected devices and sensors that are loaded with protocols and interfaces such as web interface, cloud interface, mobile interface. In a scenario where multiple devices are interconnected, each needs to be controlled via a separate application [28–30]. The next section presents potential threats to IoT.

6 Potential Threats to IoT Technology

Given all this, physical threats are always present and only an aware and smart use of devices being used by many could lower the potential fatalities of physical access to a device [31]. To ensure closed-door security in the deployment of IoT devices, a few requirements ought to be considered. Some of these are illustrated below in Fig. 7.

- *Authentication*: The process of validating the identity of someone, if the user is whom he claims himself to be. This is one of the biggest challenges faced in the security of multiple connected devices due to the effective deployment of bots by malicious users in today's scenario. Hence, along with the "who" question, the "what" question is of imminent significance and worth due consideration. There are various entities involves in an IoT technology such as devices, people, services, processing units, service providers and authenticating the entities for interactions is of utmost importance. Though it is not an easy job to authenticate all the devices but it is critical to prevent compromised enterprise networks [32–34].
- *Availability*: The data should be available to IoT device users whenever and wherever they need it. Resiliency to sustain availability to users at desired levels coupled with a guarantee of a certain level of performance as requested by the users is the need of the hour.
- *Integrity*: The deployment of effective technologies that ensure the modification of data only by the authorized users ensures integrity. Since IoT involves the communication of data across various devices; there is a need to ensure data protection from unauthorized access. This also includes protection from sabotage. Fault tolerance capabilities and robustness of the IoT system are key factors involved in integrity. Password-based solutions are one effective way to ensure integrity in connected devices [35].
- *Privacy*: Maintaining the user information in safe locations with strong protection mechanisms prevents attacks from malicious entities thereby ensuring privacy. With the growth in IoT over the years, various devices do not offer all desired

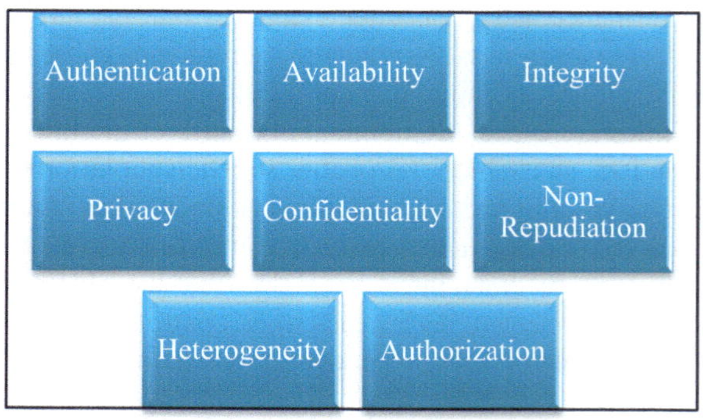

Fig. 7 IoT security requirements

solutions and this is of major concern against the backdrop of growing threats and new techniques deployed almost daily to gain access to information somehow.

- *Confidentiality*: The process of ensuring the availability of data only and only to the authorized users is known as confidentiality. These need not essentially be the people but can be devices, machines, services, objects, etc. The most effective way to ensure confidentiality is via encryption techniques.
- *Non-Repudiation*: This involves proving the involvement of an entity in a data communication to ensure that the person sending the data and receiving the same is authorized and they only have participated in the information exchange and none other. This could be done via digital signatures, digital contracts, email messages etc. The use of confirmation services gets the deal done via timestamps, etc. [36, 37].
- *Heterogeneity*: The complexity of IoT technology can be understood from the fact that it involves the connection between different entities that are equipped with different capabilities; the devices use different interfaces, designed specifically for different functions. Thus, there arises a need for protocols that work in different situations and also in different devices. In an IoT technology, each connected device communicates differently from another one. The topic of heterogeneity becomes even more important because loopholes in this issue could compromise the privacy, integrity, and identification as well [38, 39].
- *Authorization*: The process of allowing certain users access to the information or resources based on his/her identity is an authorization. A host of security systems deployed these days come equipped with a two-step process that involves authentication as the first and authorization as the second [40, 41].

Out of all layers in IoT, the security in the perception layer is indeed very important because devices in this layer usually do not have enough memory for complete security technology [42–44]. The attacks are deployed for external sources at the nodes, wherein the data is passed on to the transport layer. There are a host of threats and vulnerabilities known to be impacting IoT devices in a big way by affecting the availability, integrity, authenticity, privacy among others. Some of them are mentioned in Table 3.

7 Conclusions

The key enabling factors that led to the growth of Industry Revolution 4.0 are introduced from the aspects of Cyber-Physical Systems (CPS) and the Internet of Things (IoT) in this chapter. Firstly, a brief introduction to IoT with its key features and applications is presented to understand IoT. A detailed discussion on wearable device is also presented as it resides at the core of IoT. The chapter also discussed key features, its characteristics, the latest advancements in CPS, especially in real applications and projects. The chapter also presented activity trackers and sensors. Finally, potential

Table 3 Threats in IoT and Countermeasures

Attack	Description	Countermeasures
Type of attack: RFID vulnerabilities		
DoS attack	The denial of service attacks allows the attacker to lock the authenticated user out of his own system and take control of the resources himself. It is achieved by flooding the targeted system with superfluous requests at such a rate that they cannot be fulfilled and end up crashing the system [32]	• Improve the connection between items and tags • Active tags should have an alarm function
Eavesdropping	A communication between two authorized users may be accessed by an unauthorized user with the aid of antennas that can also record the said communication [33]	• Encryption techniques
Skimming	The attacker aims at observing the exchange of information between legitimate tags and readers. After the observation, the attacker attempts to make a clone of the tag. Usually, the data stored in credit cards and passports are at maximum risk due to skimming. This leads to the compromise on data without the users' information [34]	• Block wallets can be used
Replay attack	The attacker gains access to the communication channel between two users'; records the communication and then replays it [35]	• Counter based schemes • Time based schemes
Side channel attack	Here, the attacker targets unconventional information such as power consumption, electronic fields, and timings information. The attacker needs to have a very deep knowledge about the internal and alongside being familiarized to complex cryptographic algorithms to carry out the attack [36]	• Dissociate connection between shared information and secret information • The amount of information released should also be regulated

(continued)

Table 3 (continued)

Attack	Description	Countermeasures
Cloning	In this, the attacker accesses the user identity information and can clone it to access his bank account or any private information [37]	
Type of attack: WSN vulnerabilities		
Wormhole	The bits of the original data get relocated in this attack. At one end of the network, the attacker receives the data packets, relocates them to some other point in the network and replaces them in the network via the point [38]	• Synchronized clocks • Directional Antennas
Sybil attack	The attacker is able to create multiple identities and replicate a single node. The said identities could be generated in two possible ways, either they are stolen or are fabricated/fake, i.e., generated randomly [38]	• Certification • Resource testing
Spoofing	It is a variation of an identity theft technique wherein an attacker executes his malicious intentions by making the user believe that the request from a trusted site, user and thereby gaining access to his private information [39]	• Data encryption • Authentication protocol
Ping flood	A variation of a DoS attack, wherein the attacker overwhelms the users ICMP by sending superfluous request packets and causing the system to crash [39]	• Firewall • Intrusion detection system
Type of vulnerability: NFC		
Phishing attack	It is a type of social engineering technique that tries to acquire users' credentials in a fraudulent manner. The attacker poses as a trusted entity in an official electronic communication [40]	• Two-way Authentication • Filtering of e-mails

(continued)

Table 3 (continued)

Attack	Description	Countermeasures
Relay attack	It is a variation of a hacking technique related to the traditional man-in-the-middle attack. The said communication between two users is initiated by the attacker himself and then relays messages between two parties without necessarily modifying or reading the data [40]	• Latency variation should be limited • Maximum Latency must be regulated
Type of vulnerability: Bluetooth		
Bluesnarfing	These attacks aims at establishing a connection with the devices, usually to gain access of all the stored data and IMEI, International mobile equipment identity that may reroute the incoming calls on the users device to the attackers system [41]	• Keep the Bluetooth device updated • Turn the device off when not in use
Bluejacking	It is much similar to the phishing attack discussed above. Unsolicited messages are sent to Bluetooth enabled devices to carry out data modification such as adding any contacts to the list etc.	• Authentication techniques • Hardware and software of devices should be updated at regular intervals
Bluebugging	By exploiting any loophole in the firmware of legacy devices, not only the devices but commands can also be accessed with this attack	• Hide the device in undiscoverable mode • Limit the availability of device by turning it off when not in use
Type of vulnerability: Wi-Fi		

(continued)

Table 3 (continued)

Attack	Description	Countermeasures
Man in the middle attack	The attacker can secretly relay or alter the communication happening between two devices. The attacker can crash the system and send specific requests to compromise the user data on the system. The encrypted packets used by IoT devices during communication are extracted by the attacker and the device would decrypt it using the old keys	• Bluetooth link layer security features may be used.
Data theft	Attackers can intercept the data across a network and use it to access the accounts of the user for financial gains or simply to spoil his reputation	• Authentication and authorization techniques need to be stringent
Malware	At times, software that is intentionally created to harm the system is planted on the user system through some files. Once planted, the system behaves in an absurd manner and can transfer user information to the attacker or simply destroy the data	• Content filtering
Wi-Fi password cracking	Wireless access points that use very old security protocols are the easiest targets to crack Wi-Fi passwords. Also, very common words and phrases that can be easily guessed are also very vulnerable to such an attack	• WPA2 technique to be deployed
Type of vulnerability: Wi-Max		
Scrambling attack	This attack affects all wireless systems by the method of precisely injecting RF interfaces during the transmissions of management messages	• Diffie-Hellman encryption technique • Use of one-way function that is key-dependent

(continued)

Table 3 (continued)

Attack	Description	Countermeasures
Masquerade attack	When the attacker assumes the identity of one of the two people involved in a communication, it is known as masquerading. The attack is carried out through either sniffing or spoofing	• Innovative algorithms
Water torture attack	A series of frames are sent by the attacker to the user system in order to consume the users' system battery	• Complex mechanism to baffle the attacker
Type of vulnerability: cloud computing		
Wrapping attack	When the users' account and passwords are duplicated in the log-in phase and during the setup phase the SOAP messages are exchanged between the web browser and the server, it is a wrapping attack	• Adding a redundant bit to the SOAP header
Flooding attack	This attack mostly happens when an attacker generates a fake data which could in turn resource some type of code that could be made to run on the application of any legitimate user.	• A group of servers may be created
Type of vulnerability: LoRa		
Jamming attack	In this attack, attackers can transmit a powerful radio signal in the proximity of application devices to disrupt the radio transmissions. These attacks require dedicated hardware for the purpose which acts as a limitation for it to be carried out in real-time or real-world devices such as IoT	• Switching the operational frequency whenever anomalous behavior is perceived
Compromised devices and keys	Authentication keys and devices are compromised in this attack	• Key management • Frame counters

threats to IoT are outlined to highlight vulnerabilities with its countermeasures in IoT technology.

References

1. Cila, N., Smit, I., Giaccardi, E., Kröse, B.: Products as agents: metaphors for designing the products of the IoT age. In: Proceedings of the 2017 CHI Conference on Human Factors in Computing Systems, pp. 448–459. ACM (2017)
2. Condoluci, M., Araniti, G., Mahmoodi, T., Dohler, M.: Enabling the IoT machine age with 5G: Machine-type multicast services for innovative real-time applications. IEEE Access **4**, 5555–5569 (2016)
3. Kishore, K., Sharma, S.: Evolution of wireless sensor networks as the framework of Internet of Things—a review. Int. J. Emerg. Res. Manag. Technol. **5**(12) (2016). ISSN: 2278-9359
4. Khan, R., Khan, S.U., Zaheer, R., Khan, S.: Future internet: the internet of things architecture, possible applications and key challenges. In: 2012 10th International Conference on Frontiers of Information Technology (FIT), pp. 257–260. IEEE (2012)
5. Borgia, E.: The Internet of Things vision: key features, applications and open issues. Comput. Commun. **54**, 1–31 (2014)
6. Ghayvat, H., Mukhopadhyay, S., Gui, X., Suryadevara, N.: WSN-and IOT-based smart homes and their extension to smart buildings. Sensors **15**(5), 10350–10379 (2015)
7. Martins, A.F., Santos, D.F., Perkusich, A., Almeida, H.O.: IEEE 11073 and connected health: Preparing personal health devices for the Internet. In: 2014 IEEE International Conference on Consumer Electronics (ICCE), pp. 274–275. IEEE (2014)
8. Zancul, E.D.S., Takey, S.M., Barquet, A.P.B., Kuwabara, L.H., Cauchick Miguel, P.A., Rozenfeld, H.: Business process support for IoT based product-service systems (PSS). Bus. Process Manag. J. **22**(2), 305–323 (2016)
9. Zanella, A., Bui, N., Castellani, A., Vangelista, L., Zorzi, M.: Internet of things for smart cities. IEEE Internet of Things J. **1**(1), 22–32 (2014)
10. Ryu, M., Yun, J., Miao, T., Ahn, I. Y., Choi, S.C., Kim, J.: Design and implementation of a connected farm for smart farming system. In: 2015 IEEE Sensors, pp. 1–4. IEEE (2015
11. Sanchez, L., Muñoz, L., Galache, J. A., Sotres, P., Santana, J. R., Gutierrez, V., Ramdhany, R., Gluhak, A., Krco, S., Theodoridis, E., Pfisterer, D.: SmartSantander: IoT experimentation over a smart city testbed. Comput. Netw. **61**, 217–238 (2014)
12. Lemke, L., Maraninchi, F., Donsez, D., Privat, G.: Control-oriented Models for a shared IoT Infrastructure in Smart Cities (2015). https://doi.org/10.13140/RG.2.2.26737.07527. Accessed link: https://www.researchgate.net/profile/Laurent_Lemke/publication/294874794_Controloriented_Models_for_a_shared_IoT_Infrastructure_in_Smart_Cities/links/56c4fd8f08aeeeffa9e5dcd3/Control-oriented-Models-for-a-shared-IoT-Infrastructure-in-Smart-Cities.pdf
13. Lee, I., Lee, K.: The Internet of Things (IoT): applications, investments, and challenges for enterprises. Bus. Horiz. **58**(4), 431–440 (2015)
14. Wurm, J., Hoang, K., Arias, O., Sadeghi, A.R., Jin, Y.: Security analysis on consumer and industrial IoT devices. In: 2016 21st Asia and South Pacific Design Automation Conference (ASP-DAC), pp. 519–524. IEEE (2016)
15. Sehgal, A., Perelman, V., Kuryla, S., Schonwalder, J.: Management of resource constrained devices in the internet of things. IEEE Commun. Mag. **50**(12) (2012)
16. Lin, Y.B., Lin, Y.W., Chih, C.Y., Li, T.Y., Tai, C.C., Wang, Y.C., Lin, F.J., Kuo, H.C., Huang, C.C., Hsu, S.C.: EasyConnect: A management system for IoT devices and its applications for interactive design and art. IEEE Internet of Things J. **2**(6), 551–561 (2015)
17. Zhang, Z.K., Cho, M.C.Y., Wang, C.W., Hsu, C.W., Chen, C.K., Shieh, S.: IoT security: ongoing challenges and research opportunities. In: 2014 IEEE 7th International Conference on Service-Oriented Computing and Applications (SOCA), pp. 230–234. IEEE (2014)

18. Arias, O., Wurm, J., Hoang, K., Jin, Y.: Privacy and security in internet of things and wearable devices. IEEE Trans. Multi-Scale Comput. Syst. **1**(2), 99–109 (2015)
19. Metcalf, D., Milliard, S.T., Gomez, M., Schwartz, M.: Wearables and the internet of things for health: wearable, interconnected devices promise more efficient and comprehensive health care. IEEE Pulse **7**(5), 35–39 (2016)
20. Lee, E.A.: Cyber physical systems: Design challenges. In: 11th IEEE Symposium on Object Oriented Real-Time Distributed Computing (ISORC), pp. 363–369. IEEE (2008
21. Lee, J., Bagheri, B., Kao, H.A.: A cyber-physical systems architecture for industry 4.0-based manufacturing systems. Manuf. Lett. **3**, 18–23 (2015)
22. Sha, L., Gopalakrishnan, S., Liu, X., Wang, Q.: Cyber-physical systems: a new frontier. In: IEEE International Conference on Sensor Networks, Ubiquitous and Trustworthy Computing, 2008. SUTC'08, pp. 1–9. IEEE (2008)
23. Tan, Y., Goddard, S., Perez, L.C.: A prototype architecture for cyber-physical systems. ACM Sigbed Rev. **5**(1), 26 (2008)
24. Lee, J., Ardakani, H.D., Yang, S., Bagheri, B.: Industrial big data analytics and cyber-physical systems for future maintenance & service innovation. Procedia CIRP **38**, 3–7 (2015)
25. Shih, P.C., Han, K., Poole, E.S., Rosson, M.B., Carroll, J.M.: Use and adoption challenges of wearable activity trackers. IConference 2015 Proceedings (2015)
26. Shih, P.C., Han, K., Poole, E.S., Rosson, M.B., Carroll, J.M.: Use and adoption challenges of wearable activity trackers. IConference 2015 Proceedings (2015)
27. Patel, S., Park, H., Bonato, P., Chan, L., Rodgers, M.: A review of wearable sensors and systems with application in rehabilitation. J. Neuroeng. Rehab. **9**(1), 21 (2012)
28. Kozlov, D., Veijalainen, J., Ali, Y.: Security and privacy threats in IoT architectures. In: Proceedings of the 7th International Conference on Body Area Networks, pp. 256–262. ICST (Institute for Computer Sciences, Social-Informatics and Telecommunications Engineering) (2012
29. Hwang, Y.H.: IoT security & privacy: threats and challenges. In: Proceedings of the 1st ACM Workshop on IoT Privacy, Trust, and Security, pp. 1–1. ACM (2015
30. Zhang, Z.K., Cho, M.C.Y., Shieh, S.: Emerging security threats and countermeasures in IoT. In: Proceedings of the 10th ACM Symposium on Information, Computer and Communications Security, pp. 1–6. ACM (2015
31. Alqassem, I., Svetinovic, D.: A taxonomy of security and privacy requirements for the Internet of Things (IoT). In: 2014 IEEE International Conference on Industrial Engineering and Engineering Management (IEEM), pp. 1244–1248. IEEE (2014
32. Farooq, M.U., Waseem, M., Khairi, A., Mazhar, S.: A critical analysis on the security concerns of internet of things (IoT). Int. J. Comput. Appl. **111**(7) (2015)
33. Nawir, M., Amir, A., Yaakob, N., Lynn, O.B.: Internet of Things (IoT): taxonomy of security attacks. In: 2016 3rd International Conference on Electronic Design (ICED), pp. 321–326. IEEE (2016
34. Grabovica, M., Popić, S., Pezer, D., Knežević, V.: Provided security measures of enabling technologies in Internet of Things (IoT): a survey. In: Zooming Innovation in Consumer Electronics International Conference (ZINC), 2016, pp. 28–31. IEEE (2016)
35. Emerson, S., Choi, Y.K., Hwang, D.Y., Kim, K.S., Kim, K.H.: An OAuth based authentication mechanism for IoT networks. In: 2015 International Conference on Information and Communication Technology Convergence (ICTC), pp. 1072–1074. IEEE (2015)
36. Babar, S., Stango, A., Prasad, N., Sen, J., Prasad, R.: Proposed embedded security framework for internet of things (IoT). In: 2011 2nd International Conference on Wireless Communication, Vehicular Technology, Information Theory and Aerospace & Electronic Systems Technology (Wireless VITAE), pp. 1–5. IEEE (2011)
37. Borgohain, T., Kumar, U., Sanyal, S.:. Survey of security and privacy issues of internet of things (2015). arXiv preprint arXiv:1501.02211
38. Pongle, P., Chavan, G.: Real time intrusion and wormhole attack detection in internet of things. Int. J. Comput. Appl. **121**(9) (2015)

39. Andrea, I., Chrysostomou, C., Hadjichristofi, G.: Internet of Things: Security vulnerabilities and challenges. In: 2015 IEEE Symposium on Computers and Communication (ISCC), pp. 180–187. IEEE (2015)
40. Gupta, B.B., Tewari, A., Jain, A.K., Agrawal, D.P.: Fighting against phishing attacks: state of the art and future challenges. Neural Comput. Appl. **28**(12), 3629–3654 (2017)
41. Alaba, F.A., Othman, M., Hashem, I.A.T., Alotaibi, F.: Internet of Things security: a survey. J. Netw. Comput. Appl. **88**, 10–28 (2017)
42. Celik, Z.B., Fernandes, E., Pauley, E., Tan, G., McDaniel, P.: Program Analysis of Commodity IoT Applications for Security and Privacy: Challenges and Opportunities (2018). arXiv preprint arXiv:1809.06962
43. Al-Emran, M., Malik, S.I., Al-Kabi, M.N.: A Survey of Internet of Things (IoT) in education: opportunities and challenges. In: Toward Social Internet of Things (SIoT): Enabling Technologies, Architectures and Applications, pp. 197–209. Springer, Cham (2020)
44. Afzal, B., Umair, M., Shah, G.A., Ahmed, E.: Enabling IoT platforms for social IoT applications: vision, feature mapping, and challenges. Future Gener. Comput. Syst. **92**, 718–731 (2019)

Web Services Security Using Semantic Technology

Firoz Khan and Lakshmana Kumar Ramasamy

Abstract For the present growing distributed Web Services (WS) communication, security is of significant concern in the context of dynamically changing entities. It is a challenge to ensure the security goals such as authentication, authorisation and policy bounded exchange that exists between dynamic web-based applications. As the WS are enhanced to next level Semantic Web Services (SWS), the syntactic based security approach are to be improvised using additional domain knowledge called semantics between the entities (web services). Therefore, the incorporation of semantics will upgrade the quality of security management and provide solutions to the security challenges. The service semantics extracted as additional knowledge describes the web services with which one can check the certainty of entity authentication. The appropriate use of the semantic in policy representation and evaluation against a request is more suitable for challengeable security management of present web information system. Overall, the chapter discusses the incorporation of semantics in all attempts of seeking security solutions that lead to enrichment of interoperability between web services in a dynamic instant.

Keywords Semantic web · Access control · Ontology · Concept · Web services · Web services security · XACML · OWL · Semantic web · Access control mechanisms

F. Khan (✉)
Higher Colleges of Technology, Dubai, UAE
e-mail: fk7@hotmail.com

L. K. Ramasamy
Hindusthan College of Engineering and Technology, Coimbatore, India
e-mail: research.laksha@gmail.com

M. Al-Emran et al. (eds.), *Recent Advances in Intelligent Systems and Smart Applications*, Studies in Systems, Decision and Control 295,
https://doi.org/10.1007/978-3-030-47411-9_22

1 Introduction

World Wide Web (WWW) for global information sharing consist of a universal database with linked documents and published information for access by consumers. The exponential growth of WWW insists on the prime importance of security issues that arise during information sharing [1].

As the e-commerce business solution and the B2B usage increased, sharing of private documents was done using Secure Sockets Layer (SSL). When the organisations improved their business with several heterogeneous business systems (Enterprise Resource Planning (ERP), Customer Resource Management (CRM) among others), linked and distributed for dynamic information sharing, they need a proprietary protocol for communication. Most of the business enterprises represent the data in the form of XML standard which is independent of any platform

With the development of XML standard, the business system faced a limitation in adaption to dynamic change and tight coupling between software applications. This limitation was overcome by Service-Oriented Architecture (SOA) which enables the designing and deployment of discrete software modules. The XML protocol provides interoperability between web services distributed business environment in using other standards such as Simple Object Access Protocol (SOAP) and Web Services Definition Language (WSDL) Web services are a software component that are self-contained and self-described modules accessible and invoke via XML message on SOAP communication over the web [2].

2 Web Service Security

The web service security is essential as the number of access points of web services has increased on the internet. The privacy of data is needed when there is an increase in sharing and accessing of data. When the sharing takes place in a distributed dynamic environment, the requestor and provider are unknown to each other. For secure communication, it is necessary to have a digital agreement on policies and constraints of each other's privacy rights. The integrity of the data is maintained by using proper encryption and decryption schemes.

Business processes and workflows use web services (Shivaram 2003). As web services are an extension of SOA component, which is dynamic and loosely coupled, security measures are very much essential for keeping the business information very safe. The five areas to be considered for security are Message privacy, Message level protection, Parameter checking, Authentication and Authorization. These security points are dealt with only at the level of application layer for it is a web service security and not network security [3]. The incorporation of semantics in these five areas is very much essential for security enhancement at the application level.

2.1 WS-Security Framework

The WS-Security Framework provides a specification of request and response for communicating with security and privacy. For example, XML-encryption gives encryption and decryption specification of messages. XML- signature specifies message authentication and thereby integrity. XACML(extensible Access Control Markup Language) is for access control model and authorisation. SAML (Security Assertion Markup Language) for authentication and securing attribute information.

2.1.1 WS-Authentication

Authentication is verification on the level of certainty of both the communicating entities. In distributed web services, the requestor and provider are unknown to each other. Both of them authenticate each other on sharing their credentials. The level of certainty is increased, and the place of authentication is vital for more sensitive information. The authentication certificate or acceptance is given by trust authority. On verifying the certificate, the provider authenticates the requestor for further continuing the communication process [4].

2.1.2 Message Privacy

The message privacy does not allow any unauthorised person to access a message. The message privacy is the confidentiality of a message. An XML signature assures the confidentiality on the header part of the message. Privacy is further secured by encrypting the signature and the message. As the message is routed in a network, this encryption and decryption are done at end-to-end intermediary points.

2.1.3 WS-Authorization

Ws-Authorization is creating specification on authorisation claim and interpretation of these claims at the endpoint. The specification is on description and management of the data and its policies. This specification is followed for Ws-Privacy and Ws-Security, as shown in Fig. 1. Ws-Authorization specification is implemented as Ws-Policy.

The authorisation gains importance as the dynamicity of communication increases because it is challenging to authorise a requestor who has just entered the network. As the Web services are enhanced to form Semantic web services which get involved in the automatic discovery and business executions, Ws-Authorization becomes more critical [5].

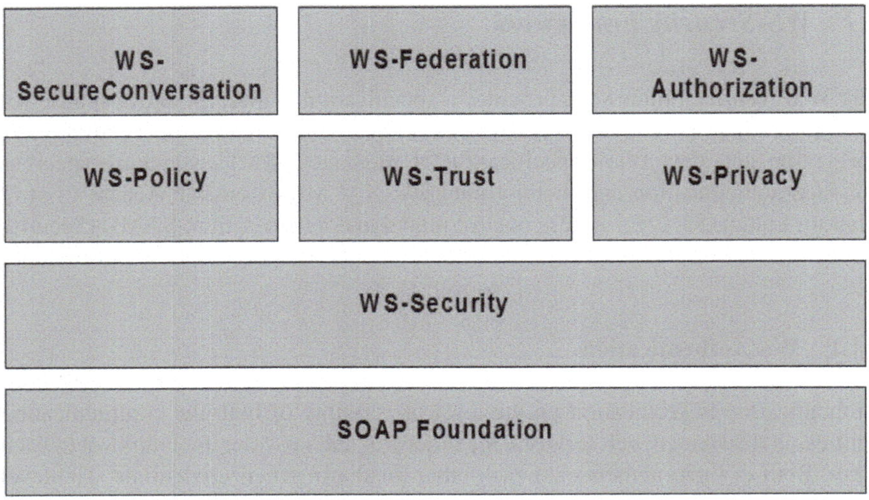

Fig. 1 Web services stack

3 Evolution of WWW to Semantic Web

The semantic web is extending of the current web with well-defined meaning [6]. It facilitates the web data to be defined and linked to be used in service discovery, service automation and service integration. The evolution of syntactic web definition into the semantic web is shown in Fig. 2.

Fig. 2 The nature of semantic Web services

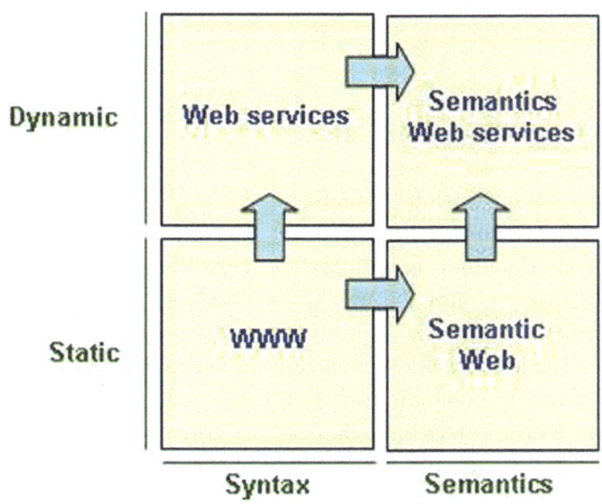

3.1 Semantic Standards

The semantic standards make the web to be a global infrastructure for sharing, to search and to reuse the documents and data efficiently and reliably. The semantic standards are the Resource Description Framework (RDF) and the Web Ontology Language (OWL). RDF is used to describe web information. OWL describes the structure of web ontology for interoperability of data. While XML speaks syntax of the web, RDF and OWL state hid semantic knowledge of the web.

3.2 Ontology Web Language

The OWL semantics of the service is used for discovery and automatic composition. The ontology of the web helps to result in the search in a very quality manner as an exact search. The high-level description of the web using OWL facilitates to have semantic-based access control and to take knowledge-based dynamic decisions during interoperability. The knowledge in OWL description is expressed as a class hierarchy of concepts with attributes as its properties, and the link between the two concepts explains the relationship between two concepts. The semantic knowledge is gained using reasoned or through the use of an inference engine [7].

3.3 Semantic Web Rule Language (SWRL)

The inference of semantic meaning based on ontology concepts is made using semantic rules expressed in Semantic Web Rule Language (SWRL). The SWRL is very expressive that can express a more complex relationship between the concept properties. The details of related concepts, its subsumption, intersection and union are easily expressed in the rule language. The context information based on which decisions differ is used for authorisation purpose. Because of this expressiveness nature, it is advantageous to integrate the access control models with semantic technology. This paper highlights various semantic-based access control models framed based on semantic knowledge. The real challenge exists in the complexity of collecting context data and semantic inference from it in a dynamic environment [8].

4 Adding Semantics to Web Services

Currently, different approaches are under research to add semantics to any resources. New standards and expressing languages are developed to bring out the hidden semantics between the concepts and links. On using this language, the information about the type of resource, interface, properties is characterised precisely.

The WWW function, like any object-oriented system, easily interacts with each other dynamically and enables discovering services. The Semantic web as an extension of WWW favours the clients to use the service dynamically when both the client and provider are unknown to each other previously. In such an environment, there exist problems like a mismatch between the providers and incompatible models of protocols. The use of semantics to describe service capability and shared semantic models helps to eradicate the identified problem [9].

Nevertheless, the implementation of such a model needs infrastructure through which each entity can interpret the published service individually and must be aware of ontologies. There must also be a provision in the architecture to deal with security authorisation and information privacy. The service providers publish the service and interface details to invoke formats to their service using XML definition and Web Services Description Language (WSDL). The service description is done in XML and has a link to the domain ontology to be referred for conceptual terms. The semantic web services consist of description about the concepts in the services so that the clients can interpret and invoke them.

Figure 3 is an example of WSDL concepts mapped to Ontology. WSDL description describes operations, messages, preconditions and effects. Operations describe the functionalities. When semantics is added, this operation is mapped to ontology pointing its functionality

Messages describe input and output parameters of operations using ontology in XML schema. The preconditions are logical conditions to be satisfied, and effects are stated after execution. The sample WSDL example is shown in Fig. 4.

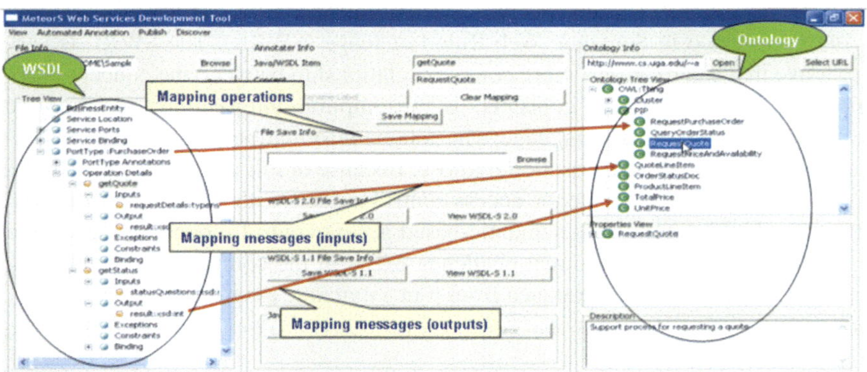

Fig. 3 WSDL mapping to ontology

```
<?xml version="1.0" encoding="UTF-8"?>
<definitions name = "StudentManagement"
targetNamespace=
"http:.../StudentManagement.wsdl20"
        xmlns="http://www.w3.org/2004/03/wsdl"
        xmlns:tns="http.../StudentManagement.wsdl20"
        xmlns:sm="http:.../StudentMng.owl#"
        xmlns:mep=http:.../TR/wsdl20-patterns>
<interface name = "StudentManagementUMA">
        <operation name = "RegisterStudent"
                pattern = "mep:in-out" >
        <action element =

"sm:RegisterStudent" />
        <input messageLabel = "student"
                        element = "sm:StudentInfo" />
        <output messageLabel = "ID"
                        element = "sm:StudentID" />
        </operation>
        <operation name = "StudentInformation"
         pattern = "mep:in-out" >

                <action element =

        "sm:StudentInformation" />
                <input messageLabel = "ID"
                                element = "sm:StudentID" />
                <output messageLabel = "student"
                                element = "sm:StudentInfo" />
        </operation>
        <operation name = "checkStatus"
         pattern="mep:in-out" >
...
        </operation>
</interface>
</definitions>
```

Fig. 4 WSDL example

4.1 Adding Semantics for Authorization

WS-Policy describes the policies of web services. It is a model and syntax which follow the specifications on WS-Trust, WS-Agreement, WS-Security, and WS-Utility within the policy. The WS-Policy provides the specifications to describe the vocabulary on domain context. Figure 5 is a healthcare example of authorisation included in requestor and service provider

The domain knowledge is used to do syntactic matching of assertions which fails. However, the request and provider information show a matching when keywords are semantically reasoned with relation and context information.

•**Service Provider:**
Must be a *Physician* working in *Emergency Service* of the *Health Services* Industry
•**Requester:**
Is an *Emergency Room Physician* working at a *Hospital* in the *Emergency Room*

Fig. 5 Authorization assertion

4.2 Condition—Extension Elements

The annotated WS-Policy have semantic representation for authorisation using the extensible element in Fig. 6. The extension elements are permission, role, subjectCategory, modelReference.

The annotations are placed within <wsp: ExactlyOne> tags. The permission element is the information to authenticate the client. The role element provides information about the authority the role possesses. The extension element subjectCategory is to indicate the subject type. For example, it may be a user and its relationship. The element modelReference is for mapping elements in a schema to ontology. Annotation comes after <All> tag. The annotations start with the namespace "wsau" and are shown in the following XML declaration example

xmlns:wsau="http://lsdis.cs.uga.edu/authorization/wsau**".**

The information in annotations describes the service provider, client policy, client details and are given to ontology inference reasoners to evaluate the authorisation. The reasoners use inference rules added to it to derive a decision [10].

5 WS-Security Policy Framing

The policies are permissions that a requestor requires to access or action on a resource. The policy can be that requestor must have an account to access that message, and Certificate Authority should trust it. Once authorised, both the parties should have a secure communication channel for sharing sensitive information after exchanging accepted security credentials. Web Service Trust Language (WS-Trust) is used to trust the credentials. It enables to exchange the security tokens for trusting each other before the exchange of credentials. For web services, policies include policies on access control, for privacy and business rules posed on two entities.

```
<wsp:Policy
xmlns:sp="http://schemas.xmlsoap.org/ws/2005/07/securitypolicy"
xmlns:wsp="http://schemas.xmlsoap.org/ws/2004/09/policy"
xmlns:base="http:/http://www.NationalEHR.com/policies"
xmlns:wsrm="http://schema.xmlsoap.org/ws/2004/03/rm"
xmlns:wsau="http://lsdis.cs.uga.edu/authorization/wsau"
xmlns:Ontology1="http://lsdis.cs.uga.edu/projects/meteor-s/wsdl-
s/ontologies/HL7_RBAC.owl">
<wsp:ExactlyOne>
<wsp:All>
<wsau:role name="Emergency_Room_Physician" wsau:ModelReference="Ontology1#
Emergency_Room_Physician "/>
<wsau:subjectCategory name="Health_Services"
wsau:ModelReference="Ontology1#Health_Services"/>
<wsse:SecurityToken>
<wsse:TokenType>
  wsse:X509v3
  </wsse:TokenType>
  </wsse:SecurityToken>
  </wsp:All>
  <wsp:All>
  <wsau:permission name="read" wsau:ModelReference="Ontology1#read"/>
  <wsau:role name="Executive_Administration"
  wsau:ModelReference="Ontology1#Executive_Administration"/>
  <wsau:subjectCategory name="Health_Services"
  wsau:ModelReference="Ontology1#Health_Services"/>
  <wsse:SecurityToken>
  <wsse:TokenType>
  wsse:X509v3
  </wsse:TokenType>
  </wsse:SecurityToken>
  </wsp:All>
  </wsp:Policy>
```

Fig. 6 Annotated WS-Policy file

5.1 Semantic Access Control Model (SAC)

The design of the model follows a semantic approach and the access conditions are on semantics. The model considers the semantic property about resource and client who respond and request. The access control system provides the semantics derived from the attribute certificates. These access control constraints are validated for correct and safety access [11].

The SAC model is implanted with a language suitable for access control and semantic integration [12]. Such a language is called Semantic Policy Language (SPL). The language expresses resource properties and context. The SPL information gives three information access criteria, allocation and resource and context properties. The relevant information is used in access control validation.

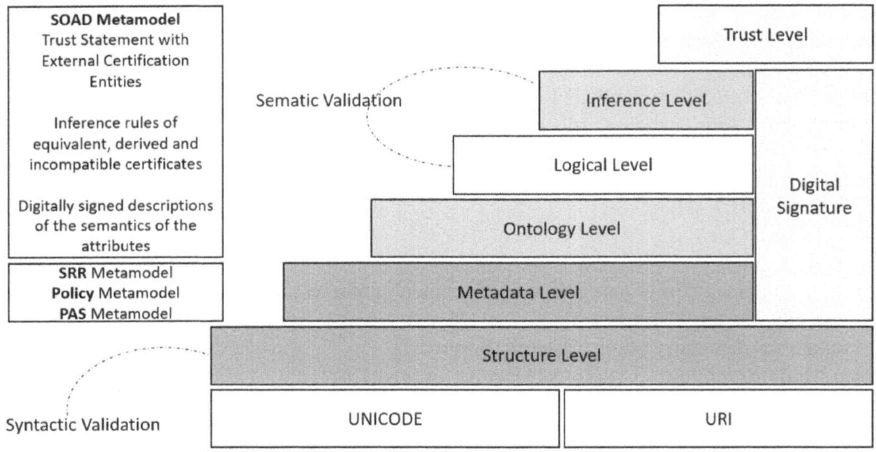

Fig. 7 Model of SAC

Figure 7 shows the underlying layers of the SAC model. Each layer includes the metamodels used for specifying various information.

The SAC metamodel is used for specifying access control criteria, validating policies and syntax. They are also used for contextual validation of policies. The certificate entities are integrated by SOAD model, which provide the necessary trust. Policy metamodel at structure level is used to represent access control criteria semantically. PAS (Policy Applicability Specification) metamodel at structure level represents policies allocation of resources. SRR (Secured Resource Representation) metamodel at the metadata level gives resources semantics. SOAD (Source of Authorization Description) metamodel represent inference rule required for validation. It is this metamodel that coordinates the process between the access control validation system and entities.

6 OWL-S Ontology and Semantic Markup for Privacy and Authentication

Ontology is a formal conceptualisation of domain knowledge. The ontology is essential for semantic information processing on the web. It gives the structural representation for the web content.

OWL-S is an ontology that is used to represent web profile, a process model of the service, and how the service can be grounded. The ontologies can be used to derive security information from web service i/p and o/p information. Semantic mark up for OWL-S is necessary for security characteristics as no encrypted data will reveal the structure of data. The structure of the data is explained with the help of mark up. The information object class (InfObj), an encrypted information object (EnclObj) and signed information object (SignInfObj) are ontology classes defined to get the input and output parameters of the services. For example, the following is class defined for a person

```
<foaf:Personrdf:ID="MarySmith">
<foaf:namexml:lang="en">Mary Smith</foaf:name>
<foaf:title>Dr.</foaf:title>
</foaf:Person>
```

Its corresponding OWL-S profile with a condition not to reveal the personal information is given as

```
<process:hasInputrdf:ID="PersonInf">
<process:parameterTyperdf:resource="EncPersonInfObj"/>
</process:hasInput>
where
<Class rdf:ID="EncPersonInfObj">
<SubClassOfrdf:resource="#EncInfObject"/>
<Restriction>
<onPropertyrdf:resource="baseObject"/>
<allValuesFromrdf:resource="&foaf;#Person"/>
</Restriction>
</Class>
```

The digital signature create for the above OWL-S profile is given as

```
<process:hasInputrdf:ID="RegInf">
<process:parameterTyperdf:resource="SigRegInfObj"/>
</process:hasInput>
<Class rdf:ID="SigRegInfObj">
<SubClassOfrdf:resource="#SigInfObject"/>
<Restriction>
<onPropertyrdf:resource="base"/>
<allValuesFromrdf:resource="&foaf;#Person"/>
</Restriction>
</Class>
```

The personal information is registered as policy and subjected to verification when requested by the requestor. The validation checks the digital signup created for the policy of personal information. The class policy in the above examples is represented in Rei ontology which is RDF based policy specification. This ontology consists of concepts about permissions (rights, prohibitions, obligations and Dispensations) and attributes called (actor, action provision and condition). The Rei is referred to in (http://www.csee.umbc.edu/~lkagal1/rei/swspolicy.owl).

There are three subclasses that Rei is represented:PrivacyPolicy for describing privacy on not sharing, AuthorizationPolicy for permission information and ConfidentialityPolicy on encryption and decryption. The following are the steps of Matchmaker for private policy check with Rei reasoner [13]

1. Matchmaker fetch OWL-S of Web service
2. Retrieve the privacy policy for privacy information given by the resource owner.
3. The Matchmaker send privacy policy and OWL-S to Rei reasoner
4. The reasoned verify for any private part and prohibition information and contradictions
5. If the check fails, send false otherwise true
6. If true, the Matchmaker send authorisation policy to Rei reasoner
7. If authorisation satisfied, reasoned returns true otherwise false to client.

7 XACML for Policy

Extensible Access Control Markup Language (XACML) is an XML language for expressing access policies defined by standard OASIS [14]. XACML is widely used in many projects. It accepts or denies the request based on attributes of the requestor. As far semantics is concerned, this approach furnishes high-level policy representation with their own semantics the main advantage of XACML is that the policies can be reasoned by any off-shelf reasoning tool for required formalisms. The other advantage is policy constructing and analysing can be done irrespective to any specific environment. XACML consists of six components as shown in its architecture in Fig. 8

1. *Policy Enforcement Point* (PEP): From SOAP request for authorisation query for request
2. Policy Decision Point (PDP): it evaluates the rules in the policy. If the policy is not available, supplied from PRP, and the decision is to send to PEP
3. Policy Retrieval Point (PRP): return requested policy to PDP
4. Policy Information Point, (PIP): it queries the predicate information,i.e. attribute information. The information is returned to PRP.
5. Policy Administration Point (PAP): create rules and combine to form policies
6. A policy store to save the policies generated at PAP.

7.1 Three Essential Part of XACML

(a) Rule: Rules are present in policies. They are validated based on content. The rules are on resources, subjects or actions. It also includes consequences as the effect
(b) Policy: a set of rules and obligations
(c) Policy Set: a set of Policies and policy element and algorithm to combine policies.

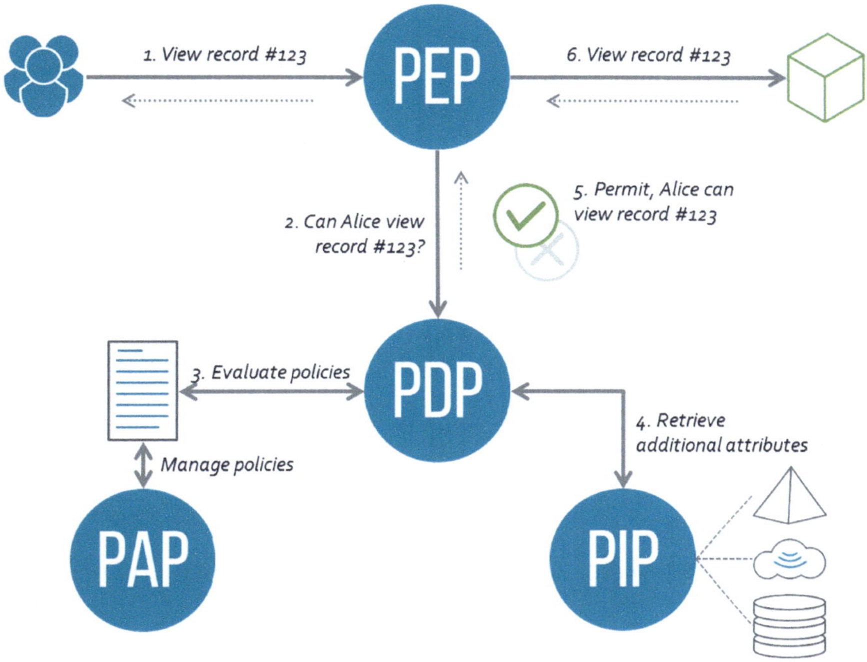

Fig. 8 Architecture of XACML and information flow

7.2 *Steps for Policy Evaluation Using XACML*

i. A request is passed to PEP to find what the user is requesting
ii. PEP request policy from PDP
iii. PDP decides which policy to supply from policy store
iv. PDP request subject and resource information from PIP to evaluate the policies
v. On policy evaluation, the result is returned to the PEP.

8 Access Control Model

The access control mechanism is defined to check the access of resources based on specified access conditions. Access control defining is essential for information security. The security policies are defined over resources to who have permission to access and validate the request to a resource for denying or accepting. The requirement for an access control mechanism is (a) it is applicable for both atomic and composite services. (b) the mechanism should satisfy complex requirement such as it may be requested functionality or provided functionality (c) consider semantic

Table 1 XACML profile elements

Core Element	XACML Profile element as
User	Subjects
Role	Subject attributes
Objects	Resource
Operations	Action
Permissions	Policy set role permission

relation between the concepts (d) the policies are crucial and certified by crypto technique (e) validate the credentials supplied to the user at the time of execution (f) as the semantic environment is open and distributed the authorisation is allowed only on proper authentication [15].

The following are the existing access control model based on semantics and uses semantic standards

8.1 Rule Based Access Control Model (RBAC) Using XACML

The XACML RBAC profile is applicable for a hierarchical role model. The roles are considered. The profile supports both dynamic separation and structural separation of action. The RPS (Role Policy set) associate the role with permission policy. The RPS define the role and permission it can have. The XACML provides a rich policy framework to be integrated into the system. The process involved in this model is given as follows

(a) The subject (user) submits a request to PEP
(b) The PEP uses XACML and recreates the request according to the attributes of the subject, resource and action. It is forwarded to PDP
(c) The PDP invokes PIP to evaluate the request according to policies in XACML
(d) The request is worked against the policies to grant or revoke
(e) The PDP answers back with permission or deny answer to PEP which is sent to the client.

The XACML profile defined for the RBAC model is given as (Table 1).

8.2 Attribute-Based Access Control Model (ABAC) with XACML

The ABAC model is for accessing the resources according to the attribute evaluation of subjects and resources. The standard XACML, when used with ABAC for a virtual organisation, supports interoperability between different organisations. It provides fine-grained authorisation and economic features. The XACML policies defined in one organisation can be used in another organisation. It can resolve conflicts in an algorithm. The limitations are XACML with ABAC do not scale and can support only a few subjects and does not consider explicit semantic relationships.

8.3 Semantic Attribute Access Control Model

Due to the limitations of ABAC model with XACML for virtual organisations with more resources and subjects the semantic-based access control model (SBAC) is combined with ABAC model which involve both the attributes and semantic in access control decision mechanism. The model is designed such that SBAC is used for virtual level and ABAC for the organisation level. In this model, the inference rule is enhanced with semantics using Semantic Web Rule Language (SWRL) [16].

8.4 Semantic Aware Context-Based Access Control Model (SCBAC) [17]

The SCBAC model uses context-based policy as the main design principle to characterise the resources, subjects, roles and actions. The context is established between the entities and permission is assigned with the resources. The components of the model are shown in Fig. 9.

(A) **Context defining**

Four types of contexts are defined

1. Subject context(SC): the context related to subjects that should hold for access to a resource or object. It may include role credential id and name.

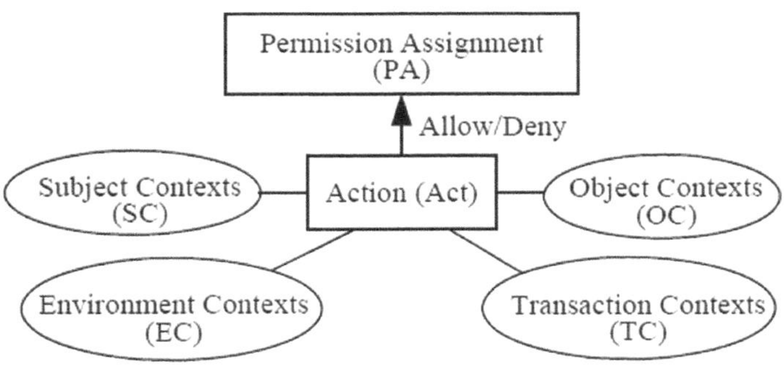

Fig. 9 SCBAC components

2. Object context(OC): the object context used to make access decisions. It provides the situation at which the object is created.
3. Transaction contexts(TC): A transaction is a particular action to be carried out in the system in a particular situation. The process gives information about transactions in future or past.
4. Environment contexts(EC): the information at the time of the transaction, like time and date.

The model is framed on the following function

The right assigned to a request based on the context is

(<Act, SC,OC,EC,TC>, P) $\in$ PS where P = (True, False) where PS is permission set which consists of permission for the contexts, Act is action to be performed for a decision.

8.5 Context-Aware Access Control Model (CAAC) Extending RBAC [18]

The RBAC model does not adapt to the changing context when requesting a resource in a dynamic environment. The model includes the context information related to the user and role and also between role and permission. It is an extension of the RBAC model. The model addresses the following assignment

(a) User-role: specify any information that characterize the state of the entity or any relationship between the entity
(b) Role-permission: consider the purpose and situation-based information, relevant entity information to a particular situation.
(c) Context/Situation specification language (CSSL): explain context related conditions and situation based conditions.

The policy framework of the model is given as

Model M = (MS, MR)

Where MS (Model based on situation) = (U,R,C_e,R_s,OP,P,S_E,P_Y)

MR(Model based on relationship) = (RH, CAUA, RSH, OPA, CAPA)

U—User

R—Role

C_e—Context expressions

R_s—Resource

OP—Operations

P—Permissions

S_E—Situation expression

P_y—Policy

R_H—Role hierarchy

CAUA—Context-aware user-role assignment

RSH—Resource hierarchy
OPA—Operation assigning
CAPA—Context-aware role-permission assignment

XACML is more prominent for policy developing Attribute-based control system (ABAC) where user id, action and resources are input as attribute values to access a resource. Role-based access control system (RBAC) can also be implemented using XACML as a specialised form of ABAC.

9 Framing and Validation of XACML Policy

Access control is a widely used security mechanism which decides the rights for a user or a process have to access a resource. To ease control management, access control policies are designed by widely used languages such as XACML. XACML is a universal language with the freedom to describe the policies. There exist several verification tools for checking the policies of XACML.

The software component Policy Decision Point (PDP) decides whether to permit or deny the request on evaluating it against the policies. When the information resources are of critically sensitive information, the policy evaluation gains more importance. So, correct implementation is based on the correctness of policy specification. The policymakers ensure the correctness of policy specification only after rigorous validation of the policies.

The essential elements of XACML are ***PolicySet, Policy, Rule, Target, and Condition***. When there is multiple sets of policy set, policy and rule, there is a combining algorithm to solve the conflicting rules for one particular request. There exist policy combining algorithm and rule combining algorithms. To find out the matching for a request, XACML initially produce conditions that match with policy set, policy and rule for a subject identifier, resource identifier and action. The attribute values of request are matched with the attribute of policy.

The requests and responses from PDP are converted to a canonical form called **XACML *context***. The PDP on checking request values and policies value takes an authorisation decision. The decision is one among ***permit (allowed), deny(not allowed), not applicable(no policy match) and indeterminate (unable to process due to error)***.

Policy is policy language that composes ***Target*****,** which are a set of capabilities (attributes) need to trace any matching policies, ***Rules*** which are facts to take access control decisions and ***Effect*** which is ***Permit or Deny***. There are combining algorithm to make decisions when there are conflicts.

A ***Target*** has four elements which are ***Subjects, Actions, Resources, and Environments***. Each of the elements contains an attribute identifier, a value and matching function with which it checks whether a policy is suitable for a request. The rule will specify ***Target and Conditions***. When the Target matches and when Conditions are satisfied, the Effect is promoted to the next level of policy. PolicySet with a collection of policies specifies an algorithm to make decisions when there are conflicts. Policy and PolicySet contain set of Obligations to do actions on receiving a response. An ***XACML Request*** is a request with attribute and value pairs for ***Subject, Action, Environment and Resource.*** The syntax of the XACML policy is given in Table 2.

The term *{Palg;Policies}* is policy set of policies and combining algorithm. The following are the results of policy evaluation.

- deny-overrides: if any policy in the set is evaluated as deny then result is deny.
- permit-overrides: permit precede over other results

Table 2 XACML policy syntax

PDPpolicies ::= {P alg ; P olicies}	(Retrieved policies)
P alg ::= only-one-applicable|Ralg	(Policy-combining alg.)
Ralg ::= deny-overrides|permit-overrides	(Rule-combining alg.)
| first-applicable	
| ordered-deny-overrides	
| ordered-permit-overrides	
Policies ::=	(Policies)
{P alg ; target :{ [T argets] } ; P olicies}	(policy set)
| <Ralg ; target :{ [T argets] } ; rules :{Rules}>	(policy)
| Policies Policies	
T argets ::= MatchId(value,name)	(Targets)
| Targets ∨ Targets	
| Targets ∧ Targets | Targets Π Targets	
MatchId ::= string-equal integer-equal	(Match functions)
| string-regexp-match	
| integer-greater-than | . . .	
Rules ::= (Effect [; target :{ Targets }] [; condition :{expression}])	(Rules)
| Rules Rules	
Effect ::= permit | deny	(Effects)

- first-applicable: the result on the evaluation of the first policy is applied to the request.
- ordered-deny-overrides/ordered-permit-overrides: the policies are evaluated in the order they occur.
- only-one-applicable: apply to policy/policyset and ensures only one applies to the request.

In a matching element, MatchId specifies the Boolean function to be used for comparing a literal value to attribute value of the given name. Moreover, the targets are built using $\vee$ as logical disjunction and $\wedge$ and Π as logical conjunction. XACML targets take the form Subjects u Resources u Actions u Environments where the subterms take the form $Subject_1 \vee Subject_2 \vee \ldots\ldots \vee Subject_n$ and each $Subject_i$ have $MatchId_1(value_1;name_1) \wedge : : : \wedge MatchId_m(value_m;name_m)$.

The logical operators are used in targets {**match; no-match; indeterminate**} where match and no match are true and false.

The context requests are represented as **request: {Attributes} where Attribute is set of (name, value) pairs** that tell the subject, resources, the actions to be performed and environmental properties.

The following is an example of policy creation for accessing e-prescription of the patient "Alice" is requested by the doctor and pharmacist.

```
PolicySet patientConsent { permit-overrides
        target: equal ( "Alice" , resource / patient-id )
        policies:
        PolicySet ePre { permit-overrides - all
                target:equal("e-Prescription",resource/type)
                policies:
                Rule writeDoc ( permit target: equal ( subject / role , "doctor" )
                        && equal ( action / id , "write" )
                        && in ("e-Pre-Write" , subject / permission )
                        && in ( "e-Pre-Read" , subject / permission ) )
                Rule readDoc ( permit target: equal ( subject / role , "doctor" )
                        && equal ( action / id , "read" )
                        && in ( "e-Pre-Read", subject / permission ) )
                Rule readPha ( permit target: equal ( subject / role , "pharmacist" )
                        && equal ( action / id , "read" )
                        && in ("e-Pre-Read" , subject / permission ) )
                obl-p:[M log ( system / time , resource / type , subject / id , action / id )
]
        }
        Rule denyRule ( deny )
        obl-p:[O compress ( ) ]
        obl-d:[ M mailTo ( resource / patient-id.mail , "Data requested by
unauthorized subject" ) ]
```

The above policy for a resource is "patientConsent" with rule combining algorithm for "permit-overrides". The target as per the syntax contains the MatchId with value "Alice" and is patient id.

The policies start with Policyset "ePre" with rule algorithm permit-overrides—all and with target MatchId as "equal" for a value "e-Prescription" which is the resource.

The policy encloses three types of rules for three permission, i.e. doctor can write, the doctor can read and pharmacist can read. All the three rules are written for the effect "permit". The target MatchId is "equal" for the three rules. For the first rule, the value of the subject role name is "doctor" and when the subject permission is "e-Pre-Write" and "e-Pre-Read", the action id is "write". For the second rule, the subject/role name is "doctor" and subject permission is "e-Pre-Read", then the action id is "write". For the third rule, the subject/role name is "pharmacist" and when subject permission is "e-Pre-Read", the actionid is "read".

Obligations are pieces of advice to take after PDP has taken the response decision. For all the three rules when the effect is permitted, the obligation expression creates obligation id "log" and message the attribute values such as system time, resource type, subject id, and action id.

The fourth rule is denyRule set the effect as "deny" when none of the preceding rules permit. In that case, the obligations of a permit are compressed. For the obligation expression obl-d, when the obligation id is "mailto", it is advised to send a message," Data requested by unauthorised subject" Let the attributes of the three requests are

Request:{ Request1
(subject/id,"Dr House")
(resource/patient-id,"Alice")
(resource/type,"e-Prescription")
(subject/role,"doctor")
(subject/permission,"e-Pre-Read","e-Pre-Write")
(action/id,"write")
}

Request:{ Request2
(subject/id,"Dr. House")
(resource/patient-id,"Alice")
(resource/type,"e-Prescription")
(subject/role,"doctor")
(action/id,"delete")
}

Request:{ Request3
(subject/id,"Dr. House")
(resource/patient-id,"Alice")
(resource/type,"e-Prescription")
(subject/role,"doctor")
(subject/permission,"e-Pre-Read")
(action/id,"read")
}

The response result for the three requests based on the policy is

```
/*
        Request1 : PERMIT
        Request2 : DENY
        Request3 : PERMIT
 */
```

The policy "patientConsent" is validated using following **XML Schema.**

```
<PolicySet xmlns="urn:oasis:names:tc:xacml:3.0:core:schema:wd-
17" xmlns:xsi="http://www.w3.org/2001/XMLSchema-
instance" xsi:schemaLocation="urn:oasis:names:tc:xacml:3.0:core:schema:wd-
17 http://docs.oasis-open.org/xacml/3.0/xacml-core-v3-schema-wd-
17.xsd" PolicySetId="AliceEPrescription" Version="1.0" PolicyCombiningAlgId
="urn:oasis:names:tc:xacml:3.0:policy-combining-algorithm:permit-
overrides" MaxDelegationDepth="1">
<Target>
<AnyOf>
<AllOf>
<Match MatchId="urn:oasis:names:tc:xacml:1.0:function:string-equal">
<AttributeValue DataType="http://www.w3.org/2001/XMLSchema#string"> Alice <
/AttributeValue>
<AttributeDesignator DataType="http://www.w3.org/2001/XMLSchema#anyURI" Mus
tBePresent="false" Category="urn:oasis:names:tc:xacml:3.0:attribute-
category:resource" AttributeId="urn:oasis:names:tc:xacml:1.0:resource:patie
nt-id"/>
</Match>
</AllOf>
</AnyOf>
</Target>
<PolicySet PolicySetId="ePre" Version="1.0" PolicyCombiningAlgId="urn:oasis
:names:tc:xacml:3.0:policy-combining-algorithm:permit-overrides">
<Target>
<AnyOf>
<AllOf>
<Match MatchId="urn:oasis:names:tc:xacml:1.0:function:string-equal">
<AttributeValue DataType="http://www.w3.org/2001/XMLSchema#string"> e-
Prescription </AttributeValue>
<AttributeDesignator DataType="http://www.w3.org/2001/XMLSchema#anyURI" Mus
tBePresent="false" Category="urn:oasis:names:tc:xacml:3.0:attribute-
category:resource" AttributeId="urn:oasis:names:tc:xacml:1.0:resource:type"
/>
</Match>
</AllOf>
</AnyOf>
</Target>
<Policy PolicyId="PolicyEnclosingRulewrite" Version="1.0" RuleCombiningAlgI
d="urn:oasis:names:tc:xacml:3.0:policy-combining-algorithm:permit-
overrides" MaxDelegationDepth="1">
<Target/>
<Rule RuleId="write" Effect="Permit">
<Target/>
<Condition>
<Apply FunctionId="urn:oasis:names:tc:xacml:1.0:function:and">
<Apply FunctionId="urn:oasis:names:tc:xacml:1.0:function:and">
<Apply FunctionId="urn:oasis:names:tc:xacml:1.0:function:and">
<Apply FunctionId="urn:oasis:names:tc:xacml:1.0:function:string-equal">
<AttributeDesignator DataType="http://www.w3.org/2001/XMLSchema#anyURI" Mus
tBePresent="false" Category="urn:oasis:names:tc:xacml:3.0:attribute-
category:subject" AttributeId="urn:oasis:names:tc:xacml:1.0:subject:role"/>
<AttributeValue DataType="http://www.w3.org/2001/XMLSchema#string"> doctor
</AttributeValue>
</Apply>
```

```
<Apply FunctionId="urn:oasis:names:tc:xacml:1.0:function:string-equal">
<AttributeDesignator DataType="http://www.w3.org/2001/XMLSchema#anyURI" Mus
tBePresent="false" Category="urn:oasis:names:tc:xacml:3.0:attribute-
category:action" AttributeId="urn:oasis:names:tc:xacml:1.0:action:id"/>
<AttributeValue DataType="http://www.w3.org/2001/XMLSchema#string"> write <
/AttributeValue>
</Apply>
</Apply>
<Apply FunctionId="urn:oasis:names:tc:xacml:1.0:function:string-subset">
<AttributeValue DataType="http://www.w3.org/2001/XMLSchema#string"> e-Pre-
Write </AttributeValue>
<AttributeDesignator DataType="http://www.w3.org/2001/XMLSchema#anyURI" Mus
tBePresent="false" Category="urn:oasis:names:tc:xacml:3.0:attribute-
category:subject" AttributeId="urn:oasis:names:tc:xacml:1.0:subject:permiss
ion"/>
</Apply>
</Apply>
<Apply FunctionId="urn:oasis:names:tc:xacml:1.0:function:string-subset">
<AttributeValue DataType="http://www.w3.org/2001/XMLSchema#string"> e-Pre-
Read </AttributeValue>
<AttributeDesignator DataType="http://www.w3.org/2001/XMLSchema#anyURI" Mus
tBePresent="false" Category="urn:oasis:names:tc:xacml:3.0:attribute-
category:subject" AttributeId="urn:oasis:names:tc:xacml:1.0:subject:permiss
ion"/>
</Apply>
</Apply>
</Condition>
</Rule>
</Policy>
<Policy PolicyId="PolicyEnclosingRuleread" Version="1.0" RuleCombiningAlgId
="urn:oasis:names:tc:xacml:3.0:policy-combining-algorithm:permit-
overrides" MaxDelegationDepth="1">
<Target/>
<Rule RuleId="read" Effect="Permit">
<Target/>
<Condition>
<Apply FunctionId="urn:oasis:names:tc:xacml:1.0:function:and">
<Apply FunctionId="urn:oasis:names:tc:xacml:1.0:function:and">
<Apply FunctionId="urn:oasis:names:tc:xacml:1.0:function:string-equal">
<AttributeDesignator DataType="http://www.w3.org/2001/XMLSchema#anyURI" Mus
tBePresent="false" Category="urn:oasis:names:tc:xacml:3.0:attribute-
category:subject" AttributeId="urn:oasis:names:tc:xacml:1.0:subject:role"/>
<AttributeValue DataType="http://www.w3.org/2001/XMLSchema#string"> doctor
</AttributeValue>
</Apply>
<Apply FunctionId="urn:oasis:names:tc:xacml:1.0:function:string-equal">
<AttributeDesignator DataType="http://www.w3.org/2001/XMLSchema#anyURI" Mus
tBePresent="false" Category="urn:oasis:names:tc:xacml:3.0:attribute-
category:action" AttributeId="urn:oasis:names:tc:xacml:1.0:action:id"/>
<AttributeValue DataType="http://www.w3.org/2001/XMLSchema#string"> read </
AttributeValue>
</Apply>
</Apply>
<Apply FunctionId="urn:oasis:names:tc:xacml:1.0:function:string-subset">
<AttributeValue DataType="http://www.w3.org/2001/XMLSchema#string"> e-Pre-
Read </AttributeValue>
<AttributeDesignator DataType="http://www.w3.org/2001/XMLSchema#anyURI" Mus
tBePresent="false" Category="urn:oasis:names:tc:xacml:3.0:attribute-
category:subject" AttributeId="urn:oasis:names:tc:xacml:1.0:subject:permiss
ion"/>
</Apply>
```

```
</Apply>
</Condition>
</Rule>
</Policy>
<Policy PolicyId="PolicyEnclosingRulepha" Version="1.0" RuleCombiningAlgId=
"urn:oasis:names:tc:xacml:3.0:policy-combining-algorithm:permit-
overrides" MaxDelegationDepth="1">
<Target/>
<Rule RuleId="pha" Effect="Permit">
<Target/>
<Condition>
<Apply FunctionId="urn:oasis:names:tc:xacml:1.0:function:and">
<Apply FunctionId="urn:oasis:names:tc:xacml:1.0:function:and">
<Apply FunctionId="urn:oasis:names:tc:xacml:1.0:function:string-equal">
<AttributeDesignator DataType="http://www.w3.org/2001/XMLSchema#anyURI" Mus
tBePresent="false" Category="urn:oasis:names:tc:xacml:3.0:attribute-
category:subject" AttributeId="urn:oasis:names:tc:xacml:1.0:subject:role"/>
<AttributeValue DataType="http://www.w3.org/2001/XMLSchema#string"> pharmac
ist </AttributeValue>
</Apply>
<Apply FunctionId="urn:oasis:names:tc:xacml:1.0:function:string-equal">
<AttributeDesignator DataType="http://www.w3.org/2001/XMLSchema#anyURI" Mus
tBePresent="false" Category="urn:oasis:names:tc:xacml:3.0:attribute-
category:action" AttributeId="urn:oasis:names:tc:xacml:1.0:action:id"/>
<AttributeValue DataType="http://www.w3.org/2001/XMLSchema#string"> read </
AttributeValue>
</Apply>
</Apply>
<Apply FunctionId="urn:oasis:names:tc:xacml:1.0:function:string-subset">
<AttributeValue DataType="http://www.w3.org/2001/XMLSchema#string"> e-Pre-
Read </AttributeValue>
<AttributeDesignator DataType="http://www.w3.org/2001/XMLSchema#anyURI" Mus
tBePresent="false" Category="urn:oasis:names:tc:xacml:3.0:attribute-
category:subject" AttributeId="urn:oasis:names:tc:xacml:1.0:subject:permiss
ion"/>
</Apply>
</Apply>
</Condition>
</Rule>
</Policy>
<ObligationExpressions>
<ObligationExpression ObligationId="log" FulfillOn="Permit">
<AttributeAssignmentExpression AttributeId="urn:oasis:names:tc:xaml:3.0:arg
ument">
<AttributeDesignator DataType="http://www.w3.org/2001/XMLSchema#anyURI" Mus
tBePresent="false" Category="urn:oasis:names:tc:xacml:3.0:attribute-
category:system" AttributeId="urn:oasis:names:tc:xacml:3.0:system:time"/>
</AttributeAssignmentExpression>
<AttributeAssignmentExpression AttributeId="urn:oasis:names:tc:xaml:3.0:arg
ument">
<AttributeDesignator DataType="http://www.w3.org/2001/XMLSchema#anyURI" Mus
tBePresent="false" Category="urn:oasis:names:tc:xacml:3.0:attribute-
category:resource" AttributeId="urn:oasis:names:tc:xacml:1.0:resource:type"
/>
</AttributeAssignmentExpression>
<AttributeAssignmentExpression AttributeId="urn:oasis:names:tc:xaml:3.0:arg
ument">
<AttributeDesignator DataType="http://www.w3.org/2001/XMLSchema#anyURI" Mus
tBePresent="false" Category="urn:oasis:names:tc:xacml:3.0:attribute-
category:subject" AttributeId="urn:oasis:names:tc:xacml:1.0:subject:id"/>
</AttributeAssignmentExpression>
```

```
<AttributeAssignmentExpression AttributeId="urn:oasis:names:tc:xaml:3.0:arg
ument">
<AttributeDesignator DataType="http://www.w3.org/2001/XMLSchema#anyURI" Mus
tBePresent="false" Category="urn:oasis:names:tc:xacml:3.0:attribute-
category:action" AttributeId="urn:oasis:names:tc:xacml:1.0:action:id"/>
</AttributeAssignmentExpression>
</ObligationExpression>
</ObligationExpressions>
</PolicySet>
<Policy PolicyId="PolicyEnclosingRuleruleDeny" Version="1.0" RuleCombiningA
lgId="urn:oasis:names:tc:xacml:3.0:policy-combining-algorithm:permit-
overrides" MaxDelegationDepth="1">
<Target/>
<Rule RuleId="ruleDeny" Effect="Deny">
<Target/>
</Rule>
</Policy>
<ObligationExpressions>
<ObligationExpression ObligationId="mail" FulfillOn="Deny">
<AttributeAssignmentExpression AttributeId="urn:oasis:names:tc:xaml:3.0:arg
ument">
<AttributeDesignator DataType="http://www.w3.org/2001/XMLSchema#anyURI" Mus
tBePresent="false" Category="urn:oasis:names:tc:xacml:3.0:attribute-
category:resource" AttributeId="urn:oasis:names:tc:xacml:1.0:resource:patie
nt-mail"/>
</AttributeAssignmentExpression>
<AttributeAssignmentExpression AttributeId="urn:oasis:names:tc:xaml:3.0:arg
ument">
<AttributeValue DataType="http://www.w3.org/2001/XMLSchema#string"> Data
request by unauthorised subject </AttributeValue>
</AttributeAssignmentExpression>
</ObligationExpression>
</ObligationExpressions>
<AdviceExpressions>
<AdviceExpression AdviceId="compress" AppliesTo="Permit"> </AdviceExpressio
n>
</AdviceExpressions>
</PolicySet>
```

10 Conclusion

The dynamic and distributed environments which are decentralised manner need more effort on access control mechanism for access to resources. The policy is designed based on semantic information of the entities either requestor or provider. The semantic rule and semantic-based access control model improve the interoperability of policies and enhance the authorisation into fine-grained authorisation model. The knowledge-driven decision taken on using semantic knowledge makes use of formal properties that helps to validate and reason the permission to access the resources. The conventional RBAC model and ABAC model are extended to semantic and context-aware model that overcome the limitations such as scalable and purely syntactic based decisions. The semantic standards such as OWL, SWRL and XACML on various SAC models facilitate the essentiality of access control mechanism for present dynamic and changing situations to achieve secure authenticated and authorised communication between web services.

References

1. Berners-Lee, T., Hendler, J., Lassila, O.: The semantic web. In: Scientific American, pp. 34–43, May 2001
2. Curbera, F., Nagy, W., et al.: Web services: why and how. In: Workshop on Object-Oriented Web Services—OOPSLA 2001, Tampa, Florida, USA (2001)
3. Patterson, R.S., Miller, J.A., Cardoso, J., Davis, M.: Bringing semantic security to semantic web services. In: Cardoso, J., Hepp, M., Lytras, M.D. (eds.) The Semantic Web. Semantic Web and Beyond Computing for Human Experience, Vol. 6. Springer, Boston, MA (2007)
4. IBM Corporation and Microsoft Corporation: Security in a Web Services World: A Proposed Architecture and Roadmap Version 1.0 (2002)
5. Verma, K., Aggarwal, R., Miller, J., Milnor, W.: Constraint driven web service composition in METEOR-S. In: Proceedings of the 2004 IEEE International Conference on Services Computing (SCC'04), Shanghai, China (September 2004), pp. 23–32 (2007)
6. Pennington, C., Cardoso, J., Miller, J.A., Patterson, R.S., Vasquez, I.: Introduction to Web Services
7. Shen, H., Cheng, Y.: A Semantic context-based model for mobile web services access control. I.J. Comput. Netw. Inf. Secur. **1**, 18–25 (2011)
8. Horrocks, I., Patel-Schneider, P.F., Boley, H.: SWRL: A Semantic Web Rule Language Combining OWL and RuleML (2004). http://www.w3.org/Submission/SWRL/
9. Burstein, M., Bussler, C., Zaremba, M., Finin, T., Huhns, M.: A semantic web services architecture. IEEE Internet Comput., Sept–Oct 2005
10. Bajaj, S., Box, D.: Web Services Policy Framework (WSPolicy). ftp://www6.software.ibm.com/software/developer/library/wspolicy
11. Lopez, V, Mana, A., Mariatroya, J., Yague, V.: Applying the semantic web layers to access control. In: IEEE International Workshop on Web Semantics (WebS03), pp. 622–626 (2003)
12. Yag¨ue, M., Troya, J.: Semantic approach for access control in web services. In: Euroweb'02, eWiC. British Computer Society (2002)
13. Kagal, L., Paolucci, M., Srinivasan, N., Denker, G., Finin, T., Sycara, K.: Authorization and privacy for semantic web services. In: AAAI 2004 Spring Symposium on Semantic Web Services, March 2004
14. Shields, B., Molloy, O., Lyons, G., Duggan, J.: Securing web services using semantic web technologies. In: Proceedings Ofiasw-2005
15. Gondara, M.K.: Access control mechanisms for semantic webservices—a discussion on requirements and future directions. IJ-ETA-ETS **4**(1), 338 (2011)
16. Amini, M., Arasteh, M.: Semantic and attribute-based access control model for virtual organizations. ISC Int. J. Inf. Secur. **7**(1), 27–45 (2015)
17. Shen, H., Cheng, Y.: Semantic context-based model for mobile web services access control. I.J. Comput. Netw. Inf. Secur. **1**, 18–25 (2011)
18. Kayes, A.S.M., Han, J., Colman, A.: A semantic policy framework for context-aware access control applications. In: 12th IEEE International Conference on TrustCom (2013)

DistSNNMF: Solving Large-Scale Semantic Topic Model Problems on HPC for Streaming Texts

Fatma S. Gadelrab, Rowayda A. Sadek, and Mohamed H. Haggag

Abstract The scalability means a time and space complexity of the topic modeling algorithms are needed to be reduced. Although the parallel algorithms on the multi-processor architecture have a small complexity for time and space, the communication costs among processors lead to a serious scalability problem. On the other hand, when the topic modeling dealing with large scale streaming data, it still suffers from the limited model capacity problem. To tackle this problem, in this paper we proposed a distributed version of the prior topic modeling algorithm (SNNMF) named DistSNNMF. The training task splits into many sub-batch tasks and distributed across multiple worker nodes, such that the whole training process is accelerated through the cooperation with the data-parallel platform. Extensive experiments conducted on real-world datasets demonstrate the usability and scalability of the proposed algorithm.

Keywords Distributed topic model · Scalability topic model · Parallel topic model · Scalable topic model

F. S. Gadelrab (✉) · R. A. Sadek
Department of Information Technology, Faculty of Computers and Artificial Intelligence, Helwan University, Cairo, Egypt
e-mail: Fatma.s.gadelrab@gmail.com

R. A. Sadek
e-mail: rowayda_sadek@yahoo.com

M. H. Haggag
Department of Computer Science, Faculty of Computers and Artificial Intelligence, Helwan University, Cairo, Egypt
e-mail: Mohamed.haggag@fci.helwan.edu.eg

M. Al-Emran et al. (eds.), *Recent Advances in Intelligent Systems and Smart Applications*, Studies in Systems, Decision and Control 295,
https://doi.org/10.1007/978-3-030-47411-9_23

1 Introduction

The real-world application of the topic modeling is limited due to the scalability issue. When the number of documents and the vocabulary size reached the million or, even the billion levels, which is quite common now in practical processing. The challenge of a topic detection model needs to improve the scalability technique by using a distributed or parallelized computing framework [1–3]. Almost all of the topic detection models depend on the sparse matrix multiplication. On the other hand, many of these algorithms are rewritten with GPU utilization, because GPU brings a massivelly parallel architecture in addition to a particularly good ratio between performance and price [4]. The scalability challenge comes in two forms: better use of available processing power (scaling in time) with making the process has a large input is feasible on computers with limited memory (scaling in space). The first problem is typically solved by devising a scheme to split the problem into smaller sub-problems to distribute the computation [5]. The parallel algorithms on the multi-processor architecture have a low time in addition to a space complexity, their communication costs among processors often scale linearly with the vocabulary size and the number of topics, leading to a serious scalability problem [6]. Both a scaling up and a scaling out are used to solve the problem. With the parallelization, there is the possibility of addressing the scaling up without either simplifying the algorithm or the task. Graphics processing units (GPUs) have enabled inexpensive and high-performance computing for general purpose applications [7]. The current application of GPU processors for parallel computing tasks show excellent results in terms of speedups compared to CPU processors. However, there is no existing framework that enables the automatic distribution of data and processing across multiple GPUs. All these elements are necessary conditions to enable users to easily perform a big data analysis to create their modules for their desired processing functionality [8].

This paper is an extension of the prior work SNNMF (Novel Semantic Tagging Detection Algorithms Based Non-Negative Matrix Factorization) [9]. The main goal of SNNMF is to present a static and flat topic model by utilizing the semantic text representation based on the knowledge-based approach to extract the term-topic matrix and the topic-document matrix for a batch dataset. In detail, SNNMF is based on document-term(V). In addition, it generates two sparse matrices, (1) term-topic (H), (2) topic-document(W). The final term-topic ($H^{(4)}$) in SNNMF is passing-by four algorithms. Each one of these algorithms generates a sparse matrix ($H^{(1)}$, $H^{(2)}$, $H^{(3)}$, $H^{(4)}$) sequentially. In more, SNNMF adapted the document-term (V) matrix for document representation by using tf-idf (term frequency-inverse document frequency) for the whole training data-set. Accordingly, one of the limitations in SNNMF is that it always requires to pass through the whole dataset and consumes considerably much space in addition to a long time, this is making it' as an intractable applying for the large-scale dataset. In other words, for a large-scale dataset, the sparse matrix cannot be processed by GPU using the common storage formats because of the memory limitation. In addition, the parallel effect is poor using general formats for the sparse matrices with extremely uneven distribution of nonzero elements, which leads to performance deterioration. SNNMF suffers from a scalability issue using a large-scale dataset. To alleviate this problem, there are various models developed to address the sparse matrix problems for a large-scale dataset. The earlier studies such as [10, 11] addressed the sparse matrix problem by decomposing the main matrix into some sub-matrices to be computed separately with parallel processes. In addition, in [12] the authors proposed a method called mainstream, that exploits a streaming technology in addition to hide the memory latency of copy between the main memory and the device memory. To fully exploit the GPU stream technology, they divided a matrix into small fixed-size units instead of performing on a compressed sparse row or, compressed sparse column format. To address tf-idf problem for a large-scale dataset, in [13, 14] the authors proposed an approximate version of tf-idf worked on the continuous data stream. In addition to the parallel implementation of the approximate tf-idf using GPU.

The above methods may help the proposed DistSNNMF to beat on the scalability in addition to the tf-idf problem for the large-scale dataset compared to SNNMF.

Considering the solutions to these problems, this paper proposes a distributed real-time semantic tagging (DistSNNMF). Besides as a distributed algorithm, it demonstrates good scalability through experiments, in addition, its text representation via tf-idf for a large-scale dataset.

In summary, DistSNNMF methodology has three major points. First, Building the data-parallel platform. Second, Building tasks-parallel distributed. Third, DistSNNMF combines between the data-parallel platform with the tasks-parallel distributed to improve the topic detection model scalability issue via the high-performance computing (HPC).

The remainder of the paper is organized as follows. In Sect. 2: background, Sect. 3: some recent related work is briefly reviewed, Sect. 4: proposed solution (Methodology), Sect. 5: result and experiments, Sect. 6: conclusion and, future work, 7: Acknowledgment, at last in Sect. 8: references.

2 Background

This section presents an overview of big data platforms and the tools are used to implement a topic detection model running over HPC.

2.1 Big Data Platforms

Currently, there are two popular types of platforms and associated technologies for conquering the needs of big data processing: (1) Hadoop, along with the related technologies such as Spark and Yarn provisioned on commodity hardware, (2) High-performance computing (HPC) platforms with or without Hadoop provisioned to them.

Pros of HPC for big data: HPC systems may spend a small number of compute cycles to compute a result by parallel processing or, concurrent processing techniques (algorithms).

Cons of HPC for a big data: HPC lacking the superb storage and I/O (data movement) capabilities important for big data processing [15, 16].

2.2 Tools

The topic modeling frameworks have been widely deployed to run over GPU servers in both academia and industry. Several popular distributed frameworks including Caffe-MPI,[1] CNTK,[2] MXNet,[3] and, TensorFlow[4] have achieved a high throughput in a single GPU with the help of cuDNN and good scalability across multiple GPUs over multiple machines. These frameworks supply an easy way for users to develop own model. In addition, try to optimize the related algorithms to achieve high throughput by using hardware platforms, such as multi-core CPU, many-core GPU, multiple GPUs, and, multiple machines. However, the different implementation methods by vendors, these tools show different performance even when training the same model on the same hardware platform [17] evaluated the performance of four distributed Deep Learning (DL) frameworks (i.e., Caffe-MPI, CNTK, MXNet and, TensorFlow) with convolution neural networks (CNNs) over GPU cluster. For a single GPU, MXNet and, TensorFlow has a better data prefetch mechanism than CNTK. TensorFlow hides the overhead of data reading. For using TensorFlow to implement the algorithm, in [18] the authors said there are exist two "versions" of TensorFlow, one for local execution on a single machine with many devices and the other one supporting a distributed execution across many machines with many devices. The second degree referred to the fact that at each machine such there was more than one device, for example, five independent GPUs and/or three CPUs. When the client requested an evaluation of a TensorFlow graph via a session's run routine, this query was sent to the primary process, which in turn delegates the task to one or, more worker processes and coordinates their execution. In addition, in [19, 20] the authors said a TensorFlow used data-flow graphs to represent the computation, shared state, and, the operations that mutate that state. It mapped the nodes of a dataflow graph across many machines in a cluster and, within a machine across multiple computing devices. TensorFlow supported both large-scale training and an inference. TensorFlow allows programmers to experiment with different parallelization schemes.

[1] https://github.com/Caffe−MPI/Caffe−MPI.github.io.

[2] https://github.com/Microsoft/CNTK.

[3] https://github.com/apache/incubator−mxnet.

[4] https://github.com/tensorflow/tensorflow.

3 Related Work

Recently, to satisfy the scalable requirements, some scholars have made attempts for the parallelization computing mechanism of the topic detection models in addition to partition the input dataset. The partitioning of the input dataset is two approaches, the offline and the online partitioning (streamed texts). This section briefly discusses the related works from both the scalable topic modeling and the topic modeling for the streaming texts.

3.1 The Scalable of the Topic Modelling

There are various models developed for topic detection. these models focus on partitioning the data (offline, online) in addition to handle the scalability issue. Such as, in [30] the authors proposed a distributed version of the online topic model algorithm named [as] DistHDP, the training task was divided into many sub-batch tasks and distributed across multiple worker nodes, such that the whole training process is accelerated. The model convergence was guaranteed through a distributed variation inference algorithm. In [21] the authors presented the CuLDA_CG model based on the delicately designed workload partition and synchronization mechanism to exploit the benefits of multiple GPUs. CuLDA_CGS pre-processing based partitioning the input dataset into multiple chunks. In [22] the authors described approach to discover the topics based on Min-Hashing, which could handle a massive text dataset with large vocabularies, using modest computer hardware and did not require to fix the number of topics in advance. In [23] the authors presented PO-sLDA to solve problems that were caused by the gradual increase of the scale of datasets such as, increasingly inefficient in addition to time-consuming and limits its applications in a very narrow range via parallel computing mechanism. PO-sLDA implemented via the MapReduce framework. PO-sLDA was proposed to promote the ability of cloud computing with big data processing. In [24] the authors proposed a ZenLDA training system that followed a generalized design for the distributed data-parallel platform. ZenLDA consisted of three main aspects: (1) it converted the commonly used serial collapsed Gibbs sampling inference algorithm to a monte-carlo collapsed Bayesian estimation method, which was embarrassingly parallel; (2) it decomposed the LDA inference formula into parts that can be sampled more efficiently to reduce computation complexity; (3) it proposed a distributed LDA training framework, which represented the dataset as a directed graph with the parameters annotated as corresponding vertices and implemented ZenLDA by other well-known inference methods based on Spark. In [25] the authors introduced empirically analyze Clustered Latent Dirichlet Allocation (CLDA) for extracting dynamic latent topics from a collection of documents. This approach based on a data decomposition in which the data was partitioned into segments, followed by topic modeling on the individual segments.

3.2 Topic Detection Models for Streaming Texts

We now turn our attention to another approach to tackle the problem of the topic detection models on the large-scale training set, it is the topic detection models on the streamed data.

In [26] the authors presented a hierarchical online NMF (HONMF) for detecting and tracking topic hierarchies in a text stream. The proposed method can dynamically adjust the topic hierarchy. The emerging documents were documents belonging to the emerging topics in the current time slot. In [27] the authors proposed a dynamic detection method for microblog topics based on the time series. Firstly, according to the law of event development, a three-level topic evolution model based on time series was proposed. The microblog text was divided into time slices. Then, a time decay function is introduced, regards it as time feature, used it, and contented features together to calculate the text similarity. Finally, the traditional Single-pass algorithm had been improved to optimize its clustering center update strategy. The experiment dataset based combined data per day.

4 The Proposed DistSNNMF Methodology

This section describes the proposed DistSNNMF methodology. The goal of this study is to improve the scale in time and the scale in space (memory) for the prior topic detection model SNNMF.

4.1 DistSNNMF Methodology Overview

The key difference between SNNMF and DistSNNMF is the advantage of online detecting topics in addition to the scalability approach. DistSNNMF has three major points shown in Fig. 1. First, Building the data-parallel platform. DistSNNMF decomposes the long documents into paragraphs (sub-documents), Then, each computational worker node owns one independent sub-documents to perform the computation over that sub-documents. Since DistSNNMF has multiple nodes scanning the sub-documents in parallel, DistSNNMF should be able to scan more sub-documents than when using a single node. This leads to an increase the throughput through distributed parallel computing. This dividing to improve the scale in space.

Second, Building the tasks-parallel distributed. The training task of SNNMF is split into many sub-batch tasks (jobs) and distributed across multiple worker nodes, such that the whole training process is accelerated. The proposed sub-tasks are function-oriented system design approach. DistSNNMF has two main categories for these tasks, the first is named the term-topic (FJ) tasks and the second named the topic-document (CJ) tasks. This partitioning to improve the scale in time per job.

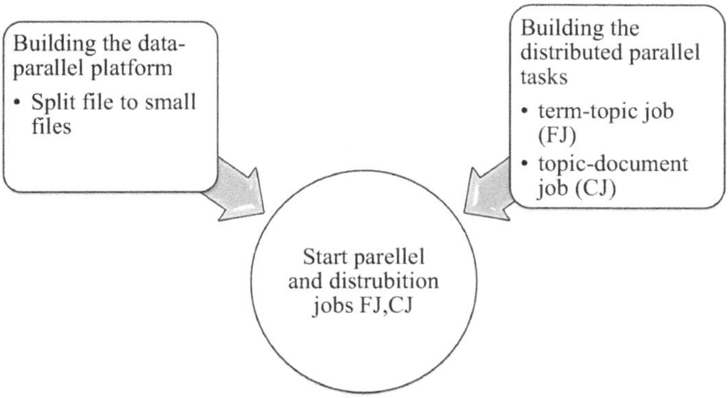

Fig. 1 The design overview of DistSNNMF

Third, DistSNNMF combines the data-parallel platform with the tasks-parallel distributed to improve the topic detection model scalability issue on high-performance computing (HPC).

4.2 *DistSNNMF Methodology Details*

This section discusses the details of DistSNNMF including (1) Term-topic jobs (FJ) and, (2) Topic-document jobs (CJ). The CJ jobs can start after at least performs all sub-jobs of FJ for one file. Each sub-job starts with a parallelism way by many nodes for the parallelism files. Each node contains CPU or, CPU + one GPU to compute this function, the output may be a file or, set of files. The relation between the input file and, the job is only one to one, the relation between the job and, the output is one to one, but, sometimes, one to many.

4.2.1 Term-Topic Jobs (FJ)

The main purpose of these groups of jobs is extracting the final topics-terms vectors per each of the streamed text file. By the end of this part, DistSNNMF can get the term-topic matrix such as $H^{(4)}$ in SNNMF, but in separated vectors form. Each vector is saved in a separate file. The term-topic group contains four sub-jobs are as the following:

- Pre-processing (FJ-1)
- Internal low-pass filter (FJ-2)
- External low-pass filter (FJ-3)
- Tag edge detection via a semantic low-pass filter (FJ-4).

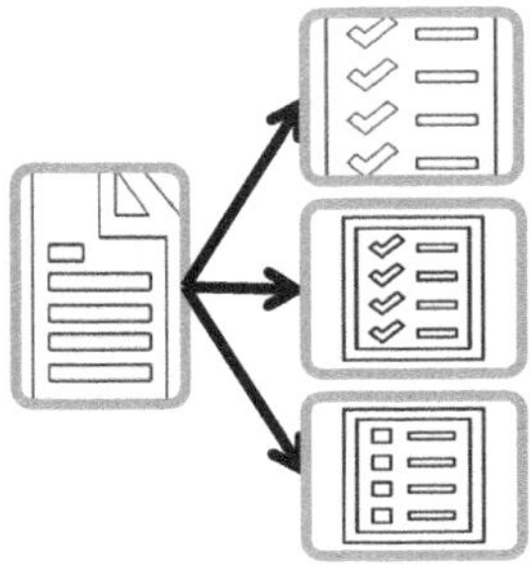

Fig. 2 Layout of pre-processing job

Pre-processing (FJ-1)

Figure 2 illustrates the pre-processing job (FJ-1) layout, the main goal of this job is to compute the term frequency (tf) and the inverse document frequency (idf) for the current text document without knowing whole dataset files. In the more, the new terms are added by the current text file to the current dataset terms. FJ-1 simulates the same steps in SNNMF pre-processing. The input is one text file. The output is a one term frequency (tf), a one inverse document frequency (idf) and, one new terms file.

$$idf_t = \log \frac{N}{df_t} \tag{1}$$

Equation 1 is designed for the inverse document frequency, where, N is the number of the current documents and the df_t is the frequency of the term (t) in the current document. With the parallel jobs, each a job adds the new file to the dataset, this leads to an increase at least the N number so, not only the current idf is calculated, but also idfs for the earlier files will be updating. This solution has pros and cons. Pros: The computation of tf and idf for each file does not need to know the whole dataset files. This solution is used via stream processing to get fresh results. It considers in-memory computing (scale in space) by loading in the memory of the two feature vectors, one for the current file and, the other for one of the earlier files. After updating the previous file, the current previous file will be closed to open the other and-so-on to keep the memory space. Cons: This job's time depends on the count of previous files and features in idf. This will lead to scaling in time issue. To beat on the long times, improve network and CPU capability. In other words, the impact factors of this job are the input/output (IO_f) in addition to the computing time of tf and idf. IO_f is the count of the previous files in addition to the tf and idf for the new file. The computing time is sensitive to the count of the feature in the recent file and the earlier files, In addition to the computing time of tf and idf for the new file and the updating for the idf previous files.

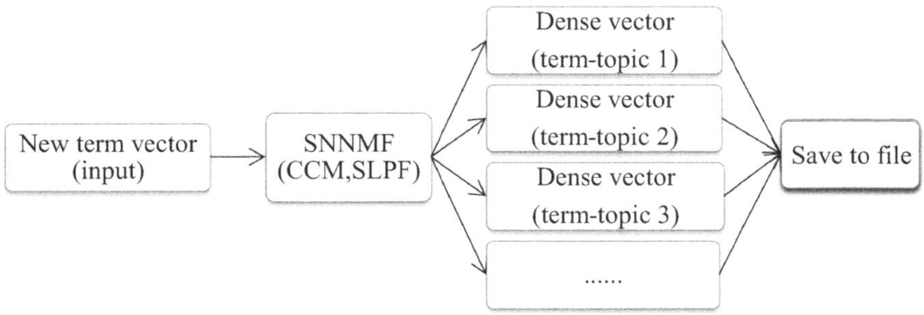

Fig. 3 Layout of internal low-pass filter job

Internal Low-Pass Filter (FJ-2)

Figure 3 illustrates the internal low-pass filter job (FJ-2) layout. The main goal of this job is to generate the term-topic vectors for each term is extracted by the output new-terms file of FJ-1. The computing of this job simulates SNNMF algorithm 1: Correlation-coefficient matrix (CCM) in addition to SNNMF algorithm 2: semantic low-pass filter (SLPF). The input of this job is a vector of new-terms that is extracted by FJ-1 and the output is many dense vectors. Each output dense vector is saved sequentially. The impact factors of this job are IO_f, in addition to the computing time of CCM and SLPF for each input feature.

IO_f is the output dense vectors, in addition to the input file "the new-terms file". The computing time is sensitive to the count of the features in the input file.

$$i = \frac{n(n + 1)}{2} \tag{2}$$

Equation 2 is designed to compute the iteration times of computing the CCM and SLPF for the input file. Where n is the count of the terms in the input file. For example, if the file has 100 terms (n = 100), this means the iteration (i) number is 100* (100 + 1)/2 = 5050. 5050 cycles run on one node to produce 100 vectors (topic files). Vectors are loaded into the memory sequentially (one by one) to keep the memory space (scale in space).

This solution has pros and cons. Pros, this job takes into consideration the scale in memory and the scale in time to compute the semantic term-topic vectors.

Cons, the time of this job depends on the count of the features (terms) in the input vector in addition to the complexity of the algorithms (CCM, SLPF) of SNNMF; In the more, the large-scale of the input vector causes the time scalability issue. To overcome this issue, the job uses the new feature only and a small bit of the original text file. To beat on a longtime, improve the network and, GPU capability.

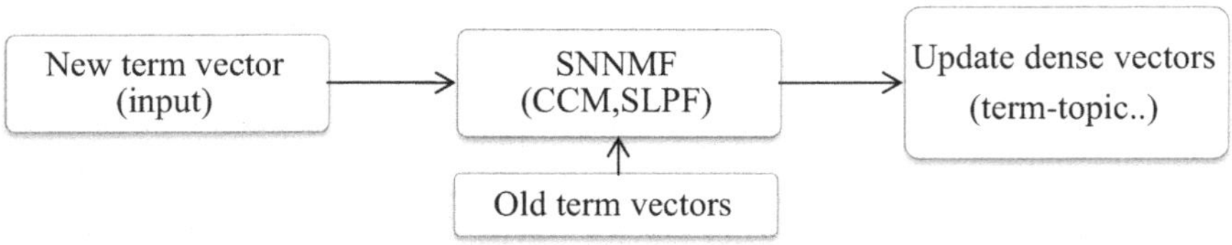

Fig. 4 Layout of external low-pass filter job

External Low-Pass Filter (FJ-3)

Figure 4 illustrates the external low-pass filter job (FJ-3) layout. This job is considered an extension of FJ-2. FJ-3 completes the term-topic computation by simulating the algorithms (CCM, SLPF) of SNNMF for the current input new terms as shown in Fig. 4 by adding the old terms vectors to the input. In detail, the input of this process is the whole old term vectors that were created before, in addition to a new term vector of FJ-2. The old vectors are loaded sequentially to keep memory space. The output is the update of the term-topic dense vectors produced by FJ-2.

For example, if the current file is index zero (job0) there is no earlier file (no execution time for job0), if the current file is index one, there is one earlier file as an input. Each an output dense vector is considered as a column of $H^{(2)}$ in SNNMF.

The impact factors of this job are IO_f in addition to the computing time of CCM and SLPF.

IO_f is the count of the output dense vectors in addition to the input files (the new-terms file and old files). The computing time is sensitive to the count of the features in the whole input files in addition to the computing time of CCM and SLPF for each input feature.

$$i = \sum_{k=1}^{n} \left((fw) \left(fo_{i=1}^{z} \right) \right) \tag{3}$$

Equation 3 is designed to compute the iteration times of computing the CCM and SLPF for the input files. Where n is a count of the terms in the input files with index k start from 1, fw is the new file (current file), fo are the old files with index i start from 1 to z is the number of the old files.

For example, the current file has 100 term (n = 100) and the previous files are two file (z = 2)contains 100,50 terms, the number of the iteration (i) = 100*100 + 100*50 = 15,000 cycle. In the first place, this job depends on the count of the features in the input vectors.

This solution has pros and cons. Pros, this job updates the output dense vector directly, not stored in the memory (scale in space). Cons, all input features (new and old) may cause the scale in time issue. To beat on the longtime, improve the network and the processors' capability.

Fig. 5 Layout of tag edge detection via semantic low-pass filter job

Tag Edge Detection via Semantic Low-Pass Filter (FJ-4)

Figure 5 illustrates the tag edge detection via semantic low-pass filter job FJ-4 layout. The main goal of this job is to generate the term-topic vectors more exact than in FJ-3. The input is one term-topic vector that was computed by FJ-3. The output is one term-topic vector that may be less than or equal to the input vector features. The process simulates SNNMF algorithm 3: Tag edge detection (TED) and SNNMF algorithm 4: Tag edge detection via a semantic low-pass filter (TEDS). Each an output dense vector is considered as a column of $H^{(4)}$ in SNNMF.

The impact factor of this job is the computing time of TED and TEDS. IO_f is not considered as an impact factor because the input is one file and the output are one file. The computing time is sensitive to the count of the features in the input file, In addition to computing the time of TED and, TEDS for each input feature.

$$i = \frac{n(n-1)}{2} \tag{4}$$

Equation 4 is designed to compute the iteration times of computing the TED and TEDS for the input file. Where n is the count of the terms in the input file. Example, if the term-topic input file has 9 terms (n = 9), The total process iterations (i) are 9(9 − 1)/2 = 36 cycles. This job depends on the first place on the count of the terms per file.

This solution has pros and cons. Pros, this job loads a one feature vector in the space (scale in space). Cons, if the input vector is a large-scale (has a huge count of features), it may cause a scale in time issue. Regarding scale in time, to beat on the complexity of the algorithm SNNMF TED and, SNNMF TEDS, improve the processors' capability and, downsize the input vector.

4.2.2 Topic-Document (CJ)

The aim of this group of jobs is extracting the topics-documents vectors based semantic. In the earlier group (Term-topic), many documents were converted to terms (tf and idf), in addition to the configuration of the terms-topics vectors for the current time slot dataset. The jobs of the topic-document group depend on the time-slot and data dependencies that already calculated at this moment (timeline). The time extended for the jobs of the topic-term group in addition to the new terms leads to emerging new topics (new term-topic vectors) and evolves current topics.

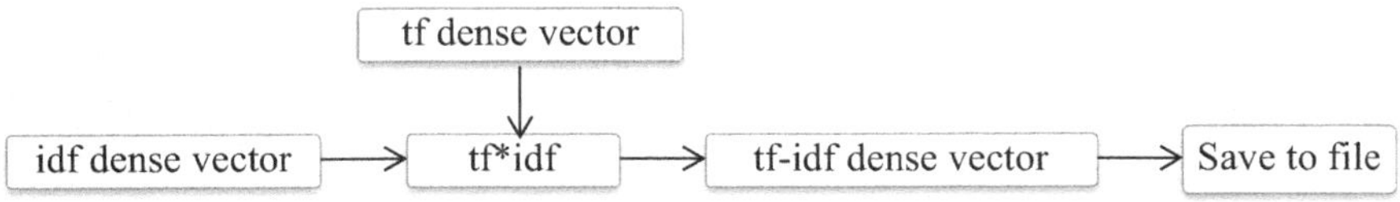

Fig. 6 Layout of tf-idf job

The emerging new topics are performed by FJ-1 and, FJ-2. The evolving of the current topics may be removing terms from the current term-topic vectors or, adding new terms to the current term-topic vectors, the evolving of the current topics is performed by FJ-3 and FJ-4.

The current DistSNNMF topic-term represents the fading of the topics by the old topic vectors that not updated. All these changes in the terms-topic vectors over time impact on the topic-document group jobs 'results.

The topic-document group has two sub jobs:

- tf-idf (CJ-1)
- Topic document (CJ-2).

Tf-Idf (CJ-1)

Figure 6 illustrates tf-idf job (CJ-1) layout. The main goal of this job is to compute tf-idf for each tf and idf generated file in the current dataset (at this moment or, at the current time slot). In the SNNMF, tf-idf was calculated in the pre-processing steps to produce the document-term sparse matrix. In DistSNNMF for a large-scale, tf-idf is a dot product between the two vectors (tf, idf) which are extracted by FJ-1 to produce the dense vectors. Each an output dense vector is considered as a column of V in SNNMF.

The impact factors of this job are IO_f, in addition to the computing time of tf-idf job. IO_f is the number of the output dense vectors in addition to the input files (tf file and, idf file). The computing time is sensitive to the number of the features in the whole input files, In addition to computing the time of tf-idf for each input feature.

$$tf - idf = tf * idf \tag{5}$$

$$i = n \tag{6}$$

Equation 6 is designed to compute the iteration times of computing the tf-idf (Eq. 5) for the input files. Where n is the count of the terms in the input file.

This solution has pros and, cons. The Pros of this process are to improve the scale in space for computing tf-idf per file and the scale in time for all current time slot dataset files. Cons: the process depends on the count of the features in tf and idf vectors, the huge count of terms may cause the scale in space issue. This job

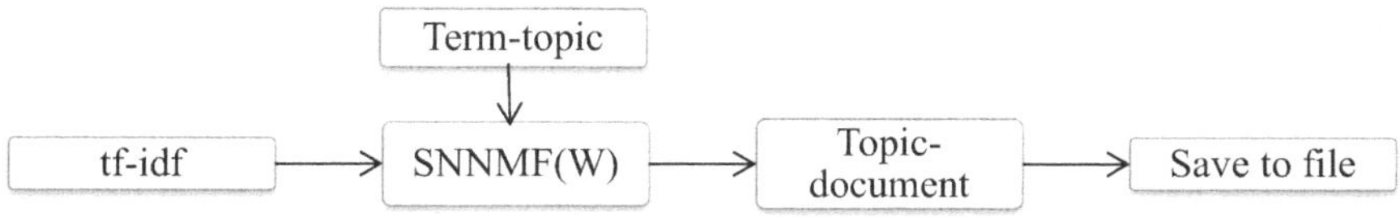

Fig. 7 Layout of topic document job

depends on IO more than the computation. To beat on the time scale, the system should improve the network capability.

Topic Document (CJ-2)

Figure 7 illustrates the computation of the topic-document (CJ-2) job layout. The main goal of this job is to generate the topic-document vector per document (tf-idf). The input is the dense vector tf-idf per the document in addition to the related dense vectors of the term-topic that were computed in FJ-4 sequentially. The output is a dense vector topic-document that will be stored in a file. The process simulates SNNMF algorithm 5: Topic-Document (W) for the input vectors. This job computes the sum scalar multiplication between tf-idf vector and term-topic vectors. Each an output dense vector is considered as a column of W matrix in SNNMF. The impact factors of this job are IOf in addition to the computing time of the same self-job.

IO_f is the number of the output dense vectors in addition to the input files (term-topic files and tf-idf file). The computing time is sensitive to the number of features in the whole input files.

$$V_i = T_{i1}, T_{i2}, T_{i3}, \ldots\ T_{in} \tag{7}$$

$$T_n = t_{n1}, t_{n2}, t_{n3}, \ldots\ t_{nx} \tag{8}$$

$$W_i = T_{in} * T_n \tag{9}$$

$$W_i = T_i * (t_{n1},\ t_{n2},\ t_{n3}, \ldots.\ t_{nx}) \tag{10}$$

Equations from 7 to 10 are designed to compute the topic-document vector for each term in tf-idf input vector. Equation 7 shows the dense document-term vector of tf-idf, where (V_i) for document index (i). V_i has many terms T_{i1} to T_{in}. Each one of these terms is considered a topic in term-topic files, suppose this topic (T_n) has index n (shown in Eq. 8). Each topic T has many terms t_{n1} to t_{nx}. SNNMF Eq. 13 is to compute the document's topics for the whole sparse matrix document-topic (W). DistSNNMF Eq. 10 shows the document-topic dense vector T_{in} is a scalar, but T_n is a vector. Equation 10 shows details of the scalar vector multiplication to generate the document-topic vector values for the document index (i) with the topic (T_{in}).

Table 1 Factors impacting on the scaling jobs

Job name	Relation (input to out)		Impact factors		Scale in space	Scale in time
	Input	Output	No. Files	No. Terms		
FJ-1	One	Three	–	–	√	√
FJ-2	One	Many	–	√	√	√
FJ-3	*Many*	*Many*	√	√	√	√
FJ-4	One	One	–	√	√	√
CJ-1	Two	One	–	√	√	√
CJ-2	Many	One	√	√	√	√

Example, document one (V_1) has 100 terms $T_{1:100}$ and, the value of $T_1 = 0.3$. In term-topic, T_1 has 5 terms ($t_1, t_2, \ldots t_5$), the values of terms are $t_1 = 2$, $t_2 = 3.5$, $t_3 = 4$, $t_4 = 5.5$ and, $t_5 = 3.1$.
$W_1 = 0.3*2 + 0.3*3.5 + 0.3*4 + 0.3*5.5 + 0.3*3.1 = 5.43$

This for every term of 100 terms ($T_{1:100}$) to finish one document. The Pros of this method are to keep a memory space by loading only 2 vectors document-term and term-topic.

This solution has cons. The time of this job depends on the count of terms in tf-idf vector and, the count of terms in each an input term-topic vector. This job depends on IO more than the computation. To beat on the time scale, the system should improve the network capability.

Table 1 illustrates the summary of the impact factors with scaling in jobs. The result section will achieve this. Based on these impact factors the results will recommend a suitable configuration.

5 Results and Experiments

This section presents the results for all jobs were discussed above. We corroborate the effectiveness of DistSNNMF by comparing it using two datasets upon different implementations for both CPU and, CPU + GPU. We compare the scalability experiment for DistSNNMF in terms of time efficiency and job performance.

Moreover, the experiments are designed to answer the following question:

Does the GPU accelerate all jobs (functions) of the distributed topic-modeling algorithms (DistSNNMF)?

Table 2 Test case of workload data sets

No. Case	No. Text files	Input terms (per-process)
TC-1	3	695
TC-2	3	37,947

5.1 Experimental Setup

5.1.1 Dataset Used

Due to the limitation of the resources, the experiments utilize two publicly different scale datasets as the following: There are two scale datasets are used in the experiments shown in Table 2: one (TC-1) has 695 term and, the other (TC-2) has 37,947 terms. These files are extracted from Wikipedia dumb[5] by this way.[6]

5.1.2 Experiments Settings and Configurations

Hardware

DistSNNMF is tested on the target high performance compute (HPC). This HPC follows the single program multiple data (SPMD) execution model, where the program data is partitioned and mapped to the different processes (compute nodes). All processes execute the same program and perform computations on the data items that they own (i.e., owner-computes rule). i.e. data items owned by each process are stored on its private (local) address space, and, when the local computations on a process involve non-local data items, these items are accessed using interposes communication.

The DistSNNMF is tested on two different types of nodes: First, Architecture x86_64, CPU op-mode (s) 32-bit, 64-bit, CPU(s) 32, Thread(s) per core 2, Core(s) per socket 8 Socket(s) 2, Memory 128 GB. Second, plus one GPU Tesla K80 | NVIDIA.

Software

Centos 6. 9, Python 3. 5, Singularity, TensorFlow 1. 4 are the software that used to train DistSNNMF. DistSNNMF is implemented by TensorFlow API. Upon invocation of the run, TensorFlow starts at the requested output nodes and work backward examining the graph dependencies. In the more, TensorFlow computes the full transitive closure of all nodes that must be executed. These nodes may then be assigned to one or many physical execution units (CPUs, GPUs, etc...) on one or many machines.

[5]https://dumps.wikimedia.org/enwiki/.

[6]http://www.evanjones.ca/software/wikipedia2text.html.

DistSNNMF implementation is based on TensorFlow for local execution on a single machine that has many devices. Devices are the smallest most basic entities in the TensorFlow execution model. All nodes in the graph must eventually be mapped to an available device to be executed. In practice, a device most often is either CPU or, a GPU.

For the comparison between the GPU-accelerated implementation and a pure CPU. TensorFlow runs session to call all functions on CPU devices by tf. Device ('/CPU: 0'). the same on GPU, Device ('/GPU: 0'). A GPU is a co-processor that performs operations requested by CPU code.

Slurm workload[7] is used to parallelized DistSNNMF jobs. SLURM job arrays offer a simple mechanism for achieving this. We did this for several reasons. First, all modern supercomputers are made up of collections of n-way dispirited memory processors, so, any code parallelization on these larger machines builds on code parallelization for a single multiprocessor node.[8] Second, job arrays are useful for submitting and, managing many similar jobs. As an example, job arrays are convenient if a user wishes to run the same analysis on 100 different files. SLURM supplies job array environment variables that allow multiple versions of input files to be easily referenced. Third, the performance of the current generation big data middleware stays unsatisfactory because several fundamental bottlenecks exist in the current software design related to IO and, hard disk as mentioned before in the background section. After SLURM starts an array of jobs. Each job calls run to start a TensorFlow session to implement the graph.

5.1.3 Scalability Evaluation of DistSNNMF

We have inspired the evaluation methods by [28–30]. In this study, we only considered the parallel implementations of DistSNNMF for parallel processing with the data parallelism. Embarrassingly parallel analysis is one of the parallel jobs in which multiple instances of the same program executes multiple data files simultaneously, with each instance running independently from others on its own allocated resources (i.e. CPUs and, memory).

It is worth mentioning that our tests do not include a performance comparison between the DistSNNMF with the other state-of-the-art [21–25, 30] which is described in the related work section. As previously stated, DistSNNMF objectives to solve the previous topic model (SNNMF) problem for the large-scale dataset. Further, these applications are based on the LDA (latent Dirichlet allocation). In addition, they are using Spark for the parallel process. In the more, HONMF [26] and, Microblog [27] did not test for the scalability.

We evaluate the scalability of DistSNNMF by scaling it to a different types of processors (CPU, GPU). The experiments record the average execution time of each tested job and utilize the time cost (TC) by ml. sec.as metrics. In addition, the

[7] https://slurm.schedmd.com/sbatch.html.

[8] See www.top500.org for a list of the world's 500 fastest supercomputers.

Table 3 Matrix of experiments

Dataset	Processors	
	CPU	CPU + GPU
TC-1	1	2
TC-2	3	4

experiments compare the execution time of all the tested jobs on two datasets, as shown in Table 3. We test each parallel job four times.

In general, in case of GPU, the input pipeline extracts data from a location, transforms it, and, loads it onto an accelerator for processing. As accelerators become faster, it is important that the input pipeline keeps up with the demand.

The measurements of experiments are inspired by [31]. The total processing time of each DistSNNMF job on GPU is the sum of transmission time, pre-treatment time, and, computing time but, there are no transmission time and pre-treatment time on the multi-core CPU.

DistSNNMF has two computing times factors impact on the computing time of the job, first, the core equation such as Eq. (3) in SNNMF and Eq. (10) in DistSNNMF. Second, the iteration count for the core equation or, the computational process (i) such as Eq. (2, 3, 4, and, 6) in this paper. Furthermore, IO file for the input and the output. IO leads to increasing the transmission time and the pre-treatment time for the job time based on a GPU.

5.2 *Results and Discussion*

We compare our model for four times (Table 3) quantitatively.

5.2.1 Quantitative Evaluation

The purpose of this section is to answer the following question "Does GPU accelerate all jobs (functions) of the distributed topic-modeling algorithms (DistSNNMF)?

To answer this question briefly, the complexity of DistSNNMF jobs time is derived from two aspects: running the time of the approach itself (computing time) and the communication cost of a job. To estimate the communication cost of the job, the following two aspects must be considered. The first aspect is the communication cost between the main processor (CPU) and the input/output (I/O) devices (IO_f), i.e., reading documents hard disk and updating documents then saving them to the hard disk. The second aspect is the communication cost between processors (CPU & GPU). In other words, this section shows a full validation of DistSNNMF job performance on its scalability by adding the systematic analysis of time complexity.

We performed different scale experiments which illustrated in Table 4. We did not directly investigate latency or, communication bandwidth effects. Nevertheless, one

Table 4 Comparing the results achieved by each strategy, the number is colored in red if its degree is a significantly outperformed, green if its degree does almost outperformed, and otherwise black

		TC-1	TC-2
FJ-1	CPU	14	27
	GPU	23	32
FJ-2	CPU	126	29,302
	GPU	414	18,182
FJ-3	CPU	30	61,069
	GPU	14	24,549
FJ-4	CPU	**27**	49
	GPU	48	57
CJ-1	CPU	9	553
	GPU	14	591
CJ-2	CPU	24	104
	GPU	31	397

could expect that if the communication time becomes long compared to the computation time, the relative time for communication versus computation also effects on the weak scaling of parallelization.

In this section, we summarize our findings regarding the behavior of all analyzed strategies. To compare the scalability. We apply the methodology introduced in Sect. 4 (jobs from FJ-1 to CJ-2) to obtain the results shown in Table 4 for DistSNNMF.

Table 4 represents the different performance between the two configurations are used (CPU or CPU + GPU).

Overall, the results obtained show GPU is significantly outperformed in FJ-3 with both dataset (TC-1,TC-2) and FJ-2 with the dataset (TC-2). In contrast, CPU is significantly outperformed in FJ-2 with TC-1 and CJ-2 with TC-2.

In detail, the experiment results find GPU is the best case (the decline in the execution time) for the processes that have a significant increase in the weight of the core equation with the iteration count such as FJ-3 with two scales (TC-1, TC-2). Although FJ-3 has a high weight of IO (many inputs and many outputs), the weight of the computing time seems to be outperformed more than the weight of IO. Similarly, FJ-2 with TC-2 only. FJ-2 has a high weight of the core equation in addition to the iteration count and the high weight of IO.

In contrast, CPU is significantly outperformed in FJ-2 with TC-1 and CJ-2 with TC-2, Both jobs have the same high weight of IO, but they are not the same in iteration weight (FJ-2 with TC-1 has a low iteration weight, CJ-2 with TC-2 has a high iteration weight) regardless the level of the weight of the core equation.

In summary, Table 5 represents the summary of recommending the configuration (CPU or GPU) against the factors per each strategy.

With these experiments, we can answer the question by saying GPU cannot accelerate all jobs of DistSNNMF. The main impact factor is the count (weight) of the

Table 5 Summary of recommend configuration, the row is highlighted in grey if its GPU is recommended

Job name	Test case	Factors			Preferred processors
		Computing time		IO_F	
		Core equation	No. of iteration (i)		
FJ-1	TC-1	Low	–	One to three (high)	CPU
	TC-2		–		CPU
FJ-2	TC-1	High (Eq. 3 SNNMF)	Low	One to many (high)	CPU
	TC-2		High		*GPU*
FJ-3	TC-1		Very high	Many to many (high)	*GPU*
	TC-2		Very high		*GPU*
FJ-4	TC-1		High	One to one (low)	CPU
	TC-2		High		CPU
CJ-1	TC-1	Low (Eq. 5)	Low	Two to one (low)	CPU
	TC-2		High		CPU
CJ-2	TC-1	Low (Eq. 10)	Low	Many to one (high)	CPU
	TC-2		High		CPU

iteration for the core equation, not the weight of the core equation itself, Next it, the weight of the core equation. In the end, the weight of IO.

6 Conclusion and Future Work

We have constructed and evaluated DistSNNMF for the scalable semantic tagging in text data. This algorithm leverages existing data parallelism with distributed tasks to increase speed and facilitate the use of large corpora. It begins by discretizing the data into disjoint segments, discretizing the algorithms into disjoint segments jobs, and applying these parallel tasks on each segment in parallel data. DistSNNMF is built using SNNMF with data parallelism and distributed task, which are efficient and simple to implement a maintain provable guarantees. The running time of these algorithms is effectively dependent on the size of the document (features or terms) and IO of documents. We found that some tasks of DistSNNMF system are performed by CPU only faster than the GPU, this is contrary to what is expected. Future work may explore alternative clustering approaches, such as, using hierarchical clustering instead of SNNMF. Future work may explore alternative task segmentation to simplify the core algorithm to beat on long running time.

Acknowledgements This work was supported by computational resources is provided by the Bibliotheca Alexandrina[9] on its High-Performance Computing (HPC) infrastructure (see HPC. Bibalex.Org).

We thank the anonymous reviewers for their constructive comments. We thank the supercomputer unite in Bibliotheca Alexandrina (BA) for supplying infrastructure and, timely support for our experiments.

References

1. Li, K.C., Jiang, H., Yang, L.T., Cuzzocrea, A. (eds.): Big Data: Algorithms, Analytics, and, Applications. CRC Press (2015)
2. Al-Drees, A., Bin-Hezam, R., Al-Muwayshir, R.: Unified retrieval model of big data. In: INNS Conference on Big Data, pp. 323–332. Springer International Publishing (2016)
3. Desarkar, A., Das, A.: Big-data analytics, machine learning algorithms and, scalable/parallel/distributed algorithms. In: Internet of Things and, Big Data Technologies for Next Generation Healthcare, pp. 159–197. Springer International Publishing (2017)
4. Platoš, J., Gajdoš, P., Krömer, P., Snášel, V.: Non-negative matrix factorization on GPU. In: Networked Digital Technologies, pp. 21–30 (2010)
5. Řehůřek, R.: Scalability of Semantic Analysis in Natural Language Processing (Doctoral dissertation, Masarykova univerzita, Fakulta informatiky) (2011)
6. Yan, J., Zeng, J., Liu, Z.Q., Yang, L., Gao, Y.: Towards big topic modeling. Inf. Sci. **390**, 15–31 (2017)
7. Bhardwaj, M.: Parallel Approach for Implementing Data Mining Algorithms (2016)
8. Chong, P.K., Karuppiah, E.K., Yong, K.K.: A Multi-GPU framework for in-memory text data analytics. In: 2013 27th International Conference on Advanced Information Networking and, Applications Workshops (WAINA), pp. 1411–1416. IEEE (2013
9. Gadelrab, F.S., Haggag, M.H., Sadek, R.A.: Novel semantic tagging detection algorithms based non-negative matrix factorization. SN Appl. Sci. **2**(1), 54 (2020)
10. Yang, W., Li, K., Li, K.: A parallel computing method using blocked format with optimal partitioning for SpMV on GPU. J. Comput. Syst. Sci. **92**, 152–170 (2018)
11. Liu, W., Vinter, B.: A framework for general sparse matrix–matrix multiplication on GPUs and heterogeneous processors. J. Parallel Distrib. Comput. **85**, 47–61 (2015)
12. Yuk, J.H.: A Large-Scale Sparse Matrix Multiplication Method based on Streaming Matrix to GPUs (Doctoral dissertation, DGIST) (2017)
13. Erra, U., Senatore, S., Minnella, F., Caggianese, G.: Approximate tf-idf based on topic extraction from massive message stream using the GPU. Inf. Sci. **292**, 143–161 (2015)
14. Zhang, Y., Mueller, F., Cui, X., Potok, T.: Data-intensive document clustering on graphics processing unit (GPU) clusters. J. Parallel Distrib. Comput. **71**(2), 211–224 (2011)
15. Ciccoti, P., Oral, H.S., Kestor, G., Gioiosa, R., Strande, S., Taufer, M., Carrington, L.: Conquering Big Data with High Performance Computing. Oak Ridge National Laboratory (ORNL), Oak Ridge, TN (United States). Oak Ridge Leadership Computing Facility (OLCF) (2016)
16. Hsu, C.H., Fox, G., Min, G., Sharma, S.: Advances in big data programming, system software and HPC convergence. J. Supercomputing **75**(2), 489–493 (2019)
17. Shi, S., Wang, Q., Chu, X.: Performance modeling and, evaluation of distributed deep learning frameworks on gpus. In: 2018 IEEE 16th Intl Conf on Dependable, Autonomic and, Secure Computing, 16th Intl Conf on Pervasive Intelligence and, Computing, 4th Intl Conf on Big Data Intelligence and, Computing and, Cyber Science and, Technology Congress (DASC/PiCom/DataCom/CyberSciTech), pp. 949–957. IEEE (2018)

[9]https://hpc.bibalex.org/.

18. Goldsborough, P.: A tour of TensorFlow. arXiv preprint arXiv:1610.01178 (2016)
19. Abadi, M., Barham, P., Chen, J., Chen, Z., Davis, A., Dean, J., Kudlur, M., et al.: TensorFlow: a system for large-scale machine learning. OSDI **16**, 265–283 (2016)
20. Abadi, M., Agarwal, A., Barham, P., Brevdo, E., Chen, Z., Citro, C., Ghemawat, S., et al.: Tensorflow: Large-Scale Machine Learning on Heterogeneous Distributed Systems. arXiv preprint arXiv:1603.04467 (2016)
21. Xie, X., Liang, Y., Li, X., Tan, W.: CuLDA_CGS: Solving Large-Scale LDA Problems on GPUs. arXiv preprint arXiv:1803.04631 (2018)
22. Fuentes-Pineda, G., Meza-Ruiz, I.V.: Topic Discovery in Massive Text Corpora Based on Min-Hashing. arXiv preprint arXiv:1807.00938 (2018)
23. Li, Y., Song, W.Z., Yang, B.: Stochastic variational inference-based parallel and online supervised topic model for large-scale text processing. J. Comput. Sci. Technol. **33**(5), 1007–1022 (2018)
24. Zhao, B., Zhou, H., Li, G., Huang, Y.: ZenLDA: large-scale topic model training on distributed data-parallel platf orm. Big Data Mining Anal. **1**(1), 57–74 (2018)
25. Gropp, C., Herzog, A., Safro, I., Wilson, P.W., Apon, A. W.: Scalable Dynamic Topic Modeling with Clustered Latent Dirichlet Allocation (clda). arXiv preprint arXiv:1610.07703. (2016)
26. Tu, D., Chen, L., Lv, M., Shi, H., Chen, G.: Hierarchical online NMF for detecting and tracking topic hierarchies in a text stream. Pattern Recogn. **76**, 203–214 (2018)
27. Zhang, D., Han, Y., Li, X.: Dynamic detection method of micro-blog topic based on time series. In: International Conference of Pioneering Computer Scientists, Engineers and, Educators, pp. 192–200. Springer, Singapore (2018)
28. Mejía-Roa, E., Tabas-Madrid, D., Setoain, J., García, C., Tirado, F., Pascual-Montano, A.: NMF-mGPU: non-negative matrix factorization on multi-GPU systems. BMC Bioinform. **16**(1), 4Stochastic3 (2015)
29. Newman, D., Smyth, P., Steyvers, M.: Scalable parallel topic models. J. Intell. Community Res. Dev. 5 (2006)
30. Li, Y., Feng, D., Lu, M., Li, D.: A distributed topic model for large-scale streaming text. In: International Conference on Knowledge Science, Engineering and Management, pp. 37–48. Springer, Cham (2019)
31. Shi, S., Wang, Q., Chu, X.: Performance modeling and, evaluation of distributed deep learning frameworks on gpus. In: 2018 IEEE 16th Intl Conf on Dependable, Autonomic and, Secure Computing, 16th Intl Conf on Pervasive Intelligence and, Computing, 4th Intl Conf on Big Data Intelligence and, Computing and, Cyber Science and, Technology Congress (DASC/PiCom/DataCom/CyberSciTech), pp. 949–957. IEEE (2018)

Predicting MIRA Patients' Performance Using Virtual Rehabilitation Programme by Decision Tree Modelling

Nurezayana Zainal, Ismail Ahmed Al-Qasem Al-Hadi, Safwan M. Ghaleb, Hafiz Hussain, Waidah Ismail, and Ali Y. Aldailamy

Abstract An effective rehabilitation procedure is required to successfully manage the disabilities caused by diseases such as stroke, spinal cord injury (SCI), traumatic brain injury (TBI), and cerebral palsy (CP). In this regard, Medical Interactive Recovery Assistant (MIRA) platform proffers virtual rehabilitation through exergames. Virtual reality therapy (VRT) has recently gained attention for upper limb rehabilitation due to its positive impacts on patients' performance. VRT is a modern interactive application that integrates computer software with hardware devices to create an interactive virtual environment when playing different types of games and exercises (exergames). The output of playing the game generates statistical features (parameters) reflecting the patients' performance. However, physiotherapists who manage the input settings of exergames according to specific movements cannot easily predict the future performance of the patients based on their observations. Thus, this study proposes a decision tree model to predict MIRA patients' future performance for three difficulty levels (easy, medium and hard) with respect to their previous/last session records. Patients' data in the previous/last session are used to

N. Zainal · I. A. A.-Q. Al-Hadi · S. M. Ghaleb · W. Ismail (✉)
Faculty of Science and Technology, Universiti Sains Islam Malaysia, Negeri Sembilan, Malaysia
e-mail: waidah@usim.edu.my

N. Zainal
e-mail: ezayana@gmail.com

I. A. A.-Q. Al-Hadi
e-mail: ismail.ahmed.alhadi@gmail.com

S. M. Ghaleb
e-mail: safwan_ghaleb@yahoo.com

H. Hussain
PERKESO Rehabilitation Centre Bandar Hijau, Hang Tuah Jaya, Melaka, Malaysia
e-mail: hafez.hussain@rehabmalaysia.com

A. Y. Aldailamy
Faculty of Computer Science and Information Technology, Universiti Putra Malaysia, Selangor, Malaysia
e-mail: a.aldailamy7@gmail.com

M. Al-Emran et al. (eds.), *Recent Advances in Intelligent Systems and Smart Applications*, Studies in Systems, Decision and Control 295,
https://doi.org/10.1007/978-3-030-47411-9_24

determine the prediction values according to the proposed model. This helps physiotherapists to monitor and also predict their patients' progress using certain prediction values. Results prove the efficiency of the proposed decision tree-based statistical tool for prediction in medical monitoring applications.

Keywords Rehabilitation · Virtual reality therapy · Exergame · Decision tree

1 Introduction

Disabilities like a sudden loss of motion are experienced by patients due to diseases such as stroke, spinal cord injury (SCI), traumatic brain injury (TBI), and cerebral palsy (CP). These disabilities are particularly frequent when such patients are not well rehabilitated [1]. For instance, stroke is a common cause of neurological and body movement impairments [2]. Recently, the effects of stroke, SCI, TBI, and CP declined because of the therapeutic training and care provided in rehabilitation centers. These centers provide long term rehabilitation programs to resuscitate patients' disabled limbs [3]. Typically, excessive ration of physical therapy and rehabilitation valuation are administered based on therapists' theoretical decisions [4]. In rehabilitation centers (such as PERKESO Tun Abdul Razak Rehabilitation Centre which provides health care services for the people who have disabilities in their limbs), the ultimate goal is to help people to get back to their normal life as soon as possible by utilizing existing regular training programs such as virtual reality therapy (VRT).

VRT is the latest technology for training patients who suffer from disease-caused disabilities that affect their movement skills [5]. Recently, VRT has become a treatment approach for upper limb rehabilitation with the advancement of virtual reality technologies [5], [6]. VRT is a modern interactive application that integrates computer software with hardware devices such as screen, computer, sensor or Kinect to create an interactive virtual environment for playing exergames (i.e., games with exercises) [7]. An exergame contains one game and one movement that help patients to train their limbs according to the rules of the game and the movement throughout the game duration. For instance, the Medical Interactive Recovery Assistant (MIRA) platform is a computer software which provides a number of exergames for patients who suffer from disabilities in their limbs or spine.

A virtual rehabilitation system (like MIRA application) provides a number of movements for the upper and lower limbs to train patients' weak limbs. In the MIRA software, the movements are performed in a set of games listed for each movement. An example is the catch game which requires body movements to catch several objects appearing on the screen. Each progressive training level needs higher velocity and acceleration to complete the game. In addition, the therapist determines the input setting values (e.g., difficulty of the game, tolerance, minimum and maximum range) for each exergame as the patient moves to the next level of the game [2]. Playing an exergame generates statistical features that are saved in a database. For example, the in-built accelerometer monitors patients' movements and detects their arm directions based on the acceleration approximation of the movement. The system then validates the correct position of the shoulder or elbow relative to the Kinect sensor in term of distance [2]. Moreover, the system detects the temporal active movement in terms of moving time, and the speed based on the duration of the exergame and the detected distance. The generated parameters of the exergame are analyzed to improve the performance of the patients' movements.

As a patient trains the weak limbs using the exergame, only a few points are awarded which reduces his motivation for playing this exergame. Also, the physiotherapist manages the input setting of exergame according to specific movements and cannot easily predict the future performance of this patient by mere observation. To address this limitation, researchers recently deployed machine learning algorithms to improve the system's performance or its predictive ability by using computer programs to explore the dataset [8]. In medical applications (e.g., disease diagnosis, disease prediction, survivability, and disease identification), several machine learning solutions have been proposed to enhance the prediction of patient's behavior [8, 9]. More recently, an efficient and popular statistical tool used in medical research for classification, prediction, interpretation, and data manipulation is the decision tree algorithm [10, 11]. Notably, prediction is one of the most important applications of the decision tree models. The tree models derived from historical data easily predict the results of future records [11]. To the best of our knowledge, this is the first study that develops and implements a prediction model that evaluates patients' rehabilitation performance in the MIRA platform using the decision tree model.

The rest of this paper is organized as follows: Sect. 2 extensively reviews works related to this study. Section 3 details the MIRA technology and decision tree technique. Section 4 describes the method along with the machine learning approach used in this paper. Section 5 discusses the results obtained while Sect. 6 concludes this work and highlights the future work.

2 Related Works

A study on determining the effect of virtual reality and computer-based biofeedback training on the balance and reaction time for older adults who suffered from several physical pains was conducted by Bisson et al. [12]. Several other works have also examined MIRA patients' performance by analyzing their rehabilitation data. In this respect, a work in [13] is proposed to evaluate the visibility of MIRA software. A self-reported questionnaire was administered to ten adult patients suffering from neurological conditions. The experiments were performed on these patients, who suffer from neurological damage, for one year after training them on using MIRA. At the end of the experiments, most of the patients were satisfied with using MIRA to improve their upper limb movement.

The MIRA software tracks the physical motion of patients by using a Kinect sensor. Mc Glinchey et al. [14] compared the amount and frequency of active upper limb exercises using MIRA. Another interesting study was carried out on a 7 years old boy having an acute motor deficit of the left upper limb after a viral respiratory infection. The study evaluates the results obtained after 4 months of recovery with MIRA exergames and Vojta therapy [15]. In this research, we propose a prediction method based on a decision tree to predict the future performance of patients using virtual rehabilitation therapy systems.

As most patients prefer Kinect for rehabilitation exercises due to its high performance and adaptability [16]. Duarte et al. proposed an application that gives feedback on patients' behavior while they are playing the exergames [17]. To facilitate the data analysis, a Kinect Serious Game for Physiotherapy (KSGphysio) was introduced. The mobile interface is a essential part used to collect the patients' data to monitor and analyze their recovery progress during the rehabilitation exercise. A new framework was described to monitor patients' health while simultaneously entertaining them [18]. During the rehabilitation session, the 3D data is captured and sent to the appropriate server. This is done using the developed web-based serious game application, named Therasoup. This application helps physiotherapists to visualize the data in their assessments. Furthermore, the framework is effective for remote physiotherapy sessions to reduce cost significantly.

Osgouei et al. [19] introduced a method that assists the performance quality of patients using MIRA software. To achieve their goal, the authors employed the rehabilitation data of patient skeleton joints trajectories recorded in advance. Hidden Markov modeling (HMM) and Dynamic Time Warping (DTW) approaches in the machine learning were used to recognize the gesture and accurately compare the motion similarity performance of the unknown patient with the human skeleton joints' reference trajectories. The authors suggested that the patient has to choose a suitable method for his exercise.

Recently, Yu et al. [20] evaluated the motion similarity of older people by developing a DTW-based algorithm to support an effective Kinect-enabled home-based virtual coaching. DWT was validated and applied to measure the similarity in movement between a virtual coach and a user at a given period of time based on eight chosen bone vectors. Results reveal that DTW effectively evaluates the performance of older users while performing home rehabilitation exercises.

A novel solution for post-stroke patients called home-based multi-scene upper limb rehabilitation training and evaluation system is proposed by Bai et al. [21]. The ultimate goal of this solution is to objectively maintain patient's commitment, especially for long term rehabilitation. This is necessary to reduce the duration of a patient's therapy. In this solution, the upper limb reachable workspace relative surface area (RSA) and Fugl-Meyer Assessment (FMA) scale are integrated to evaluate the rehabilitation process. The analysis results indicate that applying this solution encouraged the patients to participate in the training process. This significantly improved their conditions.

Despite the benefits gained by employing the MIRA software, physiotherapists are still unable to predict their patients' future performance because of the huge information produced by the system. Distinctively from the aforementioned studies, this study implemented the decision tree model on the MIRA virtual therapy system to monitor and predict patients' rehabilitation performance for the future session.

To specifically manage the sudden loss of motion, we considered the following five parameters: time, average acceleration, distance, moving time and average speed. The proposed decision tree model predicts patients' performance according to a percentage (between 10% and 100%) following a defined weightage formula.

3 MIRA Technology and Decision Tree Technique

In this section, we further explain the MIRA technology and decision tree technique deployed in this study.

4 MIRA Technology

The Kinect sensor recognizes the 3D location of the body joints and allows the MIRA platform to detect the generated features or the kinematic movements. MIRA platform provides a set of exergames which are suitable for training the upper limbs, lower limbs, and the full body. It generates five significant parameters which are time, average acceleration, average speed, distance and moving time. Note that these parameters are quite common and there are other variable parameters depending on the exergame. Clearly, the average acceleration acts as the mean of the positive variant rate of velocity distributed on the time of the exergame. The average speed is the outcome of dividing the distance covered by the duration of the movement while the distance refers to the overall distances covered by a definite joint. The moving time is the total time of active movement (correctly or incorrectly) throughout the time of the exergame. Besides, there are other useful statistical parameters generated by MIRA when a patient plays the exergame. These include the number of repetitions, the average accuracy and the total score points.

MIRA contains thirty video games and thirty-three movement exercises for various parts of the body. These parts can be divided into four sets: upper limb, lower limb, spine and the full body. The exergames involve movements such as elbow flexion, elbow abduction, knee flexion, and shoulder abduction. Each movement is carried out using various types of games such as Izzy the Bee (a game involving a bee that collects and deposits pollen in the hive while avoiding bugs), Piano (a player needs to hit the highlighted keys to play the tune they hear at the beginning of the game), Atlantis (a player has to study artefacts that appear while avoiding underwater mines that cross their path), and Jugger (a player has to help a frog to jump from lily pad to another pad while avoiding falling in the water). Figure 1 shows examples of MIRA exergames. At the end of the exergame, a total of twenty-six variable parameters, representing the results and performance of the patient, are generated. Only five of the variables (i.e., time, average acceleration, distance, moving time and average speed) are considered in this study.

Fig. 1 Screenshots of example MIRA exergames: Izzy the Bee (top left), Piano (top right), Atlantis (bottom left) and Jugger (bottom right)

The decision tree is a well-known supervised classification technique. It involves the division of a complex problem into several sub-problems, and repeating this process recursively through a tree generation procedure [2]. It characterized by a hierarchical structure consisting of nodes and branches [22]. The internal node represents an input variable, the branch of an internal node is a subset of the values of the corresponding input variable, and a leaf node is associated with the values of the output variables [22]. The decision tree algorithm begins with a set of cases or examples and creates a tree data structure that is used for the classification of new cases [23]. It is a numerical analysis procedure which accounts for all or most of the possible alternatives and outcomes of a future course of action requiring different other decisions [24]. Moreover, it is commonly used to identify the optimal strategy in decision analysis, research operation, etc. The decision tree model is proposed to motivate patients' indulgence in specific movements and to help them achieve the predicted values targeted at specific parameters.

5 Methods

In this section, we will explain the methodology of this work in details. Moreover, this section deeply demonstrate the data and the patients' sample used in this paper.

5.1 Patients'Sample

The study considered 28 patients (20 males and 8 females) at PERKESO Tun Abdul Razak Rehabilitation Centre, Malacca with various impairments caused by Traumatic Brain Injury (TBI), Spinal Cord Injury (SCI), Cerebral Palsy (CP), and humerus fracture. These patients play the exergames with various movements based on the impairments or injuries. The data were collected over 5-months period (October-March) for 2018 until 2019. Each patient have to do exercises using MIRA within this period. The patient's details includes name, gender, date of birth, diagnosis and weak side of limb will be recorded. The patients' rehabilitation performance data were used to implement the decision tree model. After that, we performed a validation testing in order to analyze the result of prediction patients' performance after using the decision tree method. Thirty patients diagnosed with several impairments were implemented prediction for their data of rehabilitation using MIRA exergames. Then, we did comparisons whether they can hit the 10% prediction values on the new session.

5.2 Decision Tree Model Framework

The framework of the decision tree model deployed is described as follows:

- Registration of patient in MIRA where patients are required to fill the following: name, gender, date of birth, diagnosis and weak side.
- Physiotherapists assist the patients to start their rehabilitation by choosing appropriate movement according to their infected limbs such as shoulder abduction and elbow flexion.
- Then, the games (i.e., exergames) for each movement are selected since different movements require different exergames.
- Patients start the rehabilitation process by playing the chosen exergames in MIRA.
- MIRA records each patient's performance in the exergames.
- Subsequently, the physiotherapist selects the movement and the side, then the system shows a list of previous movements that have been performed by the patient.
- The system also displays the latest record of the patient based on the selected movement and the side.
- The system indicates the current values of the target parameters that are needed to predict future values based on the proposed model.

Figure 2 shows the flow processes of the patient's rehabilitation using MIRA, as described above. While Fig. 3 shows a graphical representation of a decision tree model for prediction of patients' performance using MIRA.

The proposed decision tree model (as shown in 3) is implemented in the MIRA therapy system to analyze the patients' performance module. The aforementioned

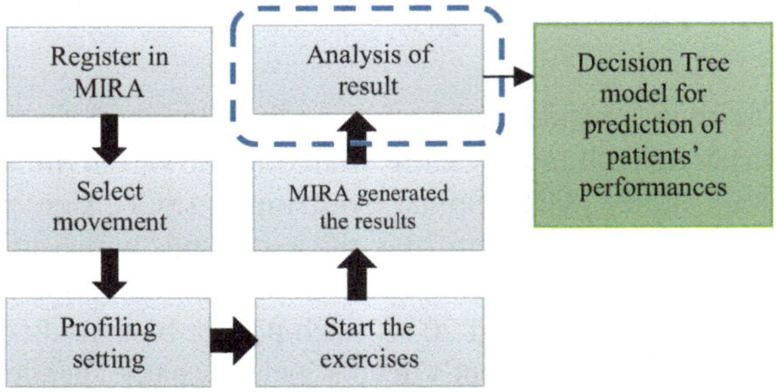

Fig. 2 Flow process of rehabilitation using MIRA

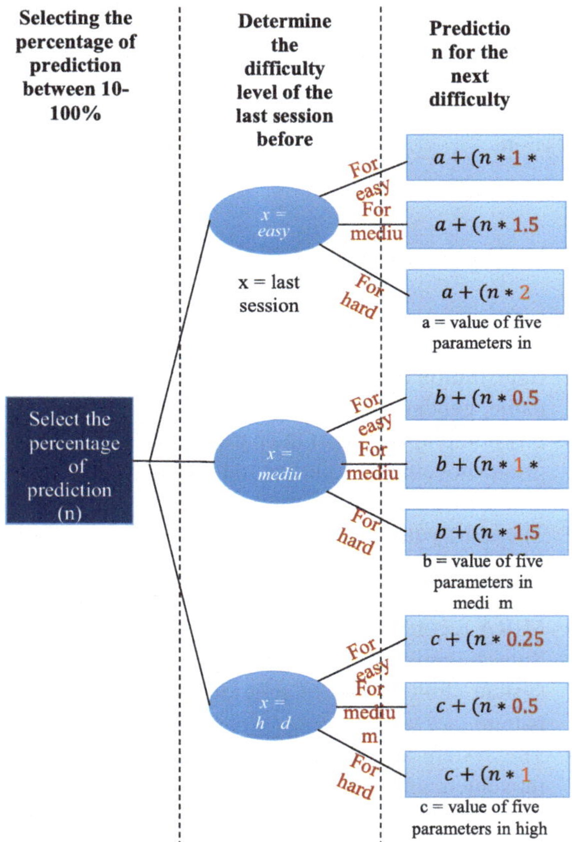

Fig. 3 Decision Tree model for predicting patient's performances

decision tree describes the prediction based on the percentage prediction for five variables; time, average acceleration, distance, moving time and average speed according to the three difficulty levels of the game. These levels of difficulty are easy (a), medium (b) and hard (c). *x* represents the patient's previous/last session which refers to a patient's last difficulty level while playing the game.

Table 1 The weights of difficulty levels

Difficulty/D weight	Easy	Medium	Hard
w^E	1	0.5	0.25
w^M	1.5	1	0.5
w^H	2	1.5	1

The prediction equations are different at all levels of difficulty as shown in Fig. 3. The value of a, b or c considers the values of five parameters in the last session. For instance, if the last session of a patient is easy, the prediction equation for an easy level is given as:

$$a + (n * w * a) \tag{1}$$

where n is the percentage of the prediction selected by the physiotherapist, a defines the value of either time, average acceleration, distance, moving time and average speed at the last session (i.e., easy). w is the weight value of difficulty level according to the threshold values in Table 1. From the observation of the patients' performance while playing exergames, we define the value of each difficulty level as 1. This value can be increased or decreased by 50% to increase or decrease the difficulty of the exergame.

6 Discussion

The decision tree model was implemented in the MIRA virtual therapy system for predicting patients' rehabilitation performance based on the time, average acceleration, distance, moving time and average speed. Figure 4 shows an example of patient's predicted values of for time, average acceleration, distance, moving time and average speed based on a 10% of prediction threshold. The prediction value might be is referred to after the patients play another session in the future. Thus, it can assists the physiotherapist to determine whether their performance is improved or not. Another comparison example of the patient's performance with the predicted values is shown in Fig. 5. It can be seen from the figure that the patient only hit the prediction value of average acceleration, distance and average speed. Furthermore, the results of patients' who played a new exergame session after the implemented 10\% prediction were reported.

Figure 6 shows the summary of the comparison of patients' performance (in terms of time, average acceleration, distance, moving time and average speed) with the set prediction values. The prediction was implemented on thirty patients diagnosed with various impairments to monitor their performance. From Fig. 6a, only two patients hit the prediction value of time during the exercise in the following session with not more than 50% improvement, while the rest did not reach the prediction threshold. Figure 6b shows that ten patients hit the prediction value of average acceleration

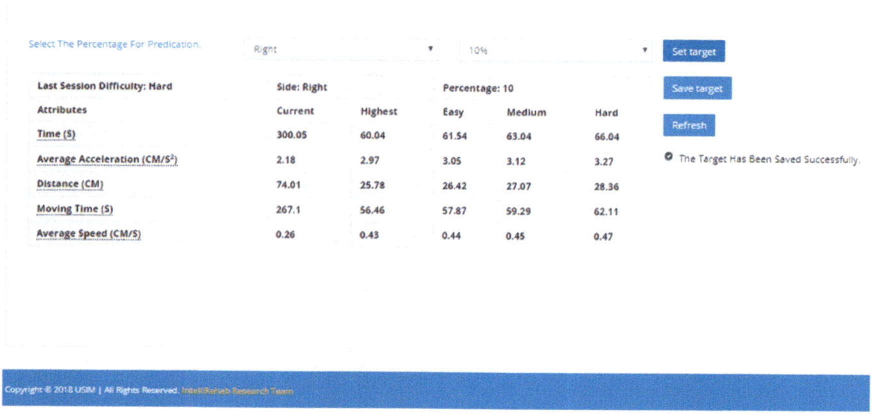

Fig. 4 Example of patient's prediction values based on 10% prediction

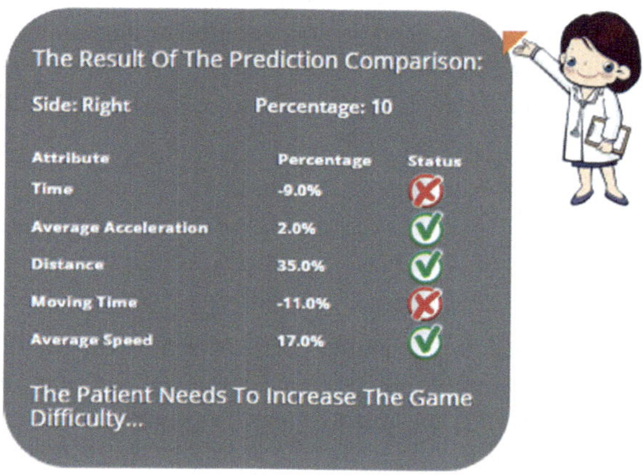

Fig. 5 An example of patient's performances compared to the predicted

by 2% (lowest) and 60% (highest) improvement. Besides, twelve patients hit the prediction value for distance with greater Range of Motion (ROM) value and one of them hit more than 100% improvement (128%) as shown in Fig. 6c. With respect to the moving time, only four patients hit the prediction value with 58% highest improvement in Fig. 6d. Finally, in Fig. 6e, eleven patients successfully hit the 10% prediction value. Interestingly, one of the patients achieved 132% improvement. However, most of the patients did not hit the 10% predicted value for average speed.

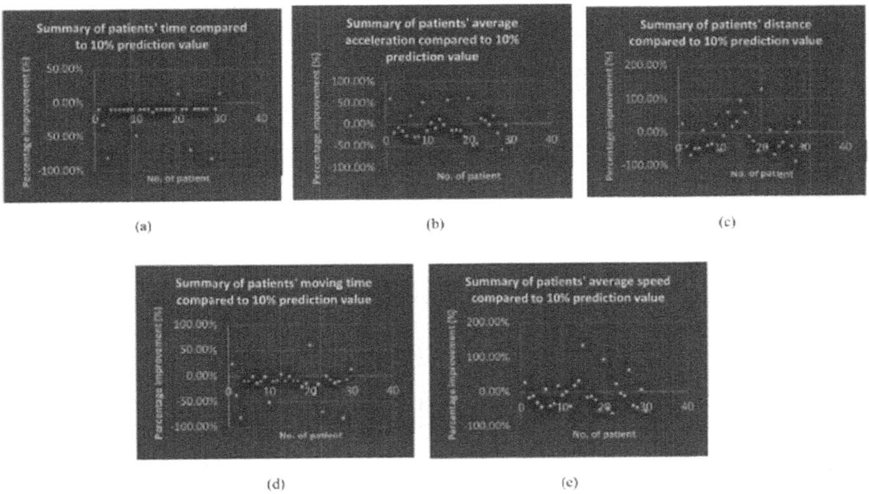

Fig. 6 Summary of patients' performance after 10% prediction: Patients' time (**a**), patients' average acceleration (**b**), patients' distance (**c**), patients' moving time (**d**) and patients' average speed (**e**)

7 Conclusion

We proposed a new decision tree model for predicting patients' performance by deploying a percentage (between 10% and 100%) that follows a defined weightage formula. The decision tree model was implemented on the MIRA virtual therapy system for predicting patients' rehabilitation future performance based on the time, average acceleration, distance, moving time and average speed. Results only showed that a few patients' performance reported hit the prediction while others are failed. In summary, the decision tree model effectively helps the physiotherapists to monitor patients' performance during the rehabilitation process. Designing an efficient algorithm to determine the most suitable MIRA setting for a more accurate patient's performance prediction procedure will be considered in the future.

Acknowledgements This work is supported by the International Grant USIM/INT-NEWTON/FST/IHRAM/053000/41616 under Newton-Ungku Omar Fund.

References

1. Krucoff, M.O., Rahimpour, S., Slutzky, M.W., Edgerton, V.R., Turner, D.A.: Enhancing nervous system recovery through neurobiologics, neural interface training, and neurorehabilitation. Front. Neurosci. **10**, 584 (2016)
2. Moldovan, I.M., et al.: Virtual rehabilitation programme using the MIRA platform, Kinect and Leap Motion sensors in an 81 years old patient with ischemic stroke. 2017 E-Health Bioeng.

Conf. EHB **2017**, 325–328 (2017)
3. Mousavi Hondori Hossein Khademi, M.: A review on technical and clinical impact of microsoft kinect on physical therapy and rehabilitation. J. Med. Eng. 16 (20140
4. Perez-Marcos, D.: Virtual reality experiences, embodiment, videogames and their dimensions in neurorehabilitation. J. Neuroeng. Rehabil. **15**(1), 113 (2018)
5. Lam, M.C., Arshad, H., Prabuwono, A.S., Tan, S.Y., Kahaki, S.M.M.: Interaction techniques in desktop virtual environment: the study of visual feedback and precise manipulation method. Multimed. Tools Appl. **77**(13), 16367–16398 (2018)
6. Lohse, K.R., Hilderman, C.G.E., Cheung, K.L., Tatla, S., der Loos, H.F.M.: Virtual reality therapy for adults post-stroke: a systematic review and meta-analysis exploring virtual environments and commercial games in therapy. PLoS ONE **9**(3), 1–13 (2014)
7. Sheehy, L., Taillon-Hobson, A.: Does the addition of virtual reality training to a standard program of inpatient rehabilitation improve sitting balance ability and function after stroke? Protocol for a single-blind randomized controlled trial. BMC Neurol. **16**(1), 42 (2016)
8. Zhu, M., Zhang, Z., Hirdes, J.P., Stolee, P.: Using machine learning algorithms to guide rehabilitation planning for home care clients. BMC Med. Inform. Decis. Mak. **7**, 1–13 (2007)
9. Yadav, G., Kumar, Y., Sahoo, G.: Predication of Parkinson's disease using data mining methods: a comparative analysis of tree, statistical and support vector machine classifiers. In: 2012 National Conference on Computing and Communication Systems, pp. 1–8 (2012)
10. Fallon, B., Ma, J., et al.: Opportunities for prevention and intervention with young children: lessons from the Canadian incidence study of reported child abuse and neglect. Child Adolesc. Psychiatry Ment. Health **7**(1), 4 (2013)
11. Song, Yan-yan, Ying, Lu: Decision tree methods: applications for classification and prediction. Shanghai Arch. Psychiatry **27**(2), 130–135 (2015)
12. Bisson, E., Contant, B., Sveistrup, H., Lajoie, Y.: Functional balance and dual-task reaction times in older adults are improved by virtual reality and biofeedback training. CyberPsychol. Behav. **10**(1), 16–23 (2007)
13. Chang, Y.-J., Han, W.-Y., Tsai, Y.-C.: A kinect-based upper limb rehabilitation system to assist people with cerebral palsy. Res. Dev. Disabil. **34**(11), 3654–3659 (2013)
14. McGlinchey, M.: Feasibility of using MIRA with adult patients presenting with upper limb motor dysfunction post neurological damage. Stud. Health Technol. Inform. **217**, 696–702 (2015)
15. Ahmed-Labib, M., Golan, J.D., Jacques, L.: Functional outcome of brachial plexus reconstruction after trauma. Neurosurgery **61**(5), 1016–1023 (2007)
16. Călin, A.D., Pop, H.F., Boian, R.F.: Improving Movement analysis in physical therapy systems based on kinect interaction. In: Proceedings of the 31st British Computer Society Human Computer Interaction Conference, pp. 87:1–87:5 (2017)
17. Duarte, N., Postolache, O., Scharcanski, J.: KSGphysio—kinect serious game for physiotherapy. In: 2014 International Conference and Exposition on Electrical and Power Engineering (EPE), pp. 606–611 (2014)
18. Cary, F., Postolache, O., Girão, P.M.: Kinect based system and serious game motivating approach for physiotherapy assessment and remote session monitoring. Proc. Int. Conf. Sens. Technol. **1**, 1–5 (2014)
19. Haghighi Osgouei, R., Soulsbv, D., Bello, F.: An objective evaluation method for rehabilitation exergames. In: 2018 IEEE Games, Entertainment, Media Conference (GEM), pp. 28–34 (2018)
20. Yu, X., Xiong, S.: A dynamic time warping based algorithm to evaluate kinect-enabled home-based physical rehabilitation exercises for older people. Sensors **19**(13), 2882 (2019)
21. Bai, J., Song, A.: Development of a novel home based multi-scene upper limb rehabilitation training and evaluation system for post-stroke patients. IEEE Access **7**, 9667–9677 (2019)
22. Lo, C.-L., Tseng, H.-T.: Predicting rehabilitation treatment helpfulness to stroke patients: a supervised learning approach. Artif. Intell. Res. **6**(2), 1 (2017)
23. Quinlan Ross, V: C4. 5: programs for machine learning. Mach. Learn. **240**, 302 (1993)
24. Mack, J.W., Xu, X., Gregory, J., Kirchain, R.: Accounting for Rehabilitation Activity Uncertainty in a Pavement Life Cycle Assessment using Probability and Decision Tree Analysis, pp. 11–13 (2015)

Segmentation of Images Using Watershed and MSER: A State-of-the-Art Review

M. Leena Silvoster and R. Mathusoothana S. Kumar

Abstract The success of watershed transform in image processing and image analysis domain is due to its ability to induce closed object boundaries and needs less computation time with reference to other segmentation methods. This review delivers a thorough study of conventional image segmentation methods like watershed algorithm, marker-controlled watershed and the Maximally Stable Extremal Region (MSER) algorithm using Magnetic Resonance Images (MRI). This study determines the major accomplishments in the performance metrics of the relevant algorithms in the three areas such as tracking of fibres, licence plate, and faces. In addition, this review examines the key outcomes and focuses on the lessons learned and thus forms the foundation for future research. The results of the literature indicate that the MSER-based procedures outperforms well as compared with the other methods. The experimental results of the applications shows that MSER provides improved speed and stability.

Keywords Watershed · Marker-controlled watershed · MSER

1 Introduction

Even though, conventional x-rays and computed tomography (CT) are employed for the diagnosis of diseases, Magnetic Resonance Imaging (MRI) has received widespread publicity because of its non-invasive nature. Segmentation of MRI is a significant process to retrieve the region of interest (ROI). To detect the ROI, certain structures like organs and tissues can be eliminated using the segmentation algorithm. The automatic detection process probably needs prior deep knowledge of normal as well as the diseased tissues and have a vital role in the diagnosing process of

M. Leena Silvoster (✉)
Computer Science Department, College of Engineering, Attingal, Thiruvananthapuram, India
e-mail: leenasilvoster@gmail.com

R. M. S. Kumar
Information Technology Department, Noorul Islam University, Tamilnadu, India
e-mail: rmsskdhujaa@gmail.com

M. Al-Emran et al. (eds.), *Recent Advances in Intelligent Systems and Smart Applications*, Studies in Systems, Decision and Control 295,
https://doi.org/10.1007/978-3-030-47411-9_25

diseases, treatment planning etc. The detection of ROI such as tumour is a challenging process because of its high spatial resolution, non-invasive characteristic and contrast of soft tissues. Various techniques have been employed for segmentation of MRI over the past years. The simplest method to segment an image is the edge-detection techniques. Despite, the disadvantages of edge detection algorithm includes its (1) inability to record some edges since it is complex and complicated, (2) it is sensitive to noise and (3) malfunctioning at corners, curves where there is a variation of gray-level values. These disadvantages could be eliminated if we are using watershed and/or MSER algorithm.

Even though there exists various techniques for the disease diagnosis process, the survey concentrates on the conventional watershed and MSER algorithm. The role of this review spans in different dimensions. First, the review includes the techniques for the segmentation of MRI. Second, a comparison of the techniques with respect to the performance measures are discussed. Third, the extension of this proposed results and the applications of relevant algorithms, future improvements are also discussed.

Systematic arrangement of the review is as follows. Section 2 discusses the related woks on MRI and some applications. In Sect. 3, basic terminology and its background is discussed, which consists of definitions and basic steps related to watershed algorithm and MSER. Section 4 discuses about the algorithm which are involved in this work. The performance that have originated from the conventional techniques and its comparisons is presented in Sect. 5. Section 6 deals with the limitations and advantages of the research findings and finally, Sect. 7 concludes the paper.

2 Related Work

Mainly watershed algorithms are categorized into two groups. First category of the algorithm implementation involves the flooding process starting from the regional minima region. Second group finds watershed points directly. Each of these category may be further grouped into three classes: sequential ones, parallel algorithms or ordered algorithms.

Watershed transform is the labelling process in which each pixel of a catchment basin assigns a label and associate a label to each points of the watershed, which is different from the label of the catchment basin. The watershed algorithm falls within region-based segmentation method. Basic principle of watershed algorithm relying on the simulation of the immersion of a landscape in a lake and construct dams to hinder the linking of diverse catchment basins. Besides, the conventional watershed algorithm uses input image or gradient of the input image and results in an over-segmented image. The reason for the widespread use of watershed is as follows: (a) For one region, it yields closed boundaries with reference to each minimum or marker. (b) Flooding starts from some minima, and thereby, reduces the number of catchment basins of the watershed of the flooded surface. After successive flooding, the collection of catchment basins results in hierarchical segmentation. (c) It is realisable to flood a surface by impose a minima at some predefined points: this induces marker based segmentation. When the segmentation considers the shape of the object, distance function is having more advantage.

The principle of watershed by immersion was introduced for detecting the contour of images and it is proposed by Beucher et al. [1] in 1979. In 1989, Vincent et al. [2] proposed a watershed transform which uses the principle of mathematical morphology. In 1991, Vincent et al. [3] presented a simulation of watersheds by immersion. Algorithm presented by Meijster et al. [4] in 1998 consists of two steps: (1) Used a FIFO-queue (First in First Out) algorithm for the transformation of a lower complete image, and (2) Used graph theory to compute the watershed and remove the previous FIFO-queue. Kostas et al. [5] introduced watershed merging method. In 2000, Lotufo et al. [6] uses shortest-path forest problem to implement watershed. This algorithm uses the same principle of the watershed algorithm using queues. Chen et al. [7] considered the issues relating to the elementary Vincent et al. algorithm such as: (1) incorrect labelling due to a point p is equidistant from more than three (or three) neighbouring catchment basins (2) redundant calculation of geodesic distance between labelled points, (3) consumption of memory, and (4) during processing, information about catchment basins is not able to retrieve. Next step, "gushing step", that in combination with the immersion step, executes in a faster fashion. Rambabu et al. [8], in 2003, presents a hill climbing simulation based algorithm which eliminates more than one scanning of the image using multiple queues. Shen et al. [9] presents a watershed transform algorithm utilizing nearest-neighbor graph (NNG) in 2005. The principle of this algorithm is to evolve the image into a NNG and then segments it by identifying geographic features. This process is repeated using the initial population result. Later, in 2008, Rambabu et al. update [8] and presented algorithm [10].

One of the major issue deals with watershed algorithms is over segmentation, which were discussed in [11, 12]. The application of the watershed method on gradient image [13] results in over segmentation of regions [3, 14, 15] and the irrelevant contours are considered as real ones. The reason of over segmentation is due to the fact that the occurrence of too many local minima due to noise or existence of other structures which is not a part of gradient image. To (overcome) decrease the over segmentation of watershed based techniques, several approaches have been proposed in the literature, we can cite for example techniques based on markers [13], region merging methods [3], scale space approaches [16], methods based on partial differential equations for image denoising or edge enhancement [17], wavelet techniques combined with a watershed transformation [18], remove irrelevant contour [19] or modify the gradient image [3], etc.

Maximally Stable Extremal Regions (MSER) is a fast and efficient detection algorithm with nearly linear time complexity to detect the affinely-invariant stable subset of extremal regions, the MSER. This algorithm is being employed for detecting high color contrast with their background and consist of homogeneous colour formation. MSER returns the set of arbitrary shaped boundaries of the region. It responds to blobs of high contrast to produce affine variant, distinctive arbitrary shape. Also, MSER is a multi-scale detector of both fine and large structure and it is susceptibility to natural illumination impacts or shadows. Maximally stable regions are those that have slight change across a wide range of intensity threshold levels. The 'maximally stable' in MSER characterize the optimized threshold selection process.

Key characteristics of MSER algorithm is regions and outer border uses intensity function over a range of thresholds. As a result, some specific regions are stable due to local binarization and possess the following properties: (1) MSER works well on images with homogeneous region having distinctive boundaries (2) It is not applicable to images with motion blur (3) Multi-resolution MSER works well on blurred images and provides an improvement of matching performance (4) Good repeatability (5) Affine invariant (6) Fastest stable local region detector

Steps of MSER includes: (1) Pre-processing - Sort the pixel in the increasing order and for each intensity level, determine the number of pixels. (2) Clustering- At each intensity level, representation of all the regions are created (3) MSER detection- For each local minima, the cardinality $|Q|$ of all extremal region and its growth rate were also recorded. (4) Display-Identify all image pixels corresponds to a detected MSER and output it. An advantage of MSER over otsu's method is its robustness if there exist inhomogeneity in the illumination. MSER is able to detect all region of interest and it is able to process an image in milliseconds.

Though the wide-baseline stereo problem is extensively studied in recent years, a new set of correspondence between a pair of images is introduced, Maximally Stable Extremal Region (MSER) detector. MSER has been used extensively in diverse areas of image processing: wide baseline stereo [20], image retrieval [21], object recognition [22], tracking [23], 3D segmentation [24], target detection [25] and text detection [26]. [27] Deals with the implementation of the MSER detector on FPGA. First implementation of MSER uses a union-find data structure and needs quasi-linear time number of pixels. The popularity of MSER is particularly due to their high availability of same regions addicted to different images in same scene: (1) Invariant to geometric transformation. (2) Invariant to illumination changes (3) Well localised in their spatial neighbourhood (4) Discriminate neighbourhood (5) Robust to occlusion.

Donoser et al. [23] implemented Maximally Stable Extremal Region (MSER) detector using component tree and obtain MSERs in quasi-linear time. Further, this work demonstrates the improvement in the detection of a single MSER by a factor of 4–10. The use of component tree improves tracking stability using weighted features and backward tracking and formulates a novel algorithm. Vedaldi [28] extended the definition of stability function for video sequence tracking by extracting the 3D region from a stack of frames. Donoser et al. [24] presented an extension of MSER detector to track the consecutive frames efficiently. They also include temporal information to improve stability. Another work of Forssen [29] uses MSER algorithm on vector-valued function. This method is similar to the work of Felzenszwalb and Huttenlocher [30]. But, the distance between two components is fixed. The "maximally stable principle" provide a region without any slight change over consecutive iterations. In another work, Forssen [29] improved the repeatability at multiple resolution.

Usually, the watershed algorithm is applied on the distance transform of the image. However, the output of the watershed algorithm forms a closed boundary and connected region, and it is highly sensitive to local minima. It also increases the computational cost of the algorithm. But the disadvantage of watershed transform is over segmentation. With the introduction of external and internal markers, watershed segmentation is very effective as compared with the traditional watershed algorithm. It also overcomes the problem of high overlap of gray level intensities of the objects. Advantage of MSER is that it is affine invariant, sensitive to image blur and can be applied to a low quality image. Each node of the component tree (data structure used in the implementation of MSER) identifies as a connected region with binary threshold images.

3 Basic Terminology and Its Background

Consider a 2D grayscale image I as topographic reliefs whose domain is $D_I \subset \mathbb{Z}^2$; where I represents discrete gray values within $[0, N]$, where N is any random positive integer

$$I\begin{pmatrix} D_I \subset \mathbb{Z}^2 \to \{0, 1, \ldots N\} \\ q \qquad\quad \mapsto I(q) \end{pmatrix} \tag{1}$$

Assume a square grid, G with 4-connectivity or 8-connectivity; $G \subset \mathbb{Z}^2\ X\ \mathbb{Z}^2$.
Notations:
$l(P)$: denotes the length of a predetermined path P
$N_G(q)$: Collection of neighbour's pixels of q, with reference to G:

$$N_G(p) = \{p' \in \mathbb{Z}^2, (q, q') \in G\} \tag{2}$$

Definition 1:
A path P through pixels u and v in an image I is a $(l + 1)$-pair of pixels $(u_0, u_1 \ldots u_{l-1}, u_l)$ thereby $u_0 = u, u_l = v$, and $\forall i \in [1, l], (u_{i-1}, u_i) \in G$; 1 is the length of the path

Definition 2:
Given an image I, minimum M at elevation h is defines as the connected plateau of pixels with the value h from which it is unable to attain a point of lower elevation without climbing:

$$\forall u \in \forall v \notin M, \text{ such that } I_v \leq \text{I(u)}$$

$$\begin{gathered} \forall P = (u_0, u_1, \ldots u_l) \text{ such that } u_0 = u \text{ and } u_l = v \\ \exists i \in [1, l] \text{ such that } I(u_i) > I(u_0) \end{gathered} \tag{3}$$

A minimum is a contiguous region with equal gray level of pixels which is darker than its neighbours such that no pixel in the region is connected to a pixel outside the region with equal or smaller grey-level value; darker pixels are having lower elevation or value. These extrema are called regional ones rather than local ones.

Let $h_{\min}$ is the smallest value in I and $h_{\max}$ is the largest in I. At level h, threshold of I, $T_h(I)$ is

$$T_h(I) = \{u \in D_I, I(u) \leq h\} \tag{4}$$

Definition 3:
The catchment basin, $C(M)$ connected with a minimum M and $C_h(M) \subset C(M)$ is the collection of pixels having an altitude equal or smaller to h:

$$C_h(M) = \{u \in C(M), I(u) \leq h\} = C(M) \cap T_h(I) \tag{5}$$

i.e., A drop of water falls at u, follows a relief, a descending path called downstream and finally it reaches M. Lines which prevent the merging of drainage basins are called catchment basins.

Definition 4:
A connected set A is a set of pixels that has a path within A between any two pixels in A.

Let $A \subset \mathbb{Z}^2$ be a connected set and y and x be two pixels of A.

Definition 5:
The geodesic distance, denoted as $d_A(y, x)$, between two pixels y and x in A is the shortest path among all paths within A which connects y and x and are totally contained in A. The lower bound of the shortest paths of length γ in A is connecting y and x.

$$d_A(y, x) = inf\{l(P), P \text{ path among two pixels that are totally contained in } A\} \tag{6}$$

Now, let $B \subseteq A$ which consists of connected components (CC)B_1, B_2, ..., B_k.

Definition 6:
The geodesic influence zone $iz_A(B_i)$ of a CC B_i of B in A is the locus of the pixels of A whose geodesic distance to B_i is minimum as comparing with the geodesic distance to any of the components of B:

$$iz_A(B_i) = \left\{p \in A, \forall j \in \frac{[I, k]}{\{\mathrm{i}\}}, d_A(p, B_i) < d_A\big(p, B_j\big)\right\} \tag{7}$$

The catchment basins of an image I represents the influence zones of its minima.

Definition 7:
Skeleton by influence zones (SKIZ) of B inside A is the pixels of A which is not associated with any of the geodesic influence zone and is denoted as $SKIZ_A(B)$:

$$SKIZ_A(B) = \mathrm{A}/IZ_A(B) \quad \text{with } IZ_A(B) = \bigcup_{i\in[1;k]} IZ_A(B) \tag{8}$$

Or in other words, the set $IZ_A(B)$ is the collection of all elements of the geodesic influence zones of the CC of B. So SKIZ contains all pixels intensities which is at equidistant to at least two nearest connected components. These set can be identified as foreground pixels.

Definition 8:
Assume $I : D \subset \mathbb{Z}^2 \rightarrow S$. The extreme regions are definable if: (1) S is completely ordered

(2) A neighborhood relation $A \subset D \times D$ is defined as $pAq \Leftrightarrow |p_x - q_x| + |p_y - q_y| \leq 1$

A region is extremal if $I(p) > I(q)$ (maximum intensity region) or $I(p) < I(q)$ (minimum intensity region) holds for all $p \in Q, q \in \partial Q$, where $Q \subset D$

Extremal region is defines as the set of pixels (connected component) of the image with higher or lower intensity pixels than the pixels of the outer boundary. Component tree is constructed using these extremal regions. Minimal region are otherwise known as MSER−. If the image is inverted and repeat the above procedure produces maximal region or MSER+.

Definition 9:
A region Q is a connected component of D, i.e. for any pair $p, q \in Q$ there is a path $p, a_1, a_2, \ldots a_n, q$ so, that $pAa_1, a_iAa_{i+1}, a_nA_q$ holds

Definition 10:
The outer border ∂Q is a subset of $D \backslash Q$ so that for any pixel there is at least one pixel $p \in Q$ with pAq

Definition 11:
Maximally Stable Extremal Region (MSER): Let $Q_1, \ldots, Q_{i-1}, Q_i, \ldots$ be a sequence of nested extremal regions, i.e. $Q_i \subset Q_{i+1}$. Extremal region Q_{i*} is maximally stable iff

$$q(i) = |Q_{i+\Delta} \backslash Q_{i-\Delta}| / |Q_i|$$

has a local minimum at i^* where $\Delta \in S$ is a free parameter.

Definition 12:
MSER Margin is defines as the count of threshold when the region becomes stable.

Definition 13:
Morphological reconstruction is a nonlinear filter based on mathematical morphology. Mathematical reconstruction can keep the information of object contours when filtering the image. Let the original image is $m(a, b)$, m is called marker, $n(a, b)$ is the mask. Size of m and n must be the same and se is the structuring element.

Morphological Opening Reconstruction:

$$\widetilde{r}(m) = r_{mose}(m) \tag{9}$$

Here, $m \circ se = (m\Theta se) \oplus se$ denotes the morphological opening for m by se.

Morphological Closing Reconstruction:

$$\widetilde{\phi}(m) = r^{*}_{m\bullet se}(m) \tag{10}$$

and $m \bullet se = (m \oplus se)\Theta se$ is the morphological closing for m by se.

The processing of reconstruction can restore those object edges that cannot be eliminated by the opening/closing and can remove the peak/valley that totally contained by the structuring element.

Morphological opening reconstruction is the process of a morphological opening followed by the closing reconstruction, that is

$$\begin{aligned} m_1 &= \tilde{r}(m) \\ m_2 &= \tilde{\phi}(m_1) \end{aligned} \tag{11}$$

The noise and the fine texture in m_2 are greatly suppressed and the true contours are restored through the reconstruction process. There will not be any loss of information and the original image gets simplified after this process.

Definition 14:
Morphological gradient is obtained after the processing of opening/closing reconstruction, the gradient image is

$$g(m_2) = (m_2 \oplus se) - (m_2\Theta \mathrm{se}) \tag{12}$$

where $m_2 \oplus se$ and $m_2\Theta se$ are the elementary dilation and erosion respectively.

4 Algorithm

In this section various region growing algorithms which uses intensity as a feature are discussed.

4.1 Watershed Algorithm

Vincent and Soile first introduced watershed by immersion in 1991. Image could be viewed as topographic surface; the gray value of a point can be treated as the elevation of the surface. The idea of immersion is analogous to pierce holes in the local minima and dip this surface in a lake. Valleys and basins of the relief represent dark regions whereas the mountain and the crest represent the light areas. Catchment basin (CB) is the region filling with water. Water starts flooding from the local minima and starts creating small lakes or basins. Eventually, at certain point, one or two lakes may coincide and begins merging each other. Construct dams at this position to prevent this overflow. Dams are called watershed lines. i.e., watershed lines divides the catchment basins in the surface. Segmentation of the terrain detects CB in the gradient image, the collection of all dams constitute the watershed lines. Finally, each minimum is enclosed by dams, which demarcate its corresponding catchment basin. The implementation of flooding uses morphological operations, geodesic skeleton and reconstruction. Dam boundaries corresponds to the connected boundaries retrieved from watershed segmentation algorithms. Algorithm is illustrated in Algorithm 1 and Algorithm 2.

Algorithm 1: Flooding Watershed process

---

for gray-level from h_{min} to h_{max}
  // Action 1: Expand current region
  Repeat ($R_{gray-level} \in R_L$) **do**
    Perform Growing [R]
  Until gray-level h
  //Action 2: Form new region
  for each (pixel $P \in$ ***level***)
    if (pixel P is not added to any region R) **then**
      Construct new region [R] in R_L
      Add pixel P to region [R]
      Growing [R] Until level
  end-for
end-for

Algorithm : **2 :** Flooding watershed [3]

```
Input : Digital grey scale image G=(I, E, f)
Output : Labelled watershed image lab on W
MASK←−2 //initial value of each level
WSHED ← 0 //label of the watershed pixels
Q ← ∅
INIT ← −1 // initial value of I
fp ← -1,1 //fictitious pixel ∉ W
current_lab ← 0 // current_lab is the current label
for each p∈ I do
W[p] ← ∞
d[p] ← 0 //dist is a work image of distances
end
// Stage I
SORT pixels of I in increasing order of graylevels such that h_min ≤ h ≤ h_max
// Stage II

for each h = h_min: h_max do
for each p∈ I with I[p] = h do
    W[p] ← MASK
    if q ∈ N_G(p) & L[q] > 0 or L[q] = WSHED then
            //initialize queue with neighbours at level h of current basins or watersheds
            d[p] ← 1
          enqueue (p, Q)
    end if
end for
    current_dist ← 1
enqueue (fp, Q)

loop( ) //extend basins
    p ← dequeue(Q)
    if ( p = fp) then
     if not empty (Q) then
       enqueue (fp, Q)
       current_dist ← current_dist+1
        p← dequeue(Q)
     end if
   endif
 for each q ∈ N_G(p) do //Scanning for the neighbours of P
     if ( [q] < current_dist ) and ( L[q] > 0) or ( L[q] = WSHED) then
         //q belongs to an existing basin or watershed
     if (L[q] > 0) then
                      if L[p] = MASK or L[p] = WSHED then
                       L[p] = L[q]
                        elseif L[p] ≠ L[q] then
                              L[q] = WSHED
                      endif
              elseif L[p] = MASK then
                      L[p] = WSHED
      endif
     elseif (L[q] = MASK) & (d[q] = 0) then //q is plateau pixel
              d[q]=current_dist+1
              enqueue (q, Q)
      endif
   end for
end_loop
```

```
//detect and process new minima at level h
for (p ∈ W) & (f[p] = h) do
        d[p] ← 0 // reset distance to zero
    if L[p] = mask then // p is inside a new minimum
        current_lab ← current_lab + 1 //create new label
        enqueue (p, Q)
        L[p] = current_lab
            while not (dequeue(Q)) do
                q ← dequeue(Q)
                for all (r ∈ N_G(q) do // inspect neighbours of q
                    if (L[r] = MASK) then
                            enqueue (r, Q)
                            L[r] ← current_lab
                    end if
                end for
            end while
    end if
  end for
end for
```

4.1.1 Watershed Transform by Topographical Distance

Here, distance from every pixel to the nearest non-zero valued pixel is calculated. Let f be a grey level image, with $f^* = f_{LC}$ the lower completion of f. Let $(m_i)_{i \in I}$ be the collection of minima of f. The basin $CB(m_i)$ of f corresponding to a minimum m_i is defined as the basin of the lower completion of f:

$$CB(m_i) = \{\ p \in D | \forall j \in I \backslash \{i\} : f^*(m_i) + T_{f^*}(p, m_i) < f^*(m_j) + T_{f^*}(p, m_j)\} \tag{13}$$

and the watershed of f is defined as

Watershed $W_{shed}(f)$ of f is the complement of $B_{h\text{max}}$ in Ω

$$W_{shed}(f) = D \cap \left(\bigcup_{i \in I} CB(m_i) \right)^c \tag{14}$$

Let W be some label, $W \notin I$. The watershed transform of f is a mapping $\lambda : D \rightarrow I \bigcup \{W\}$,such that $\lambda(p) = i$ if $p \in CB(m_i)$, and $\lambda(p) = \text{W}$ if $p \in W_{shed}(f)$

Thus, watershed algorithm on f aims to label each pixels of D in such a way that catchment basins are uniquely labelled and a special label W is allocated to all the pixels of watershed of f.

4.1.2 Watershed Transform by Gradient Method

Watershed transform by gradient method uses gradient image as the input. Along the object edges, the gradient magnitude is higher and all other pixels are having lower values. The outcome of watershed transform is watershed ridge lines along object edges. In some cases a homogeneous region may be represented as a large number of regions and as a result false contours may appear. A solution to this problem is to threshold the gradient [31, 32]

4.2 Marker-Controlled Watershed Segmentation

Conventional watershed algorithm employ pixel intensity of the image of the gradient image. Hence, to incorporate the shape information and to eliminate the over segmentation problem of watershed, marker based method is introduced which uses markers. This method is robust and flexible to segment objects accurately with closed contours.

A marker is a connected component which is a part of an image. Markers may transform the gradient image. The initial (user) parameters are internal and external marker. First, the pixels belonging to any object is marked as the object markers or foreground markers or internal markers. The external markers constitute the background and are formed using distance transform of the internal markers. The pixels which are not the part of the object constitute the background markers. The distance transform of an image is defined as the distance from every pixel to the closest nonzero pixel value. Usually, euclidean metric is used for calculating the distance. At the end of segmentation, the boundaries of the watershed regions are formed along the ridges, thus isolating object from its neighbors. Even if the boundaries of the object are not clear, the boundaries are represented as the ridges between the markers. Then apply the watershed transform. In conventional approach, initial markers represents all local minima of the image. Thus, each of the marked elements, or the local minima, constitute the center of the catchment basin, but some of the minima are less important.

4.3 Maximally Stable Extremal Region

MSER uses intensity-based thresholds to detect stable regions that are brighter or darker than its boundary pixels. Standard Maximally Stable Extremal Region (MSER) method work the same way as that of the watershed segmentation profile of the grey scale image and image can be viewed as landscape elevation. Standard immersion watershed algorithm starts with piercing a hole in minimum of the image and thus enables the flow of water into the local minima and this will continue until level is equal in all parts of the image (until the whole landscape is immersed). MSER represents regions that are distinguishing, stable and having invariant property. Such connected component areas of data-dependent shape, are known as distinguished regions, plays an active role in object recognition. These set of distinguished regions are otherwise known as extremal regions. The properties of such regions are listed below. The set is closed under continuous (and thus perspective) transformation of image coordinates and, secondly, it is closed under monotonic transformation of image intensities. During flooding component tree is constructed (external tree region) using union-find forest, obtaining region stability criteria and cleaning up.

Assume a threshold t based on intensity and partition the set of pixels into 2 categories: D (darker or black pixels) and B (brighter or white pixels). Initially, all the pixels in D and B are assumed as empty. We can view thresholded image, I_t, as a movie with t as the threshold. First image of the movie is a white one whereas the last frame is black. Consider all possible threshold values of image intensities. The pixel intensity which are lesser than a threshold are assumed as 'black' and which are having equal or above are assumed as 'white'. Subsequently, as the threshold decreases, white spots represents the local intensity minima starts to grow. At a certain point, regions with two local maxima may merge. Finally, when the threshold reaches a minimum intensity, all of these regions will merge and the final image will be white (all pixels are in B and D is empty). Finally, the frames of the movie contains set of all connected components regions corresponds to the set of all maximal regions; invert the image and then find the minimal regions and procedure is repeated. Minimal regions are otherwise known as MSER-. If the image is inverted and repeat the above procedure produces maximal region are MSER+. Finally, threshold is set as the local minima of the rate of change of the area and thus, retrieves MSER. Each MSER represents the threshold as well as the location of a local intensity maximum (or minimum). Even though, an extremal region is maximally stable, it might be discard if (1) max area is too large (2) too small or if there is a parameter variation (3) it is analogous to the parent.

MSER algorithm retrieves a number of co-variant regions called MSERs from an image I. During flooding, component tree (extremal tree) is constructed using union-find forest, obtain region stability criteria and cleaning up. This algorithm uses a component tree for preserving the area of the connected region (extremal regions) as a function of gray-level values. Connected component evolves a component tree. As we trace through the grey level, when a new component is found and minimum intensity pixels is considered as the leaf of the tree. By increasing the threshold, two or several components are merged into single component and the new component is allocated a new node and assigned as the parent of the previous nodes. This procedure is repeated until the entire image becomes a single component, which represents as the root node. Output of MSER is represented as location of all local minima (maximum) and a threshold corresponding to it.

5 Comparison of the Performance of the Related Works

Method	Advantage	Disadvantage	Computational time and memory requirement
Watershed	Works well when the region homogeneity is uniform. The outcome of watershed algorithm is a connected closed boundary. These boundaries represents the contours of objects. Final output is the combination of all such regions	Region growing algorithm has an impact on the selection of seed points. Usually these transforms are semi-automatic and the variation of intensity or noise may lead to over segmentation. Also, low contrast regions and thin structures are poorly detected	Computationally expensive in terms of time and memory. Time Complexity: O(n)
Marker-controlled watershed	Compared with the watershed transform, this method are more effective for segmenting objects with closed boundaries. Even though the contours are poorly defined, it is represented as ridges between two markers. As a result of the merging of the catchment basins, this process is quite expensive in terms of computation than watershed transformation. This in turn reduces the running time of the entire process	User must have a thorough knowledge of the image so that user can define the markers effectively. The output of the algorithm depends on the definition of the internal as well as the external marker	Time Complexity: O(n)

(continued)

(continued)

Method	Advantage	Disadvantage	Computational time and memory requirement
MSER	Extremal region can be enumerated in linear time	Low repeatability score in blurred sequence and textured sequences. Blur has an impact on the number of regions generated is affine invariant and can be applied to low quality images and has an efficient implementation	$O(n\alpha(n))$ where α (n) is the Ackermann function; better than previous algorithms

6 Discussion

The three major algorithms used in the past few decades are marker-controlled [33, 34], watershed [35], MSER [36]. Also, this review considers three applications such as tracking of fibers, license plates and faces and are explained in Donoser [23]. Differently, Chevrefils et al. [37] presents a learning algorithm for the classification of disc of spine MRI slices using watershed algorithm. Some works of spine MRI concentrated on spinal canal also utilises the watershed algorithm [38] for the segmentation purpose. But many of these lacks performance evaluation. Grau [41] presented a watershed algorithm and the marker-controlled watershed to segment image and obtain a good kappa statistics. Abdelfadeel [39] and Khan [40] presents MSER algorithm and obtain good results. And the result shows that MSER outperforms all the other methods.

Author	Method	Dice (%)	Type of image
Mohammed [39]	MSER	88	Cardiac MRI
Grau [41]	Watershed	65	Knee Cartilage, Brain MRI
Khan [40]	Marker-controlled Watershed	90	Brain MRI
	MSER	98	

7 Conclusion

The segmentation of medical images is a challenging process since it involves complex and complicated structures. Among the state-of-art-of literature of image segmentation, some algorithm based on mathematical morphology is considered. In this review, various region extraction techniques such as watershed, marker-controlled watershed and MSER are discussed. Even if the watershed algorithm is simple and robust, the problem with it is over-segmentation. One of the solution is to use the marker-controlled watershed algorithm. However, this algorithm needs some prior information about the image, to introduce the internal marker and the external marker. Also, it is to be noted that there exists no standard algorithm for the marker detection. The selection of markers and its detection varies from one modality to another. To eliminate this, MSER algorithm is introduced. Based on these discussions we can conclude that MSER has better performance. This automated procedure would be helpful for a physician for the complex decision process.

One of the drawback of this study is its application is confined on a small set of spine image. Future work deals with the 3D implementation of watershed algorithm, marker-controlled watershed and MSER algorithm and applies it to a spine image.

References

1. Beucher, S., Lantuéjoul, C.: Use of Watersheds in Contour Detection. Workshop Published (1979)
2. Vincent, L., Beucher, S.: The morphological approach to segmentation: an introduction.Technical report, School of Mines, CMM, Paris (1989)
3. Vincent, L., Soille, P.: Watersheds in digital spaces: an efficient algorithm based on immersion simulations. IEEE Trans. Pattern Anal. Mach. Intell. **13**(6), 583–598 (1991)
4. Meijster, A., Roerdink, J.B.T.M.: A disjoint set algorithm for the watershed transform. In: Proceedings EUSIPCO '98, IX European Signal Processing Conference, pp. 1665–1668 (1998)
5. Haris, K., Efstratiadis, S.N., Maglaveras, N., Katsaggelos, A.K.: Hybrid image segmentation using watersheds and fast region merging. IEEE Trans. Image Process. **7**(12), 1684–1699 (1998)
6. Lotufo, R., Falcao, A.: The ordered queue and the optimality of the watershed approaches. Math. Morphol. Appl. Image Signal Process. **18**, 341–350 (2000)
7. Chen, T.: Gushing and Immersion Alternative Watershed Algorithm, pp. 246–248 (2001)
8. Rambabu, C., Rathore, T., Chakrabarti, I.: A new watershed algorithm based on hill climbing technique for image segmentation. In: TENCON 2003. Conference on Convergent Technologies for Asia-Pacific Region, Vol. 4, pp. 1404–1408 (2003)
9. Shen, W.C., Chang, R.F.: A nearest neighbor graph based watershed algorithm, pp. 6300–6303 (2005)
10. Rambabu, C., Chakrabarti, I.: An efficient hill climbing-based watershed algorithm and its prototype hardware architecture. J. Signal Process. Syst. **52**(3), 281–295 (2008)
11. Beucher, S.: The watershed transform applied to image segmentation. In: Proceedings of the Pfefferkorn Conference on Signal and Image Processing in Microscopy and Microanalysis, pp. 299–314 (1991)
12. Haris, K., Efstratiadis, S.N., Maglaveras, N., Katsaggelos, A.K.: Hybrid image segmentation using watersheds and fast region merging. IEEE Trans. Image Process. **7**(12) (1998)

13. Meyer, F., Beucher, S.: Morphological segmentation. J. Visual Commun. Image Representation **1**(1), 21–46 (1990) (Academic Press)
14. Grau, V., Mewes, A.U.J., Alcaniz, M., Kikinis, R., Warfield, S.K.: Improved watershed transform for medical image segmentation using prior information. IEEE Trans. Med. Imaging **23**(4), 447–458, 0278-0062 (2004)
15. Tek, F.B., Dempster, A.G., Kale, I.: Noise sensitivity of watershed segmentation for different connectivity: experimental study. Electron. Lett. **40**(21), 1332–1333 (2004). 0013-5194. Dept. of Electron. Syst., Univ. of Westminster, London, UK
16. Jackway, P.T.: Gradient watersheds in morphological scale space. IEEE Trans. Image Proc. **5**, 913–921 (1999)
17. Weickert, J.: Efficient image segmentation using partial differential equations and morphology. Pattern Recogn. **34**, 1813–1824 (2001)
18. Jung, C.R., Scharcanski, J.: Robust watershed segmentation using wavelets. Image Vision Comput. **23**, 661–669 (2005)
19. Cates, J.E., Whitaker, R.T., Jones, G.M.: Case study: an evaluation of user-assisted hierarchical watershed segmentation. Med. Image Anal. ITK Open Science-Combining Open Data Open Source Softw. Med. Image Anal. Insight Toolkit **9**(6), 566–578 (2005)
20. Matas, J., Chum, O., Urban, M., Pajdla, T.: Robust wide-baseline stereo from maximally stable extremal regions. BMVC **2002**, 384–396 (2002)
21. Sivic, J., Zisserman, A.: Video google: a text retrieval approach to object matching in videos. Int. Conf. Comput. Vision **2**, 1470–1477 (2003)
22. Obdrzalek, S.J.: Object recognition using local affine frames on distinguished regions. In: British Machine Vision Conference, Vol. 1, pp. 113–122 (2002)
23. Donoser, M., Bischof, H.: Efficient maximally stable extremal region (mser) tracking. In: IEEE Conference on Computer Vision and Pattern Recognition (CVPR), pp. 553–560 (2006)
24. Donoser, M., Bischof, H.: 3d segmentation by maximally stable volumes (msvs). In: ICPR 2006: Proceedings of the 18th International Conference on Pattern Recognition, pp. 63–66 (2006)
25. Chen, H., Tsai, S.S., Schroth, G., Chen, D.M., Grzeszczuk, R., Girod, B.: Robust text detection in natural images with edge-enhanced maximally stable extremal regions. In: 118th IEEE International Conference on Image Processing (ICIP) (2011)
26. Gao, Y., Shan, X., Hu, Z., Wang, D., Li, Y., Tian, X.: Extended compressed tracking via random projection based on MSERs and online LSSVM learning. Pattern Recogn. **59**, 245–254 (2016)
27. Kristensen, F., MacLean, W.J.: Fpga real-time extraction of maximally-stable extremal regions. In IEEE International Symposium on Circuits and Systems (2007)
28. Vedaldi, A.: An Implementation of Multi-dimensional Maximally Stable Extremal Regions. Technical Report, Feb 7 (2007)
29. Forssen, P.: Maximally stable colour regions for recognition and matching. In: IEEE Conference on Computer Vision and Pattern Recognition (2007)
30. Felzenszwalb, P.F., Huttenlocher, D.P.: Efficient graph-based image segmentation. Int. J. Comput. Vision **59**, 167–181 (2004)
31. Hafts, S.N., Maglaveras, E.N., Pappas, C.: Hybrid image segmentation using watersheds. In: SP1E Proceedings of Visual Communications and Image Processing '96, Vol. 27, pp. 1140–1151, Orlando, Florida, U.S.A. (1996)
32. Salembier: Morphological multiscale segmentation for image coding. Signal Process. **38**, 359–386 (1994)
33. Kaleem, M., Sanaullah, M., Hussain, M.A., Jaffar, M.A., Choi, T.S.: Segmentation of brain tumor tissue using marker controlled watershed transform method. Commun. Comput. Inf. Sci. **281**, 222–227 (2012)
34. Shafarenko, L., Petrou, M., Kittler, J.: Automatic watershed segmentation of randomly textured color images. IEEE Trans. Image Process. **6**(11), 1530–1544 (1997)
35. Moga, A.N., Gabbouj, M.: Parallel image component labelling with watershed transformation. IEEE Trans. Pattern Anal. Mach. Intell. **19**(5), 441–450 (1997)

36. Zhu, H., Sheng, J., Zhang, F., Zhou, J., Wang, J.: Improved maximally stable extremal regions based method for the segmentation of ultrasonic liver images. Multi-media Tools Appl
37. Chevrefils, C., Cheriet, F., Aubin, C.-E., Grimard, G.: Texture analysis for automatic segmentation of intervertebral disks of scoliotic spines from MR images. IEEE Trans. Inf. Technol. Biomed. **13**(4), 608–620 (2009)
38. Chevrefils, F.C.: Watershed segmentation of intervertebral disk and spinal canal from MRI images. In: Proc. Int. Conf. Image Anal. Recognit., no. 3, pp. 1017–1027 (2007)
39. Abdelfadeel, M.A., ElShehaby, S., Abougabal, M.S.: Automatic segmentation of left ventricle in cardiac MRI using maximally stable extremal regions. In: Biomedical Engineering Conference (CIBEC), pp. 145–148 (2014)
40. Khan, M.A., Lali, I.U., Rehman, A., Ishaq, M., Sharif, M., Saba, T., Zahoor, S: Brain tumor detection and classification: a framework of marker-based watershed algorithm and multilevel priority features selection. Microsc. Res. Tech. **82**(6), 909–922 (2019)
41. Grau, V., Mewes, A.U.J., Alcañiz, M., Kikinis, R., Warfield, S.K.: Improved watershed transform for medical image segmentation using prior information. IEEE Trans. Med. Imaging **23**(4), 447–458 (2004)

Block Chain Technology: The Future of Tourism

Aashiek Cheriyan and S. Tamilarasi

Abstract Digitalization in every field has created a remarkable development in one way or another. The changes due to digitalization is evident across the industries and one such recent technological development happened across the world is that of "Blockchain"; which have generated a significant excitement across the industries. Most of the industries have adopted the blockchain technology for its ease of use. Tourism industry is considered to be one of the promising industries which doesn't have a retirement. The impact of digitalization has reflected in tourism sector too; but the question is; "Whether the tourism industry copes up with the emerging technologies?". Sadly, the answer is a big NO. Gradually, the tourism industry is planning to introduce Blockchain Technology into its operations to create a revolution. The introduction of Blockchain Technology into Crypto Currencies reaped huge success in the past. The same level or a little higher level of success is expected from this technology; but, the aversion and dilemma towards the new technologies makes the tourism sector to wait for its adoption. The future milestones of the tourism industry will definitely witness Blockchain technology. The tourism industry gets benefitted from Block chain technology through many ways, including secure payment, identification of the customers, management of the baggage, customer rewards management and the ratings of business. This paper aims to study the development of blockchain technology in the tourism sector, its applications and its challenges, and its future perspectives on the tourism industry.

Keywords Digitalization · Tourism sector · Blockchains

A. Cheriyan (✉) · S. Tamilarasi
Department of Commerce, Faculty of Science and Humanities, SRM Institute of Science and Technology, Kattankulathur, Tamil Nadu, India
e-mail: aashiek.666cheriyan@gmail.com

S. Tamilarasi
e-mail: tamilars2@srmist.edu.in

M. Al-Emran et al. (eds.), *Recent Advances in Intelligent Systems and Smart Applications*, Studies in Systems, Decision and Control 295,
https://doi.org/10.1007/978-3-030-47411-9_26

1 Introduction

Digitalization has been benefited almost all the industries in one way or another. The third industrial revolution paved the way for mind blowing revolutions which resulted in n number of startups coming up with new ideas. New terminologies were known to the common people. The IoT (Internet of Things), AI (Artificial Intelligence) etc.... is stimulating the industries around the world. The internet of communications will drive the whole cities into smart cities by the year 2021.

As per [1], tourism constitutes one of the fastest growing economy and it includes many new technologies and advancements in digitalization that helps the tourists and other interested stakeholders to exploit the tourist destinations in a potential manner. The era of digitalization helps the tourism industry to utilize its opportunities to its maximum. The competition in the tourism industry raises in numbers day by day. In order to withstand with the competitions, the companies in the tourism industry should keep pace with the digitalization or else they will fade into history. The number of travelers is increasing year by year. Even new tourist destinations are been discovered as the year passes on. The tourism industry has faced so many changes in the past. 57% of the tickets for travelling are booked through online. 65% of the tourists books the hotel for the same day over the internet. Even though the figures are a positive sign of growth of digital travel industry; it needs to push a bit more forward to position itself among the top industries. Tourism industry is a promising sector where it contributes 9.9% of the total employment in the world. The tourism industry alone contributes 10.4% to the Global GDP [2]. Even in this digitalization era there are so many travel companies who still stick on to the conventional methods; but in the near future due to the competition their existence is at sting. But, most of the travel companies are gradually shifting from the conventional era to the digitalized era. The digitalization is done keeping the stakeholders in the mind. Because each stakeholder is unique and the services must be provided according to their desires, cultures, language etc.... so that the digitalization should be done in such a manner that it includes all the stakeholders.

Blockchain technology is making a revolution in the way we store the data and the usage of it and other sources of information. It has generated significant excitement within many industries and fields. This is possible only because the Blockchain technology has the potential to drastically change the way in which the information or data is stored and used. With the advancement of technology, it will revolutionize our whole life in the near future. As suggested by Firica [3], Blockchain enabled projects are mostly initiated by the Banking and IT companies. By the year 2025, 10% of the global Gross Domestic Product will be stored on Blockchain technology as per the statistics provided by the World Economic Forum. The main aim of this technology is to improve the transparency and it enables high security in the transactions. The working of a blockchain technology can be simply depicted in Fig. 1 as shown.

The blockchain technology stores all the transactions publicly that a network produces. Every new user needs to request for a transaction and then the validity of the user is checked through P2P networks. Each verified transaction involves a crypto

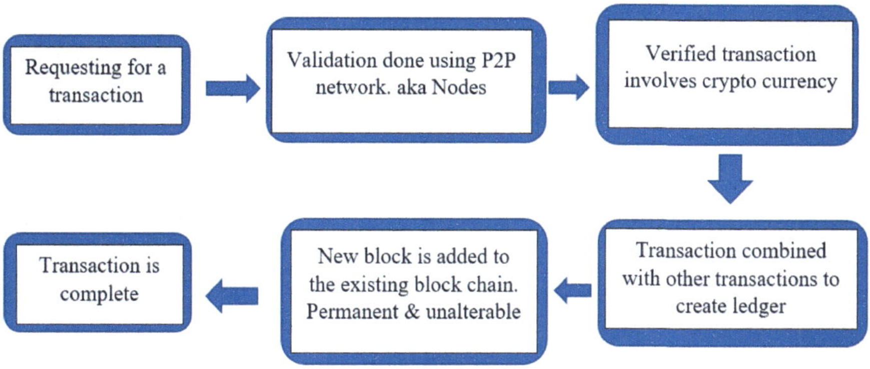

Fig. 1 Working of a blockchain technology

currency. After the verification the transaction is combined with other transactions to create a ledger. The new block gets added to the existing blockchain and it goes on as it is. The blockchain is permanent and it cannot be altered by anyone else. The data stored in a blockchain is highly secured, it can be traced and it is transparent in nature.

There are many new technological trends including Internet, Cloud Computing, gamification, new digital travelers etc.... which all together can be coined as "Digital Tourism". But the blockchain technology can be described as "a revolutionary technology that in the future will transform financial transactions and will strongly influence the tourism industry". After the implementation of the blockchain technology the travel companies will provide the unique experiences for the consumers, will benefit in personalized interactions and will have so many loyalty programs. The tourism industry is still in the transition period. The influence of blockchain technology will give new wings to the tourism industry and that's not too far to achieve.

1.1 Objectives of the Study

- To study about the blockchain technology and the digitalization of the tourism sector.
- To what extent will blockchain technology benefit the tourism industry and in what ways.
- To know about the challenges if blockchain technology is applied in tourism industry.

The objectives are framed in such a manner to determine whether the blockchain technology is helping the tourism industry. The application of blockchain technology in the tourism sector and how it is benefited both for the consumer as well as for the business people. The challenges raised in implementing the blockchain technology is also analyzed in this paper. Since tourism industry is considered one of the promising industries and the number of travelers increasing day by day had made the researcher to frame the objectives accordingly.

2 Literature Review

2.1 *Blockchain Technology*

A database which is shared among its users and allowing them to transact a valuable asset without having any intermediary in between the transactions is known as a Blockchain [4]. Blockchain technology was first applied in the Cryptocurrency Bitcoin. It is an electronic cash system which is an alternate to the traditional cash. It was induced into the economy in the year 2008, since its introduction it has gained a warm response across the globe. After the success of the cryptocurrency it has been used for commercial applications and many companies are trying to adopt blockchain technology for a smoother and healthier transactions [5].

It is a composition of cryptographic algorithms, databases and a decentralised consensus mechanism. The data are stored cryptographically in inter connected blocks. These blocks gets its validity from the P2P networks also known as Nodes. The users of a blockchain technology can interact among themselves without having an extending hand from the central authority [6]. Blockchain technologies setup is independent in nature, so that even the smart contracts are built upon this technology [7]. The decentralization is considered as one of the major advantages of the Blockchain technology. Apart from that transparency, and the valid historical log adds up to the benefits of blockchain [8]. The above advantages help in writing complex contracts without any difficulties and enables the information to be shared [9]. Even though blockchain is beneficial it lacks from the limitations too. The decentralized data can be retrieved easily since it is built on historical data. So, the protection of privacy becomes a greater challenge. The explicit interventions are necessary in preparing the smart contracts they cannot develop by its own [4]. The blockchain technology is still in its fantasy and it has so many hurdles which needs to be overcome. Understanding the implementations and applications of blockchain technology is a bit tedious task for the practitioners and users alike [4].

2.2 *Digitalization in Tourism*

The world tourism started witnessing the era of digitalization when Thomas Cook started its first organized trip in the nineteenth century and later on it developed and added a lot more into its basket by allowing the tourists to create custom tour packages as per the convenience of the stake holder [10]. ICT driven and supported innovations in the tourism sector has made a drastic change, which enables the traveler to organize his/her trips before they step out from the house/office. The developments in the transportation sector has given immense support for the development of the tourism industry [11]. Wherever a traveller wishes to go the first and foremost thing he does is he just Google it up and see whether that place has been visited by someone and if the reviews and the directions to reach the destination, nearby attractions, mode of transportation, hotels to be booked etc.... all these things happens over the internet or in the digitalized platforms. After visiting a particular tourist site each traveller becomes the "digital ambassador" of that particular destination. The traveller might come back again to the same destination to cover the missed-out places or he may accompany with his group of friends to the same destination. The digital tourism promotes a wide variety of destinations including the undiscovered museums, rallies, zoos etc.... [12]. When the London hosted Olympics in the year 2012, the AR (Augmented Reality) was the centre of attraction of tourism industry [13].

95% of the Online Travel Agencies (OTAs) are controlled by the two groups viz, Priceline Group and Expedia Inc. The Priceline group has a concentrated structure in the travel distribution which in turn benefits the company to gain around 40% growth before EBITA [14]. Even though the competition in the tourism industry is higher these companies are still hesitant to shift towards an updated technological era because the dominant players in the tourism industry are very few. But these companies still lack in updating to the latest technologies because there are only few dominant players. These companies even use worse tactics to generate increased revenue. Most of the OTAs tend to use rate parity when booking a hotel or airlines. The hotels have to pay around 25% as commission to get listed in the OTAs website for booking a room by the traveler [15].

The third parties whom the OTAs trust to store the data of their customers always had some security holes. The hackers used to retrieve the information of the consumers easily because with one master password the hackers are able to gain the entire information which ranges from millions or billions of dollars of information [16]. The security implications have made the tourism industry to restrain from further developments. But, the exact power of technology and its usage has not yet been discovered by the travel companies. With the blockchain technology all the hurdles faced by the tourism industry can be solved up to an extent [17]. Even though technology is much beneficial to the tourism industry it should never detract the traveller from experiencing a tourist destination or a hotel or any modes of transportation [18].

3 Application of Blockchain Technology in Tourism Industry

Blockchains are developed with the main intention of eliminating the middle men from the transactions. The tourism industry when joined with blockchain technology do have the potential to make a revolution in the industry. The technology can provide more security and transparency at each point of transaction. As the world grows the commutation between the countries increases. A traveler's information is passed on to different hands from the minute he books a ticket online, then to the travel agencies, the payment gateways, the banks, then to the airports, governments, border agencies and others. When such information is passed on to different parties over the internet it is highly exposure to risk. But, in blockchain technology the whole network is secured and no one can access. The tourism industry while using blockchain technology must note only open and permission less blockchains which increases the security and reduces the risk of getting hacked.

When using the blockchain technology, the suppliers and sellers are connected to a single market place. The suppliers can put their information into a database and it can be discovered by sellers and if it is feasible then the seller can purchase the service from the supplier. All these transactions are designed without any human intervention and it is performed automatically. The possible uses of blockchain in a tourism industry can be pointed out as follows:

- **Easy, safe and traceable payments**—The payments in a blockchain technology can be done using cryptocurrencies which is much easier, safer and it can be traced out. No third part intervention. Verified blocks and algorithms, network of computers etc., can help in transferring the money easily without any delay [3]. In the near future, while we travel abroad, its not necessary to exchange the currencies since we are using cryptocurrencies.
- **Better Coordination and Management**—As blockchain technology uses decentralization as a platform it saves time for the large firms. The same block of chain can be used and it can be accessed at all times reducing the managements time to check after it at each and every moment.
- **Baggage Management**—Travelling around the world in airlines is a part of the tourism. Every year the number of tourists travelling across the world increases enormously. As the number of tourists increases it is common for the airlines to lose the luggage of the passengers. The passengers don't have an actual idea of what happens to their luggage immediately after giving it at the check—in counters. And in turn it affects the passenger and the airlines company in terms of time and revenue [19]. The introduction of block chain technology acts as a creative platform wherein the passengers gets an update of his/her luggage at each checkpoint and he or she can have a tighter control and visibility. The longer ques for the check in baggage and its payment for extra weight can all be avoided with the blockchain technology [20]. The blockchain technology traces the baggage between the companies as it is stored in the block of chain.

- **Ratings**—Review and Ratings are the two unavoidable factors in the tourism industry. At present the ratings and review available in different websites differ and no one knew who the real author is, whether it is valid or not etc. Based upon the review, only the customers do decide upon a travel purchase. Depending heavily on the prior customer reviews is considered as a study done by the customer on a travel purchase. It is difficult for a person to identify whether the review existing in the online is a valid one or a fake one, the credibility matters [21]. But, through blockchain technology, it is possible to introduce a decentralized, trustworthy, unbiased and a transparent review system. After updating a review using the blockchain technology it is not possible for an updation or revision of an existing review. It is possible because all the parties who enter a review do have a unique private key which assures that the transaction has been raised from a particular user. Thus, it enables the customers to peacefully go through the reviews and purchase a travel decision with the help of blockchains [22].
- **Loyalty programs**—Usually all the companies give rewards to its customer for attracting them towards the business. But, the existing reward systems can be used only in the specified business and it cannot be exchanged for anything else other than the product or service issued by the company. Introducing Blockchain technology to the rewards systems benefits the customers in many ways. The customer can use the rewards across the countries and across the business without losing its original value. Such reward systems help the customer to enjoy freedom, privacy, autonomy and more personalized service offerings [23].

4 Challenges of Blockchain Technology in Tourism Industry

Adopting blockchain technology in the tourism industry is still in its budding stage. Even though the potential of blockchain technology is huge, when it comes to practical application it does faces some exciting challenges. That can be pointed down as follows:

- **Capacity**—The blockchain technology which is prevalent today is not possible to handle all the transactions within a specified speed and time. Only 7 transactions can be taken place in a second and moreover the size per transaction is comparatively higher than VISA and Master Cards [3]. Whereas Ethereum can handle 10–20 transactions per second. Better holding capable blockchains are under development but its entry to the present world is not known till date.
- **Security**—Even though the blockchain technology is better since it doesn't have any third parties the question of security arises. The lack of a Central Authority makes the Blockchain technology much more difficult. If the transactions have been hacked a complaining authority is lacking which in turn affects the customers confidence [3]. Even though each data gets stored into a block of chains it can

be accessed if the data is not properly managed by the companies. Resulting in a huge flow of sensitive information from blockchain to the hackers.

- **Standardization**—Standardization and Interoperability is lacking which in turn makes the Blockchain technology difficult to reach its potential heights of success [3]. Even though blockchain creates a revolution, all the companies which uses blockchain technology doesn't follow the same standard which in turn makes the systems difficult to perform according to the commands of the traveler.
- **Knowledge**—Educating the people what is blockchain technology and its application in the tourism industry is considered to be a tedious task which will be a great challenge for the companies who are willing to implement this technology.

5 Future Perspectives of Blockchain Technology in Tourism Industry

The adoption of blockchain technology in the tourism industry can be traced out to a couple of years back. When the cryptocurrencies were introduced in the year 2008, the commercial applications of blockchain technology has been discussed from then onwards. The blockchain technology is undergoing a great revolution which will obviously influence our lives in the near future as similar to AI or IoT. Even though the blockchain technology and its application in the tourism industry is not much familiar to India; the roots of blockchain technology in the tourism industry has been planted abroad. There are nearly 3 real life tourism companies who have adopted the blockchain technology. They are:

- **Winding Tree**—A unique travel experience with the help of blockchain technology and excluding the middle men is what it is offered by the company. The company do assures that the prices will drop down by 20% while booked through their website with the help of the new technology. The company do have its own cryptocurrency named "Lif Token" to make the transactions secure. The company is developed by a group of engineers who were working with the travel industry for decades. The problems faced during the tenure of employment made them to develop Winding Tree with the help of blockchain technology. The Winding Tree search engine is available from the 4th quarter of 2018 onwards [24].
- **Trippki**—"You travel we reward you". A loyalty based website especially concentrating on the tourism industry is the main of Trippki. Trippki gets revealed to the world in this January, 2019. It helps the customers and companies in the tourism field to come in for a direct contact. They do assign tokens to the customers for any travelling purpose, the traveler gets registered and he can redeem the points at any time he wishes. They do accept 8 types of cryptocurrencies and fiat [25].
- **Sho Card and SITA**—It is a patented digital library platform which works with the help of blockchain technology. It allows the users and enterprises to establish

their identities in a secured manner. It can be used wherever we go and it reduces the risk of taking identity cards with us while we travel.

The year 2019 will be witnessing strong and upcoming blockchain technologies in the tourism industry. Even the Amadeus group the successor group in the tourism industry will implement blockchain technology. There are many companies who comes up with this technology either directly or in directly influencing the tourism industry.

6 Conclusion

We believe that the blockchain technology and the tourism industry will emerge in the near future and it will become a revolution similar to that of the cryptocurrencies. It is an upcoming field of study where a greater number of researches can be done. Since the blockchain technology and its application in the tourism industry is in its earlier stages going into a deeper research is not possible at this stage. Educating the common man about this technology is an essential matter before the industries turn on to the blockchain. The technological advances always get accepted at a slower pace. The current level of knowledge about Blockchain and its potential implications for the tourism industry is very low. The progress in this field and the media hypes can be avoided by the researches in this field. We do hope that the future researches will be done based upon the blockchain technology in the tourism industry with working companies.

References

1. Pilkington, M., Crudu, R.: Blockchain & Bitcoin as a Way to Lift a Country out of Poverty Tourism 2.0 and E-Governance in the Republic of Moldova (2017)
2. WTTC: WTTC, December 2018 (Online). Available: http://www.wttc.org/-/media/files/reports/economic-impact-research/cities-2018/city-travel-tourism-impact-2018final.pdf. Accessed 5 Oct 2019
3. Firica, O.: Blockchain technology: promises & realities of the year 2017. Qual Access Success 51–58 (2017)
4. Glaser: A taxonomy of decentralized consensus systems. In: European Conference on Information Systems (2017)
5. Wörner, D., Von Bomhard, T., Schreier, Y.-P., Bilgeri, D.: The bitcoin ecosystem: disruption beyond financial Services? In: ECIS Proceedings (2016)
6. Gipp, B., Meuschke, N., Gernandt, A.: Decentralized trusted timestamping using the crypto currency bitcoin. In: I Conference (2015)
7. Xu, J.J.: Are blockchains immune to all malicious attacks? Financ. Innov. **2**(25) (2016)
8. Beck, R., Stenum, C.J., Lollike, N., Malone, S.: Blockchain—the gateway to trust free cryptographic transactions. In: Twenty-Fourth European Conference on Information Systems (2016)
9. Notheisen, B., Cholewa, J., Shanmugam, A.P.: Trading real-world assets on blockchain. Bus. Inf. Syst. Eng. 425–440 (2017)

10. Keller, Marketing: Management. Pearson Education, Singapore (2006)
11. Stipanovic, C., Rudan, E.: Tourism product club in generating the value chain. Pol. J. Manag. Stud. **14**(2) (2016)
12. Durrant, A., Gloembewski, M., Kirk, D., Benford, S., Fischer, J., Rowland, D., McAuley, D.: Automics: souvenir generating photoware for theme parks. In: Annual Conference on Human Factors in Computing Systems (2011)
13. Watanabe, A.: Inside World's First Augmented Reality Hotel. Aust. J. (2012)
14. Forbes: NASDAQ, 2018 (Online). Available: www.forbes.com. Accessed 2 Nov 2019
15. TnooZ: Travelport aims to raise up to $480 M with its IPO in US. TnooZ, US (2018)
16. Economist, T.: The worlds largest online travel company. The Economist (2018)
17. Winding Tree: Winding Tree, 2018 (Online). Available: www.windingtree.com. Accessed 26 Oct 2019)
18. Benyon, D., Quigley, A., O'Keefe, B., Riva, G.: Presence and digital tourism. AI & Soc, pp. 521–529 (2014)
19. Schumacher, M.: Trust in tourism via blockchain technology: results from a systematic. In: Information and Communication Technologies in Tourism 2019: Proceedings of the International Conference in Nicosia, Cyprus (2019)
20. Ali, A., Frew, A.J.: Information and Communication Technologies for Sustainable Tourism. Routledge, Abingdon (2013)
21. Yoo, K.H., Gretzel, U.: Comparison of deceptive and truthful travel reviews. In: Information and Communication Technologies in Tourism, pp. 37–47. Springer, Berlin (2009)
22. Önder. I., Treiblmaier, H.: Blockchain and tourism: three research propositions. Ann. Tour. Res. 180–182 (2018)
23. Kowalewski, D., McLaughlin, J., Hill, A.: Blockchain will transform customer loyalty programs. Harv. Bus. Rev. (2017)
24. Winding Tree: Winding Tree (Online). Available: www.windingtree.com
25. Trippki: Trippki (Online). Available: www.trippki.com

Biomedical Corpora and Natural Language Processing on Clinical Text in Languages Other Than English: A Systematic Review

Mohamed AlShuweihi, Said A. Salloum, and Khaled Shaalan

Abstract Natural Language Processing (NLP) applications on real-life textual content require a suitable fit for purpose corpora, which can accommodate the ambiguity of the domain. Researchers in the field managed to synthesize gold-standard corpora in many domains and for varying tasks, assisted by domain experts and linguists. The wealth of information buried in free-text electronic documents in healthcare systems presents itself as a leading contender for NLP applications. In this literature review, the efforts to come up and utilize a clinically annotated corpus in a particular healthcare information extraction task are explored. Those efforts can be more pronounced when done on a new language with limited existing gold-standard clinical references. A great number of people around the globe interact with healthcare systems in languages other than English. Advancing the Clinical NLP research in their languages will propel the general progress in the field and potential healthcare advantages considerably. For the purposes of this review, we considered three major world languages: Spanish, Italian, and Chinese. This led to considering the research question to be about the viability of the creation or utilization of a gold-standard clinical corpus in a language other than English and how it can contribute in performing a complex clinical language mining task. The implementations reviewed in these languages considered varying approaches to overcome complexities in biomedical NLP in these languages. This study managed to highlight novel solutions to complex tasks and found that efforts in these languages can be highly successful if a non-English medical corpus is created from scratch, off-the-shelf tools are used or machine translation is considered to bridge the gap in biomedical NLP domain-specific lingual resources in these languages.

Keywords Natural language processing · Healthcare · Gold-standard clinical corpus

M. AlShuweihi · S. A. Salloum (✉) · K. Shaalan
Faculty of Engineering and IT, The British University in Dubai, Dubai, UAE
e-mail: ssalloum@sharjah.ac.ae

S. A. Salloum
Research Institute of Sciences and Engineering, University of Sharjah, Sharjah, UAE

M. Al-Emran et al. (eds.), *Recent Advances in Intelligent Systems and Smart Applications*, Studies in Systems, Decision and Control 295,
https://doi.org/10.1007/978-3-030-47411-9_27

1 Introduction

The effectiveness of healthcare is inductive to the wellbeing of societies and a significant indicator of its development [1, 2]. Governments and organizations realized that data and information derived from healthcare operations are valuable factors for continuous improvement and optimization of the provided services [3, 4]. To harness the power of data in the sector, healthcare organizations sought the adoption of Electronic Medical Records (EMR) as a potential solution [2, 5–10]. The adoption of EMRs reached the levels anticipated by the policymakers in the field in the past decade or so. Such prevalence of EMR solutions introduced clinicians, statisticians, and administrators to a deluge of data, but also introduced a difficult conundrum. These systems not only generated the expected tabular and structured data but also captured a significant amount of unstructured data, in the form of free-text clinical notes, forms, and reports.

Data scientists worked diligently to find sustainable solutions to complete the picture of healthcare data and get the most from the unstructured free-text data. This is due to the huge efforts of Natural Language Processing (NLP) techniques [11–17]. Due to its popularity, many research articles were conducted to employ NLP in different applications [18, 19]. NLP presented itself as a highly viable technology to manage unstructured data, generated from free-text and electronically captured human natural language. Clinicians and data scientists ventured in exploring varied tasks, which are made possible only through NLP. Tasks such as public-health surveillance of epidemics based on events in social media and other textual internet sources [20]. Numerous applications in information extraction using NLP from clinical electronic artifacts and assessing early detection and diagnostics [21, 22] were researched. Clinical safety was also studied for potential areas for improvement based on the NLP analysis of documentation [23, 24], addressing aspects such as patients' falls and drug adverse reaction. The research in the medical domain for NLP led to more specialized research in Biomedical Language Processing (BLP), which mainly focuses on advancing the field and reaching the state-of-the-art performance in a variety of tasks in the domain.

To say that NLP is a straightforward solution to solve the unstructured clinical data puzzle is a gross oversimplification. NLP application on real-life task is complex in most of the cases, but this complexity can increase exponentially in the clinical domains. The main hurdle in most cases is the scarcity of an annotated clinical corpora [24]. Although advances were made on the genomic NLP, due to the existence of relevant corpora, ethical and legal constraints hindered the progress of clinical corpora development [25]. Further to this complexity, researching NLP in healthcare tasks is made more difficult when dealing with other world languages than English as most of the research in the field is heavily focused on English. This added scarcity of clinical corpora in world languages other than English makes the research in the development and the application of these corpora a worthwhile research question to pursue. Research in clinical NLP in world languages other than English was conducted in varying levels of depth and tasks. Clinical NLP applications were researched in Chinese [26, 27], Swedish [28], Spanish [24], and many other languages. Assessing the approaches to develop the clinical corpora in those languages and juxtaposing the specifics of these techniques is in the core of what this paper attempts to study.

This paper attempts to systematically review published work about NLP and its application on healthcare tasks, with a concentration on research to develop clinical corpora in different world languages in differing use-cases. The researched work is selected based on structured inclusion criteria that will be detailed in the coming sections. The main research question is how viable NLP and the development of clinical corpora as an effective technique to mine clinical unstructured textual artifacts for valuable information. In general, the focus will be on work relevant to natural language understanding. The included work will be analyzed based on the motivation, the study design, the solution presented, and results. A discussion of the reviewed work will attempt to focus on the contribution of the results to the researched question and the limitations that can be extended in future research.

2 Systematic Review Methodology

2.1 Motivation

NLP in the biomedical domain or Biomedical Language Processing (BLP) established itself as a credible candidate to uncover insights in unstructured medical data. State-of-the-art performance and rates demonstrated the viability of BLP to interject itself in critical healthcare outcomes both in the public and individual health. To the center of the research of the field are biomedical and domain-specific corpora. Such representations in the domain were extensively researched in the English language and in a way that established a solid foundation and baseline for researching non-trivial language processing tasks in the field. Motivated by the great proliferation of HIS and EMRs rich with unstructured data in non-English speaking regions, this research examines efforts to mitigate gaps in the availability of non-English clinically-annotated corpora. The review researched high-quality work to identify methods that can propel forward the BLP research in these languages and overcome the mentioned restrictive lack of clinical lingual resources. Novel work to research BLP in major world languages such as Chinese, Spanish, and Italian were recognized the methodologies and results of the research were analyzed and compared. The analysis and synthesis of the researched work in this specific domain contributed in defining the research question to consider the viability of conducting BLP and clinical annotation in languages other than English. Complex medical lingual tasks can benefit from the researched methodologies. The outcome of this work can be generalized for other domains taking into consideration the tools, methodologies, and workflows proposed by the researched work.

Table 1 Inclusion/exclusion criteria

Inclusion criteria	Exclusion criteria
• Article language: English	• Article language: others
• Published in the last five years (2014–2019) • Peer-reviewed: yes	• Peer-reviewed: no
• Search terms: KM (clinical) and KM (corpus) and KM (natural language processing) • Format: article	

2.2 Eligibility Criteria and Analysis Methodology

This systematic review follows the same approach employed in previous systematic reviews conducted in different domains [29–35]. The articles used in this review are mainly sourced from the British University in Dubai Library, which uses the electronic portal worldcat.org. The search was done on three main keywords: Clinical, Corpus and Natural Language Processing. The researched content was limited to articles. The screening of the retrieved articles went through preliminary filtering that limited the articles to English articles, published in the past five years and has to be peer-reviewed. Further examination eliminated most of the duplicate titles and limited the result to full-text articles. A preliminary screening of the titles and abstracts of the articles examined the articles against the research question. A quality check of the publishing source limited the selected articles to quartile 1 (Q1) journals according to Scimago Journal Ranking (SJR). A final screening took into consideration novelty and originality of the research on clinical NLP in the chosen language the published article has as of the writing date and finally fine-tuning to further fit the researched topic. A detailed breakdown of the eligibility criteria used and the databases to the source are listed in Tables 1, 2 and Fig. 1.

3 Literature Analysis

3.1 On the Creation of a Clinical Gold Standard Corpus in Spanish: Mining Adverse Drug Reactions

3.1.1 Motivation and Background

Disease and drug interactions and annotations of pharmacological entities are increasingly becoming a good research area of NLP. However, the research in the field still being hindered by the lack of gold-standard corpora. The existence of such corpora can facilitate NLP application efficacy in areas, such as machine learning-powered drug adverse reactions. Oronoz et al. [24] attempted to fill this gap in the field by

Table 2 Initial search across the databases

Databases
WorldCat.org
ABI/INFORM Global
Business Source Complete
ScienceDirect
ArticleFirst
Taylor and Francis Journals
Academic Search Complete
Computers & Applied Sciences Complete
Emerald eCase Collection
SAGE Journals
Electronic Books
Electronic Collections Online
Emerald Group Publishing Limited
ERIC
Directory of Open Access Journals

creating a gold standard pharmacology corpus in Spanish. The corpus is considered as a basis to conduct machine learning-powered automated extraction of adverse drug reactions from electronic health records written in Spanish. Oronoz et al. [24] proposed IxaMed-GS as the gold standard corpus, which is manually annotated by pharmacology and pharmacovigilance experts. As gold standard corpora require significant manual input in the form of domain experts annotation, Oronoz et al. [24] ventured further to develop an automated annotation tool based on a seed of manual annotation. This task is done using a tool called FreeLing-Med. The main focus for [24] corpus is treatment adverse effects and more specifically adverse drug reactions.

Adverse drug reactions are defined as "unavoidable or difficult-to-avoid disorders, with or without injury, produced when drugs are used in an appropriate way" [24, p. 319]. According to [24], documentation of adverse drug reactions is critical to improving preventive health. This documentation takes the form of varying events and relations, such as drug-drug interactions and relations between substance-allergy. Other documentation about disease and drug and disease and symptoms also critical in improving health outcomes and reducing adverse drug reactions [24]. A study in Spain showed that adverse drug reactions constitute 37.4% of adverse effects [24]. Oronoz et al. [24] affirms the impact of adverse drug reactions by reporting that the cost associated with adverse drug reactions annually in the United States is estimated at $75 billion per year. Taking this reasoning into consideration, improving the quality of health outcomes by implementing NLP and machine learning in the information extraction of adverse drug reactions can yield significant positive health outcomes and cost-efficiency.

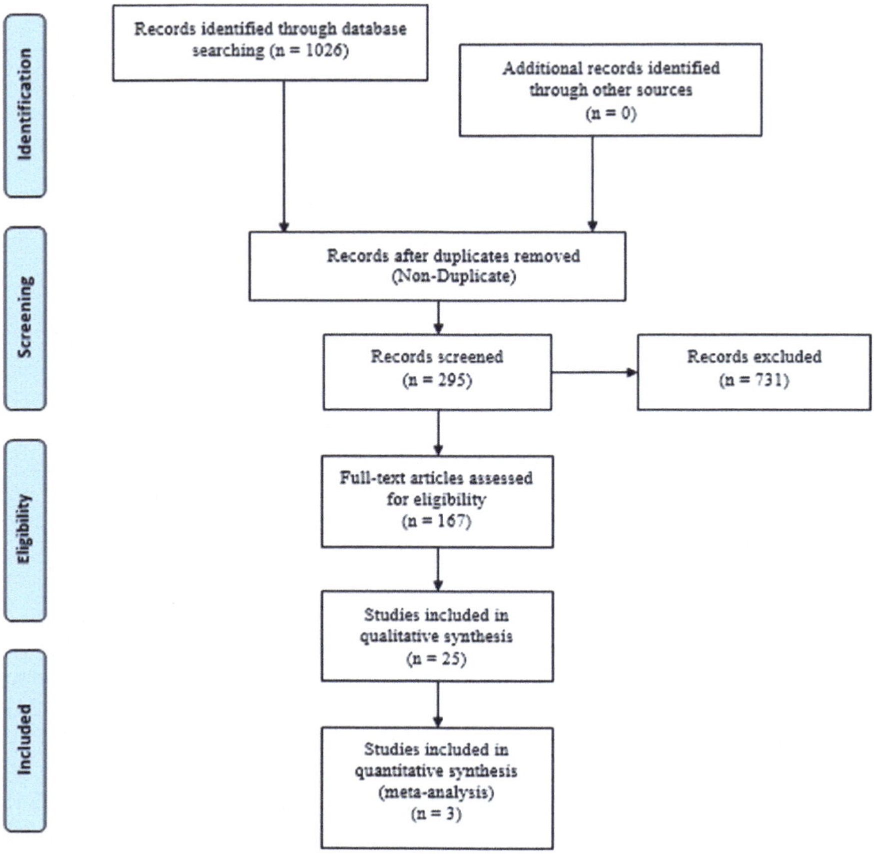

Fig. 1 PRISMA Flowchart

3.1.2 Methodology and Solution

One of the main contributions in [24] work is the creation of a gold standard adverse drug reactions corpus called IxaMed-GS. The corpus took a year to be completed and was manually annotated by four doctors and pharmacists working in pharmacology and pharmacovigilance in a Spanish hospital. Five researchers from the University of the Basque Country were on the team too. The corpus annotation objective was "the detection of adverse drug reactions in discharge reports" [24, p. 320]. The entities to be annotated and tagged were diseases, procedures, and drugs and how they are related. Only events were the drug caused the adverse reaction to the patient were annotated. The annotation tool used was the Brat rapid annotation tool.

Oronoz et al. [24] adapted FreeLing, the linguistic analyzer by augmenting it lexically with medical terms, resulting in FreeLing-Med. FreeLing-Med automatic annotation was validated by the domain experts against the manually annotated documents and modifications were introduced if required. FreeLing-Med is used to contribute to the automation of the clinical corpus in Spanish and as a potential replacement to time and effort consuming manual corpus annotation. FreeLing-Med uses SNOMED CT to add medical named entity recognition, Yetano and Alberola dictionary for medical abbreviations and BOTPLUS drug database. The annotated documents are presented in Kyoto Annotation Format (KAF).

3.1.3 The Model, Data Collection and Analysis

The resulting corpus IxaMed-GS consists of 51,061 words with 690 drug names, 735 substances, 891 diseases, 1387 symptoms, and 381 procedures. This corpus is based on 142,154 discharge reports written in Spanish generated in the period between 2008 and 2012 from outpatient consultations. All records were pre-anonymized by the medical facility before being processed. The usefulness of the corpus and the automated analyzer FreLing-Med were assessed using Random Forest (RF) as an example classifying method. The features of the model were separated into a cause, effect, and relational that measure the interaction between the classifying features. Filters in Weka were used to reduce sample imbalance and skewness. 10-fold cross-validation was used to improve the accuracy of the model training.

3.1.4 Discussion and Limitations

Inter-Annotator Agreement (IAA) was used to measure the annotation agreement level and the results [24, p. 328] are demonstrated in Table 3. The automatic analyzer performance is measured by means of classifying based on events across different sets [24, p. 328] as per the results in Table 4. Oronoz et al. [24] presented a very novel contribution in BLP by creating a unique adverse drug reaction corpus, intensively annotated on many levels including entities and most common relations between annotated entities. The corpus was evaluated using IAA and achieved rates of 90.53% agreement on the terms level and 82.86% agreement on the level of the event. This indicates that the work is done on this corpus and the guidelines implemented by

Table 3 IAA results

Results of the annotation process					
	Terms			Events	
Phase	#Agreement	#Disagreement	#Agreement	#Agreement	#Disagreement
50 documents	2251	284	88.63	104	72
25 documents	1282	134	90.53	58	12

Table 4 Automatic adverse drug reactions events classifying performance

Performance on the positive class of the automatic ADR event classifier inferred and evaluated on different sets

Inferences set	Evaluation set	Class			
		Percision	Recall	F-Measure	
Train	Train	1.000	0.892	0.943	1
	10f-CV	0.222	0.027	0.048	2
	Dev	1.000	0.019	0.038	3
$Train_{resampled}$	$Train_{resampled}$	0.998	1.000	0.999	4
	10f-CV	0.934	0.996	0.964	5
	Dev	0.300	0.346	0.321	6
$(TrainUdev)_{resampled}$	$(TrainUdev)_{resampled}$	0.999	1.000	0.999	7
	10f-CV	0.943	1.000	0.971	8
	Test	0.228	0.325	0.268	9
	Test o filter	0.619	0.325	0.426	10

the authors can be a sound indicator of how a task-specific clinical corpus can be designed in a language other than English.

The automation of the annotation was tested and provided a certain level of support to the manual annotation as measured by the classifying strategy and experts' usage. However, the application of the automatic annotation is considered as a semi-automated as opposed to a full automation tool. The corpus can be improved by increasing the source documents and expanding the adverse effects use cases. The corpus can be extended due to its extensive capabilities based on the captured entities, which are not limited to adverse drug reactions.

3.2 Use of "off-the-Shelf" Information Extraction Algorithms in Clinical Informatics: A Feasibility Study of MetaMap Annotation of Italian Medical Notes

3.2.1 Motivation and Background

Research in NLP and clinical corpora annotation does not fare better in the Italian language than Spanish, and it is still scarce in comparison to biomedical NLP applications in English. Chiaramello et al. [36] tackled the information extraction of clinical concepts from clinical notes written in the Italian language. However, instead of coming up with a completely new architecture to map extracted concepts from the clinical text, an existing knowledge source and an "of the shelf" existing tool was used. Chiaramello et al. [36] proposed using MetaMap, a well-established and broadly used information extraction tool to map concepts in clinical notes to the Unified Medical Language System (UMLS). UMLS is developed by the US National Library of Medicine (NLM), which has been initiated to advance the BLP research across the world languages. It contains a wide variety of biomedical and health vocabularies and semantic groups. These semantic categories are organized in Metathesaurus databases that aggregate terms and concepts from over 100 medical vocabularies.

Chiaramello et al. [36] proposed two main research questions to test the efficacy of MetaMap as a viable tool to annotate and map clinical concepts from Italian clinical notes. The first question attempts to check the possibility to adequately map medical concepts in clinical notes associated with the "Disorders" semantic group in UMLS to the Italian UMLS knowledge sources. The second research question is related to test MetaMap as a viable tool to extract medical concepts from Italian clinical notes. To check this question, Chiaramello et al. [36] designed three experiments that differ in scope and structure, which will be described in the following section. Chiaramello et al. [36] asserts that if implemented properly, BLP techniques and algorithms can have a direct impact on individual and public health. They concede that computer-assisted BLP in clinical decision support systems, bio-surveillance and pharmacovigilance correlation detection can improve health outcomes significantly.

3.2.2 Methodology and Solution

Chiaramello et al. [36] designed three experiments to answer the two proposed research questions as mentioned in the previous section. These experiments were designed to test three main tasks related to the extraction and annotation of Medical concepts from clinical notes. EXP1 attempts to identify how many medical concepts from the "Disorder" semantic group in UMLS can be found in clinical notes written in Italian and can be mapped to UMPS Italian knowledge sources. EXP2 assesses the MetaMap processing step, which is English dependent suitability to be applied to Italian text and map the Italian clinical notes to UMLS clinical Italian sources in the process. EXP3 tests feasibility of using MetaMap on automatically translated Italian clinical notes to English using Google Translator, to map the translated text to English UMLS medical sources.

Generally, the research by Chiaramello et al. [36] is to measure how feasible it is to use an English based linguistic algorithm with minimal to no adaptation for a similar task in another language other than the one it was designed for. Chiaramello et al. [36] laid out the MetaMap algorithm process in six main steps, to pinpoint the areas where the language dependence can impede the effectiveness of the tool in processing content from another language. The steps are tokenization, parsing, variant generation, candidate retrieval, candidate evaluation, and mapping construction. Out of those steps, parsing and variant generation are English-specific.

3.2.3 The Model, Data Collection and Analysis

The data corpus used in the research by Chiaramello et al. [36] consists of 3462 unstructured sentences retrieved from 100 clinical notes written in Italian. The notes are from five ambulatory domains: cardiology, diabetology, hepatology, nephrology, and oncology. These source notes were manually anonymized and a level of normalization to eliminate acronyms and spelling errors was done. This effort resulted in the "ITA-TXT" set, which was translated using Google translator to produce the English version "ENG-TXT". For the sake of the proposed experiments, the two sets are going to be mapped for medical concepts against two knowledge sources from UMLS, "ITA-Metathesaurus" and "ENG-Metathesaurus".

3.2.4 Results, Discussion, and Limitations

Chiaramello et al. [36] experiments to use an "off the shelf" linguistic algorithm designed for another language to map it to UMLS and apply on clinical notes in the Italian language is unprecedented as far as the authors know. For EXP1, IAA was used to gauge agreement between the automated extraction of medical concepts and manual annotation by two domain experts. Alternatively, the results for EXP2 and EXP3 were validated using recall, precision, and F-measure. Interestingly, the results for the proposed experiments were highly promising with documented variance, which was mostly traced to errors in the automated generic translation of Google translator of the clinical notes in Italian. The results for EXP1 achieved fewer rates in the identification of Italian medical concepts from "ITA-TXT" than the identification of English terms in "ENG-ITA" [36, p. 28]. However, the identification rates of medical concepts from "ITA-TXT" were very good none the less. The average identification for the concepts in Italian was at around 91% across the five medical domains as evident from Table 5.

Table 5 Result to EXP1 to identify the rate of medical concepts from "ITA-TXT" and "ENG-TXT" in "ITA-Metathesaurus" and "ENG-Metathesaurus"

Number (and %) of concepts identified in the "ITA-TXT" and "ENG-TXT" datasets found in the "ITA-Metathesaurus" and "ENG-Metathesaurus" knowledge source			
#Annotated noun phrases		#Annotated noun phrases	
		ITA-Metathesaurus	ENG-Metathesaurus
Cardiology	422	382 (90.6%)	416 (98.6%)
Diabetology	488	439 (90.0%)	481 (98.6%)
Hepatology	402	367 (91.3%)	396 (98.5%)
Nephrology	517	477 (92.3%)	517 (100%)
Oncology	248	223 (89.9%)	246 (99.2%)
Total	2077	1888 (90.9%)	2056 (99.0%)

Table 6 Result to EXP1 to identify the rate of medical concepts from "ITA-TXT" and "ENG-TXT" in "ITA-Metathesaurus" and "ENG-Metathesaurus"

Mean recall, precision, and F-measure obtained in EXP2 and EXP3, without and with the normalization

	EXP2		EXP3	
	Without normalization	With normalization	Without normalization	With normalization
Recall	0.51	0.53	0.73	0.75
Precision	0.95	0.98	0.9	0.93
F-measure	0.66	0.69	0.8	0.83

The result of EXP2 to measure the ability of MetaMap to successfully map medical concepts in Italian text to a matching description in a UMLS Italian Metathesaurus achieved 0.53 recall, which is good considering that two steps in MetaMap process are English specific. Also, the precision was higher than expected at 98%. In EXP3, where MetaMap was applied to the translated "ITA-TXT" to English, the mapping achieved high rates in the recall, precision, and F-measure, which are in line with the literature baselines. Applying MetaMap on "ENG-TXT", achieved 0.75 recall and 0.83 F-measure as evident from Table 6. The authors attributed much of the failures in the correct annotation to errors in Italian to English translation as 62% of the failures were due to this [36, p. 31]. In general, Chiaramello et al. [36] concluded that using MetaMap in its original form is adequate to process translated clinical text to English. The authors proceeded to investigate the reasons for the failure in both "ITA-TXT" and "ENG-TXT" annotation and came up with the reasons charted in Fig. 2.

Chiaramello et al. [36] can be improved on and extended by implementing techniques that can improve the translation by using a specialized medical translation tool that can eliminate a substantial amount of the measured annotation failure. Also, this work limited the annotation to a single semantic group in UMLS, which is the "Disorders" group. The work can be extended to accommodate further semantic groups such as diagnosis, adverse effects and other biomedical vocabularies that exist in UMLS. Finally, the manual annotation can be improved on as the current number of annotators of one cannot be adequate if the clinical corpus is to be expanded and more domain complexity is introduced. Improving the manual annotation capabilities will improve the IAA performance and further validate the outcomes of the experiments.

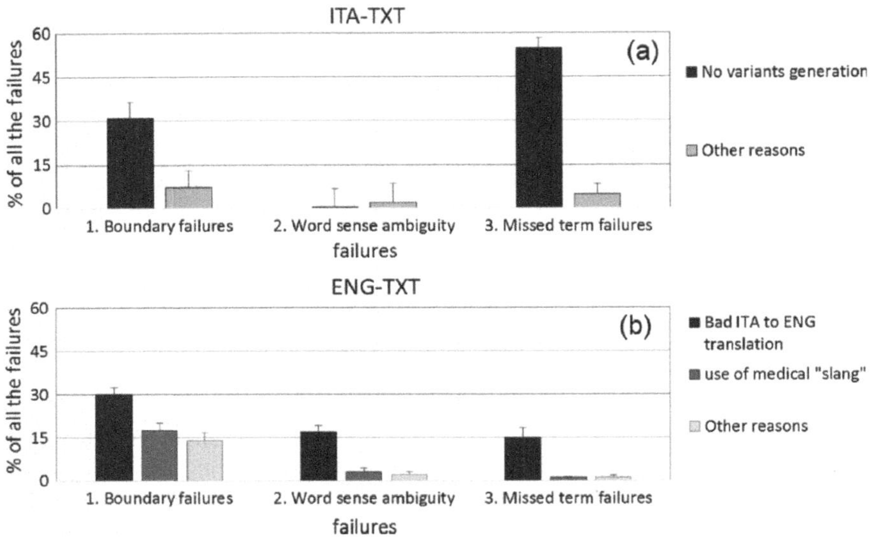

Fig. 2 Identified reasons for annotation failures across the two clinical sets in Italian and English

3.3 Building a Comprehensive Syntactic and Semantic Corpus of Chinese Clinical Texts

3.3.1 Motivation and Background

China has the largest population in the world and consequently, the popularization of EMR based electronic documentation of clinical data witnessed a rapid increase recently [27]. The amount of data, which populated the semi-structured clinical documents presented itself as a rich source of information about the entities and relations of the healthcare provided to the patients. He et al. [27] identified the gap in biomedical NLP application to clinical text written in Chinese and proceeded to develop an unprecedented syntactic and semantic clinical Chinese corpus. The corpus annotation training was iterative. This endeavor was based on previous work mostly done in English, but the lack of consistent guidelines for clinical text annotation in general and in Chinese in particular, drove the authors to venture further in establishing their own low and higher-level NLP tasks guidelines. The ultimate goal from the produced corpus was to offer a suitable baseline for NLP and BLP research on Chinese clinical text and improve greatly the efficiency and quality of healthcare service provided to millions of people.

This effort was augmented by developing the Chinese Clinical Text Processing and Information Extraction System (CCTPIES), to perform major NLP tasks on the generated corpus [27]. CCTPIES is constructed of multi-task modules to handle major NLP tasks. The system consists of word segmenter, Part-Of-Speech (POS) tagger, shallow parser, full parser, named entity recognizer and relation extractor modules [27]. Out of this POS tagger, shallow parser and full parser modules are introduced for the first time in clinical Chinese NLP, adding to the value of the work studied.

3.3.2 Methodology and Solution

He et al. [27] adopted a structured methodology to iteratively annotate a syntactic and semantic biomedical Chinese corpus, which covered POS tags, syntactic tags, entities, assertions, and relations of clinical free-text elements. He et al. [27] resorted to building the corpus as clinical Chinese text contains sublanguage elements, which general Chinese corpora are inadequate to handle. He et al. [27] research aimed to produce two complementing outputs: a comprehensive clinical Chinese annotated corpus and a multi-module NLP system. The construction of the corpus went through an iterative guideline comprised of three stages. The first stage takes into consideration characteristics of Chinese clinical texts and existing annotation guidelines to produce a draft of annotation guidelines. The second stage is the core of the iterative annotation guideline that begin by randomly sampling unannotated sentences and apply the double annotation to them. The annotation is evaluated using IAA and if the IAA is stable then the annotated text is added to the corpus. In case the IAA was not stable, then the annotation is discussed and the guideline is updated and applied to another sampled sentence.

As the aim by He et al. [27] to develop a comprehensive clinical corpus, guidelines for both low-level and high-level NLP tasks were established. The corpus is semantically annotated based on the low-level tasks of word segmentation and POS tagging and parsing. The corpus is then annotated for higher-level tasks represented by entities and assertions annotation and relations among entities. Based on the constructed corpus, a system consisting of trained models including word segmenter, POS tagger, shallow parser full parser, named entity recognizer and relation extractor. These models were trained using the sequence-labeling method and trained by CRF++, which is an open-source implementation of the conditional random fields' algorithm [27] with the reasons charted in Fig. 3.

3.3.3 The Model, Data Collection and Analysis

The corpus construction is based on clinical text from semi-structured discharge summaries and progress notes from a single hospital in China. The annotation is done by two physicians and eight computational linguists. The low-level tasks annotation was based on 72 discharge summaries and 66 progress notes including 2612 parsing trees. The higher-level tasks annotation was done from text from 500 discharge summaries and 492 progress notes including 39,511 entities and one-to-one relations. The annotated entities types are Diseases, Symptoms, and Treatments. The entities and relations between them and their respective rates are listed in Tables 7 and 8.

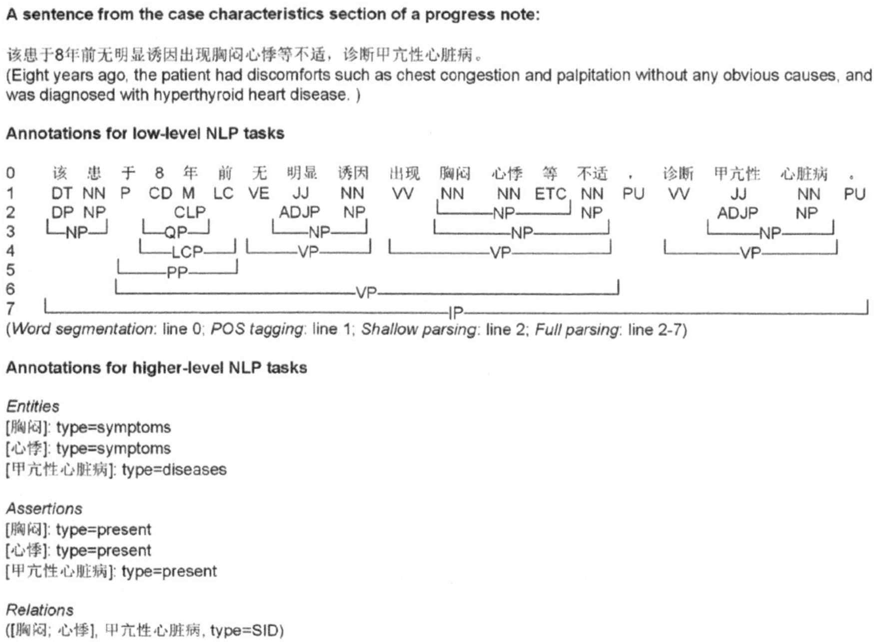

Fig. 3 An example of the annotation of a sentence that demonstrates low and higher-level NLP tasks

Table 7 IAA of the corpus entities annotation in the training iterations and corpus construction stages

Inter-annotator agreement in the latest three annotator training iterations and corpus construction stage (F1 measure)				
	IAA			
	Training [-3]	Training [-2]	Training [-1]	Corpus construction
Word segmentation	0.965	0.979	0.983	–
POS tagging	0.893	0.952	0.956	–
Shallow parsing	0.956	0.969	0.970	–
Full parsing	0.805	0.840	0.865	–
Entity (span, type, assertion)	0.848	0.920	0.927	0.922
Relation (entity group preserved)	0.765	0.781	0.843	0.772
Relation (one-to-one)	0.742	0.774	0.805	0.755

"–" means not evaluated

IAA inter-annotator agreement; *POS* past-of-speech

Table 8 Performance of CCTPIES modules trained on the annotated corpus

Performance of system modules trained on our annotated clinical text			
Module	Precision	Recall	F1
Word segmentation	0.981	0.979	0.980
POS tagging	0.966	0.964	0.965
Shallow parsing	0.946	0.949	0.948
Full parsing	0.845	0.841	0.843
Entity recognizer	0.923	0.902	0.912
Relation extractor	0.784	0.691	0.735

3.3.4 Results, Discussion, and Limitations

He et al. [27] work is a significant contribution in the area of NLP application to Chinese clinical text. The performance of the corpus and the models trained on it showed application-level performance. The corpus construction, which benefited from performance improvement based on the iterative update of annotation guidelines on low and higher-level annotation. Based on IAA, the F-measure rating for the main tasks performed on the corpus demonstrated improved results over the training iterations. Word segmentation reached 0.983 and POS tagging 0.956. The system modules showed excellent performance based on the results demonstrated in Table 8. Relation extractor showed relatively lower performance than the other modules.

Although [27] research is highly novel and can be considered as a baseline for future research in the domain and language, it can benefit from varying the sources of the clinical text. The corpus is developed based on two types of clinical documents, which are the discharge summaries and progress notes. Furthermore, the notes were from two departments in the hospital. These two aspects can result in lower performance in the relational and segmentation annotation. Also, introducing more robust automated annotation techniques can serve in improving the efficiency of the annotation process as the iterative process can prove taxing to the limited manual annotation resources and improve the cost and effort efficiency.

4 Discussion and Information Synthesis

As demonstrated in the previous sections, NLP and BLP are advancing steadily in the clinical domain in languages other than English. Research in NLP application in the clinical domains is vital with the rapid rate in the proliferation of EMRs in the world. The unstructured data in these records contain a wealth of information, which is an ideal candidate for NLP low and higher-level tasks. Many challenges exist in the face of research in other languages, such as lesser access to language-specific clinical text datasets, but this is improving steadily. The performance of the reviewed research showed production level results based on IAA evaluations mostly in low-level tasks and at varying rates in more complex relational higher-level tasks. In conclusion, NLP in the clinical domain in general and languages other than English is a highly viable domain with promising potential to improve individual and public health goals.

The previous section provided an overview of the novel and unprecedented contributions in NLP and clinical and biomedical corpus development in languages other than English. The paper covered experiments on Spanish, Italian and Chinese clinical text to develop robust and comprehensive biomedical corpora. The researched work attempted to develop internal annotation guidelines to handle the manual and automated annotation. The contrast between the approaches used to develop the corpora and their scopes offer a unique exploration of how specialized domain tasks are annotated and the methodologies used to achieve state-of-the-art performance in these tasks. Oronoz et al. [24] targeted a very specialized area of biomedical NLP, which was the annotation of a Spanish adverse drug reaction corpus, while [27] aimed toward a comprehensive clinical corpus in the Chinese language that annotated entities, assertions and relations in a complex language that necessitates working with sublanguage features.

The purpose of targeting efforts in languages other than English is vital due to the increased proliferation of EMR based clinical systems in newly developed regions and parts of the world with other languages than English. The techniques can vary based on available resources such as manual annotators and knowledge sources such as UMLS. The field of studying NLP in the biomedical domain is advancing steadily, due to increased research maturity and availability of datasets to develop language-specific corpora [37]. The majority of researched work in BLP in languages other than English is concerned with information extraction [38]. Other tasks widely researched in English clinical NLP early detection systems [39] and mental health sentiment analysis [40, 41] experienced limited exploration in non-English NLP research.

5 Conclusion

As demonstrated in the previous sections, NLP and BLP are advancing steadily in the clinical domain in languages other than English. Research in NLP application in the clinical domains is vital with the rapid rate in the proliferation of EMRs in the world. The unstructured data in these records contains a wealth of information, which is ideal candidate for NLP low and higher-level tasks. Although the researched work in this paper is English, research in the native languages of the discussed lingual BLP use-cases may exist. To mitigate this, the main research in this chapter considered novel and unprecedented work, which was not attempted as far as the authors know. Future work should consider advancement in the researched field, which may have been published in these languages with added focus in establishing gold-standard biomedical corpora in these languages. Many challenges exist in the face of research in other languages, such as lesser access to language-specific clinical text datasets, but this is improving steadily. The performance of the reviewed research showed production level results based on IAA evaluations mostly in low-level tasks and at varying rates in more complex relational higher-level tasks. In conclusion, NLP in the clinical domain in general and languages other than English is a highly viable domain with promising potential to improve individual and public health goals.

Acknowledgements This work is a part of a project undertaken at the British University in Dubai.

References

1. Alhashmi, S.F.S., Salloum, S.A., Abdallah, S.: Critical success factors for implementing artificial intelligence (AI) projects in Dubai Government United Arab Emirates (UAE) health sector: applying the extended technology acceptance model (TAM). In: International Conference on Advanced Intelligent Systems and Informatics, pp. 393–405 (2019)
2. Alhashmi, S.F.S., Salloum, S.A., Mhamdi, C.: Implementing artificial intelligence in the United Arab Emirates healthcare sector: an extended technology acceptance model. Int. J. Inf. Technol. Lang. Stud. **3**(3) (2019)
3. Ghannajeh, A., et al.: A qualitative analysis of product innovation in Jordan's pharmaceutical sector. Eur. Sci. J. **11**(4), 474–503 (2015)
4. Alshurideh, M.: The factors predicting students' satisfaction with universities' healthcare clinics' services: a case-study from the Jordanian Higher Education sector. Dirasat: Adm. Sci. **41**(2), 451–464 (2014)
5. Aburayya, A., Alshurideh, M., Albqaeen, A., Alawadhi, D., Ayadeh, I.: An investigation of factors affecting patients waiting time in primary health care centers: an assessment study in Dubai. Manag. Sci. Lett. **10**(6), 1265–1276 (2020)
6. Alshurideh, M.: Pharmaceutical promotion tools effect on physician's adoption of medicine prescribing: evidence from Jordan. Mod. Appl. Sci. **12**(11) (2018)
7. Salloum, S.A., Al-Emran, M., Khalaf, R., Habes, M., Shaalan, K.: An innovative study of E-payment systems adoption in Higher Education: theoretical constructs and empirical analysis. Int. J. Interact. Mob. Technol. **13**(6) (2019)
8. Habes, M., Salloum, S.A., Alghizzawi, M., Alshibly, M.S.: The role of modern media technology in improving collaborative learning of students in Jordanian universities. Int. J. Inf. Technol. Lang. Stud. **2**(3), 71–82 (2018)
9. Al Kurdi, B., Alshurideh, M., Salloum, S.A., Obeidat, Z.M., Al-dweeri, R.M.: An empirical investigation into examination of factors influencing university students' behavior towards E-learning acceptance using SEM approach. Int. J. Interact. Mob. Technol. **14**(02), 19–41 (2020)
10. Al-Maroof, R.S., Salloum, S.A., AlHamadand, A.Q., Shaalan, K.: Understanding an extension technology acceptance model of Google Translation: a multi-cultural study in United Arab Emirates. Int. J. Interact. Mob. Technol. **14**(03), 157–178 (2020)
11. Salloum, S.A., Al-Emran, M., Monem, A., Shaalan, K.: A survey of text mining in social media: Facebook and Twitter perspectives. Adv. Sci. Technol. Eng. Syst. J. **2**(1), 127–133 (2017)
12. Salloum, S.A., AlHamad, A.Q., Al-Emran, M., Shaalan, K.: A survey of Arabic text mining. In: Studies in Computational Intelligence, vol. 740, Springer, Berlin (2018)
13. Salloum, S.A., Al-Emran, M., Shaalan, K.: A survey of lexical functional grammar in the Arabic context. Int. J. Comput. Netw. Technol. **4**(3), 141–147 (2016)
14. Salloum, S.A., Al-Emran, M., Monem, A.A., Shaalan, K.: Using text mining techniques for extracting information from research articles. In: Studies in Computational Intelligence, vol. 740, Springer, Berlin (2018)

15. Mhamdi, C., Al-Emran, M., Salloum, S.A.: Text mining and analytics: a case study from news channels posts on Facebook, vol. 740 (2018)
16. Salloum, S.A., Al-Emran, M., Shaalan, K.: Mining text in news channels: a case study from Facebook. Int. J. Inf. Technol. Lang. Stud. **1**(1), 1–9 (2017)
17. Salloum, S.A., Al-Emran, M., Abdallah, S., Shaalan, K.: Analyzing the Arab gulf newspapers using text mining techniques. In: International Conference on Advanced Intelligent Systems and Informatics, pp. 396–405 (2017)
18. Al Emran, M., Shaalan, K.: A survey of intelligent language tutoring systems. In: Proceedings of the 2014 International Conference on Advances in Computing, Communications and Informatics, ICACCI 2014, pp. 393–399 (2014)
19. Al-Emran, M., Zaza, S., Shaalan, K.: Parsing modern standard Arabic using Treebank resources. In: 2015 International Conference on Information and Communication Technology Research, ICTRC 2015 (2015)
20. Velasco, E., Agheneza, T., Denecke, K., Kirchner, G., Eckmanns, T.: Social media and internet-based data in global systems for public health surveillance: a systematic review. Milbank Q. **92**(1), 7–33 (2014)
21. Afzal, N., et al.: Natural language processing of clinical notes for identification of critical limb ischemia. Int. J. Med. Inform. **111**, 83–89 (2018)
22. Roch, A.M., et al.: Automated pancreatic cyst screening using natural language processing: a new tool in the early detection of pancreatic cancer. Hpb **17**(5), 447–453 (2015)
23. Patterson, B.W., et al.: Development and validation of a pragmatic natural language processing approach to identifying falls in older adults in the emergency department. BMC Med. Inform. Decis. Mak. **19**(1), 138 (2019)
24. Oronoz, M., Gojenola, K., Pérez, A., de Ilarraza, A.D., Casillas, A.: On the creation of a clinical gold standard corpus in spanish: mining adverse drug reactions. J. Biomed. Inform. **56**, 318–332 (2015)
25. Cohen, K.B., Demner-Fushman, D.: Biomedical Natural Language Processing, vol. 11. John Benjamins Publishing Company, Amsterdam (2014)
26. Wang, H., Zhang, W., Zeng, Q., Li, Z., Feng, K., Liu, L.: Extracting important information from Chinese Operation Notes with natural language processing methods. J. Biomed. Inform. **48**, 130–136 (2014)
27. He, B., et al.: Building a comprehensive syntactic and semantic corpus of Chinese clinical texts. J. Biomed. Inform. **69**, 203–217 (2017)
28. Grigonyte, G., Kvist, M., Velupillai, S., Wirén, M.: Improving readability of Swedish electronic health records through lexical simplification: first results. In: Proceedings of the 3rd Workshop on Predicting and Improving Text Readability for Target Reader Populations (PITR), pp. 74–83 (2014)
29. Al-Emran, M., Mezhuyev, V., Kamaludin, A.: Technology acceptance model in M-learning context: a systematic review. Comput. Educ. **125**, 389–412 (2018)
30. Al-Emran, M., Mezhuyev, V., Kamaludin, A., Shaalan, K.: The impact of knowledge management processes on information systems: a systematic review. Int. J. Inf. Manag. **43**, 173–187 (2018)
31. Al-Qaysi, N., Mohamad-Nordin, N., Al-Emran, M.: A systematic review of social media acceptance from the perspective of educational and information systems theories and models. J. Educ. Comput. Res. **57**(8), 2085–2109 (2020)
32. Al-Saedi, K., Al-Emran, M., Abusham, E., El-Rahman, S.A.: Mobile payment adoption: a systematic review of the UTAUT model. In: International Conference on Fourth Industrial Revolution (2019)
33. Saa, A.A., Al-Emran, M., Shaalan, K.: Factors affecting students' performance in higher education: a systematic review of predictive data mining techniques. Technol. Knowl. Learn. (2019)
34. Salloum, S.A.S., Shaalan, K.: Investigating students' acceptance of E-learning system in Higher Educational Environments in the UAE: applying the extended technology acceptance model (TAM). The British University in Dubai (2018)

35. Salloum, S.A., Alhamad, A.Q.M., Al-Emran, M., Monem, A.A., Shaalan, K.: Exploring students' acceptance of E-learning through the development of a comprehensive technology acceptance model. IEEE Access **7**, 128445–128462 (2019)
36. Chiaramello, E., Pinciroli, F., Bonalumi, A., Caroli, A., Tognola, G.: Use of 'off-the-shelf' information extraction algorithms in clinical informatics: a feasibility study of MetaMap annotation of Italian medical notes. J. Biomed. Inform. **63**, 22–32 (2016)
37. Névéol, A., Dalianis, H., Velupillai, S., Savova, G., Zweigenbaum, P.: Clinical natural language processing in languages other than english: opportunities and challenges. J. Biomed. Semant. **9**(1), 12 (2018)
38. Campillos, L., Deléger, L., Grouin, C., Hamon, T., Ligozat, A.-L., Névéol, A.: A French clinical corpus with comprehensive semantic annotations: development of the Medical Entity and Relation LIMSI annOtated Text corpus (MERLOT). Lang. Resour. Eval. **52**(2), 571–601 (2018)
39. Doan, S., et al.: Building a natural language processing tool to identify patients with high clinical suspicion for Kawasaki disease from emergency department notes. Acad. Emerg. Med. **23**(5), 628–636 (2016)
40. Carson, N.J., et al.: Identification of suicidal behavior among psychiatrically hospitalized adolescents using natural language processing and machine learning of electronic health records. PLoS ONE **14**(2), e0211116 (2019)
41. McCoy, T.H., Castro, V.M., Cagan, A., Roberson, A.M., Kohane, I.S., Perlis, R.H.: Sentiment measured in hospital discharge notes is associated with readmission and mortality risk: an electronic health record study. PLoS ONE **10**(8), e0136341 (2015)

A Proposed Context-Awareness Taxonomy for Multi-data Fusion in Smart Environments: Types, Properties, and Challenges

Doaa Mohey El-Din, Aboul Ella Hassanein, and Ehab E. Hassanien

Abstract this paper presents a new taxonomy for the context-awareness problem in data fusion. It interprets fusing data extracted from multiple sensory datatypes like images, videos, or text. Any constructed smart environment generates big data with various datatypes which are extracted from multiple sensors. This big data requires to fuse with expert people due to the context-awareness problem. Each smart environment has specific characteristics, conditions, and roles that need to expert human in each context. The proposed taxonomy tries to cure this problem by focusing on three dimensions classes for data fusion, types of generated data, data properties as reduction or noisy data, or challenges. It neglects the context domain and introduces solutions for fusing big data through classes in the proposed taxonomy. This taxonomy is presented from studying sixty-six research papers in various types of fusion, and different properties of data fusion. This paper presents new research challenges of multi-data fusion.

Keywords Data fusion · Big data · Telemedicine · Internet-for-Things · Smart environment · Data visualization

1 Introduction

A smart environment is defined by the simulation of physical systems in the real system that is based on the connection between multiple sensors through the Internet. The extracted data has several characteristics: big volume, various formats types (such as text, video, audio, or image), and veracity of changes [1, 2]. It refers to

D. M. El-Din (✉) · A. E. Hassanein · E. E. Hassanien
Information Systems Department, Faculty of Computers and Artificial Intelligence, Cairo University, Cairo, Egypt
e-mail: d.mohey@alumni.fci-cu.edu.eg

A. E. Hassanein
e-mail: aboitcairo@gmail.com

E. E. Hassanien
e-mail: E.Ezat@fci-cu.edu.eg

M. Al-Emran et al. (eds.), *Recent Advances in Intelligent Systems and Smart Applications*, Studies in Systems, Decision and Control 295,
https://doi.org/10.1007/978-3-030-47411-9_28

a highly adaptive environment for various targets in sensor fusion [3]. The fusion process has happened when integrating the extracted data with various formats or various objectives. The fusion can interpret the data into useful values. Data fusion is classified further as low, intermediate and high corresponding to data-level, feature-level, and decision-level fusions [4, 5]. Researchers usually to provide and introduce a Multi-modal for sensor fusion to be more familiar for the human-computer interaction research field that used the input and delivering output involving multiple sensor modalities [6, 7].

The context-aware relies on the user's targets and domain [8]. For any smart environment, Multi-sensor data fusion requires to answer the mention questions in the following:

1. *What is the target of the fusion?*
2. *How many users can deal with the outcomes data?*
3. *How can we make fusion?*
4. *When can automate the fusion?*
5. *Who gets the data from them?*

Multi-model or Multi-sensor fusion still faces various challenges into fusion in different format types, meaningful interoperation, and specific domain [9, 10]. The adaptive context-aware relies on learning has problems in the accuracy rate, performance evaluation, and explanation reasoning of fusion. Multi-sensor fusion targets reaching the information heterogeneity of multi-sources. The raw data to analyze might be generated by a big number of heterogeneous sensors and extensive research effort has been devoted to coherently and accurately combine the resulting data.

The standardization of fusion-context types that is still the biggest problem in basic fusion is often targeted specific domain [11–13]. Any fusion model requires the expert user to follow the process of the fusion from scratch. However, the adaptive model of fusion targets getting benefits from the features to improve the fusion adaptation for multi-context.

This paper proposes a new standardization taxonomy for multiple types of context data fusion problem. The proposed taxonomy provides a new classification of context-awareness based on several dimensions: the type of context, the reduction process type, the data noisy, and the time of streaming this data. It presents a solution for the context-awareness problem by focusing on three dimensions classes for data fusion, types of generated data, data properties as reduction or noisy data, or challenges. It neglects the context domain and introduces solutions for fusing big data through classes in the proposed taxonomy.

The rest of the paper is organized as the following: Sect. 2, discusses the literature review of the context problem in data fusion, Sect. 3, the proposed taxonomy of data fusion, Sect. 4, Discussion, Sect. 5, open research challenges of data fusion and sensor fusion. Finally, Sect. 6 targets the conclusion outlines.

2 Literature Review

This section discusses the importance of multi-data fusion in smart environments. It also examines the context-awareness problem in data fusion and presents several taxonomies for fusion classification. Smart Environment refers the simulation systems for real environments based on connecting sensors via internet. It is based on internet-of-things and artificial intelligent. These data record each step in each second, that gets huge data with variant formats, various objectives or supplemental information with specific conditions and properties. Data fusion means the collecting big data from multiple sources to achieve the main target. The importance of data fusion is founded in the hardness in fusing huge data with various data properties or types. It improves making decisions and provides the accuracy results.

Context is defined by any data that can be utilized to depict the situation of an entity. An entity is defined by any object or people that interacts between a user and an application, including the user and applications themselves. The awareness is defined by the responsibilities of the system whether from users or conditions. Context-Awareness is still a big challenge of data fusion due to the specific domain properties, targets, and conditions. For each context domain, there are some data that consider complementary data for reaching the full vision of objective or idea. The main picture of context includes any factor affected on the environment that includes location, identity, environment, activities, and time. The context requires to inference the sensory data from various factors in one or various types.

Previous researches present several taxonomies for context-Awareness problem that are based on a study of context-aware frameworks. Researchers in [14], presents a new taxonomy based on eight categories to discuss the implementation criteria for implementing a context-aware system. These categories are system types, used sensors, context abstraction level, context model, context storage, architecture, privacy and security, and application domain. Researchers in [15], presents two main dimensions of context analysis in a proposed taxonomy for the mobile guides' literature. It creates multiple criteria for dealing with complexity of the mobile guides application space. It consists of five classifications for the mobile applications, context awareness system, user interface and requirements, social relationships, service tasks, and environment whether spatial or temporal. Researchers in [16], presents all data fusion taxonomies based on six dimensions that includes the relationship between multiple resources, the fusion abstraction level, the type of input and output, data level fusion, the data type fusion, user's requirements in fusion process. Researchers in [17], presents a main data fusion taxonomy that is classified into five classes for improving the usability of data fusion systems. It includes five classes, relationships between applications, the level of abstraction for the input and the output, JDL data fusion framework, the architecture type of data fusion.

However, there is a main obstacle in all previous taxonomies that appears in how to deal with various context (as shown in Table 1). In other words, how can construct any fusion model without expert people. That requires the study of data, data types interpretation, features, and conditions check. So, there is a need to take at

Table 1 A comparison between previous taxonomies of data fusion

Paper No.	Taxonomy target	Categories	Benefits	Limitations
[14]	Observation of the implementation systems of context-aware	Eight	Improving the implementation system methodology Managing the fusion context control	Lack of context storage, less caring for data in any system Lack of the criteria to secure data and secure the network
[15]	Introduce the taxonomy classification for mobile guides	Five	It can classify context to improve the mobile guides	Lack of technology evolution, lack of hard devices, less performance
[16]	Survey of all data fusion taxonomies	Six	Improving saving power, benefit from data redundancy. It discusses the previous data fusion taxonomies	No grantee spatial and temporal in fusion process concurrently
[17]	It classifies the proposed taxonomy based on applications strategies	Five	It proposes a data fusion taxonomy to classify fusion strategies to prevent ambiguously. Powerful taxonomy	Interpretation of fusion is highly complex to create fusion applications

the occasional look of any applications. There are several conditions, features, and data changes that must consider in any data fusion construction application as the comparison in Table 2.

In the following, a comparison between several data fusion applications in various contexts.

Previous comparison between twelve researches finds a big obstacle in variant context in context features, conditions, and properties (as shown in Table 2). Each environment requires creating a specific application for it, that becomes very hard in implementation and needs experiment people. All limitations in mention applications are shown in need choosing suitable size of a dataset, improving performance or accuracy results to be more effective or flexible system. The conclusion is no adaption fusion application for various contexts due to various datatypes, conditions and properties for each system.

Table 2 A comparison between data fusion applications

Paper No.	Context domain	Data size	Time stream type	Conditions	Data type	Advantages	Disadvantages
[19]	Telemedicine	1585 article for patients	Offline	Array of features of disease, cardiovascular disease (N = 37) or diabetes (N = 18) aged $\geq$60 years	Audio or video	Monitoring remotely by 87%	There is hardness in adaption framework with several systems special national healthcare system Can't handle n 26 of 68 cases due to the lack of expert data for them
[20]	Daily living activities	Wearable dataset	Real-stream and offline	Automatic recognition of sedentary uses smart clothes (shirts)	Signal	Monitoring patients remotely by 95% accuracy results	Can't interpret all activities. and requires authorized users for recording activities
[21]	Telehealth	Hand and fingerprint in 30 frames per second for five people in six times (Records 360)	Real-time	Color-based method for determination of hand location, and hand contour analysis	Image dataset	Improves for tracking and recognition system	Requires improving system by analyze contour for fingers and hands to be more flexible

(continued)

Table 2 (continued)

Paper No.	Context domain	Data size	Time stream type	Conditions	Data type	Advantages	Disadvantages
[23]	Image fusion quality	Two different image types: anatomic (MRI) and functional (SPECT, positron emission tomography [PET])	Offline	Identifying tumor masses	Images	Good fusion for two images with more details	Takes a long time that requires improving the performance and taking less cost of time
[30]	Sonar videos	Sequenced data in 3 days in the underwater environments	Offline	Target tracking for images and record videos	Videos	Improving tracking	Requires tracking more scenarios to cover multiple cases
[31]	Video surveillance	Optical images	Real-time	Tracking objects to measure confidence and adoption	Images optical and infrared (IR) sensors	High accuracy	Finding localization errors and incorrect segmentation
[33]	Speech	Speech dataset	Offline	Diversity text-independent closed-set speaker identification	Speech	Improve performance 94.1%	Requires improving performance

(continued)

Table 2 (continued)

Paper No.	Context domain	Data size	Time stream type	Conditions	Data type	Advantages	Disadvantages
[38]	Social media	Social media people's reviews (getty imags, flick, twitter)	Offline (getty image: dataset containing more than 500,000 image and text pairs, for twitter 20,000 weakly, And Flickr 312,796 weakly)	Requires understanding English words, grammar, or keywords and comparing with the evaluation of images	Image-text	Improving prediction model for images and text)	Requires improving the effective due to the lack of datasets
[44]	Audio-visual data	Offline	Emotions samples of 830 frames in 6 emotions	Interpreting the semantic meaning of emotions in two data types	Emotion recognition for face and audio data	Fusing in semantic level for improving multi-model recognition, improving accuracy by 15% to reach 55.9%	Requires bigger dataset and improving accuracy results
[57]	Speech	7 males and 3 females	Offline	Measurements and simulation various SNR levels	Audio and visual	Improve decision fusion level and improve decision making with high accuracy	Improving results by some noisy data in testing

(continued)

Table 2 (continued)

Paper No.	Context domain	Data size	Time stream type	Conditions	Data type	Advantages	Disadvantages
[71]	Outlier detection data	Has noisy data	Offline	Classifying outliers whether event or error	Video-audio-text	Increase accuracy by 14% by detecting outliers	Appling in various datasets and requires improving the performance
[77]	Multi context with several Attributes	Variant contexts	Real dataset for real sensors	Based on multi attitude theory	Sensors	Applies on variant contexts	Improving performance

3 The Proposed Taxonomy of Data Fusion

In recent years, a lot of attention has been paid to include context information in the data fusion process in order to reduce ambiguities and errors. The goals of fusion just only determine the type of fusion context. Our observation classifies the previous motivations that contain two types of categorization of fusion: based on the type of fusion and based on the type of data offline or real-time. These Categories targets reaching the semantic meaning of various contexts fusion.

The proposed taxonomy provides a new classification of context-awareness based on several dimensions: the type of context, the reduction process type, the data noisy, and the time of streaming this data, as shown in Fig. 1. A context type refers to four categories: one same type: text only, video, only, audio only, or image only, the second category: includes two context types only and the third contains three context types, and the fourth type: more than three types of context that has several motivations to be dynamic or adaptive recently.

3.1 Fusion Type One: Based on A Context-Type Fusion

This type refers to the same type of context with different meanings such as several texts, several images and so on. This category depends on four Categorizations' classification as the following:

1. **Category one: The same Context Type**:

(1) Text-Context Fusion

There is a mutual idea between the 'Text' and 'context', but the 'Context' is more complex. The process of understanding the context and convert into text that refers to the semantic-pragmatic and right meaning. The text-context may be short words,

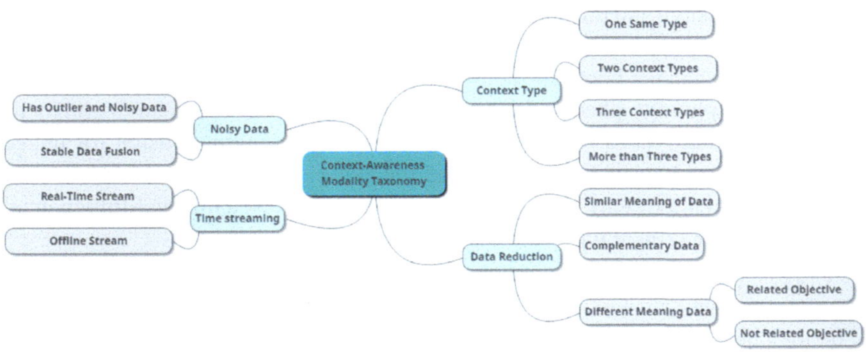

Fig. 1 The proposed taxonomy of classification of context-awareness modality

numbers, or long documents. That is very useful for several research in machine translation to identify the culture filter dynamically for improving accuracy. The definition of the context often refers to everything around the text [17]. The fusion between various texts may cause of the new definition mentions entitled 'intertextuality' that text's relation with other texts. The text usually focuses on the syntax of the sentence but the context and intertextuality focus on the semantic meaning of the text, in other words, the core of the text.

This type is very powerful in sentiment analysis when collecting the various sentiments via multi-sources about something. The telemedicine has also a portion of this area, there are motivations which cover the ECG, blood pressure about patients [18–21]. These observation records in every second to manage remotely and medicine patients that can serve the patients in saving time, cost and their life's. On another hand, supporting the doctors to observe a high number of patients at the same time and making decisions on their cases concurrently. Telemedicine becomes very important for saving people life's and elderly people who stand alone.

(2) Image-Context fusion

The image fusion of context includes some motivations such as flat image, 2 Dimensions, and 3 Dimensions, and X-rays images types [22–27]. The image expresses the human emotion, disease type or analysis image, pattern identification such as fruits. All these fusion processes on different images use for enhancing the information quality. Not only has the type of image influenced the results but also the resolution of each image that can improve the classification. Colorized images, resolutions, pixels are factors that should identify from the first of the fusion process. That also can determine the anomaly detection of each fusion domain and environment [28]. X-ray and MRI images have several research to improve accuracy and classification [29]. That can support the Type and shape of tumors from the x-rays. There is another goal of fusion for recognizing the pattern classification that can support the size or weight of something.

Multiview fusion of images from the same modality and taken at the same time but from different viewpoints. Multimodal image fusion is a result of extracting data from multi-sensors (visible and infrared, CT and NMR, or panchromatic and multispectral satellite images) [30]. A multi-temporal fusion of images taken at different times in order to detect changes between them or to synthesize realistic images of objects which were not photographed in the desired time. Other types of image fusion focus on a 3D scene that picked repeatedly with several focal lengths. Fusion for image restoration, fusion two or more images of the same scene and modality, each of them blurred and noisy, may lead to a de-blurred and de-noised image. Multichannel deconvolution is a typical representative of this category.

(3) Video-context fusion

There are a huge number of videos only online and offline, the new research targets to identify some patterns, emotions and actions in the video [31]. So, fusion can support various classification in the same time. That can serve the telemedicine of remotely monitoring patients and their actions in a real-time stream of television channels, news, Announcer, signer and actors online [19].

(4) Audio-context fusion

Audio data fusion data refers to the audio data from various sensors [32]. Not only is it healthy, but it is also very important for people with disabilities. There are several motivations of research in converting the speech-to-text and still, the fusion between various audios is an obstacle of influencies modelo. The Speed, length, type of the audio are the essential three factors will impact on analyzing and fusing the data. They influence a Sequence of Overlapping Frames.

2. **Category two: based on two fusion context-types**

(1) Text-Speech

Previous research convert speech into text (Speech-to-text) to enhance pattern recognition and accuracy [33, 34]. The artificial intelligent algorithms provide several solutions for improving the accuracy of the meaning, pattern identification, and feature frequently. They often focus on the language's grammar, verb, adverb, adjective or noun. This part-of-speech method can support the reliability of the system and enhance time performance. Lexicon-based is another method for text parsing and processing that can support by stop words list, each word has value and polarity. That has a good benefit for each domain individually. Machine learning [34] can classify several categorizations and predict the results. The recent researches [35, 36] benefit from the reliability of the deep learning algorithms to identify the features of text and enhance the accuracy of fusion. This type has some challenges into accent, language, requires translation or not, and convert usually the speech into text to unify the topology structure of variant context.

(2) Text-Image

There are three types of image and text fusion,

- *the image targets determining the text from it* [37],
- *the image should be expressed into text* [38] *to complete the fusion type or reversal situation* [39].
- *The image is an emotional feeling of sentiment opinion* [40], *the fusion mostly tries to convert text into an image to fuse images.*

(3) Text-Video

Researchers provide various solutions for identification of the text from the real-stream video which can improve the fusion levels and decide [40–42]. The main targets of their researchers are converting videos into storyboards. That requires to track information, determine the time, and a group of actions. That is considered multi-intelligent activities or actions which can detect some activates, and actions on them in spatiotemporal. The fusion process between video and text is proof in some applications, for example, the production of multimedia (e.g., video archiving and understanding). The challenge of this type often how to convert the actions in the video into interpreted text.

(4) Video-Image

The motivations refer to split the streams of video into image frames. Although This fusion type is easier than previous fusion types, the performance time is very important to can classify, predict and fusion especially in real-stream of the videos. Motion estimation and warping functions are the main building blocks of the proposed framework [43]. The timestamp for each frame is very important for any video-image fusion. The researchers proposed [44] a software based on a hybrid distribution from Gaussian and uniform that was developed to be highly safe and reliable for the observations.

The main problem of this type demonstrates how to identify objects, activates into the video and interprets with the image meaning. Another question of this problem show in that is there any redundant data.

(5) Video-Audio

While the audio is a major source of speech information, the visual component is a valuable supplementary information source in noisy environments because it remains unaffected by acoustic noise. Many studies have shown that the integration of audio and visual features leads to more accurate speaker identification even in noisy environments [45]. Audio-visual integration can be divided into three categories: feature fusion, decision fusion, and model fusion [46]. In feature fusion, multiple features are concatenated into a large feature vector and a single model is trained [47]. However, this type of fusion faces a problem in the hard representation of losing timing simultaneously for audio-visual features. In decision fusion, audio and visual features are processed separately to build two independent models, which completely ignore the audio-visual correlations. In model fusion, several models have been proposed, such as multi-stream hidden Markov model (HMM), factorial HMM, coupled HMM, mixed DBN, etc. [48–51]. This type utilizes for tracking multiple speakers. Other research in this area targets determining the speech from the video. The authors aimed at detecting instances of aggressive human behavior in public environments based on Dynamic Bayesian Network [52]. The goal of visual-audio fusion improves the accuracy of video only or audio only and enhances the features' classification quality. There is a new approach to audio-visual fusion is focused on the modeling of audio and video signals. It targets decomposing each modality into a small set of functions representing the structures that are inherent in the signals. The audio signal is decomposed into a set of atoms representing concentrations of energy in the spectrogram (sounds) and the video signal is concisely represented by a set of image structures evolving through time, i.e. changing their location, size or orientation. As a result, meaningful features can be easily defined for each modality, as the presence of a sound and the movement of a salient image structure. Finally, the fusion step simply evaluates the co-occurrence of these relevant events. This approach is applied to the blind detection and separation of the audio-visual sources that are present in a scene.

In contrast, other methods presented used basic features and it is more focused on the fusion strategy that combines them. This approach was based on a nonlinear diffusion procedure that progressively erodes a video sequence and converts it into an audio-visual for the video sequence, where only the information that is required in applications in the joint audio-visual domain is kept. For this purpose, they define a diffusion coefficient that depends on the synchrony between video motions and audio energy and preserves regions moving coherently with the presence of sounds. Thus, the regions that are least diffused are likely to be part of the video modality of the audio-visual source, and the application of this fusion method to the unsupervised extraction of audio-visual objects is straightforward. The challenge of this type demonstrates how to convert and understanding the input videos and support the full idea of audios meanings. Another big obstacle faces the researchers here in the accent and language detection of the speech.

3. **Category Three: Three Fusion types**

The main challenge of this category which type should select to convert all contexts into it. There is no standard model or method to support researchers in choosing suitable topology. So, this process takes a long time to search about the results of methods to check which converted topology that can hold the highest accuracy with good performance time. Most researchers make a great effort at searching for convenient algorithms to their objective. But they still face a big obstacle in selecting the suitable topology that can improve the classification, reduction, and pattern recognition.

In the following, a brief of several types of this type.

(1) Text, audio, Image

The researchers presented a fusion method, based on deep neural networks, to predict personality traits from audio, language, and appearance [53]. They observed three modalities that included a signal relevant for personality prediction, that using all three modalities combined greatly outperforms using individual modalities, and that the channels interact with each other in a non-trivial fashion. By fusing the last network layers and fine-tuning the parameters we have obtained the best result, average among all traits, of 0.0938 Mean Square Error, which is 9.4% better in the performance. Video frames (appearance) are slightly more relevant than audio information (i.e. non-verbal parts of speech).

(2) Text, Image, video

The evolution of this fusion type relies on the semantic meaning of the context. The keywords play a vital role to identify the domain and indexing the context's features. The authors [54] introduced a probabilistic framework for semantic video indexing based on learning probabilistic multimedia representations of semantic events to represent keywords and key concepts. Ellis [55] presents a framework for detecting sources of sounds in audio using such cues as onset and offset. The proposed solution [56] to use machine learning in fusion between text, image and video contexts based on defining a dictionary for semantic concepts and using data retrieval. The authors [57] investigated the correlations between audio and visual features. A new audio-visual correlative model (AVCM) based on a dynamic Bayesian network (DBN) was proposed, which describes both the inter-correlations and the loose synchronicity between audio and visual streams. The experiments on the audio-visual bimodal speaker identification show that the AVCM model enhances the accuracy results.

(3) Text, video, audio

The researchers presented a learning-based approach to the semantic indexing of multimedia content using cues derived from audio, visual, and text features. They used semantic labeling as a machine learning problem. They created a lexicon and used statistical methods for making the fusion between various contexts [58, 59]. A Proposed solution [58] is a middleware for context-aware applications that dynamically learns associative rules among context attributes. The main challenge of this type of how to reach the unified meaning of variant contexts according to each context feature.

4. **Category four: based on more than three types of fusion**

This type faces the category three challenge also, which refers to the standardization of selecting the suitable topology to convert all various contexts into one topology automatically and dynamically with the highest accuracy.

Recently, the smart environment is a hot area of research and industry that includes huge data invariant topologies structures that require to analyze concurrently to support users in decision making in real-time. Any Smart environment has several sensors that hold the data about an object in one context type or more. Few motivations in this area to support some sensors or data fusion invariant context such as mobile sensors data. The recent researches target dynamic models or adaptive models to be convenient with some types of contexts [60–62]. The changing of the environment may be constant, but the system requires to be adaptive and dynamic [61] to each individual user and situation, that needs to automatic its behavior modifications simultaneously. Dynamic Bayesian networks (DBNs), in addition to current sensory readings, consider the past belief of the system, thus capturing the dynamicity of the phenomena under observation. This research used [62] to develop adaptive systems for getting automated inferences of the surrounding environment, the context-aware concept is adopted by the computing world in combination with data fusion. The research [63] provided a contribution tool that uses for data fusion based on features classification into twelve defect categories and using deep Convolutional Neural Networks (CNN) for surface defect detection in manufacturing processes. The researchers observed the effect on the result detection and compare between them as shown in Table 3.

The evolution of context fusion in this type reaches creating a new adaptive model [64] for raw sensor data that can make extracting the data and features directly based on deep neural network and reinforcement learning. The researchers' approach [65] uses the convolutional neural networks (CNN) for improving the classification, features recognition with reliable performance but that requires to weakly supervised

Table 3 The observed comparison between data mining in the present methods and past ways

	Traditional mining	Recent mining
Domain	Single domain	Usually multi-domain
Volume of data	Small	Big
Modality	One context model	Multi-models of contexts
Datasets	Often unified	Diversity data
Distribution	Not different	Different
Data representation	Not different	Different
Data interconnection	Not requires	Paramount
Data noisy	Ignore it	Deal with outliers (error or event)

object localization. The experiment relies on the benchmark data set of PASCAL VOC 2007 and 2012 that demonstrates a new context-aware approach the significantly enhances weakly supervised localization and detection. The observation shows the comparison between data mining [66] past and present.

Recent researches [66, 67] discuss the gap in the extracted meaning from various contexts in internet-of-things (IoT) environments. Each environment has a target, domain and many objects which require to track them and know all activities on it with respect to the essential dimension known as 'Time'. This dimension can support these environments in prediction the actions in the next time. The authors [67] introduced a new life cycle for the context in tracking objects and the design of contexts. That was considered a management framework for context-aware. But still, the tracking objects and objects management is a challenge that requires some motivation to improve accuracy and performance.

3.2 Fusion Type Two: Based on Data Reduction

The core of the difference between data integration [68, 69] and data fusion the data reduction [6, 70] process. The data integration targets to combine all data with others without neglect any data or avoid similar or complementary data. So, data integration requires making cleaning for integrated data and normalized. But the data fusion targets combine the data with some data or attributes reduction. Not any data can reduce or avoid it in fusion process. The fused data often interprets into three categories: data holds the same meaning, complementary data that includes some data in different attributes refer to the one objective and full idea of meaning, and the different data that can't compensate instead of it (as shown in Fig. 2). The target of the fusion process that only can control which the data can apply the reduction

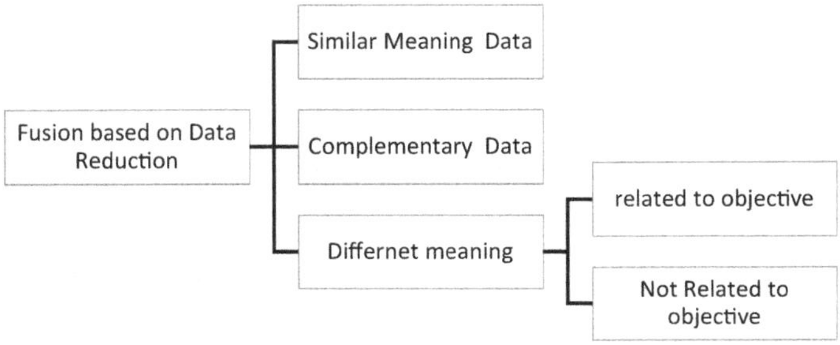

Fig. 2 The datatypes classification architecture

of it. So not all the different data is important but there are some data related to the objective and user's targets and some data not important that causes to headache on the fusion process.

3.3 Fusion Type Three: With Respect to Noisy or Outliers

Anomalies, or outliers [71], can be a serious issue when training machine learning algorithms or applying statistical techniques. Errors and events cause of fault measurements and conditional reading that causes of faulty decisions. Outlier detection [72] refers to any strange phenomena with the same sample data. There are four types of outliers [73]: numeric, DBScan, Z-score and Isolation forest.

3.4 Fusion Type Four: Based on Data Time Stream Classification

1. Type 1: Real-Time Stream

The recent motivations target the adaptive multi-model for context-aware but still have a big obstacle in the accuracy and performance measurements [74]. The essential challenge of how can make the data fusion process between variant topologies in real-time stream [75]. That requires huge data analysis and analytics simultaneously. This data analysis is very important in the decision making and saves lives in emergency cases in a smart environment such as smart city and smart health.

2. Type 2: Offline data or near to real-time

Traditional mining researches of fusion targets offline analysis or near the real-time analysis. This process requires powerful computers and servers and good cloud and data integrity to guarantee no manipulation in the data from sensors.

Previous research targeted [76] constructing a context framework based on the hierarchical aggregation that deals with a broad spectrum of contexts, from personal (e.g., the activities of individuals) to city-wide (e.g., locations of groups of people and vehicles) and world-wide (e.g., global weather and financial data) [77]. Defined a formal model capable of representing context and situations of interest and developed a technique that exploits multi-attribute utility theory for fusing the modeled information and thereby attain situation awareness [78]. Smart Healthcare Applications and Services: Developments and Practices, an extensive framework to mediate ambiguous contexts in pervasive environments is presented in.

An extensive framework [79] to mediate ambiguous contexts in pervasive environments is presented in. Several works adopt [80] DBNs to perform adaptive data fusion for different applications, such as target tracking and identification and user presence detection [81].

It is a simulation of multiple-model [82] target tracking, data up-link, and missile state observer that used for terrain following navigation, transfer alignment, and pre-launch studies involving air-launched munitions from various fighter aircraft. Kalman Filter. A static schema of exclusion [83] of most costly sensors are often emery hungry that focus on the reduction of their consumption.

The proposed solution is constructed based on the occasional look of smart environment that for avoid the expert for each expectation. This solution targets to check properties, types, and challenges for multiple contexts. The proposed context system aims to construct the framework of any context with respect to data types, data properties, and various challenges. It is better than previous taxonomies [14–17] in multiple context properties. The taxonomy in [14], requires discussing the abstraction level and architecture level. The taxonomy in [15], focuses on the complexity of context system. The taxonomy in [16], includes relationship between multiple resources and fusion levels. The taxonomy in [17], improves the usability of the fusion system. It differs in the properties construction system, better in neglecting the context properties. It prevents current the limitations of expert people for each domain, datatype formats, and context properties. The main measurements to select this taxonomy on how to fuse multiple sensory data with variant datatypes.

4 Discussion

Data fusion relies on fusing some multi-source with variant context types that holds some data mining processes that include the preprocessing data and identify patterns and requires visualizing these data (as shown in Fig. 3).

The basic architecture of data fusion [84] or sensor fusion taxonomy introduced three levels of fusion (as shown in Fig. 4).

The proposed data fusion taxonomy architecture shows in Fig. 1. That passes on all features of data fusion and mining processes as Fig. 3. The proposed architecture takes care of all affected features (inner and outer) on the data fusion process to reach the highest accuracy of fusion. The good fusion reaches good decision making, we

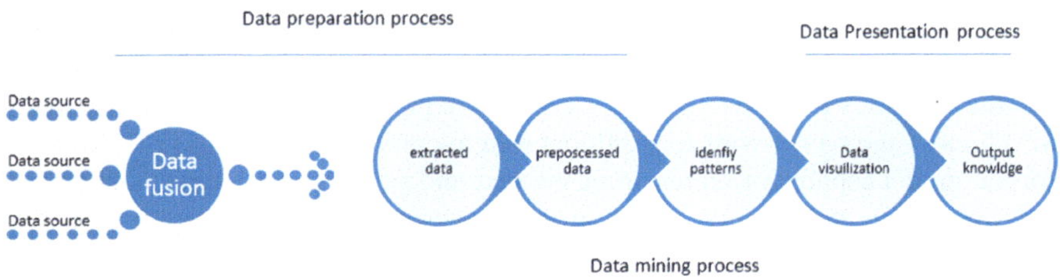

Fig. 3 Data fusion processes

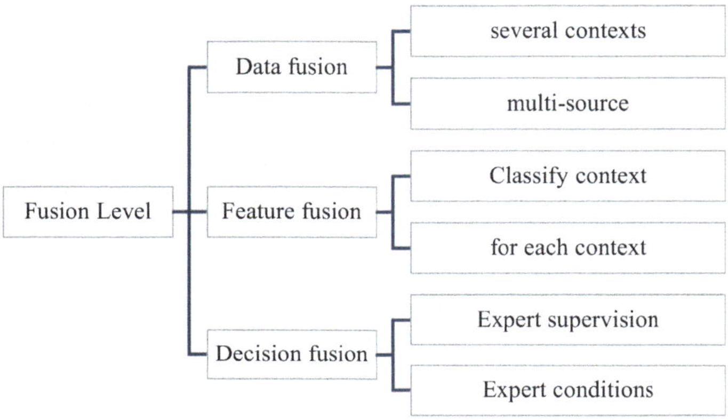

Fig. 4 The basic architecture of data fusion

can express it in a direct correlation as Eq. (1).

$$Good\ Fus. \propto Good\ Dec.M, \tag{1}$$

The basic architecture still faces problems in understanding the level of data reduction, the outliers happen, and time streaming observation that effects on the analytics and users' requirements.

Our observation recommends using deep learning algorithms for improving accuracy and performance meaning because there are a big number of effected features in the smart environment. The smart environments require to decide concurrently and tracking several objects.

Each context-aware type in fusion process has some features that should consider in the fusion process at any fusion level (as shown in Table 4). A Context is a set of objects or characteristics or features for identifying simple situations or complex events. A "context" has a necessary role at any level of Fusion. The fusion relies on two essential concepts, (1) Context is just collecting the characteristics of the data environment, (2) Context is a representation of physical environment features.

The main challenge of all fusion types is the methodology or model that targets to combine several topology structures into one type. But other properties have a big

Table 4 The observed features of data fusion for context-aware types

Fusion context type	Features
Text	Detect language, syntax, semantic, and grammar
Audio	Speed, length, and type of speech
Image	Dimension type, resolution of the image, and image type
Video	Number of frames, real-stream or offline

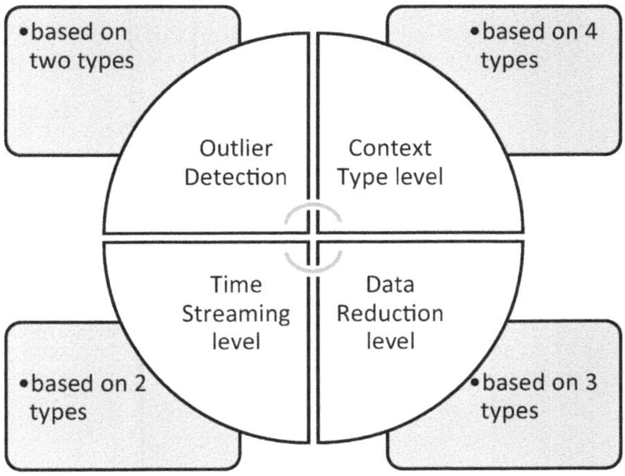

Fig. 5 The summary of the proposed taxonomy

effect in fusion in a smart environment that often in a real-time stream and have noisy. However, the level of reduction relies on the user's requirements and the number of users known as Social Internet-of-Things as Eqs. (2) and (3).

$$Reduc.Level \propto User\ Req, \tag{2}$$

$$No.Users \propto Reduc.level, \tag{3}$$

The fusion taxonomy makes a full picture of the fusion properties in smart environments. They have become an integral part of the importance of the details of the fusion process (as shown in Fig. 5).

5 Open Challenges and Research Areas

The summary of the previously mentioned challenges (as shown in Fig. 6),

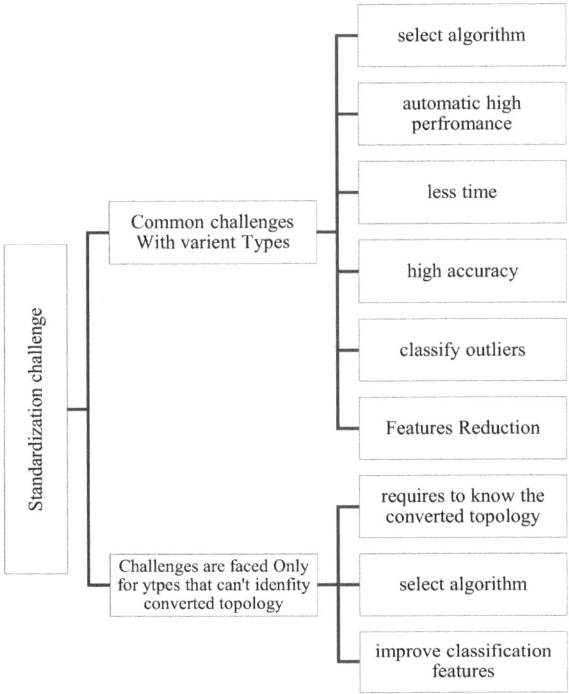

Fig. 6 The challenges in standardization data fusion based on context types

5.1 A Standardization Challenge

The main standardization challenge is multi-model context-awareness data fusion. It depends on artificial intelligence, user interfaces. The fusion of context-aware requires the expert human in the specific domain. The previous categories refer to the types that can know the converted topologies automatically and the second category that refers to the types can't know the converted topology automatically. There are two types of this challenge based on context: common challenges between the mentioned categories and different challenges between them as Fig. 6.

5.2 Expert Users

Each smart environment has one or more users that will be an expert to follow the fusion process and write features and thresholds of the proposed system. No way to know can automate the fusion with high accuracy and performance. Smart environment still faces a problem with user supervision and social Internet-of-things.

The user supervision refers to the expert user that can provide the system control. Social Internet-of-Things (SIoT) refers to the variant users' requirements that cause of making some fusion levels of data reduction.

5.3 Context Features Identification

Each context-aware type has several features but not all data have the features. There are some features missed and others must observe manually. The description of the data often is not enough to extract features of each context.

5.4 Choosing Topology

It is not an easy process to select the convenient topology that will convert into it such as (category 3, 4 in type 1 of the taxonomy). The recent researches have some types of conversions with a variant in the accuracy results. The comparison between results and algorithms that is not easy to choose the suitable topology for each fusion environment.

5.5 Data Integrity

The data integrity becomes a hot area of research that should guarantee data on the network. Measuring the quality of big data is significant to secure the data about sensors and objects. Data integrity is imposed within a system at its design stage using standard rules and procedures and is maintained using error checking and validation routines.

6 Conclusion and Future Work

Although data fusion is not a new field, the combination in the smart environment generates new requirements and challenges in data fusion. Smart environment relies on an interconnected set of sensors and objects that connected via internet that relies on the Internet-of-things domain. These sensors hold big data with variant contexts types. This paper demonstrates an observation classification taxonomy for data fusion that relies on types of contexts and some recent properties have a big effect on the smart environments. These properties area (data reduction, time streaming, and noisy data). This paper presents a new taxonomy for a context-awareness problem that is classified for multiple formats of data fusion. It introduces a new classification taxonomy based on classified types, properties, and challenges. The future work goes forwards to the new multi-model of context-awareness in any type to make a standardized model for reaching the convenient meaning automatically and identifying the features of data. This model targets automating classification, new features recognition and outlier detection. The purpose of the proposed application is to investigate proprieties from neuroscience providing insight into neural processing mechanisms involved in multisensory fusion for contextual awareness known to be present in humans.

References

1. Thapliyal, R., Patel, R.K., Yadav, A.K., Singh, A.: Internet of Things for smart environment and integrated ecosystem. In: International Conference on Advanced Research in Engineering Science and Management At: Dehradun, Uttarakhand (2018)
2. Bhayani, M., Patel, M., Bhatt, C.: Internet of Things (IoT): in a way. In: Proceedings of the International Congress on Information and Communication Technology, Advances in Intelligent Systems and Computing (2016)
3. Bongartz, S., Jin, Y., Paternò, F., Rett, J., Santoro, C., Spano, L.D.: Adaptive User Interfaces for Smart Environments with the Support of Model-Based Languages. Springer, Berlin (2012)
4. Ayed, S.B., Trichili, H., Alimi, A.M.: Data fusion architectures: a survey and comparison. In: 15th International Conference on Intelligent Systems Design and Applications (ISDA) (2015)
5. Chao, W., Jishuang, Q., Zhi, L.: Data fusion, the core technology for future on-board data processing system. Pecora 15/Land Satellite Information IV/ISPRS Commission I/FIEOS 2002 Conference Proceedings (2002)
6. Kalyan, L.O.: Veeramachaneni, Fusion, Decision-Level, Hindawi Publishing Corporation The Scientific World Journal Volume 2013, Article ID 704504, 19 pages
7. Lahat, D., Adal, T., Jutten, C.: Multimodal data fusion: an overview of methods, challenges and prospects. In: Proceedings OF THE IEEE (2015)
8. Jaimes, A., Sebe, N.: Multimodal human computer interaction: a survey. Comput. Vis. Image Underst. **108**(1), 116–134 (2007)
9. Kashevnika, A.M., Ponomareva, A.V., Smirnov, A.V.: A multi-model context-aware tourism recommendation service: approach and architecture. J. Comput. Syst. Sci. Int. **56**(2), 245–258 (2017). (ISSN 1064-2307)
10. Lahat, D., Adali, T., Jutten, C.: Multimodal data fusion: an overview of methods, challenges, and prospects. Proc. IEEE **103**(9) (2015)
11. Hall, D.L., Llinas, J.: An introduction to multi-sensor data fusion. Proc. IEEE **85**(1) (1997)
12. Hofmann, M.A.: Challenges of model interoperation in military simulations. Simulation **80**(12), 659–667 (2004)
13. El-Sappagh, S., Ali, F., Elmasri, S., Kim, K., Ali, A., Kwa, K.-S.: Mobile Health Technologies for Diabetes Mellitus: Current State and Future Challenges, pp. 2169–3536 (2018)
14. Žontar, R., Heričko, M., Rozman, I.: Taxonomy of context-aware systems. Elektrotehniški Vestnik **79**(1–2), 41–46 (2012). (English Edition)
15. Emmanouilidis, C., Koutsiamanis, R.-A., Tasidou, A.: Mobile guides: taxonomy of architectures, context awareness, technologies and applications. J. Netw. Comput. Appl. **36**(1), 103–125 (2013)
16. Almasri, M., Elleithy, K.: Data fusion in WSNs: architecture, taxonomy, evaluation of techniques, and challenges. Int. J. Sci. Eng. Res. **6**(4) (2015)
17. Biancolillo, A., Boqué, R., Cocchi, M., Marini, F.: Data fusion strategies in food analysis (Chap. 10). In: Data Fusion Methodology and Applications, vol. 31, pp. 271–310 (2019)

18. Ferrin, G., Snidaro, L., Foresti, G.L.: Contexts, co-texts and situations in fusion domain. In: 14th International Conference on Information Fusion Chicago, Illinois, USA (2011)
19. den Berg, N., Schumann, M., Kraft, K., Hoffmann, W.: Telemedicine and telecare for older patients—a systematic review. Maturitas **73**(2) (2012)
20. Kańtoch, E.: Recognition of sedentary behavior by machine learning analysis of wearable sensors during activities of daily living for telemedical assessment of cardiovascular risk. Sensors (2018)
21. Kang, S.-K., Chung, K., Lee, J.-H.: Real-time tracking and recognition systems for interactive telemedicine health services. Wireless Pers. Commun. **79**(4), 2611–2626 (2014)
22. Gite, S., Agrawal, H.: On context awareness for multisensor data fusion in IoT. In: Proceedings of the Second International Conference on Computer and Communication Technologies, pp. 85–93 (2015)
23. Deshmukh, M., Bhosale, U.: Image fusion and image quality assessment of fused images. Int. J. Image Process. (IJIP) **4**(5) (2010)
24. Moravec, J., Šára, R.: Robust maximum-likelihood on-line LiDAR-to-camera calibration monitoring and refinement. In: Kukelová, Z., Skovierovă, J.: (eds.) 23rd Computer Vision Winter Workshop, Český Krumlov, Czech Republic (2018)
25. De Silva, V., Roche, J., Kondoz, A.: Robust fusion of LiDAR and wide-angle camera data for autonomous mobile robots. Sensors (2018)
26. Ghassemian, H.: A review of remote sensing image fusion methods. Inf. Fusion **32**(part A) (2016)
27. Palsson, F., Sveinsson, J.R., Ulfarsson, M.O., Benediktsson, J.A.: Model-based fusion of multi- and hyperspectral images using PCA and wavelets. IEEE Trans. Geosci. Remote Sens. **53**(5) (2015)
28. Kim, Y.M., Theobalt, C., Diebel, J., Kosecka, J., Miscusik, B.: Sebastian, multi-view image and ToF sensor fusion for dense 3D reconstruction. In: IEEE 12th International Conference on Computer Vision Workshops, ICCV (2009)
29. Choia, J., Radau, P., Xubc, R., Wright, G.A.: X-ray and magnetic resonance imaging fusion for cardiac resynchronization therapy. Med. Image Anal. **31** (2016)
30. Krout, D.W., Okopal, G., Hanusa, E.: Video data and sonar data: real world data fusion example. In: 14th International Conference on Information Fusion (2011)
31. Snidaro, L., Foresti, G.L., Niu, R., Varshney, P.K.: Sensor fusion for video surveillance. Electr. Eng. Comput. Sci. **108** (2004)
32. Heracleous, P., Badin, P., Bailly, G., Hagita, N.: Exploiting multimodal data fusion in robust speech recognition. In: IEEE International Conference on Multimedia and Expo (2010)
33. Boujelbene, S.Z., Mezghani, D.B.A., Ellouze, N.: General machine learning classifiers and data fusion schemes for efficient speaker recognition. Int. J. Comput. Sci. Emer. Technol. **2**(2) (2011)
34. Gu, Y., Li, X., Chen, S., Zhang, J., Marsic, I.: Speech intention classification with multimodal deep learning. Adv. Artif. Intell. (2017)
35. Zahavy, T., Mannor, S., Magnani, A., Krishnan, A.: Is a picture worth a thousand words? A deep multi-modal fusion architecture for product classification in E-commerce. Under Review as a Conference Paper at ICLR 2017
36. Gallo, I., Calefati, A., Nawaz, S., Janjua, M.K.: Image and encoded text fusion for multi-modal classification. Published in the Digital Image Computing: Techniques and Applications (DICTA), Australia (2018)
37. Viswanathan, P., Venkata Krishna, P.: Text fusion watermarking in medical image with semi-reversible for secure transfer and authentication
38. Huang, F., Zhang, X., Zhao, Z., Xu, J., Li, Z.: Image-text sentiment analysis via deep multimodal attentive fusion. Knowl.-Based Syst. (2019)
39. Blasch, E., Nagy, J., Aved, A., Pottenger, W.M., et al.: Context aided video-to-text information fusion. In: 17th International Conference on Information Fusion (FUSION) (2014)
40. Video-to-Text Information Fusion Evaluation for Level 5 User Refinement,18th International Conference on Information Fusion Washington, DC, 6–9 July 2015

41. Jain, S., Gonzalez, J.E.: Inter-BMV: Interpolation with Block Motion Vectors for Fast Semantic Segmentation on Video, arXiv:1810.04047v1
42. Gidel, S., Blanc, C., Chateau, T., Checchin, P., Trassoudaine, L.: Non-parametric laser and video data fusion: application to pedestrian detection in urban environment. In: 12th International Conference on Information Fusion Seattle, WA, USA, 6–9 July 2009
43. Katsaggelos, A.K., Bahaadini, S., Molina, R.: Audiovisual fusion: challenges and new approaches. Proc. IEEE **103**(9) (2015)
44. Datcu, D., Rothkrantz, L.J.M.: Semantic audio-visual data fusion for automatic emotion recognition, recognition. Emot. Recognit. 411–435 (2015)
45. O'Conaire, C., O'Connor, N.E., Smeaton, A.: Thermo-visual feature fusion for object tracking using multiple spatiogram trackers. Mach. Vis. Appl. **19**(5–6), 483–494 (2008)
46. Kumar, P., Gauba, H., Roy, P.P., Dogra, D.P.: Coupled HMM-based multi-sensor data fusion for sign language recognition. Pattern Recogn. Lett. **86** (2017)
47. Chen, C., Liang, J., Zhao, H., Tian, J.: Factorial HMM and parallel HMM for gait recognition. IEEE Trans. Syst. Man Cybern. Part C (Appl. Rev.) **39**(1), 114–123 (2009)
48. Cetin, O., Ostendorf, M. and Bernard, G.D.: Multi-rate coupled hidden markov models and their application to machining tool-wear classification. IEEE Trans. Signal Process. **55**(6) (2007)
49. Eyigoz, E., Gildea, D., Oflazer, K.: Multi-rate HMMs for word alignment. In: Proceedings of the Eighth Workshop on Statistical Machine Translation, Bulgaria, pp. 494–502 (2013)
50. Zajdel, W., Krijnders, J.D., Andringa, T., Gavrila, D.M.: CASSANDRA: audio-video sensor fusion for aggression detection. In: IEEE International Conference Advanced Video and Signal Based Surveillance (AVSS), London, UK (2007)
51. Kampman, O., Barezi, E.J., Bertero, D., Fung, P.: Investigating audio, video, and text fusion methods for end-to-end automatic personality prediction. In: Proceedings of the 56th Annual Meeting of the Association for Computational Linguistics (Short Papers), pp. 606–611 (2018)
52. Ji, C.B., Duan, G., Ma, H.Y., Zhang, L., Xu, H.Y.: Modeling of image, video and text fusion quality data packet system for aerospace complex products based on business intelligence (2019)
53. Xiong, Y., Wang, D., Zhang, Y., Feng, S., Wang, G.: Multimodal data fusion in text-image heterogeneous graph for social media recommendation. In: International Conference on Web-Age Information Management WAIM, Web-Age Information Management (2014)
54. Naphade, M., Kristjansson, T., Frey, B., Huang, T.S.: Probabilistic multimedia objects (multi-jects): a novel approach to 9 video indexing and retrieval in multimedia systems. In: Proceedings of IEEE International Conference on Image Processing, vol. 3, pp. 536–540, Chicago, USA (1998)
55. Ellis, D.: Prediction-driven computational auditory scene analysis. Ph.D. thesis, MIT Department of Electrical Engineering and Computer Science, Cambridge, Mass, USA (1996)
56. Adams, W.H., Iyengar, G., Lin, C.-Y., Naphade, M.R., Neti, C., Nock, H.J., Smith, J.R.: Semantic indexing of multimedia content using visual, audio, and text cues. EURASIP J. Appl. Signal Process. (2003)
57. Wu, Z., Cai, L., Meng, H.: Multi-level fusion of audio and visual features for speaker identification. In: International Conference on Biometrics ICB 2006: Advances in Biometrics (2006)
58. Yurur, O., Labrador, M., Moreno, W.: Adaptive and energy efficient context representation framework in mobile sensing. IEEE Trans. Mob. Comput. **13**(8) (2014)
59. De Paola, A., Gaglio, S., Re, G.L., Ortolani, M.: Multi-sensor fusion through adaptive Bayesian networks. Congress of the Italian Association for Artificial Intelligence AI*IA 2011: AI*IA 2011: Artificial Intelligence Around Man and Beyond (2011)
60. Hossain, M.A., Atrey, P.K., El Saddik, A.: Learning multi-sensor confidence using a reward-and-punishment mechanism, integrate machine-learning algorithms in the data fusion process. IEEE Trans. Instrum. Meas. **58**(5), 1525–1534 (2009)
61. Gite, S., Agrawal, H.: On context awareness for multi-sensor data fusion in IoT. In: Proceedings of the Second International Conference on Computer and Communication Technologies (2016)

62. Malandrakis, N., Iosif, E., Prokopi, V., Potamianos, A., Narayanan, S.: DeepPurple: lexical, string and affective feature fusion for sentence-level semantic similarity estimation. In: Second Joint Conference on Lexical and Computational Semantics (*SEM), Volume 1: Proceedings of the Main Conference, and the Shared Task. ACM (2013)
63. Barzilay, R., McKeown, K.R.: Sentence fusion for multidocument news summarization. Comput. Linguist. **31**(3) (2005)
64. Durkan, C., Storkey, A., Edwards, H.: The context-aware learner. In: ICLR 2018
65. Weimer Ariandy, D., Benggolo, Y., Freitag, M.: Context-aware deep convolutional neural networks for industrial inspection. In: Australasian Conference on Artificial Intelligence, Canberra, Australia, Volume: Deep Learning and its Applications in Vision and Robotics (Workshop) (2015)
66. Brenon, A., Portet, F., Vacher, M.: Context feature learning through deep learning for adaptive context-aware decision making in the home. In: The 14th International Conference on Intelligent Environments, Rome, Italy (2018)
67. Kantorov, V., Oquab, M., Cho, M., Laptev, I.: ContextLocNet: context-aware deep network models for weakly supervised localization. ECCV 2016, Oct 2016, Amsterdam, Netherlands. Springer, pp. 350–365 (2016)
68. Savopol, F., Armenakis, C.: Merging of heterogeneous data for emergency mapping: data integration and data fusion? In: Symposium of Geospatial Theory, Processing and Applications (2002)
69. Dong, X.L., Naumann, F.: Data fusion: resolving data conflicts for integration. J. Proc. VLDB **2**(2) (2009)
70. Zhu, Y., Song, E., Zhou, J., You, Z.: Optimal dimensionality reduction of sensor data in multisensor estimation fusion. IEEE Trans. Signal Process. **53**(5) (2005)
71. Nesa, N., Ghosh, T., Banerjee, I.: Outlier detection in sensed data using statistical learning models for IoT. In: 2018 IEEE Wireless Communications and Networking Conference (WCNC) (2018)
72. Chandola, V., Banerjee, A., Kumar, V.: Outlier detection: a survey. ACM Comput. Surv. **41**(3), Article 15 (2009)
73. Aggarwal, C.C.: Outlier Analysis, 2nd edn. Springer, Berlin (2016)
74. Tonjes, R., Ali, M.I., Barnaghi, P., Ganea, S., et al.: Real Time IoT Stream Processing and Large-scale Data Analytics for Smart City Applications (2014)
75. Bonino, D., Rizzo, F., Pastrone, C., Soto, J.A.C., Ahlsen, M., Axling, M.: Block-based realtime big-data processing for smart cities. According to Eurostat, IEEE 2016
76. Cho, K., Hwang, I., Kang, S., Kim, B., Lee, J., Lee, S., Park, S., Song, J., Rhee, Y.: HiCon: a hierarchical context monitoring and composition framework for next-generation context-aware services. IEEE Netw. **22**(4) (2008)
77. Padovitz, A., Loke, S.W., Zaslavsky, A., Burg, B., Bartolini, C.: An approach to data fusion for context awareness. In: International and Interdisciplinary Conference on Modeling and Using Context, Modeling and Using Context (2005)
78. Roy, N., Das, S.K., Julien, C.: Resolving and mediating ambiguous contexts in pervasive environments. In: User-Driven Healthcare: Concepts, Methodologies, Tools, and Applications, IGI Global disseminator of knowledge (2013)
79. Roy, N., Das, S.K., Julien, C..: Resource-optimized quality-assured ambiguous context mediation framework in pervasive environment. IEEE Trans. Mob. Comput. **11**(2) (2012)
80. De Paola, A., La Cascia, M., Lo Re, G., Ortolani, M.: User detection through multi-sensor fusion in an AmI scenario. In: 2012 15th International Conference on Information Fusion (FUSION) (2012)
81. Roy, N., Pallapa, G.V., Das, S.K.: A middleware framework for ambiguous context mediation in smart healthcare application, user activity recognition. In: Third IEEE International Conference on Wireless and Mobile Computing, Networking and Communications, WiMob 2007, White Plains, New York, USA, 8–10 Oct 2007
82. Nwe, M.S., Tun, H.M.: Implementation of multi-sensor data fusion algorithm. Int. J. Sens. Sens. Netw. (2017)

83. Rahmati, A., Zhong, L.: Context-based network estimation for energy-efficient ubiquitous. IEEE Trans. Mob. Comput. **10**(1) (2011)
84. Klein, L., Mihaylova, L., El Faouzi, N.E: Sensor and data fusion: taxonomy challenges and applications. In: Pal, S.K., Petrosino, A., Maddalena, L. (eds.) Handbook on Soft Computing for Video Surveillance. Taylor & Francis. Sensor and Data Fusion: Taxonomy Challenges and applications. Chapman & Hall/CRC (2013)

Systematic Review on Fully Homomorphic Encryption Scheme and Its Application

Hana Yousuf, Michael Lahzi, Said A. Salloum, and Khaled Shaalan

Abstract Security and efficiency of transferred data is the main concern in big data and cloud computing service. Fully Homomorphic Encryption is one of the most recent solutions that ensure the security and confidently. This paper expected to lead precise writing on the subject of Fully Homomorphic Encryption and its Application. The primary point of the audit is to pick up knowledge of the Fully Homomorphic Encryption and to locate the best way to deal with executing it. First, Fully Homomorphic Encryption Algorithms were discovered since 2009. After this year, different approaches were followed in order to enhance the scheme such as Fully Homomorphic Encryption based on Integer (DGHV) Scheme, Fully Homomorphic Encryption based on BGV schemes, Multi-key Fully Homomorphic Encryption Scheme and Fully Homomorphic Encryption based on GSW. This paper uses a systematic review of Fully Homomorphic Encryption technologies. The technique is performed to lead this methodical writing included utilizing the examination inquiries or research questions to determine keywords which were then used to search for bits of peer-reviewed papers, articles or book at scholastic directories. Through introductory hunts, 250 papers and scholarly works were found and with the assistance of choice criteria and PRISMA methodology, the number of papers reviewed decreased to 24. Jadad scale used to measure the quality of selected publications. Most of the papers developed new schemes on Fully Homomorphic Encryption to enhance efficiency and reduce noise. Other papers developed a new algorithm to improve the security of the transferred data. Last Thematic modeling analysis presented in this paper to ensure that all the articles were related to Fully Homomorphic Encryption and cloud computing.

Keywords Fully homomorphic encryption · Lattices · Integer · LWE · Cloud computing

H. Yousuf · M. Lahzi · S. A. Salloum (✉) · K. Shaalan
Faculty of Engineering and IT, The British University in Dubai, Dubai, UAE
e-mail: ssalloum@sharjah.ac.ae

S. A. Salloum
Research Institute of Sciences and Engineering, University of Sharjah, Sharjah, UAE

M. Al-Emran et al. (eds.), *Recent Advances in Intelligent Systems and Smart Applications*, Studies in Systems, Decision and Control 295,
https://doi.org/10.1007/978-3-030-47411-9_29

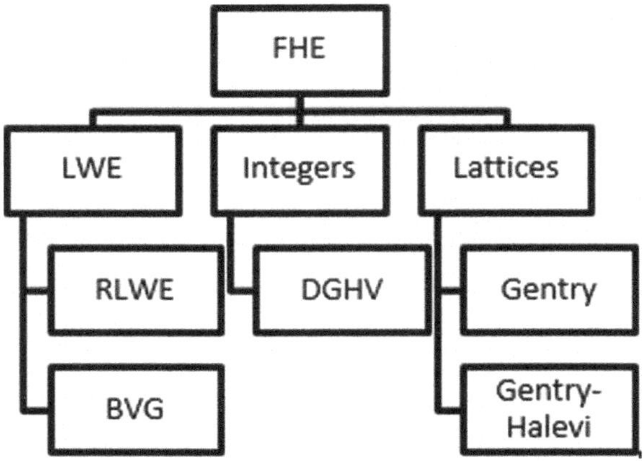

Fig. 1 The three types of fully homomorphic encryption

1 Introduction

Governments and organizations cannot trust any translation service provider unless the service is fully owned by them [1]. Moreover, many technologies popped up, such as cloud computing and big data, make the dataset difficult to capture, manage and analyze using typical data analysis tools [2–4]. Consequently, these key issues led to guide experts and developers to work toward solutions, which is compliant with this trend of technologies [5, 6]. Homomorphic Encryption is a modern cryptographic technique proposed by Gentry [7] based on lattices. It was the first encryption to add and multiply ciphertexts without decrypting. Fully Homomorphic Encryption is divided into three types. The first type is based on the lattice which constructs a Somewhat Homomorphic Encryption (SWHE) on the ideal of various rings, then compresses the decryption circuit to reduce polynomials, and finally complete the Fully Homomorphic encryption under the assumption of cyclic security through bootstrapping technology [8]. The second type is Fully Homomorphic Encryption based on Integer. The third is Fully Homomorphic Encryption based on Learning With Error (LWE) or Learning With Errors over Ring (RLWE). This Encryption uses a non-linearization encryption scheme such as BGV. Figure 1 illustrates the three types of Fully Homomorphic Encryption.

2 Literature Review

Cloud computing considered a Quantum leap in the field of Information Technology. Cloud computing brings great challenges in the field of data security and privacy protection [9].

Zhao and Geng [8] explained that the Fully Homomorphic Encryption algorithm refers to an encryption algorithm that has the characteristics of additive homomorphism and multiplicative homomorphism and can perform any multiple addition and multiplication operations.

The first time the data security problem addressed with a multiplication homomorphism solution was in 1977 by the creators of the RSA system it is called Pre-Fully Homomorphic Encryption. Gentry [7] proved the principle of encryption (Homomorphic Encryption) with some limitations in decryption. To solve the decryption limitations Gentry founded the bootstrapping method (Fully Homomorphic Encryption) which led to the increment of ciphertexts weight. In 2010, Coron and Mandal made the public key of the Fully Homomorphic Encryption system shorter, they reduce the public key size of the Homomorphic scheme from O (λ^{10}) down to O (λ^{7}). Brakerski and Vaikuntanathan [10] develop Fully Homomorphic Encryption system without bootstrapping, with this optimization, the per-gate computation of FHE schemes that follow the blueprint from O(λ^{6}) down to O(λ^{3}), but it is still inefficient [9].

Several studies addressed the Fully Homomorphic Encryption different application. For example, Wang et al. [11] concerns were to provide data security and privacy protection in big data. So they chose Fully Homomorphic Encryption to fulfill their requirements. In their article, they proposed to use Somewhat Homomorphic Encryption with shorter public keys based on DGHV scheme by encrypting pseudo-random number generator using a cubic form instead of linear form. Gentry [7] proposal is to design a system model for Fully Homomorphic Encryption scheme in big data to allow the user to encrypt private data and connect it with the real value of the data at the same time.

3 Fully Homomorphic Encryption Algorithms

3.1 The Early Lattice-Based Cryptosystems

Gentry constructed an encryption scheme using ideal lattices. Lattice-based cryptosystems typically have decryption algorithms with low circuit complexity, often dominated by an inner product computation that is in NC1 (Nick's Class, it is class of language which all Boolean functions use circuits of linear depth). Also, ideal lattices provide both additive and multiplicative homomorphisms [7]. The encryption scheme can be broken down into three steps: (1) construct the encryption that handles low order polynomial while maintaining its homomorphisms, (2) Apply a technique to squash the decryption circuit that reduces the complexity of decryption scheme, and (3). Bootstrap the result by a process to cipher the text for the decryption circuits and their extended circuit in order to obtain the Fully Homomorphic Encryption scheme. The security of the scheme is based on the computational difficulty of bounded distance coding on ideal lattices and the computational difficulty of sparse subsets [8].

3.2 Fully Homomorphic Encryption Based on Integer (DGHV) Scheme

In [12], the author proposed a new algorithm for the Fully Homomorphic Encryption scheme based on Gentry's 2009 scheme. The scheme uses an integrated operation. It constructs a Fully Homomorphic scheme from boots trappable by using addition and multiplication over the integers (Symmetric/asymmetric encryption scheme) instead of using ideal lattices over a polynomial ring. The main disadvantage of this scheme is that the public key is long and cannot be implemented on any system. This algorithm is used particularly in optimization schemes.

3.3 Fully Homomorphic Encryption Based on BGV Schemes

The research by Alhashmi et al. [6] and Zhao and Geng [9] proposed two Fully Homomorphic Encryption schemes based on the hardness of Learning with Error (LWE). LWE encrypts large integer into a single ciphertext. Both encryption schemes differ in the switching phase. BGV considered a first scheme to be practical in real-life applications. The disadvantage of this scheme that the linearization and switching techniques need to be shared between the data owner and the service provider.

3.4 Multi-key Fully Homomorphic Encryption Scheme

López-Alt et al. [13] Proposed a multi-key encryption scheme based on NTRU (Nth degree Truncated polynomial Ring Units) cryptosystem. This scheme is used to perform multiple Homomorphic calculations with multiple public keys by using re-lineralization and modulo reduction techniques. NTRU is a cryptosystem based on the polynomial ring. The safety of this scheme relies on the shortest vector question (SVP). Relative to the open secret system such as discrete logarithm or decomposition of large numbers, it has many advantages [14].

3.5 *Fully Homomorphic Encryption Based on GSW*

Gentry et al. [15] improved BGV scheme by building a Fully Homomorphic Encryption scheme using an approximate eigenvalue method. This scheme provides no evaluation key which was necessary for pervious schemes to obtain a Homomorphic evaluator. All the operations are done by Homomorphic evaluator with no need to know the user's public key which reduces the length of the public key and improves the space efficiency. The disadvantage is this scheme is not as efficient in computing as the schemes based on RLWE.

4 Advances and Application of Fully Homomorphic Encryption

Most of the reviewed studies and papers that discussed Fully Homomorphic Encryption were within cloud security, big data, and health care applications. Zhao et al. [16] established a privacy-preserving service for cloud systems that can be used in election services. They used Fully Homomorphic Encryption for QoS (Quality-of-Service) Encryption to ensure security and enhance the performance; they proposed the MapReduce model for parallel execution. Martins and Sousa [17] Proposed an automatic and methodical technique to approximate a wide range of functions Homomorphically. They proposed two ways in representation: stochastic and fixed-point approaches.

A unique establishment of levelled Fully Homomorphic signature scheme is proposed by Luo et al. [18]. The scheme needs to call the signing algorithm twice for generating signatures for all messages in a dataset into two matrices. Luo et al. [18] proved that their construction is secure against fully adaptive chosen-message attacks under the standard small integer solution (SIS) assumption.

A new scheme for Fully Homomorphic Encryption was proposed by Kim et al. [19]. They proposed a new encryption scheme for Fully Homomorphic Encryption over integer which enables scale-invariant multiplication and all evaluation over encrypted messages can be performed with partial information.

For example, it can be public that the message space M is a set of positive Integers less than some integer g, but the exact value of g will be provably hidden from all public information under a reasonable assumption. This property called informally message space hid-ability. Message space hid-ability has an immediate and interesting application, namely, Homomorphic commitment to large integers [19].

Patil et al. [20] proposed system that avoids the loss of confidential data. The system's data are kept to be encrypted before transmitting over the network. The arithmetic computation is carried out as server-side without the need for decryption [20]. The state-of-art is driven by Xu et al. [21] and used to construct Fully Homomorphic Encryption using based Merkle Tree (FHMT) which is a novel technique for streaming authenticated data structures (SADS) to achieve the streaming verifiable computation. Xu et al. [21] presented in their paper dynamic FHMT (DFHMT) and divided the achieved algorithm into five phases: initialization, insertion, tree expansion, query, and verification. Xu et al. [21] chose dynamic FHMT instead of static to remove the drawbacks of the static FHMT. The result from this encryption showed that DFHMT balanced the performance between the client and server which considered an advantage for lightweight devices.

4.1 Cloud Computing

Cloud computing provides a new way to shift computing and storage capabilities to external service providers with on-demand provisioning of software and hardware at any level of granularity. In recent years, the privacy of cloud data and the computations thereon has emerged as an important research area in the domain of cloud computing and applied cryptographic research [22].

The other application of Fully Homomorphic Encryption is the field of cloud computing. Gai et al. [23] Stated that the Fully Homomorphic Encryption in cloud computing is an effective scheme to protect the data in the cloud system. However, they highlighted that Fully Homomorphic Encryption is not yet utilized to meet the practical demand because of inaccuracy rate or insufferable latency time. To solve the inaccuracy rate and inefficiency time, Gai et al. [23] developed a new scheme designed for operating real numbers named Fully Homomorphic Encryption over Real Numbers (FHE-RN). They proved by experimental evaluations that their scheme has an outstanding performance in both accuracy and efficiency. Oppermann et al. [24] presented in their paper a developed communication protocol to ensure authenticity, integrity, and privacy of measurement data across a distributed measuring system within the cloud computing architecture [24].

Umadevi and Gopalan [25] developed a new way to keep the data private and secure with limitation of access based on the data owner preferences called Fully Homomorphic Encryption using Q_p^n Matrices with Enhanced Access Control (SFHE-ACC). This scheme transformed the data into a square matrix, the encryption and decryption are performed based on the symmetric key (Smith Normal Form). The developed scheme enhances the security, confidentiality and access control.

5 Research Methodology

The principle of research methodology is developed based on a systematic review methodology. "A Systematic reviews have foremost been developed within medical science as a way to synthesize research findings in a systematic, transparent, and reproducible way and has been referred to as the gold standard among reviews" [26].

"A systematic review aims to identify all empirical evidence that fits the pre-specified inclusion criteria to answer a particular research question or hypothesis. By using explicit and systematic methods when reviewing articles and all available evidence, bias can be minimized, thus providing reliable findings from which conclusions can be drawn and decisions made" [27].

The reason for selecting the systematic review for Fully Homomorphic Encryption scheme that no systematic review focuses on Fully Homomorphic Encryption scheme, algorithm, and applications. Moreover, this methodology able us to collect, evaluate, analyze and summarize the most current knowledge level, algorithm, and applications of Fully Homomorphic Encryption. This study follows the systematic review procedures provided by Kitchenham and Charters [28] and other systematic reviews [29–32]. The systematic review methodology accomplished by following the five steps listed below:

Step 1: Problem statement formation: The problem statement in our research was that there was no systematic review that clearly addresses the development and progress of the Fully Homomorphic Encryption scheme.

Step 2: Identification of relevant work and search for studies via multiple subsequent databases. The most relevant research articles for our subject were collected using keywords that contained Fully Homomorphic Encryption combined with algorithms, schemes, technology, and application. The search for these studies was on October 2019.

Step 3: Assessment of collected the research articles. The critical appraisal and quality assessment are required and revised in every step of systematic review: the research articles that are collected multiple subsequent databases revised and the quality of each paper assed based on our research criteria.

Step 4: Data synthesis and usage of statistical methods to explore the differences between the research articles and combine their effects (meta-analysis).

Step 5: The results (Interpreting the findings). All the findings that satisfy the research objective presented in the results section.

5.1 PRISMA Framework

In 1999 a group of international people developed a guideline called QUOROM which stands for (Quality of Reporting of Meta-analyses). QUOROM focused on the reporting of meta-analyses of randomized controlled trials. By 2009, the QUOROM framework was updated to include conceptual and practical advances in the science of systematic reviews and renamed as PRISMA (Preferred Reporting Items of Systematic reviews and Meta-Analyses) [27]. PRISMA defined as an evidence-based framework utilized for systematic reviews and meta-analyses. PRISMA framework provides researchers the ability to conduct a critical appraisal for the published articles. In our research the framework of PRISMA was applied as follow:

- The PRISMA checklist was downloaded from PRISMA official site and filled up based on Fully Homomorphic Encryption articles, see Fig. 2.
- The search strategies for Fully Homomorphic Encryption articles were performed by searching in the subsequent database using keywords added up with Boolean operators (AND) and (OR). The search results retrieved 250 articles using the research keywords.
- Eliminating the duplicated articles: After retrieving 250 articles, it was evaluated and filtered out in terms of duplication in algorithms, schemes, technology, and application. Twenty-five articles were founded as duplicated items and filtered out.
- Screening the titles and abstracts of the remaining 24 articles using the inclusion and exclusion criteria in order to identify the most relevant articles for our systematic review.

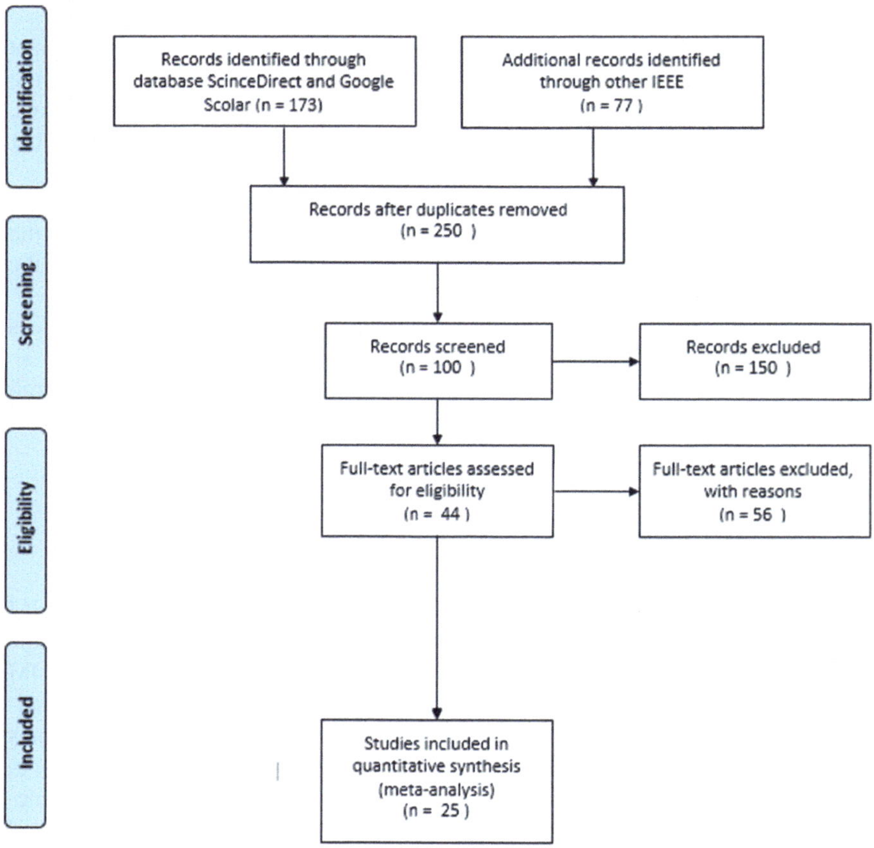

Fig. 2 Article review framework used in the study

Table 1 Inclusion and exclusion criteria

Inclusion criteria	Exclusion criteria
Articles published in peer-reviewed journals and subsequent databases such as IEEE, ScienceDirect, Springer, and Google Scholar	Articles are written in another language
Articles published in 2015 and afterword	Includes Homomorphic encryption scheme in general
Articles should be written in English	The articles keywords include Fully Homomorphic Encryption without considering it in their research paper
Articles contain Fully Homomorphic Encryption AND (algorithm) AND (Schemes) OR (Cloud Computing) OR (Technology) OR (Application)	

5.2 Inclusion and Exclusion Criterion

The Inclusion and exclusion criterion for our systematic review in Fully Homomorphic Encryption is simple and straightforward, as described in Table 1.

5.3 Data Search and Extraction

As mentioned previously the articles search was conducted via subsequent databases, peer-reviewed journals and periodicals using the keywords and Boolean operators "AND" and "OR". Accordingly, an initial review was observed, which are related to the Fully Homomorphic Encryption and are associated with new algorithms, cloud computing, and applications of the scheme. For the mentioned reason, the keywords for the search were constructed around (Fully Homomorphic Encryption) AND (algorithm) AND (Schemes) OR (Cloud Computing) OR (Technology) OR (Application).

The research articles were narrowed to be for articles published in 2015 and afterward. Most of the papers retrieved from the Institute of Electrical and Electronics Engineering (IEEE) journals, ScienceDirect and Springer. Figure 2 illustrates steps that have been used for filtering the research articles using the PRISMA framework. Firstly, we performed the search on the subsequent database and the number of articles includes the keywords recorded. Secondly, removed the duplicated articles. Thirdly, the remaining articles were screened based on the PRISMA checklist by excluding the articles that miss any item. Finally, the remaining articles were fully evaluated and sorted based on its contribution to our systematic review of Fully Homomorphic Encryption.

6 Discussion

Out of 250 retrieved articles, 24 of them met the inclusion-exclusion criteria with prioritization of Fully Homomorphic Encryption string in publication title and content. 25 articles included in the systematic review, 10 articles were from ScienceDirect, and Google Scholar and the rest retrieved from IEEE journals. The IEEE articles publication was as follows: 15 articles presented in different IEEE conferences. Figure 3 summarizes the leading countries of Fully Homomorphic Encryption publication, most of the authors' articles were from China and India with a total number of six and five numbers of publications. There is no surprise that China and India are leading the publication of security and cryptography subjects. Figure 3. The peak year for Fully Homomorphic Encryption articles is the year 2017, which should be normally since all the security issues are important topics, whereas all the data were transferred via the cloud and wirelessly and easy to be threatened by others. The theoretical development of Fully Homomorphic Encryption was by Gentry in 2009. However, the first function of the encryption was in 2011. Fully Homomorphic Encryption is an efficient encryption scheme in cloud computing, provides high confidentiality, which was the main issue discussed in those articles with different solutions for handling the noise (Fig. 4).

El-Yahyaoui and Ech-Chrif El Kettani [33] presented a new way they called it a Verifiable Fully Homomorphic Encryption scheme (VFHE). The scheme considered to be a free-noise fully Homomorphic encryption scheme without the usage of any noise management technique. The scheme does an infinity number of operations on the same ciphertext with no noise growing during the operation. VFHE considered to be faster than FHE but it suffers from security issues as this scheme is cryptanalysis.

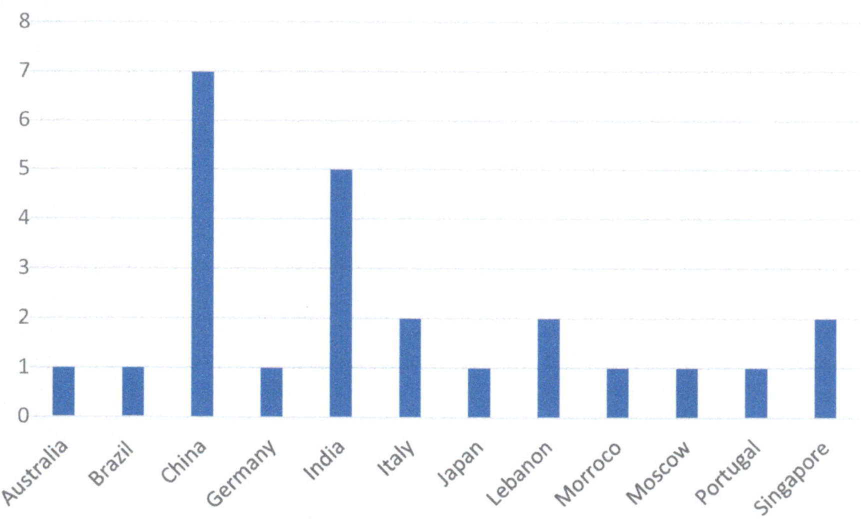

Fig. 3 Distribution of studies by countries

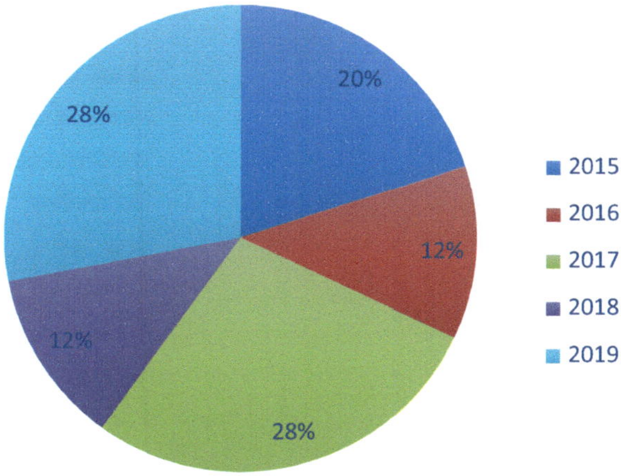

Fig. 4 Distribution of articles published between 2015 and 2019

Another scheme was approved theoretically to solve efficiency and privacy concerns proposed by Gai and Qiu [34] called An Optimal Fully Homomorphic Encryption (O-FHE) "this scheme can achieve non-noise computations over the cipher-texts and retrieve the plain-text results from the decryption within an adaptive execution time period. The proposed approach is designed to achieve Homomorphic mathematical computations, including multiplications and additions, which utilizes the law of the Kronecker Product (KP). The solution can be applied in either a vector space or a real number field" [34].

From Fig. 5, it is noted that the most articles resources from IEEE conferences which included different novel approaches to increase the efficiency and/or security of Fully Homomorphic Encryption Scheme in cloud computing and big data. The quality measurement of retrieved articles is done based Oxford Quality Scoring System (Jadad Scale) which we used it to measure the quality of the articles. Jadad scale uses three features to capture the quality: randomization, masking, withdrawal, and dropout. The quality average of the selected articles using the Jadad scale is 72%. The highest score of the measured features was the publication of Fully Homomorphic Encryption from China followed by India. From 2009 with Gentry's proposal of FHE, the rapid development of the scheme and its application occurs with different algorithm approaches in order to enhance cloud computing and big data security.

6.1 Thematic Analysis

Thematic analysis is a qualitative research approach that identifies and interprets the theme within the selected data. For our systematic review, we built a word cloud with the most common words used among the articles of Fully Homomorphic Encryption. Figure 6 represents the most common appearance is encryption around 1770 times, followed by Homomorphic (1690), data (1676), Scheme (1246) and Encrypted

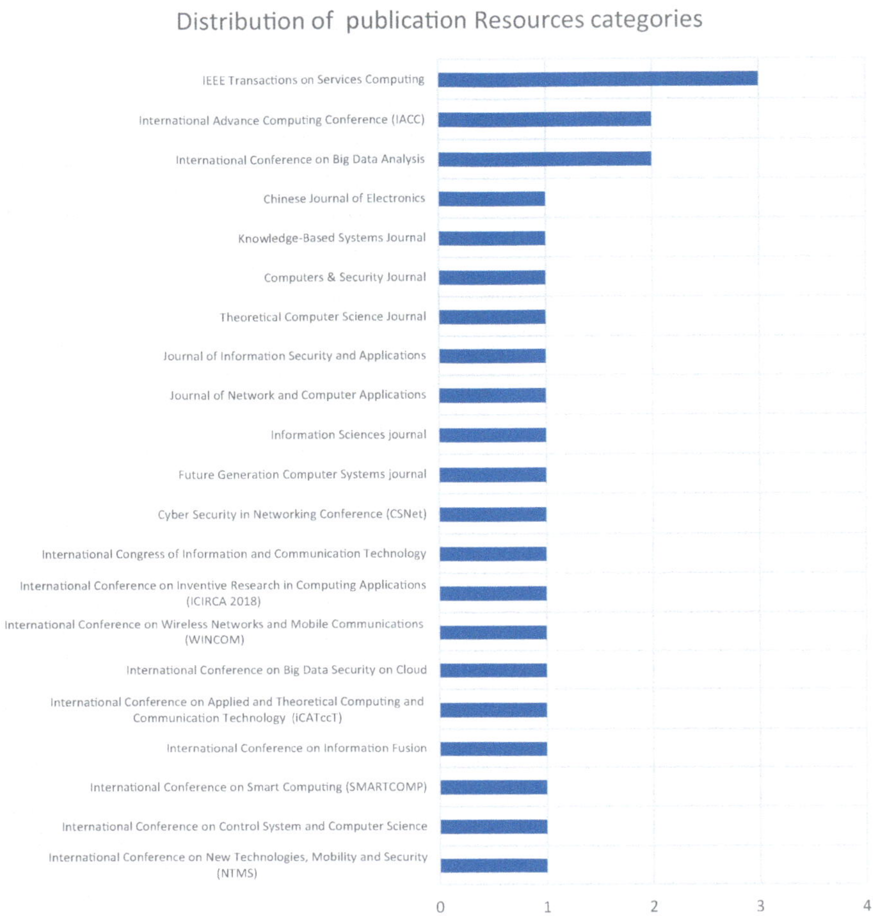

Fig. 5 Distribution of publication resources

(1073). Fully Homomorphic Encryption has been rising with added algorithms to enhance the efficiency and security of the scheme [19–24].

7 Conclusion and Future Work

In conclusion, this paper aimed to conduct a systematic literature review on the topic of Fully Homomorphic Encryption and its application. The main aim of the review is to gain insight into the Fully Homomorphic Encryption Schemes and to find the best and latest approaches to implement it. Fully Homomorphic Encryption Algorithms founded in 2009 by Gentry. After this year, different approaches were followed in

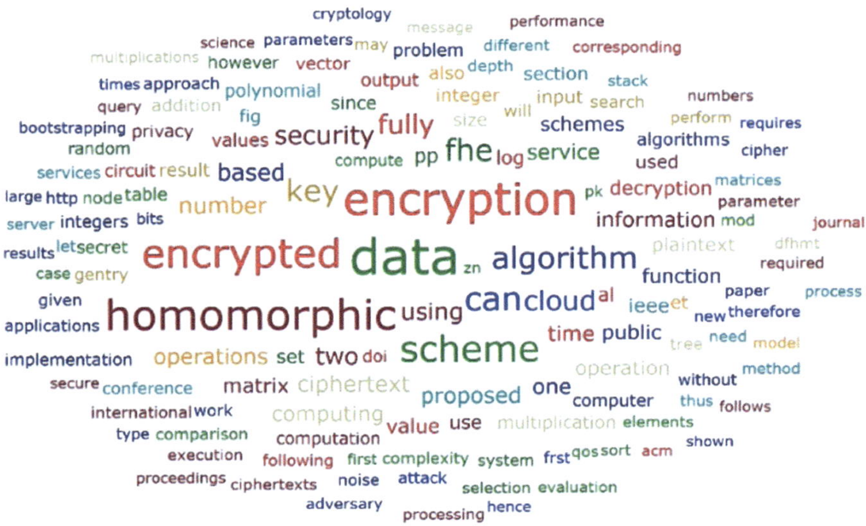

Fig. 6 Thematic modelling for fully homomorphic encryption articles

order to enhance the scheme such as Fully Homomorphic Encryption based on Integer (DGHV) Scheme, Fully Homomorphic Encryption based on BGV schemes, Multi-key Fully Homomorphic Encryption Scheme and Fully Homomorphic Encryption based on GSW. The research question derived for the purposes of the literature review were encompassing the applications of Fully Homomorphic Encryption Schemes, their advantages, and disadvantages, as well as the best implementation approach for them. The procedure done to conduct this systematic literature review included using the research questions to derive keywords which were then used to look for pieces of peer-reviewed works on academic directories. Through initial searches, 250 papers and academic works were found and with the help of selection criteria and PRISMA procedure, the number of papers reviewed in this paper was reduced to 24. Each of the 24 papers was categorized by their contribution to each research question and they were analyzed. All the selected research papers underwent a quality assessment using the Jadad scale tool in order to capture the quality of the publication selected in the systematic review. Finally, a thematic analysis conducted for Fully Homomorphic Encryption reviewed articles.

Acknowledgements This work is a part of a project undertaken at the British University in Dubai.

References

1. Gaid, M.L.: Secure translation using fully homomorphic encryption and sequence-to-sequence neural networks, Oct, pp. 1–4 (2018)
2. Al-Emran, M., Arpaci, I., Salloum, S.A.: An empirical examination of continuous intention to use m-learning: an integrated model. Educ. Inf. Technol. (2020)
3. Salloum, S.A., Al-Emran, M.: Factors affecting the adoption of E-payment systems by university students: extending the TAM with trust. Int. J. Electron. Bus. **14**(4), 371–390 (2018)
4. Al Kurdi, B., Alshurideh, M., Salloum, S.A., Obeidat, Z.M., Al-dweeri, R.M.: An empirical investigation into examination of factors influencing university students' behavior towards E-learning acceptance using SEM approach. Int. J. Interact. Mob. Technol. **14**(02), 19–41 (2020)
5. Alhashmi, S.F.S., Salloum, S.A., Abdallah, S.: Critical success factors for implementing artificial intelligence (AI) projects in Dubai Government United Arab Emirates (UAE) health sector: applying the extended technology acceptance model (TAM). In: International Conference on Advanced Intelligent Systems and Informatics, pp. 393–405 (2019)
6. Alhashmi, S.F.S., Salloum, S.A., Mhamdi, C.: Implementing artificial intelligence in the United Arab Emirates healthcare sector: an extended technology acceptance model. Int. J. Inf. Technol. Lang. Stud. **3**(3) (2019)
7. Gentry, C.: Fully homomorphic encryption using ideal lattices. Proc. Annu. ACM Symp. Theory Comput. 169–178 (2009)
8. Zhao, E.M., Geng, Y.: Homomorphic encryption technology for cloud computing. Procedia Comput. Sci. **154**, 73–83 (2019)
9. Song, X., Wang, Y.: Homomorphic cloud computing scheme based on hybrid homomorphic encryption. In: 2017 3rd IEEE International Conference on Computer and Communications ICCC 2017, vol. 2018-Janua, pp. 2450–2453 (2018)
10. Brakerski, Z., Vaikuntanathan, V.: Efficient fully homomorphic encryption from (standard) lwe. SIAM J. Comput. **43**(2), 831–871 (2014)
11. Wang, D., Guo, B., Shen, Y., Cheng, S.J., Lin, Y.H.: A faster fully homomorphic encryption scheme in big data. In: 2017 IEEE 2nd International Conference on Big Data Analysis, ICBDA 2017, pp. 345–349 (2017)
12. Van Dijk, M., Gentry, C.: Advances in cryptology—EUROCRYPT 2010. Advances in Cryptology—EUROCRYPT 2010, vol. 6110, pp. 24–43 (2010)
13. López-Alt, A., Tromer, E., Vaikuntanathan, V.: On-the-fly multiparty computation on the cloud via multikey fully homomorphic encryption. In: Proceedings of the Forty-Fourth Annual ACM Symposium on Theory of Computing, pp. 1219–1234 (2012)
14. Liu, J., Han, J.L., Wang, Z.L.: Searchable encryption scheme on the cloud via fully homomorphic encryption. In: 2016 Sixth International Conference on Instrumentation & Measurement, Computer, Communication and Control, IMCCC 2016, pp. 108–111 (2016)
15. Gentry, C., Sahai, A., Waters, B.: Homomorphic encryption from learning with errors: conceptually-simpler, asymptotically-faster, attribute-based. Lecture Notes in Computer Science (LNCS), including its subseries Lecture Notes in Artificial Intelligence (LNAI) and Lecture Notes in Bioinformatics (LNBI), vol. 8042 LNCS, no. PART 1, pp. 75–92 (2013)
16. Zhao, F., Li, C., Liu, C.F.: A cloud computing security solution based on fully homomorphic encryption. In: 16th International Conference on Advanced Communication Technology, ICACT, pp. 485–488 (2014)
17. Martins, P., Sousa, L.: A methodical FHE-based cloud computing model. Futur. Gener. Comput. Syst. **95**, 639–648 (2019)
18. Luo, F., Wang, F., Wang, K., Chen, K.: A more efficient leveled strongly-unforgeable fully homomorphic signature scheme. Inf. Sci. (NY) **480**(2318), 70–89 (2019)
19. Kim, J., Kim, S., Seo, J.H.: A new scale-invariant homomorphic encryption scheme. Inf. Sci. (Ny) **422**, 177–187 (2018)

20. Patil, T.B., Patnaik, G.K., Bhole, A.T.: Big data privacy using fully homomorphic non-deterministic encryption. In: 2017 IEEE 7th International Advance Computing Conference, IACC 2017, pp. 138–143 (2017)
21. Xu, J., Wei, L., Zhang, Y., Wang, A., Zhou, F., Zhi Gao, C.: Dynamic fully homomorphic encryption-based Merkle Tree for lightweight streaming authenticated data structures. J. Netw. Comput. Appl. **107**, 113–124 (2018)
22. Slamanig, D., Rass, S.: Cryptography for Security and Privacy in Cloud Computing. Artech House Publishers, Norwood (2013)
23. Gai, K., Qiu, M., Li, Y., Liu, X.Y.: Advanced fully homomorphic encryption scheme over real numbers. In: 2017 IEEE 4th International Conference on Cyber Security and Cloud Computing, CSCloud 2017 3rd IEEE International Conference on Cyber Security and Cloud, SSC 2017, pp. 64–69 (2017)
24. Oppermann, A., Grasso-Toro, F., Yurchenko, A., Seifert, J.P.: Secure cloud computing: communication protocol for multithreaded fully homomorphic encryption for remote data processing. In: 2017 IEEE International Symposium on Parallel and Distributed Processing with Applications and 2017 IEEE International Conference on Ubiquitous Computing and Communications (ISPA/IUCC), pp. 503–510 (2018)
25. Umadevi, C.N., Gopalan, N.P.: Matrices with enhanced access control. In: 2018 International Conference on Inventory Research Computer Applications, no. Icirca, pp. 328–332 (2018)
26. Davis, J., Mengersen, K., Bennett, S., Mazerolle, L.: Viewing systematic reviews and meta-analysis in social research through different lenses. Springerplus **3**(1), 1–9 (2014)
27. Moher, D., et al.: Preferred reporting items for systematic reviews and meta-analyses: the PRISMA statement. PLoS Med. (2009)
28. Kitchenham, B., Charters, S.: Guidelines for performing systematic literature reviews in software engineering. In: Software Engineering Group, School of Computer Science and Mathematics, Keele University, pp. 1–57 (2007)
29. Al-Saedi, K., Al-Emran, M., Abusham, E., El-Rahman, S.A.: Mobile payment adoption: a systematic review of the UTAUT model. In: International Conference on Fourth Industrial Revolution (2019)
30. Saa, A.A., Al-Emran, M., Shaalan, K.: Factors affecting students' performance in higher education: a systematic review of predictive data mining techniques. Technol. Knowl. Learn. (2019)
31. Al-Qaysi, N., Mohamad-Nordin, N., Al-Emran, M.: A systematic review of social media acceptance from the perspective of educational and information systems theories and models. J. Educ. Comput. Res. **57**(8), 2085–2109 (2020)
32. Al-Emran, M., Mezhuyev, V., Kamaludin, A., Shaalan, K.: The impact of knowledge management processes on information systems: a systematic review. Int. J. Inf. Manag. **43**, 173–187 (2018)
33. El-Yahyaoui, A., Ech-Chrif El Kettani, M.D.: A verifiable fully homomorphic encryption scheme to secure big data in cloud computing. In: Proceedings of 2017 International Conference on Wireless Networks and Mobile Communications, WINCOM 2017 (2017)
34. Gai, K., Qiu, M.: An optimal fully homomorphic encryption scheme. In: Proceedings of 3rd 2017 IEEE 3rd International Conference on Big Data Security on Cloud (Bigdatasecurity), IEEE International Conference on High Performance and Smart Computing (HPSC), and 2nd IEEE International Conference on Intelligent Data and Security (IDS), pp. 101–106 (2017)

Ridology: An Ontology Model for Exploring Human Behavior Trajectories in Ridesharing Applications

Heba M. Wagih and Hoda M. O. Mokhtar

Abstract The increasing adoption of social network applications has been an important source of information nowadays. The analysis of human behaviors in social networks has been brought to the forefront of several studies. Location-Based Social Networks (LBSN) are one of the possible means that allow the prediction of human behaviors through the efficient analysis of user's mobility patterns. Despite the remarkable progress in this research direction, however, LBSN is still hindered by the lack of literature defining the semantic aspects of the user's mobility. This research presents a contribution to the latest knowledge representation languages and Semantic Web technologies. We focus on studying human behavior mobility which is the core in location recommendation systems. Bringing to the ridesharing context, an ontology model with its underlying description logics to efficiently annotate human mobility is presented. Finally, experimental results, performed on two location-based social networks, namely, Brightkite (https://snap.stanford.edu/data), and BlaBlaCar (https://www.blablacar.co.uk/) are presented.

Keywords Location-based social networks · Sematic web technologies · Ontology modeling · Description logics

1 Introduction

The broad usage of social network applications has proved a notable impact on the internet systems as well as people's daily life. The popularization of such applications has generated a huge amount of data that introduced new research opportunities and challenges. Several current problems in social networks are still under investigation

H. M. Wagih (✉) · H. M. O. Mokhtar (✉)
The British University in Egypt, Cairo, Egypt
e-mail: heba.wageeh@bue.edu.eg

H. M. O. Mokhtar
e-mail: h.mokhtar@fci-cu.edu.eg

Cairo University, Giza, Egypt

M. Al-Emran et al. (eds.), *Recent Advances in Intelligent Systems and Smart Applications*, Studies in Systems, Decision and Control 295,
https://doi.org/10.1007/978-3-030-47411-9_30

including social network modeling and analysis, recommendation systems in social networks, managing big data in social networks, and many other research challenges.

Location-based social network (LBSN) has made remarkable strides in the last few years and gained a remarkable attraction to millions of users. LBSN enables users to share information not only about their physical location but also their everyday interests and activities. Users share their points of interest via Check-In feature, and the routes they choose to reach their destinations.

Ridesharing applications are a new field of study that has emerged under the umbrella of location-based social networks [1] and utilizes the concept of spatio-temporal trajectories of moving objects. Ridesharing has become a popular mean of transportation especially when it comes to a cheap way of traveling. It allows two or more persons traveling from the same source to the same destination to share a ride. Several benefits to using ridesharing, among them, (1) decrease in the number of vehicles in rush hours and busy locations, (2) decrease in gas expenses, (3) a probable channel for new social connections to be developed.

Semantic Trajectory Analysis is another major contribution that has recently emerged in the field of social networking. Semantic trajectories are annotated trajectories where additional information about the application context is added. Such information helps in understanding the human mobility and provides insights about human behaviors. Annotated trajectories are commonly represented using ontologies.

Ontologies were previously known as a branch of philosophy that represents the world reality. Ontologies represent a structural model of world objects that are classified into categories based on their similarities. In [2] the first formal definition for ontology was introduced where the author defined ontology as a "formal, explicit specification of a shared Conceptualization". Ontology allows data exchange over web independently from any application domain and has the same structural components (individuals, classes, properties, axioms and assertions) regardless the language used for its development.

Human Behavior Trajectory (also known as Activity Trajectory) is a special type of semantic trajectories where a new time dimension is added to the raw trajectory to represent the points of Interest for a user. Detecting these points of interest is extremely useful in analyzing human daily behavior trajectory and predicting or recommending new activities to be practiced. Human Behavior Trajectories are used in many applications especially those related to the activity recommendation systems and location recommendation systems.

Moving from raw trajectory representation level to a higher level of semantic abstraction and understanding is the main engine behind our research. The semantic abstraction representation combines both spatio-temporal patterns of trajectories and knowledge extracted from the object's dynamics. Many previous researches had studied the problem of semantic enrichment of raw trajectories; however, the understanding of human behaviors using semantic trajectories is still an area of investigation. The core objective of this chapter is introducing an ontology model named Ridology with its underlying description logics formulation for understanding and analyzing human behaviors and activities extracted from their daily mobility.

The main contributions of this chapter are as follows:

1. An ontology-based model named Ridology for exploring human behavior trajectory in ridesharing applications is proposed.
2. A description logics formulation for axioms governing the human behavior trajectory model is presented.
3. A protégé based query formulation that is capable of querying the proposed model is presented.

The rest of this chapter is organized as follows: Sect. 2 presents some related works to location-based social networks, ridesharing applications, and ontology development. Section 3 overviews the main technical concepts used in the chapter and present our human behavior ontology in ridesharing applications. Section 3.2 presents the experimental results on a real case study. Finally, Sect. 4 concludes and proposes directions for future work.

2 Related Work

Human mobility behavior has been considered as an important source of information that supports many applications such as recommendation systems. Recommendation Systems play an important role in providing better customized user experiences. Different recommendation systems have been developed to provide users with personalized recommendations on items such as movies, books, news and others.

A new generation of recommendation Systems based on locations had emerged known as Location-Based Recommendation Systems (LRSs). Location-based recommendation systems utilizes user's locations (Checks-in), points of interest, user ratings and social ties extracted from location-based social networks. In [3] the authors investigated the human motion using three data sets, two from a location-based social networks and a one from a cell phone location data. A Periodic and Social Mobility Model for predicting user mobility was introduced. The model was built to monitor people mobility, where people tend to have a constant dynamic behavior in working days while social relations affect the user location in weekends and long travel distances. The model presented shows 40 percent correctness in determining the exact user location as well as high correlation in results between the cell phone and the location-based social network check-in extracted data. Based on these results, the authors found that human mobility is affected by three constraints; the geographic movement, temporal behavior and the social network and their mobility towards friends' location is twice stronger than other cases in weekends.

Another contribution in LBSN is presented in [4] where a location-based social networking system named Sindbad was introduced. The system consists of three main components; the location-aware news feed, the location-aware news ranking, and the location-aware recommendation and three types of data were pumped to the system; spatial messages, user profiles, and spatial ratings. This system provided users with geo-tagged messages submitted by his or her friends to recommend a certain place within a specific region.

Ridesharing is a new type of applications that can be combined with social networking. Ridesharing has gained significant focus in the research in the past few years due to its importance in both social and economic sides. Many works have been introduced to enhance the performance of rideshare applications as in [5, 6]. In [5], the authors introduced a dynamic carpooling system with social network-based filtering. The introduced system gives the users the freedom to choose whether their trip to be private or public to share it with other users or friends. The proposed system is integrated with social network applications and real locations are identified using GPS.

In [7, 8], the authors employed network flow techniques and a mathematical programming method to develop a ridesharing application. The model presents the different requests to available cars as an integer multiple commodity network flow problem. Despite the good results achieved by this model, however it still lacks the dynamic matchmaking between user requests. Later, the problem of dynamic matching between user requests has been addressed in [9] where the authors proposed a mobile android system that supports dynamic ridesharing.

Although those contributions seem interesting, however raw data (spatio-temporal points) acquired from heterogeneous sources (e.g. sensors, monitors, GPSs, etc.) are exponentially increasing with difficulty to analyze and manage leading to a demand of annotated data sets.

Semantic trajectory is a newly growing approach that provides applications with knowledge about objects' dynamics where it enriches data collected from sources as location based social network applications with contextual data. Semantic trajectories are used to define the sequence of points of interest to user along with metadata as what type of point did the user visit?, what means of transportation did he use?, and other data that enriches raw spatial trajectory information. Many researches as in [10, 11] have explored the conceptual and ontology modeling approaches for semantic trajectories and defining mobility patterns for developing semantic annotations. However, one important aspect that still needed to be highly addressed is data management in semantic trajectories.

In [12] the authors highlighted some challenges related to data management in semantic trajectories such as storage and accessibility of semantic trajectories. The authors presented a new data model that represents different types of semantic trajectories. They presented semantic information in symbolic form consists of labels that represent the movement of an entity during a time interval. They also introduced a new pattern language that queries their symbolic trajectories.

In [13] the authors presented a new model called semantic model and a computation and annotation platform that converts raw mobility data into semantic trajectories enriched with annotations. The authors focused on developing a model that is independent from any application, integrated data from heterogeneous sources and used a generic annotation algorithm to provide a good performance especially over complex zone. The model encapsulated raw spatio-temporal trajectory data and then converted these data into higher-level of abstraction for semantic representations based on the stop-move concept.

A special type of semantic trajectories was introduced and is currently known as Human Behavior trajectories (also known as Activity trajectories). Human Behavior trajectories are annotated trajectories that provide knowledge about activities the user had done in particular places during specific period of time. Human Behavior trajectories are useful in activity recommendations as well as trip planning. Due to its fast pace emergence, interpreting such trajectories is definitely an important field of research. Several contributions were introduced to propose models and tools that support semantic enrichment, data analysis and querying [14, 15].

In [16] the authors proposed a Semantic-Enriched Knowledge Discovery Process to represent expressive patterns for human behaviors. The proposed approach handled the syntactic and semantic complexity of human dynamics through integrating inductive reasoning (movement pattern discovery) and deductive reasoning (human behavior inference). Their significant contribution was the development of a mobility behavior ontology that is composed of two components, the core component and the application component. The Core Ontology component represented the human behavior concepts independently from any application context while the Application Ontology component focused on the concepts of human behavior with respect to a particular application domain. The implemented ontology was based on stop-move concepts with three defined types of trajectories namely, measured, syntactic, and semantic trajectory.

Semantic annotations for raw data trajectories are more useful when based on a shared vocabulary. Ontology design patterns are gaining a remarkable focus in the geospatial semantics community [17, 18] and they are more suitable for semantic and activity trajectory representation. Among the recent contributions in this direction is the work in [19]. In this work the authors developed a Geo-Ontology design pattern for semantic trajectories based on the Stop-Move concepts. They used Description Logics to show the logical formalism used for the ontology design pattern development. Description logics (DL) [20, 21] is an essential step in defining, integrating, and maintaining ontologies. Semantic Web technology is heavily based on Description Logics.

Following the above contributions and motivated by the importance of efficiently representing activity and semantic trajectories, in the rest of this chapter we propose a ridesharing ontology based on human behavioral trajectory representation using the Protégé Ontology Development tool along with Description Logics.

3 Ridology: Proposed Ontology for Rideshare Applications

Ridology is the proposed ontology that shows the categorization of concepts regarding the rideshare domain, also the restrictions and axioms of these concepts. The proposed ontology consists of four main entities, users, trips, drivers, and vehicles.

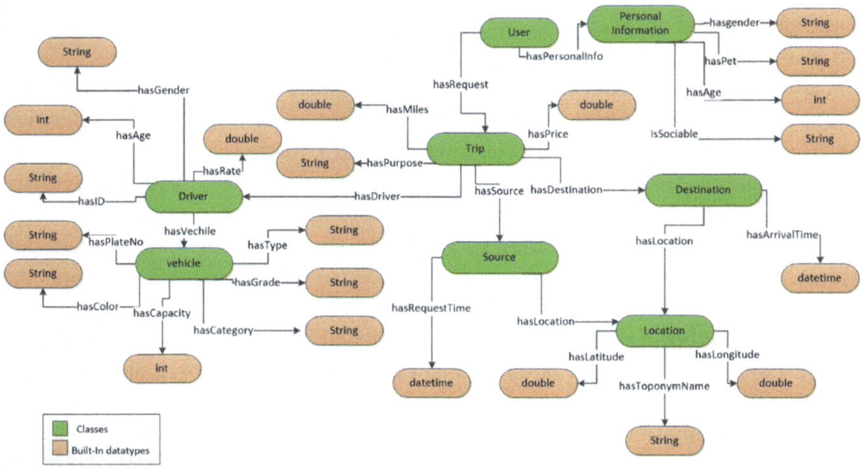

Fig. 1 Schema description of the proposed ontology model

Each of these entities is characterized by an OWL annotation which represents relevant properties. Users' attributes include their relevant preferences (e.g., is sociable, has pet) as well as their demographic features (age, gender). Each user can request a number of trips with different start and end locations at different times. Trips are handled by different drivers where each driver possess a vehicle that has a certain capacity and category. Figure 1 shows the proposed ontology schema.

The following set of axioms formally describe the proposed model in description logics.[1]

Axiom 1 User$\sqsubseteq$ ∃hasPersonalInfo.UserInformation ∏ ∃hasRequest.Trip

- hasPersonalInfo: An ObjectProperty of Domain Class User and Range Class UserInformation
- hasRequest: An ObjectProperty of Domain Class User and Range Class Trip

Axiom 2 PersonalInfo$\sqsubseteq$ ∃hasgender.string ∏ ∃hasAge.int ∏ ∃hasPet.string ∏ ∃IsSociable.string

- hasgender: A DataProperty of Domain Class PersonalInfo and Range DataType string
- hasAge: A DataProperty of Domain Class PersonalInfo and Range DataType int
- hasPet: A DataProperty of Domain Class PersonalInfo and Range DataType string
- IsSociable: A DataProperty of Domain Class PersonalInfo and Range DataType string

Axiom 3 Trip$\sqsubseteq$ ∃hasSource.Location ∏ ∃hasDestination.Location ∏ ∃hasDriver. Driver ∏ ∃hasPurpose.string ∏ ∃hasMiles.double ∏ ∃hasPrice.double

[1]Following SROIQ description logic notations introduced in [22].

- hasSource: An ObjectProperty of Domain Class Trip and Range Class Location
- hasDestination: An ObjectProperty of Domain Class Trip and Range Class Location
- hasDriver: An ObjectProperty of Domain Class Trip and Range Class Driver
- hasPurpose: A DataProperty of Domain Class Trip and Range DataType String
- hasMiles: A DataProperty of Domain Class Trip and Range DataType double
- hasPrice: A DataProperty of Domain Class Trip and Range DataType double

Axiom 4 Driver$\sqsubseteq$ $\exists$hasID.string $\sqcap$ $\exists$hasGender.string $\sqcap$ $\exists$hasAge.int $\sqcap$ $\exists$hasRate.double $\sqcap$ $\exists$hasVehicle.Vehicle

- hasID: A DataProperty of Domain Class Driver and Range DataType string
- hasGender: A DataProperty of Domain Class Driver and Range DataType string
- hasAge: A DataProperty of Domain Class Driver and Range DataType int
- hasRate: A DataProperty of Domain Class Driver and Range DataType double
- hasVehicle: A DataProperty of Domain Class Driver and Range Class Vehicle

Axiom 5 Source$\sqsubseteq$ $\exists$hasLocation.Location $\sqcap$ $\exists$hasRequestTime.DateTime

- hasLocation: An ObjectProperty of Domain Class Source and Range Class Location
- hasRequestTime: A DataProperty of Domain Class Source and Range DataType DateTime

Axiom 6 Destination$\sqsubseteq$ $\exists$hasLocation.Location $\sqcap$ $\exists$hasArrivalTime.DateTime

- hasLocation: An ObjectProperty of Domain Class Destination and Range Class Location
- hasArrivalTime: A DataProperty of Domain Class Destination and Range DataType DateTime

Axiom 7 Location$\sqsubseteq$ $\exists$hasLongitude.double $\sqcap$ $\exists$hasLatitude.double $\sqcap$ $\exists$hasToponymName.string

- hasLongitude: A DataProperty of Domain Class Location and Range DataType double
- hasLatitude: A DataProperty of Domain Class Location and Range DataType double
- hasToponymName: A DataProperty of Domain Class Location and Range DataType string

Axiom 8 Vehicle$\sqsubseteq$ $\exists$hasCategory.string $\sqcap$ $\exists$hasPlateNo.string $\sqcap$ $\exists$hasCapacity.int $\sqcap$ $\exists$hasGrade.string $\sqcap$ $\exists$hasType.String $\sqcap$ $\exists$hasColor.string

- hasCategory: A DataProperty of Domain Class Vehicle and Range DataType string
- hasPlateNo: A DataProperty of Domain Class Vehicle and Range DataType string
- hasCapacity: A DataProperty of Domain Class Vehicle and Range DataType int
- hasGrade: A DataProperty of Domain Class Vehicle and Range DataType string
- hasType: A DataProperty of Domain Class Vehicle and Range DataType string
- hasColor: A DataProperty of Domain Class Vehicle and Range DataType string

3.1 Analyzing Human Behavior Trajectories

Analyzing human behavior trajectories has played an important role in enhancing ridesharing services. User' preferences can be known by detecting users' mobility from data collected from Social networking and hence offering rideshare with other users who best suite one's request. It enables different users to share vehicles with other passengers to the same or nearby destination.

Consider the following example, we assume that the user has requested three different trips with three different locations (Home, Office, Mall) where the user spends a certain time in each location. The time dimension in our model represents the start time and the estimated arrival time for each requested trip. The human behavior trajectory can be viewed as follows:

- Trip1 (Home, Start Time 7:55 a.m.—Arrival Time 8:15 a.m., Office)
- Trip2 (Office, Start Time 14:30 a.m.—Arrival Time 14:55 p.m., Mall)
- Trip3 (Mall, Start Time 20:00 p.m.—Arrival Time 20:25 p.m., Home)

We used both DL Query Language [23] and SPARQL Query Language [24, 25] in Protégé 5.0 beta [23] to query the proposed ontology and assure the validity of our ontology model through expressing real word queries. We compared the results retrieved from both query languages and due to some reasoning limitations in DL Query Language, SPARQL Query Language has proven to be more efficient and accurate.

Given the following queries, Query 1 and Query 2 retrieved correct results using both query languages; however, for Query 3, DL Query Language could not support these types of queries.

Query 1: *Retrieve the personal information for all users requesting trips.*
DL Query:
User and hasPersonalInfo PersonalInformation
SPARQL Query:
SELECT ?User ?PersonalInformation
WHERE ?User onto:hasPersonalInfo ?PersonalInformation.
Query 2: *Retrieve all the trips that are requested by certain user.*
DL Query:
User and hasRequest Trip
SPARQL Query:
SELECT ?User ?Trip
WHERE ?User onto:hasRequest ?Trip.
Query 3: *Retrieve all the locations that are visited by certain user*
SPARQL Query:
SELECT ?User ?Location
WHERE ?User onto:hasRequest ?Trip.
?Trip onto:hasSource ?Source.
?Trip onto:hasDestination ?Destination.
?Source onto:hasLocation ?Location.
?Destination onto:hasLocation ?Location.

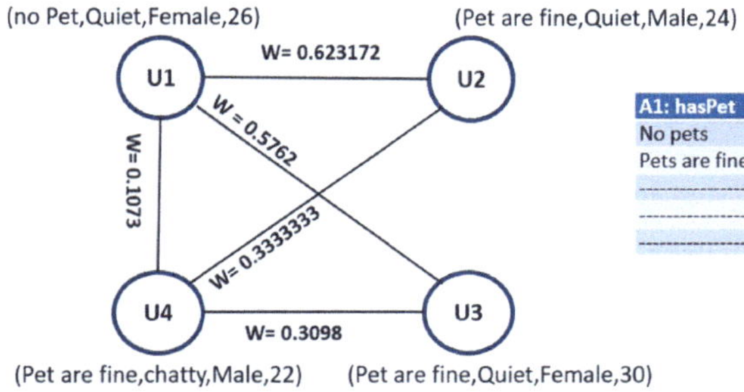

A1: hasPet	A2: isSociable	A3: hasGender	A4: hasAge
No pets	Quiet	Female	<15
Pets are fine	Chatty	Male	15-18
---------------	Very Chatty	---------------	19-22
---------------	---------------	---------------	23-40
---------------	---------------	---------------	>40

Fig. 2 Weighted relations between users in social network

For best matching between ridesharing users, we assume a certain weight value between every pair of users based on the users' attributes as well as the start location, the final destination and the time the request was made. The higher the weight value, the most probable that both users will share the same ride.

Weight: for each pair of users (user_i and user_j) in the social network, a weight value is computed. The weight is assigned the correlation value between the users' attributes ($\text{user}_{\text{iattr}}$ and $\text{user}_{\text{jattr}}$) using Pearson Correlation Coefficient (PCC) [26], and is given by,

$$\textbf{Weight}\left(\textbf{user}_i, \textbf{user}_j\right) = \textbf{PCC}\left(\textbf{user}_{iattr}, \textbf{user}_{jattr}\right) \quad (1)$$

$$\textbf{PCC}\left(\textbf{user}_{iattr}, \textbf{user}_{jattr}\right) = \frac{n\sum\left(\textbf{user}_{iattr}, \textbf{user}_{jattr}\right) - \left(\sum \textbf{user}_{iattr}\right)\left(\sum \textbf{user}_{jattr}\right)}{\sqrt{n\left(user_{iattr}^2\right) - \left(\sum user_{iattr}\right)^2}\sqrt{n\left(user_{jattr}^2\right) - \left(\sum user_{jattr}\right)^2}} \quad (2)$$

Example: consider a social network with relations between four users A, B, C and D. The Weight of the relations is the correlation value between the attributes of each friend. Figure 2 shows an example of weighted relation social network.

3.2 *Case Study*

To better explain the proposed approach and highlight its benefits, we used two mobility datasets namely, BlaBlaCar and Gowalla. BlaBlaCar is a rideshare application that was previously introduced in 2006. The idea behind BlaBlaCar is to allow drivers to offer their empty seats to people who cannot afford traveling by other transportation means. Ridesharing has many advantages on both economic and social sides. For this research, data for rideshares in 2016 was collected using BlaBlaCar's application programming interface (API). On analyzing these collected data, we noticed some important observations that help in improving the ridesharing services. First, some locations have a higher visiting frequency than others as shown in Fig. 3. Second, the number of requests is noticeably higher in rush hours especially from 01:00 p.m. till 06:00 p.m. as shown in Fig. 4 as well as in certain months as shown in Fig. 5. Counting on these observations, ridesharing companies can offer buses besides the cars for ridesharing which can increase the financial income and decrease the traffic congestion in such areas, ridesharing vehicles can be assigned parking spots in such areas. Third, the correlation value between certain users using ridesharing service might be relatively high as shown in Fig. 6, thus recommending such users to share the same ride might lead to a great riding experience and new social connections might evolve.

Gowalla is another location-based social network that is available for public use. Gowalla dataset has 196,591 users with 6,442,890 check-ins in 1,280,956 different

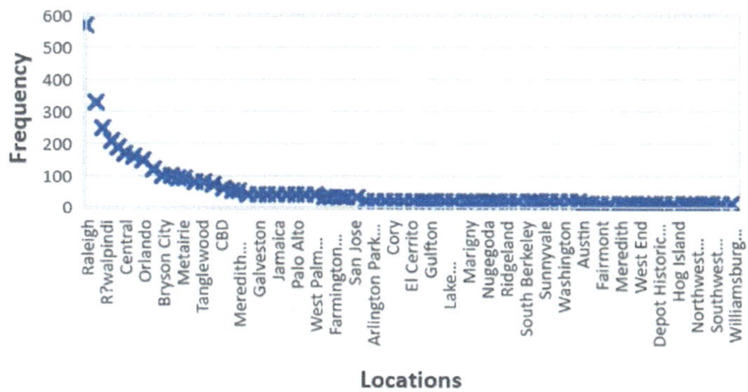

Fig. 3 Visiting frequency for different locations

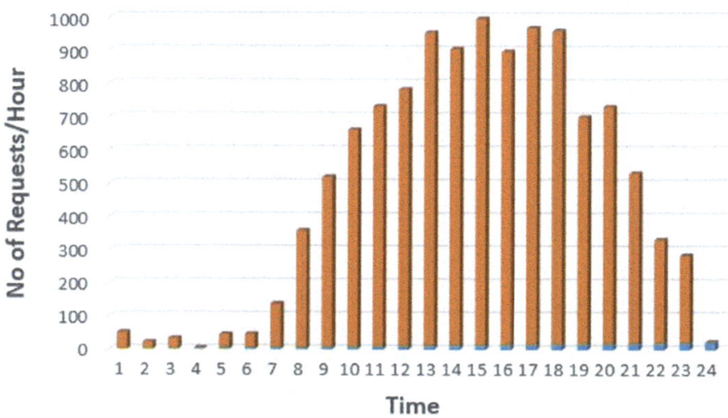

Fig. 4 Number of users' requests per hour

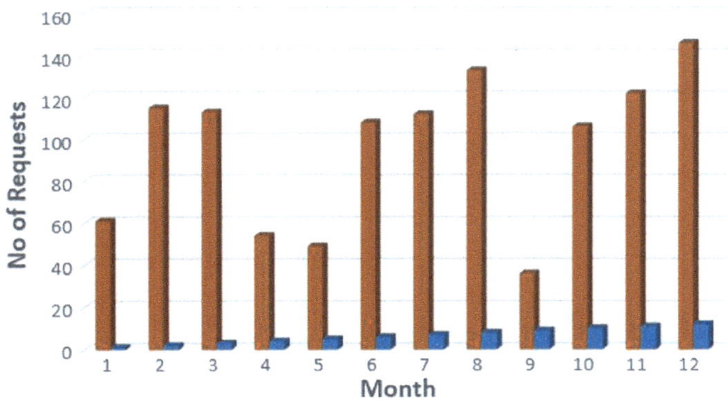

Fig. 5 Number of users' requests per month

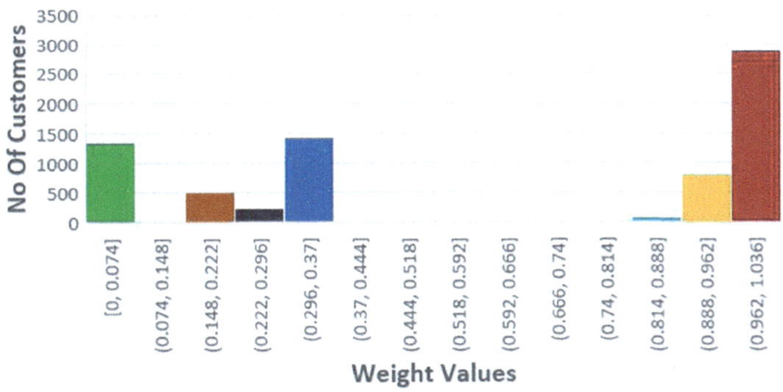

Fig. 6 Weight values between rideshare users

visited location. The check-ins consist of the user identification, the physical coordinates (longitude and latitude) of the check-in location and check-in time. Since data available in Gowalla dataset has no semantic information to define the physical coordinates, thus we use the approach introduced in [27] to provide semantic translation for each spatial coordinate corresponding to the user's check-in. In our experiments, we used the Gowalla dataset to check the visiting frequency for different locations in four countries namely, Germany, United Kingdom, United States, and Japan. We found that most of the users are distributed close to certain locations as shown in Fig. 7 which could be a target for either recommending more cars to be available for ridesharing services or to favor users with more services or offers to those locations at specific times.

(a) Germany

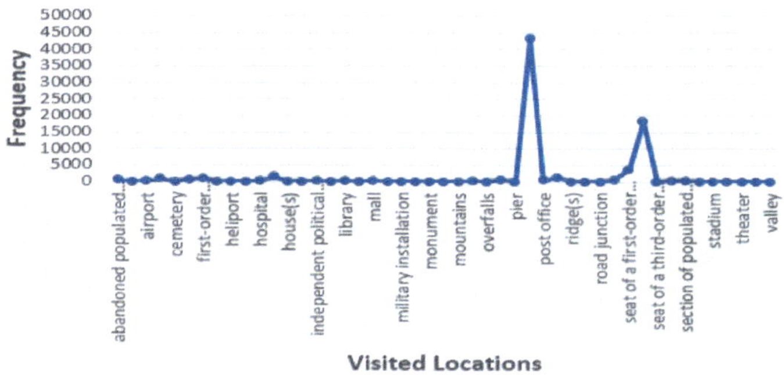

(b) United States

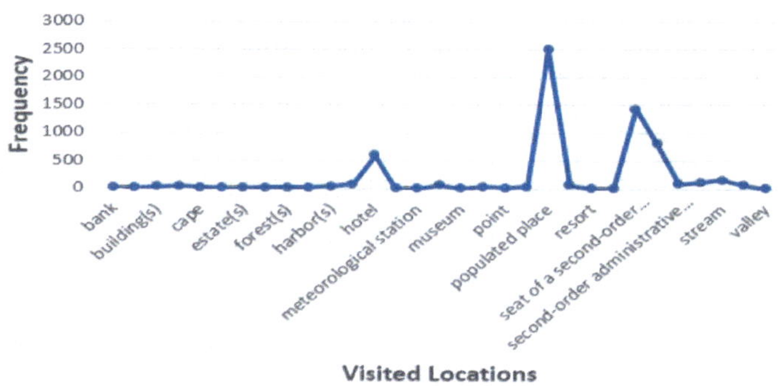

(c) United Kingdom

Fig. 7 Friend of friend recommendation

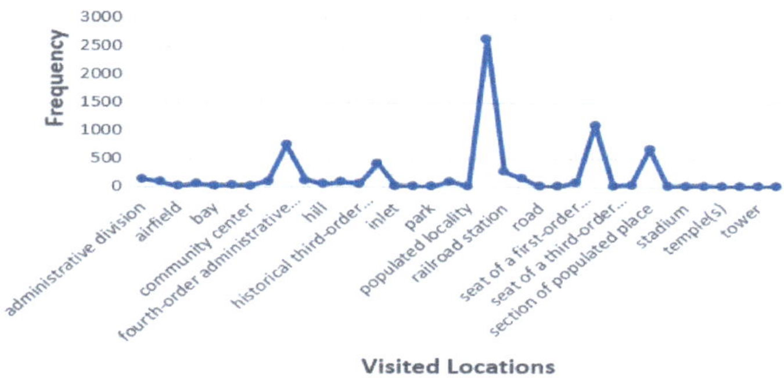

(d) Japan

Fig. 7 (continued)

4 Conclusion and Future Work

Social networking has gained a great attention in the past few years leading to a huge change in the internet ecosystem. Many research studies had been done to investigate Social Networks, among these studies is the location-based social network and its developed applications. Location-based social networks can serve as a prediction tool to enhance a ridesharing services by providing more insights, predictions and recommendation for human mobility. Human behavior trajectories are great asset in location based social network applications where an annotated version of the human dynamics data (spatio-temporal points) is provided. These annotated trajectories are used to understand and predict the behavior of moving objects. In this chapter we have introduced an ontological approach to represent the human behavior trajectories in ridesharing context. Our approach is based on a 2-dimensional space; however we intend to integrate the 3-dimensional space and more complicated environmental surfaces in our future work. Combining more than one ontology to provide a complete description for the behavior trajectory is also among our future work plan.

Acknowledgements This research is an extension to the previously published paper [28] in the 11th International Conference on Semantics, Knowledge and Grids (SKG) conference in 2015.

References

1. Schreieck, M., Safetli, H., Siddiqui, S., Pflugler, C., Wiesche, M., Krcmar, H.: A matching algorithm for dynamic ridesharing. Transp. Res. Procedia 272–285 (2016)
2. Gruber, T.R.: A translation approach to portable ontology specifications. Knowl. Acquis. **5**(2), 199–220 (1993). (Special Issue: Current Issues in Knowledge Modeling Archive. Academic Press Ltd., London, UK)

3. Cho, E., Myers, S.A., Leskovec, J.: Friendship and mobility, user movement in location-based social networks. In: Proceedings of the 17th ACM SIGKDD International Conference on Knowledge Discovery and Data Mining, San Diego, CA, USA, 21–24 Aug 2011, pp. 1082–1090
4. Sarwat, M., Bao, J., Eldawy, A., Levandoski, J.J., Magdy, A., Mokbel, M.F.: Sindbad: a location-based social networking system. In: Proceedings of the ACM SIGMOD International Conference on Management of Data, SIGMOD 2012, Scottsdale, AZ, USA, 20–24 May 2012, pp. 649–652
5. Sasikumar, C., Jaganathan, S.: A dynamic carpooling system with social network based filtering. Res. J. Eng. **8**(3), 263–267 (2017)
6. Aissaoui, R., Houssaini, O.I.: CARPOOLING APPLICATION? KwiGo, Available via http://www.aui.ma/ssecapstone-repository/pdf/CARPOOLING-APPLICATION-KwiGo.pdf, Apr 2015
7. Yan, S., Chen, C.Y.: A model and a solution algorithm for the carpooling problem with pre-matching information. Comput. Ind. Eng. **61**(3), 512–524 (2011)
8. Yan, S., Chen, C.Y., Chang, S.C.: A carpooling model and solution method with stochastic vehicle travel times. IEEE Trans. Intell. Transp. Syst. **15**(1), 7–61 (2014)
9. Nagare, D.B., More, K.L., Tanwar, N.S., Kulkarni, S., Gunda, K.C.: Dynamic car-pooling application development on android platform. Int. J. Innov. Technol. Explor. Eng. (2013)
10. Parent, C., Spaccapietra, S., Renso, C., Andrienko, G., Andrienko, N., Bogorny, V., Damiani, M.L., Gkoulalas-Divanis, A., Macedo, J., Pelekis, N., Theodoridis, Y., Yan, Z.: Semantic trajectories modeling and analysis. ACM Comput. Surv. **45**(4), 42:1–42:32 (2013)
11. Fileto, R., Krüger, M., Pelekis, N., Theodoridis, Y., Renso, C.: Baquara: a holistic ontological framework for movement analysis using linked data. In: 32nd International Conference on Conceptual Modeling, Hong-Kong, China, 11–13 Nov 2013, pp. 342–355
12. Damiani, M.L., Güting, R.H., Valdés, F., Issa, H.: Moving objects beyond raw and semantic trajectories. In: Proceedings of the 3rd International Workshop on Information Management for Mobile Applications, Riva del Garda, Italy, 26 Aug 2013, p. 4
13. Ya, Z., Chakraborty, D., Parent, C., Spaccapietra, S., Aberer, K.L.: Semantic trajectories: mobility data computation and annotation. ACM TIST **4**(3), 49 (2013)
14. Bogorny, V., Renso, C., Ribeiro de Aquino, A., Siqueira, F.D.L., Alvares, L.O.: CONSTAnT—a conceptual data model for semantic trajectories of moving objects. Trans. GIS (2013)
15. Zheng, K., Shang, S., Yuan, N.J., Yang, Y.: Towards efficient search for activity trajectories. In: 29th IEEE International Conference on Data Engineering, Brisbane, Australia, 8–12 Apr 2013, pp. 230–241
16. Renso, C., Baglioni, M., Fernandes de Macêdo, J.A., Trasarti, R., Wachowicz, M.: How you move reveals who you are: understanding human behavior by analyzing trajectory data. Knowl. Inf. Syst. **37**(2), 331–362 (2013)
17. Carral, D., Scheider, S., Janowicz, K., Vardeman, C., Krisnadhi, A.A., Hitzler, P.: An ontology design pattern for cartographic map scaling. In: 10th International Conference on The Semantic Web: Semantics and Big Data, Montpellier, France, 26–30 May 2013, pp. 76–93
18. Martínez, D.C., Janowicz, K., Hitzler, P.: A logical geo-ontology design pattern for quantifying over types. In: SIGSPATIAL 2012 International Conference on Advances in Geo-graphic Information Systems (formerly known as GIS), Redondo Beach, CA, USA, 7–9 Nov 2012, pp. 239–248
19. Hu, Y., Janowicz, K., Carral, D., Scheider, S., Kuhn, W., Berg-Cross, G., Hitzler, P., Dean, M., Kolas, D.: A geo-ontology design pattern for semantic trajectories. In: 11th International Conference on Spatial Information Theory, Scarborough, UK, 2–6 September 2013, pp. 438–456
20. Baader, F., Calvanese, D., McGuinness, D.L., Nardi, D., Patel-Schneider, P.F.: The Description Logic Handbook: Theory, Implementation, and Applications. Cambridge University Press, Cambridge (2003)
21. Baader, F., Sattler, U.: An overview of tableau algorithms for description logics. Stud. Logica **69**(1), 5–40 (2011)
22. Horrocks, I., Kutz, O., Sattler, U.: The even more irresistible SROIQ, pp. 57–67

23. The Protégé website, Available via http://protege.stanford.edu
24. Sirin, E., Parsia, B.: SPARQL-DL: SPARQL Query for OWL-DL, In 3rd OWL Experiences and Directions Workshop (OWLED) (2007)
25. Kollia, I., Glimm, B., Horrocks, I.: SPARQL query answering over OWL ontologies. In: Proceedings of the 8th Extended Semantic Web Conference on The Semantic Web: Research and Applications, Heraklion, Crete, Greece, 2011, pp. 382–396
26. Archdeacon, T.J.: Correlation and Regression Analysis: A Historian's Guide. University of Wisconsin Press, Madison (1994)
27. Wagih, H.M., Mokhtar, H.M.O., Ghoniemy, S.S.: SIMBA: a semantic-influence measurement based algorithm for detecting influential diffusion in social networks. In: The 13th IEEE International Conference on Semantic Computing, 2019, pp. 178–182
28. Wagih, H.M., Mokhtar, H.M.O.: HBTOnto: an ontology model for analyzing human behavior trajectories. In: 11th International Conference on Semantics, Knowledge and Grids (SKG), 2015, pp. 126–132

Social Networks

Factors Affecting the Adoption of Social Media in Higher Education: A Systematic Review of the Technology Acceptance Model

Noor Al-Qaysi, Norhisham Mohamad-Nordin, and Mostafa Al-Emran

Abstract Numerous review studies were conducted in the past to examine the technology acceptance model (TAM) on the one hand, and the social media adoption on the other hand. However, understanding the factors affecting the adoption of social media in higher education through the lenses of TAM is still not yet examined. Therefore, this study aims to systematically review the TAM-based social media studies for identifying the most dominant factors. This study also attempts to determine the most dominant publication venues through which the analyzed studies were published. Out of 539 articles collected, a total of 57 studies were found to meet the inclusion criteria, and hence, were included in the analysis phase. The results indicated that "perceived enjoyment", "subjective norm", "self-efficacy", "perceived playfulness", "perceived critical mass", and "openness" were the most frequent factors affecting the adoption of social media in higher educational institutions. The identification of these factors would assist the decision-makers of these institutions in making informed decisions regarding the employment of social media platforms. Limitations and future work were also discussed.

Keywords Adoption · Social media · Higher education · Systematic review · TAM

N. Al-Qaysi (✉) · N. Mohamad-Nordin
Faculty of Art, Computing & Creative Industry, Universiti Pendidikan Sultan Idris, Tanjung Malim, Malaysia
e-mail: noor.j.alqaysi@gmail.com

N. Mohamad-Nordin
e-mail: norhisham@fskik.upsi.edu.my

N. Mohamad-Nordin
College of Economics, Management and Information Systems, University of Nizwa, Nizwa, Oman

M. Al-Emran
Department of Information Technology, Al Buraimi University College, Al Buraimi, Oman
e-mail: mustafa.n.alemran@gmail.com

M. Al-Emran et al. (eds.), *Recent Advances in Intelligent Systems and Smart Applications*, Studies in Systems, Decision and Control 295,
https://doi.org/10.1007/978-3-030-47411-9_31

1 Introduction

Social media has altered the landscape through which information can be easily shared among users [1–3]. Social media term refers to the Internet-based applications that are built on the conceptual and technological foundations of Web 2.0 to allow the creation and exchange of user-generated content [4]. According to the Webster online dictionary, the term social media also refers to the electronic communication websites (e.g., "Twitter", "Facebook", "Google", "blog", and "LinkedIn"), where users can create online communities to share information, ideas, and personal messages among many others.

Technology adoption by users marks an important step prior to the use of any technology [5]. Considering the education sector, social media acceptance has received a considerable attention from the perspective of many researchers [6–11]. However, the decision to adopt social media still needs further investigation [12]. This can be accomplished through the understanding of the factors affecting its adoption. Those factors could be individual factors, technological factors, or social factors [13].

There are several information systems (IS) theories and models that were developed to explain the users' adoption of any technology. Amongst these models and theories, the "Technology Acceptance Model (TAM)" is considered as one of the most influential IS models [14]. TAM is a simple, adaptable, and sound model that has been frequently employed to analyze the acceptance of new technology [15]. Likewise, the validity of TAM has been confirmed across many application areas [16]. The popularity of TAM stems from several features, including parsimony, verifiability, and generalizability [17]. First, parsimony means simplicity; it is used as a guideline to develop a successful information system [18]. Second, verifiability means supported by data. Third, generalizability means the ability to predict the acceptance of new technology across multi-contexts [19]. The original TAM has been extensively modified to improve its validity and applicability to several technologies [14, 20]. Concerning social media, TAM has been recently found as the most frequent theoretical model used to explain the users' acceptance of social media [21].

In line with the existing literature, a number of review studies were conducted to study the TAM on the one hand, and the social media adoption on the other hand. Nevertheless, understanding the factors affecting the adoption of social media through the lenses of TAM is still not yet examined. Therefore, this study aims to systematically review the TAM-based social media studies for identifying the most dominant factors. This study also attempts to determine the most dominant publication venues through which the TAM-based social media studies were published. More specifically, the following research questions were formulated:

RQ1: What are the most frequent factors affecting social media adoption?

RQ2: What are the most dominant publication venues through which the TAM-based social media studies were published?

2 Literature Review

People and societies seem unable to escape the existence of the growing phenomenon "social networking sites" in their everyday lives. Examining the acceptance of these sites is increasingly attracting the scholars' attention [22, 23]. In order to understand the acceptance of social media, there is a need to examine the factors affecting its use [24].

An increasing number of theoretical models were developed to explain individuals' rejection or acceptance of new technology [25–27]. These theories and models include "Theory of Reasoned Action (TRA)" [28], "Theory of Planned Behavior (TPB)" [29], "Innovative Diffusion Theory (IDT)" [30], "Technology Acceptance Model (TAM)" [27], "Unified Theory of Acceptance and Use of Technology (UTAUT)" [26], "Cognitive Load Theory (CLT)" [31], and "Social Cognitive Theory (SCT)" [32]. Among these theories and models, the TAM is considered as one of the best effective models for predicting the factors that influence technology acceptance [14, 26, 27, 33]. Besides, the TAM has been extensively adopted in different contexts for predicting the attributes that affect the individuals' acceptance of technology [12, 14, 34–40].

Technology acceptance could be indirectly affected by numerous external variables through two main constructs, namely "perceived usefulness" and "perceived ease of use" [18, 41]. The original TAM has been modified for improving its applicability and validity to numerous technologies [14, 20, 42, 43]. Therefore, several review studies were carried out in the past to analyze the TAM and its application areas, as illustrated in Table 1.

Table 1 TAM-based review studies

Source	Study type	Number of analyzed studies	Results
[44]	Meta-analytic review	145	TAM is considered as a basic model to be utilized for the purposes of recognizing gaps and providing clear guidelines for management implementation
[45]	Quantitative meta-analytic review	95	The main focus of the TAM empirical investigations was on modeling intention for its impact on self-reported usage behavior
[36]	Concept-centric literature review	85	Several areas of the TAM application still need to be further investigated

With respect to the higher educational context, the significance of social media acceptance has been widely studied [6–8, 46, 47]. Higher educational institutions are required to investigate the determinants that affect the students' acceptance of social media as a prior step to the implementation process. In line with this trend, an increasing number of review studies were carried out to investigate the social media use in different educational fields, as shown in Table 2.

3 Method

The literature review is defined as a method to highlight the applicable sources to the topic under study and contribute to the research relevance [58]. A systematic literature review (SLR) is a powerful step-by-step approach that helps researchers to clearly depict the research findings [59]. SLR approach also helps researchers to collect, identify, and synthesize evidence based on specific guidelines [60], and find gaps in the research domain [57]. The present SLR follows the systematic review guidelines provided by Kitchenham and Charters [61], and other relevant systematic reviews [62–65]. The following subsections present the steps through which the current SLR follows.

3.1 Inclusion and Exclusion Criteria

The inclusion and exclusion criteria have been formulated to ensure the relevance of the collected studies in the analysis stage. In that, each study should be critically analyzed to meet the inclusion and exclusion criteria. Figure 1 describes the inclusion and exclusion criteria for the present SLR.

3.2 Data Sources and Search Strategies

The surveyed articles in this systematic review were collected from a wide range of online databases namely, "Emerald", "Wiley", "IEEE", "ScienceDirect", "Springer", "Taylor & Francis", and "Google Scholar". The search keywords include the string [("TAM" OR "Technology Acceptance Model") AND ("Social Media" OR "Facebook" OR "Twitter" OR "Instagram" OR "WhatsApp" OR "YouTube" OR "Wikis" OR "Blogs")]. The search for these articles was undertaken in September 2017. The search results showed that the total number of the collected studies is 539. A total of 45 studies were removed as duplicates. In that, a total of 494 studies remained. Based on the inclusion and exclusion criteria, the total number of the included articles in the analysis stage becomes 57. Figure 2 represents the systematic review process and the number of articles determined at each stage.

Table 2 A list of review studies in social media

Source	Study type	Field	Number of analyzed studies	Results
[48]	Systematic review	Pharmacy education	24	Students were positive towards using Facebook, Twitter, and Wikis as instructive applications in the field of pharmacy education
[49]	Systematic review	Medical education	6	Using social media in psychiatric training is still in its early stages
[50]	Meta-analysis and systematic review	Medical education	10	The majority of students used social networking sites, yet the minority used such sites for sharing educational information
[51]	Systematic review	Medical education	9	Educators admitted using SNSs (Facebook and Twitter), and a custom-made website (MedicineAfrica) was established to reach their goals
[52]	Integrative literature review	Nursing education	14	Identifying the benefits of using Facebook, Ning, Twitter, and Myspace by students, educators, professionals, and institutions in the process of teaching and learning
[53]	Integrative literature review	Nursing education	25	Wikis improved students' learning outcomes and helped them to learn several skills and gain new knowledge

(continued)

Table 2 (continued)

Source	Study type	Field	Number of analyzed studies	Results
[54]	Systematic review	Special education	11	Social media increased deaf or hard of hearing students' interaction, learning motivation, support, and feedback
[55]	Preliminary systematic review	E-learning	37	Analyzing the interactions among students in several collaborative settings. Such interactions mainly result from the communication among students via chats, e-mails, blogs, and wikis
[56]	Systematic review	Information-seeking behavior	71	The information-seeking behavior was highlighted as an information source with respect to social media. Moreover, the students' information needs were identified and classified by the roles in which social media played
[57]	Systematic review	Environmental sustainability awareness	82	Students need to fully leverage the social media use for environmental sustainability practices like recycling, water and electricity consumption reduction, and paper reduction to enjoy green higher educational settings

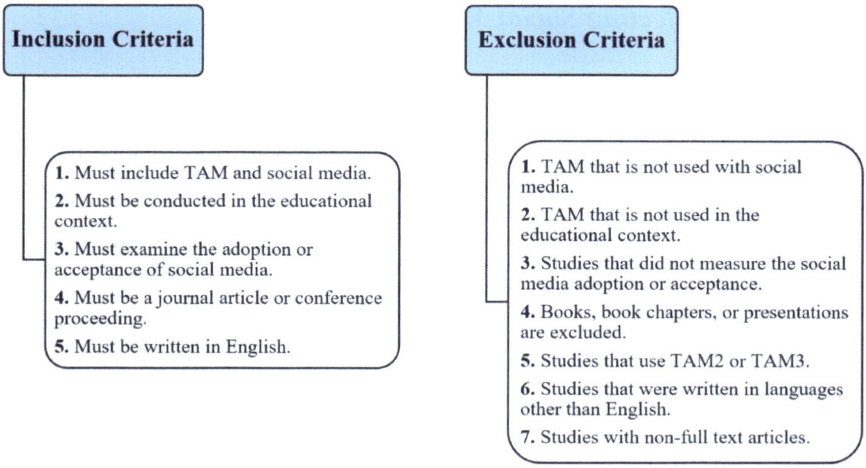

Fig. 1 Inclusion and exclusion criteria

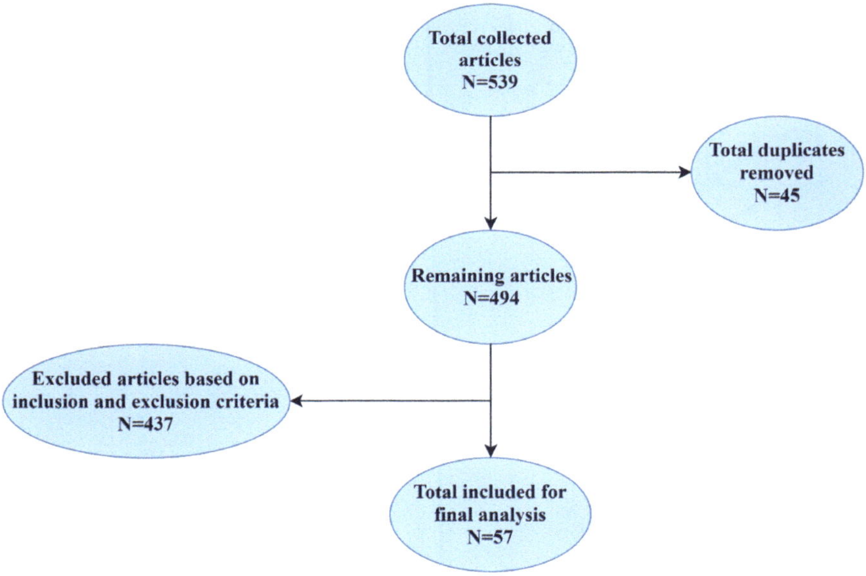

Fig. 2 Systematic review process

4 Results and Discussion

The answers to the research questions were approached under the following subsections.

4.1 Factors Affecting Social Media Adoption

For identifying the most frequent factors affecting the adoption of social media, we have analyzed the external factors to TAM. In that, the frequency and significance of each external factor have been counted. It is important to report that we have only considered the factors that achieved significant results. Also, the factors that appeared only once in the analyzed studies were discarded. Figure 3 illustrates the most frequent factors affecting the adoption of social media. These factors were appeared at least twice in the analyzed studies and achieved significant results. It can be noticed that "perceived enjoyment", "subjective norm", "self-efficacy", "perceived playfulness", "perceived critical mass", and "openness" were the most frequent factors that significantly influenced the social media adoption. It is imperative to report that "perceived enjoyment" also refers to "perceived playfulness" [66].

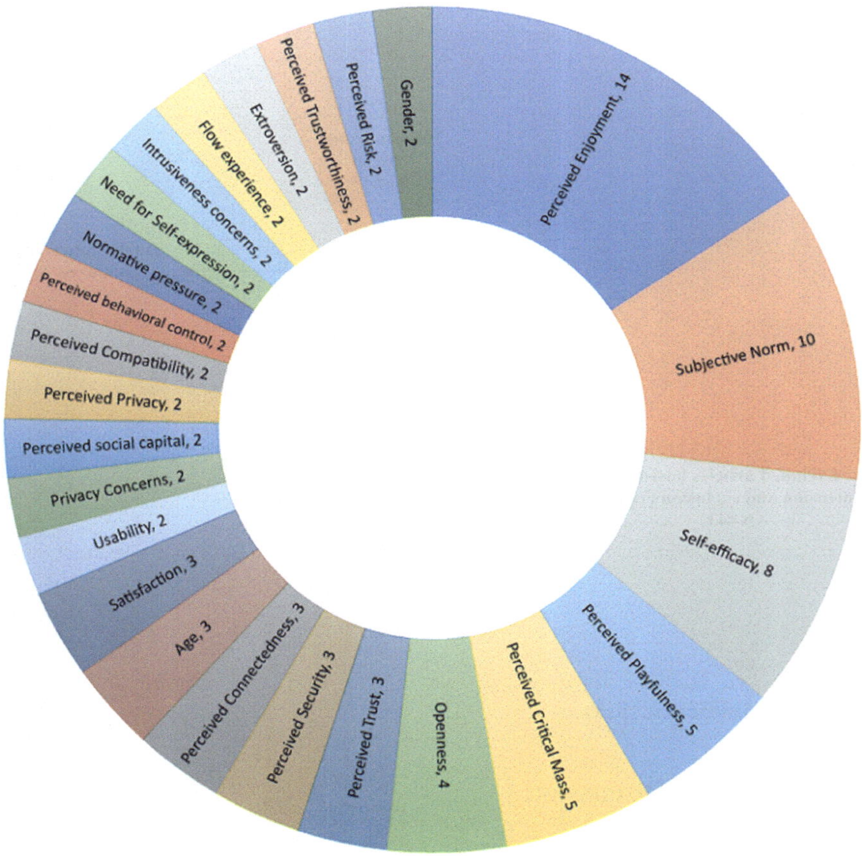

Fig. 3 Factors affecting social media adoption

4.2 Publication Venues

Identifying the most frequent publication venues (e.g., journals and conferences) can assist researchers in finding the appropriate venue to publish their TAM-based social media studies. Table 3 exhibits the most frequent publication venues through which the analyzed studies were published. It can be seen that conferences represent the first category (N = 4) through which the analyzed studies were published. Concerning the scientific journals, it is obvious that the most active journals that published the TAM-based social media studies are "Computers in Human Behavior" (N = 3), "International Journal of Human–Computer Interaction" (N = 3), "Education and Information Technologies" (N = 2), "Interactive Learning Environments" (N = 2), "Journal of Computer Information Systems" (N = 2), "Journal of Direct, Data and Digital Marketing Practice" (N = 2), "Journal of Research in Interactive Marketing" (N = 2), and "Telematics and Informatics" (N = 2).

5 Conclusion

5.1 Research Contributions

Through the lenses of TAM, the current systematic review contributes to the existing literature by determining the most frequent factors affecting the adoption of social media in higher education. The identification of these factors would assist the decision-makers of the higher educational institutions to make informed decisions regarding the employment of social media platforms. In addition, further research could also rely on the identified external factors in examining the adoption of social media in contexts other than education. This research also determined the most frequent scientific journals through which the TAM-based social media studies were published. This would assist scholars in selecting the appropriate journal for publishing their perspective research.

5.2 Limitations and Future Work

This systematic review throws up some limitations which need to be considered in future trials. First, the present study only considered analyzing the factors affecting the adoption of social media and the publication venues through which the analyzed studies were published. Further research could extend this systematic review by considering other attributes, such as educational level, participants (i.e., undergraduates and postgraduates), and active countries among many others. Second, this systematic review only considered analyzing the external factors to TAM in determining the factors affecting the adoption of social media. Further trials might consider examining

Table 3 Most frequent publication venues

Publication venue	Frequency
Conference	4
"Computers in Human Behavior"	3
"International Journal of Human–Computer Interaction"	3
"Education and Information Technologies"	2
"Interactive Learning Environments"	2
"Journal of Computer Information Systems"	2
"Journal of Direct, Data and Digital Marketing Practice"	2
"Journal of Research in Interactive Marketing"	2
"Proceedings of the Association for Information Science and Technology"	2
"Telematics and Informatics"	2
"Aslib Journal of Information Management"	1
"Behaviour & Information Technology"	1
"British Journal of Educational Technology"	1
"Communication Education"	1
"Computers & Education"	1
"Family and Consumer Sciences Research Journal"	1
"Information and Communication Technologies in Tourism"	1
"Information Systems Frontiers"	1
"Innovations in Education and Teaching International"	1
"Interdisciplinary journal of contemporary research in business"	1
"International Journal of Business and Management"	1
"International Journal of Educational Technology in Higher Education"	1
"International Journal of Retail & Distribution Management"	1
"Internet Research"	1
"Journal of Big Data"	1
"Journal of Consumer Behaviour"	1
"Journal of Current Issues & Research in Advertising"	1
"Journal of Interactive Advertising"	1
"Journal of King Saud University – Computer and Information Sciences"	1
"Journal of Marketing Management"	1
"Management Research Review"	1
"Multimedia Tools and Applications"	1
"On the Horizon"	1
"Online Information Review"	1
"Performance Improvement Quarterly"	1
"Procedia-Social and Behavioral Sciences"	1

(continued)

Table 3 (continued)

Publication venue	Frequency
"Psychology & Marketing"	1
"The Electronic Library"	1
"The International Journal of Information and Learning Technology"	1
"The Social Science Journal"	1
"Universal Access in the Information Society"	1
"VINE Journal of Information and Knowledge Management Systems"	1
"Journal of Enterprise Information Management"	1

the role of moderating variables in explaining the adoption of social media in higher educational institutions.

References

1. Salloum, S.A., Al-Emran, M., Abdallah, S., Shaalan, K.: Analyzing the Arab Gulf newspapers using text mining techniques. In: International Conference on Advanced Intelligent Systems and Informatics, pp. 396–405 (2017)
2. Mhamdi, C., Al-Emran, M., Salloum, S.A.: Text Mining and Analytics: A Case Study from News Channels Posts on Facebook, Vol. 740 (2018)
3. Salloum, S.A., Al-Emran, M., Shaalan, K.: Mining text in news channels: a case study from Facebook. Int. J. Inf. Technol. Lang. Stud. **1**(1), 1–9 (2017)
4. Kaplan, A.M., Haenlein, M.: Users of the world, unite! The challenges and opportunities of Social Media. Bus. Horiz. **53**(1), 59–68 (2010)
5. Al-Emran, M., Mezhuyev, V., Kamaludin, A.: Technology acceptance model in M-learning context: a systematic review. Comput. Educ. **125**, 389–412 (2018)
6. Akram, M.S., Albalawi, W.: Youths' social media adoption: theoretical model and empirical evidence. Int. J. Bus. Manag. **11**(2), 22 (2016)
7. Pinho, J.C.M.R., Soares, A.M..: Examining the technology acceptance model in the adoption of social networks. J. Res. Interact. Mark. **5**(2/3), 116–129 (2011)
8. Hartzel, K.S., Marley, K.A., Spangler, W.E.: Online social network adoption: a cross-cultural study. J. Comput. Inf. Syst. **56**(2), 87–96 (2016)
9. M. Alshurideh, Salloum, S.A., Al Kurdi, B., Al-Emran, M.: Factors affecting the social networks acceptance: an empirical study using PLS-SEM approach. In: 8th International Conference on Software and Computer Applications, pp. 414–418 (2019)
10. Al-Qaysi, N., Mohamad-Nordin, N., Al-Emran, M., Al-Sharafi, M.A.: Understanding the differences in students' attitudes towards social media use: a case study from Oman. In: 2019 IEEE Student Conference on Research and Development (SCOReD), pp. 176–179 (2019)
11. Al-Qaysi, N., Mohamad-Nordin, N., Al-Emran, M.: An empirical investigation of students' attitudes towards the use of social media in Omani higher education. In: International Conference on Advanced Intelligent Systems and Informatics, pp. 350–359 (2019)
12. Rauniar, R., Rawski, G., Yang, J., Johnson, B.: Technology acceptance model (TAM) and social media usage: an empirical study on Facebook. J. Enterp. Inf. Manag. **27**(1), 6–30 (2014)
13. Al-Saedi, K., Al-Emran, M., Abusham, E., El-Rahman, S.A.: Mobile payment adoption: a systematic review of the UTAUT model. In: International Conference on Fourth Industrial Revolution (2019)

14. Venkatesh, V., Davis, F.D.: A theoretical extension of the technology acceptance model: four longitudinal field studies. Manage. Sci. **46**(2), 186–204 (2000)
15. King, W.R., He, J.: A meta-analysis of the technology acceptance model. Inf. Manag. **43**(6), 740–755 (2006)
16. Venkatesh, V., Bala, H.: Technology acceptance model 3 and a research agenda on interventions. Decis. Sci. **39**(2), 273–315 (2008)
17. Lee, Y., Kozar, K., Larsen, K.: The technology acceptance model: past, present, and future. Commun. … **12**(1), 752–780 (2003)
18. Venkatesh, V., Davis, F.D.: A model of the antecedents of perceived ease of use: development and test. Decis. Sci. **27**(3), 451–481 (1996)
19. Chintalapati, N., Daruri, V.S.K.: Examining the use of YouTube as a learning resource in higher education: Scale development and validation of TAM model. Telemat. Informatics **34**(6), 853–860 (2017)
20. Teo, T., Lee, C.B., Chai, C.S.: Understanding pre-service teachers' computer attitudes: applying and extending the technology acceptance model. J. Comput. Assist. Learn. **24**(2), 128–143 (2008)
21. Al-Qaysi, N., Mohamad-Nordin, N., Al-Emran, M.: A systematic review of social media acceptance from the perspective of educational and information systems theories and models. J. Educ. Comput. Res. **57**(8), 2085–2109 (2020)
22. Hester, V.: Innovating with organizational wikis : factors facilitating adoption and diffusion of an effective collaborative knowledge management system. Knowl. Creat. Diffus. Util., 161–163 (2008)
23. Al-Qaysi, N., Mohamad-Nordin, N., Al-Emran, M.: What leads to social learning? Students' attitudes towards using social media applications in Omani higher education. Educ. Inf. Technol. (2019)
24. Teo, T.: Modelling technology acceptance in education: a study of pre-service teachers. Comput. Educ. **52**(2), 302–312 (2009)
25. Rogers, E.M.: Diffusion of Innovations Theory, p. 5th edn. New York Free Press (2003)
26. Venkatesh, V., Morris, M., Davis, G., Davis, F.: User acceptance of information technology: toward a unified view. MIS Q. **27**(3), 425–478 (2003)
27. Davis, F.D.: Perceived usefulness, perceived ease of use, and user acceptance of information technology. MIS Q. **13**(3), 319–340 (1989)
28. Fishbein, M., Ajzen, I.: Belief, attitude, intention and behaviour: an introduction to theory and research. Read. MA Addison Wesley, no. August, p. 480 (1975)
29. Ajzen, I.: The theory of planned behavior. Organ. Behav. Hum. Decis. Process. **50**(2), 179–211 (1991)
30. Rogers, R.W.: Cognitive and physiological processes in fear appeals and attitude change: a revised theory of protection motivation. In: Social Psychophysiology: A Sourcebook, pp. 153–176 (1983)
31. Sweller, J.: Cognitive load during problem solving: effects on learning. Cogn. Sci. **12**(2), 257–285 (1988)
32. Compeau, D.R., Higgins, C.A.: Computer self-efficacy: development of a measure and initial test. MIS Q. **19**(2), 189–211 (1995)
33. Lederer, A.L., Maupin, D.J., Sena, M.P., Zhuang, Y.: The technology acceptance model and the World Wide Web. Decis. Support Syst. **29**(3), 269–282 (2000)
34. Lu, J., Yu, C., Liu, C., Yao, J.E.: Technology acceptance model for wireless Internet. Internet Res. **13**(3), 206–222 (2003)
35. Birch, D., Burnett, B.: Advancing e-learning policy and practice: influences on academics' adoption, integration and development of multimodal e-learning courses. In: Institutional Transformation through Best Practices in Virtual Campus Development: Advancing E-learning Policies, IGI Global, pp. 65–80 (2009)
36. Marangunić, N., Granić, A.: Technology acceptance model: a literature review from 1986 to 2013. Univers. Access Inf. Soc. **14**(1), 81–95 (2015)

37. Al-Emran, M., Mezhuyev, V., Kamaludin, A.: Towards a conceptual model for examining the impact of knowledge management factors on mobile learning acceptance. Technol. Soc. (2020)
38. Al-Emran, M., Arpaci, I., Salloum, S.A.: An empirical examination of continuous intention to use m-learning: an integrated model. Educ. Inf. Technol. (2020)
39. Al-Emran, M., Teo, T.: Do knowledge acquisition and knowledge sharing really affect e-learning adoption? An empirical study. Educ. Inf. Technol. (2019)
40. Salloum, S.A., Al-Emran, M.: Factors affecting the adoption of E-payment systems by university students: extending the TAM with trust. Int. J. Electron. Bus. **14**(4), 371–390 (2018)
41. Legris, P., Ingham, J., Collerette, P.: Why do people use information technology? A critical review of the technology acceptance model. Inf. Manag. **40**(3), 191–204 (2003)
42. Moon, J.W., Kim, Y.G.: Extending the TAM for a World-Wide-Web context. Inf. Manag. **38**(4), 217–230 (2001)
43. Park, Y., Son, H., Kim, C.: Investigating the determinants of construction professionals' acceptance of web-based training: An extension of the technology acceptance model. Autom. Constr. **22**, 377–386 (2012)
44. Yousafzai, S.Y., Foxall, G.R., Pallister, J.G.: Technology acceptance: a meta- analysis of the TAM: Part 1. J. Model. Manag. **2**(2007), 251–280 (2007)
45. Yousafzai, S.Y., Foxall, G.R., Pallister, J.G.: Technology acceptance: a meta-analysis of the TAM: Part 2. J. Model. Manage. **2**(3), 281–304 (2007)
46. Dhume, S.M., Pattanshetti, M.Y., Kamble, S.S., Prasad, T.: Adoption of social media by business education students: application of Technology Acceptance Model (TAM). In: 2012 IEEE International Conference on Technology Enhanced Education (ICTEE), pp. 1–10 (2012)
47. Srinivasan, N., Damsgaard, J.: Tensions between individual use and network adoption of social media platforms. In: International Working Conference on Transfer and Diffusion of IT, pp. 261–278 (2013)
48. Benetoli, A., Chen, T.F., Aslani, P.: The use of social media in pharmacy practice and education. Res. Soc. Admin. Pharm. **11**(1), 1–46 (2015)
49. O'Hagan, T.S., Roy, D., Anton, B., Chisolm, M.S.: Social media use in psychiatric graduate medical education: where we are and the places we could go. Acad. Psychiatry **40**(1), 131–135 (2016)
50. Guraya, S.Y.: The usage of social networking sites by medical students for educational purposes: a meta–analysis and systematic review. North Am. J. Med. Sci. **8**(7), 268–278 (2016)
51. Cartledge, P., Miller, M., Phillips, B.: The use of social-networking sites in medical education. Med. Teach. **35**(10), 847–857 (2013)
52. Kakushi, L.E., Évora, Y.D.M.: Social networking in nursing education: integrative literature review. Rev. Lat. Am. Enfermagem. **24**(0) (2016)
53. Trocky, N.M., Buckley, K.M.: Evaluating the impact of wikis on student learning outcomes: an integrative review. J. Prof. Nurs. **32**(5), 364–376 (2016)
54. Toofaninejad, E., Zaraii Zavaraki, E., Dawson, S., Poquet, O., Sharifi Daramadi, P.: Social media use for deaf and hard of hearing students in educational settings: a systematic review of literature. Deaf. Educ. Int., 1–18 (2017)
55. Cela, K.L., Sicilia, M.A., Sanchez, S.: Social network analysis in E-Learning environments: a preliminary systematic review. Educ. Psychol. Rev. **27**(1), 219–246 (2015)
56. Hamid, S., Bukhari, S., Ravana, S.D., Norman, A.A., Ijab, M.T.: Role of social media in information-seeking behaviour of international students: a systematic literature review. Aslib J. Inf. Manag. **68**(5), 643–666 (2016)
57. Hamid, S., Ijab, M.T., Sulaiman, H., Anwar, RMd, Norman, A.A.: Social media for environmental sustainability awareness in higher education. Int. J. Sustain. High. Educ. **18**(4), 474–491 (2017)
58. Baker, M.J.: Writing a literature review. Mark. Rev. **1**(2), 219–247 (2000)
59. Kitchenham, B.: Procedures for performing systematic reviews. Keele Univ. Natl. ICT Aust. (2004)
60. Kitchenham, B., Pearl Brereton, O., Budgen, D., Turner, M., Bailey, J., Linkman, S.: Systematic literature reviews in software engineering—a systematic literature review. In: Information and Software Technology (2009)

61. Kitchenham, B., Charters, S.: Guidelines for performing systematic literature reviews in software engineering. Softw. Eng. Group, Sch. Comput. Sci. Math. Keele Univ., 1–57 (2007)
62. Wu, W.-H., Wu, Y.-C.J., Chen, C.-Y., Kao, H.-Y., Lin, C.-H., Huang, S.-H.: Review of trends from mobile learning studies: a meta-analysis. Comput. Educ. **59**(2), 817–827 (2012)
63. Sung, Y.-T., Chang, K.-E., Liu, T.-C.: The effects of integrating mobile devices with teaching and learning on students' learning performance: a meta-analysis and research synthesis. Comput. Educ. **94**, 252–275 (2016)
64. Saa, A.A., Al-Emran, M., Shaalan, K.: Factors affecting students' performance in higher education: a systematic review of predictive data mining techniques. Technol. Knowl. Learn. (2019)
65. Al-Emran, M., Mezhuyev, V., Kamaludin, A., Shaalan, K.: The impact of knowledge management processes on information systems: a systematic review. Int. J. Inf. Manage. **43**, 173–187 (2018)
66. Dumpit, D.Z., Fernandez, C.J.: Analysis of the use of social media in Higher Education Institutions (HEIs) using the technology acceptance model. Int. J. Educ. Technol. High. Educ. **14**(1) (2017)

Online Social Network Analysis for Cybersecurity Awareness

Mazen Juma and Khaled Shaalan

Abstract The importance of analyzing social networks lies in enabling researchers to understand better the relationships among network actors over time through social data mining and patterns examining social network growth. This paper aims to analyze the topology structure and critical influences on Facebook for employees' group of cybersecurity awareness. The hypotheses were tested and analyzed following the quantitative research methodology through conducting appropriate statistical tests, such as correlations, t-tests, ANOVA, and regression. The independent variables were gender, age, education, seniority, and cyber compromising; the dependent variable was cybersecurity awareness. Data collection for the employee's group acquired online by Facebook. The sample includes 1033 active employees out of 9264 employees representing the study population. The results provided several pieces of evidence that online communications by young employees were more efficient than older ones, and cybersecurity awareness of males was better than females. Nevertheless, there was a strong correlation between seniority employees, and the effectiveness of their cybersecurity awareness, as well as the topology structure, had profound effects on the ability of the employees' group to fulfill their objectives. Employees with higher education had more influence than the others; but in contrast, there was no apparent difference between the influence of employees whereas compromised or non-compromised.

Keywords Online social network · Social network analysis · Cybersecurity awareness

---

M. Juma (✉) · K. Shaalan
Faculty of Engineering and Information Technology, British University in Dubai, Dubai International Academic City, Block 11, 1st and 2nd Floor, Dubai, UAE
e-mail: mazen.juma@ieee.org

K. Shaalan
e-mail: Khaled.Shaalan@buid.ac.ae

M. Al-Emran et al. (eds.), *Recent Advances in Intelligent Systems and Smart Applications*, Studies in Systems, Decision and Control 295,
https://doi.org/10.1007/978-3-030-47411-9_32

1 Introduction

The increasing popularity of the Online Social Network (OSN) was a stemmed from a massive number of users acquired in a short amount of time; some social networking services now have gathered hundreds of millions of users, e.g., Facebook, Myspace, and Twitter. The growing accessibility of the Internet, which is available to most users 24/7, through several media, encourages them to build a reliable online interconnection of relationships [1]. As OSNs become the tools of choice for connecting people, their structure gradually mirrored their real-life society and relationships. For example, Facebook, with an estimated 13 million transactions per seconds at the peak, is one of the most challenging computer science artifacts, posturing several optimization and robustness challenges [2]. The analysis of online social network connection has become a scientific interest that requires investigations at multiple levels. For example, large-scale studies of models reflecting a real community structure and its behavior were impossible before. Moreover, data defined by some structural constraints, usually provided by the online social network structure itself, concerning real-life relations are often hardly identifiable [3]. A social network is a social structure made up of people or actors called nodes, which tied by one or more specific types of interdependent relationships or interactions between these actors, called edges or links, such as friendship, common interest, financial exchange, relationships of beliefs, and knowledge [4]. Social networks and the techniques to analyze them existed for decades. There can be several types of social networks, such as an email or telephone network and collaboration network. Recently, online social networks like Facebook, Twitter, LinkedIn, and Myspace have developed such that they gained popularity within a short amount of time and gathered a large number of users [5]. Facebook serves more than 1.7 billion users worldwide in 2020. The field of social networks and their analysis have evolved from graph theory, statistics, and sociology and now used in several other fields of information science, business application, communication, and economy [6]. Analyzing a social network is similar to the analysis of a graph because a social network forms the topology of a graph. Graph analysis tools have been there for decades. However, they not designed for analyzing a social network graph, which has sophisticated properties. An online social network graph may be enormous. It may contain millions of nodes and edges [7]. Analysis tasks of social networks include discovering the structure of the social network, finding communities in the social network, and various attribute values for the network, such as diameter, centrality, Betweenness, shortest paths, density, and visualizing the whole or part of the social network [8]. For instance, Franchi et al. [9] studied the use of social networks analysis for online collaboration and team formation identification in firms and organizations. Froehlich [10] used mixed methods approaches to social networks analysis. Asim and Malik survey [11] is concerned with the access control techniques through social network analysis for disaster management. Alotaibi et al. worked on the analysis of the Facebook social network [12]. There are two kinds of social network analysis, egocentric and complete network analysis. Egocentric network analysis is concerned with the analysis of individual nodes. A network can have as many egos as nodes in the graph. Egos can be persons, organizations, or a whole society. In egocentric network analysis, individual behavior and its variation have been described [13]. Complete network analysis is concerned with the analysis of all the relationships among a set of nodes. Techniques such as subgroup and equivalence analysis and measures like centrality, closeness, and degree all require entire networks [14].

2 Problem Statement

The problem statement of this paper is twofold. The first problem is to analyze the topology structure and critical influences on Facebook for an employee group of cybersecurity awareness. Appropriate tools, such as the graph visualization and manipulation software (Gephi) used [15], to measure the primary statistical metrics of the acquired social network dataset including average path length metric to rank employees based on their influence over this employee group. Besides, the Betweenness centrality metric used to set minimum and maximum influences in the employee group. It is also required to measure the modularity metric to show the clustering of employees that are more densely connecting compared to the rest of the social network [16]. The second problem is to analyze, report, and interpret the results of the extracted dataset from Facebook of this group of employees, as a non-profit group, to conduct statistical tests, such as frequencies and descriptive to find the relevant variables for the subsequent tests. Bivariate statistics reported too, for example, correlation and ANOVA analysis, t-test and chi-square test for selected variables. Regression analysis for the selected group depends on variables and its associated independent variables to the prediction of binary outcomes [17]. Appropriate statistical packages, such as the IBM Statistical Package for the Social Sciences (SPSS) used [18].

3 Questions and Hypotheses

This paper seeks to answer three questions:

Q1 How does the online topology structure of an employee group affect its ability to fulfill its objectives for cybersecurity awareness successfully?

Q2 What kind of employees exerts the most influence within the online employee group?

Q3 Who are the employees that have efficient communications more than the rest of the group?

This paper has defined six hypotheses by its questions:

Q1 Hypotheses:

1H Topology structure affects the ability to fulfill the objectives of cybersecurity awareness.
2H Seniority of employees and active cybersecurity awareness are related.

Q2 Hypotheses:

3H Higher-education employees can predict their influences.
4H Compromised employees likely exert more influence on cybersecurity awareness than those non-compromised.

Q3 Hypotheses:

5H Cybersecurity awareness of males is better than the cybersecurity awareness of females.
6H Efficient communications between the employees differ based on their ages.

4 Methodology

This paper used quantitative research methods for sampling from the data source and extracting the required information, using the social network mechanisms for data collection and proceeding appropriate approaches for data analysis. There are several schemes used for OSN sampling that shows the average density of the social network. For example, the proportion of real connections and measure of the amount of interaction within a population. This paper used the uniform sampling-based algorithm of the topology filter that was provided by Gephi to get the sample size of 1033 employees of the dataset that are most influential employees on Facebook group out of 9264 employees which are the whole population for cybersecurity awareness. Online social network data collection requires deciding on a period for the relationships of interest that take on a given cross-sectional realization through joining as a member to the Facebook group of the employees for cybersecurity awareness that is targeted to analyze. This paper permanently will not mention any critical information about the employees who joined this Facebook group such as names and identifications to avoid the violation of their privacy and take into account that the data collected from this group merely used for research purposes. By using Netvizz tool, it was possible to get an interchange geographic data as a geographic data file generated from Facebook employees group and to provide designated statutes for data aggregate and demonstration. A broad range of standard relationships attributes and features for the active employees of this group, technically named as nodes, and the connections between them, technically named as edges, which form the cybersecurity awareness. This paper analyzes the observations about population parameters with various analytic procedures through applying a systematic process of formal testing that used to accept or reject the statistical hypotheses and describe, illustrate, condense, recap, and evaluate the collected data. The dataset visualizes the graph of the Facebook employee group that targeted to analyze [18]. The essential visualization features of online social network dataset, such as force atlas layout, pushes the best-connected hubs separately from each other and then determines the positions of the nodes that are interconnected within the clusters around these central hubs to produce the most precise view of this community of employees. The calculation of the primary metrics for the graph like average path length and ranking the nodes based on their size. Moreover, using the Betweenness centrality metric to set minimum and maximum sizes. The nodes ranged based on the Betweenness centrality metric, which measures the influence of each node in the online social network and shows the shortest path that appears randomly between dually selected nodes through the employee group. If a node has a high-level score of Betweenness centrality, it will be the account able node, which connects different clusters through exerting the highest influence within the overall topological structure. The modularity metric calculated to determine the most connected clusters, which have the dense nodes linked together more than the rest of the online social network. In addition, it identifies the faster clusters in the employee group that communicate cybersecurity awareness and spread messages to the remaining nodes within the employee group with the analysis of the statistical tests conducted on the selected depending variable with associated independent variables. The overall total of the sample was 1033 group employees (nodes) and 81,086 connections (edges) among them. The primary metrics for the graph has calculated automatically through statistics pane in Gephi, the average path length (an average number of steps along the shortest paths for all possible pairs of network nodes) is 1.2. It is relatively short and means that clusters of employee group are well connected and have high efficiency in transporting mass of information and cybersecurity awareness within the social network.

Figure 1, Network diameter (a length of longest of all computed shortest paths between all pairs of nodes in the network) is 1.5. Average degree (number of connections for each employee in the sample) is 9.7, which means that everyone in the group has Facebook friends with an average of 10 people who are also a part of that group.

Figure 2, Betweenness centrality (the number of shortest paths between any two nodes that pass through the particular node) is 3.3, meaning that every member in the group can reach any other member within three steps. Moreover, closeness centrality (the shortest path between the node and all other nodes within a close distance network) is 1.1 i.e., there is a short average distance to all other nodes in the network, as shown in Fig. 3.

From Fig. 4, eccentricity (distance between a node and the node that is furthest from it) is 0.98, which is relatively small. That means the furthest away nodes are quite close.

In the same way, Fig. 5 shows the resolution of the modularity class (the strength of division of a network into groups, clusters, or communities) with 0.578 for the six most massive clusters constitute close to 68% of employees of the group. It shows that

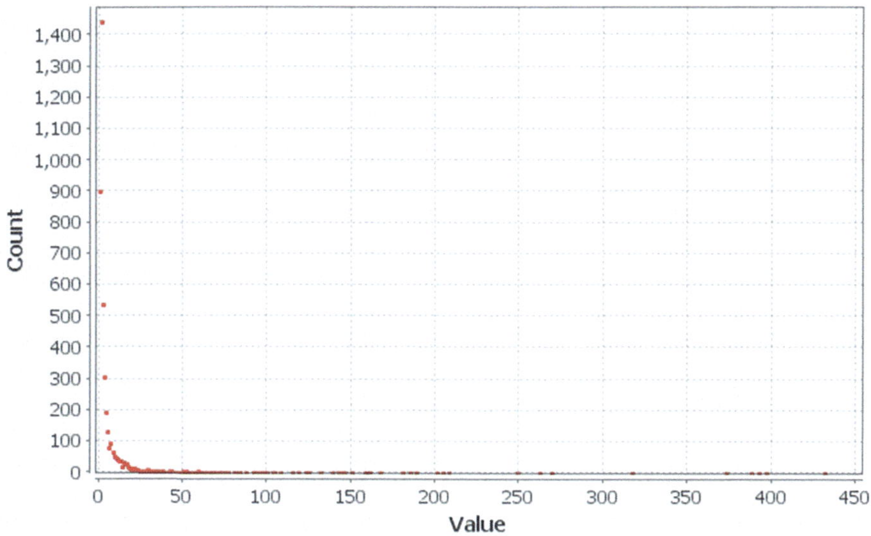

Fig. 1 Diameter

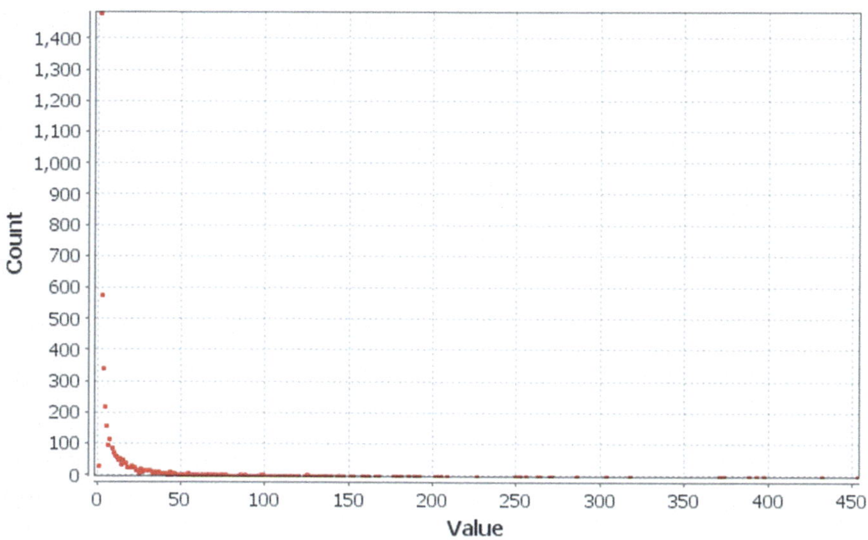

Fig. 2 Betweenness centrality

there are very dense connections between the nodes within clusters and community structure is optimizing to take faster actions for spreading incoming cybersecurity awareness to the rest of the group communities.

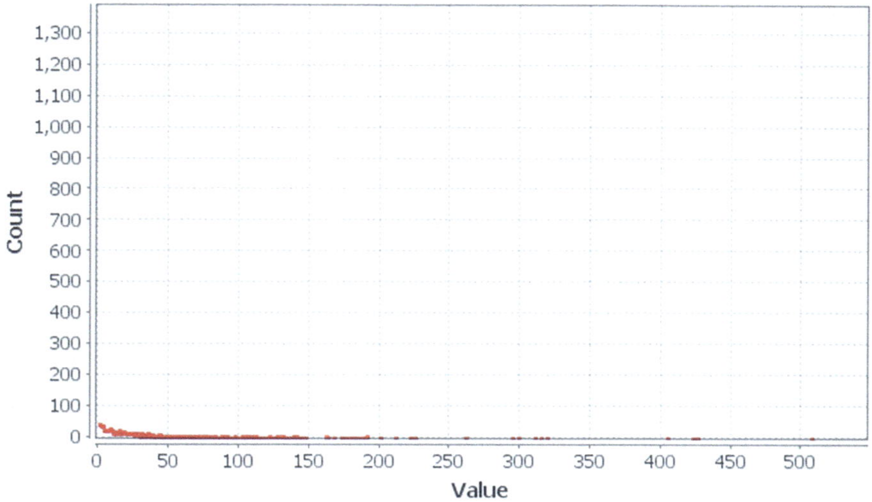

Fig. 3 Closeness centrality

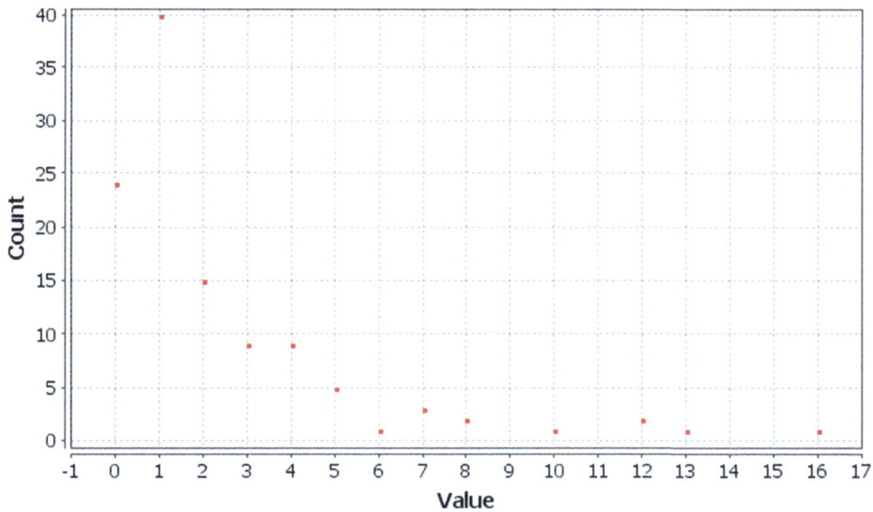

Fig. 4 Eccentricity

In Fig. 6, clustering coefficient (whole neighborhoods of the node over all the nodes in the entire network) is 0.419. Graph density (complete graph of the close network that has all possible edges and density equal to 1) is 0.088, which is high.

It means that the group employees have strong interconnection, better in propagating information and the lifetime of their cybersecurity awareness would be longer than the other groups in the completely social network.

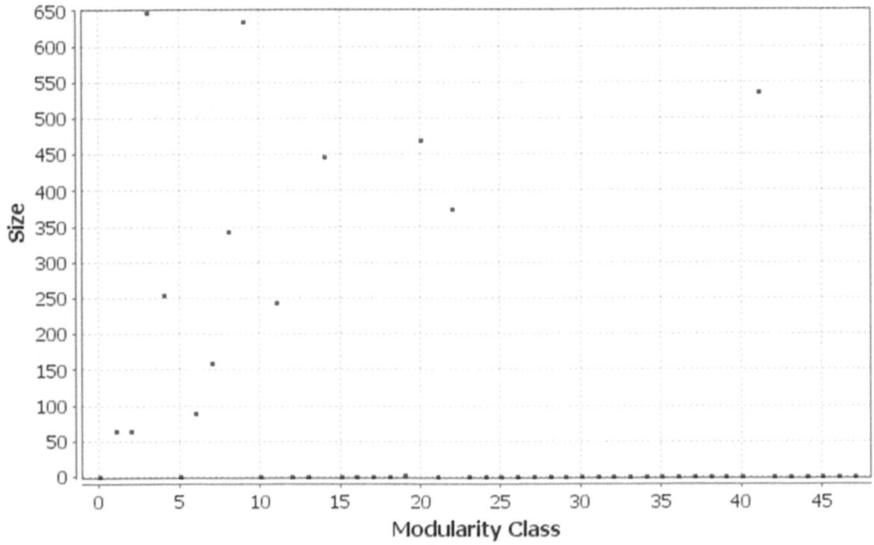

Fig. 5 Modularity

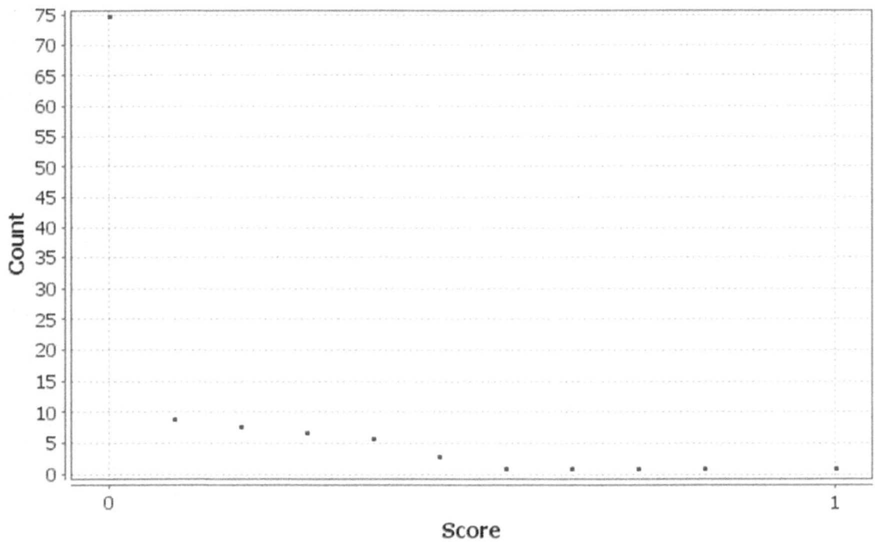

Fig. 6 Clustering coefficient

5 Results

The research data of Facebook group processed statistically through quantitative methods using Gephi and SPSS software for inspecting, cleaning, transforming, and modeling. In addition, the results were subject for reporting, interpreting, discussing, and collating with cross-referencing with the objectives of this paper to find out the valuable information to answer the questions, propose conclusions, and recommend helping decision-making.

5.1 Descriptive Statistics for Categorical Variables

Table 1 describes descriptive statistics for categorical variables in the sample; there are three categorical independent variables (compromising, gender, and education level) with an average of edges ($M = 402.3$, $SD = 218.85$) giving 1033 nodes as the total.

Table 2 describes the malicious cyber compromising that shows that the 585 or 56.6% compromised in terms of cybersecurity, and 448 or 43.4% were non-compromised.

Table 3 shows that 575 or 55.7% were males, and 458 or 44.3% were females in the sample.

Table 4 shows four levels of education: bachelor level was the highest with 568 or 55% of a bachelor degree while other levels were 188 or 18.2% high schools, 195 or 18.9% diploma, and 82 or 7.9% postgraduate.

Table 1 Statistics

	Edges	Compromising	Gender	Education
N	1033	1033	1033	1033
Mean	402.30	1.43	1.44	2.53
Standard deviation	218.849	0.496	0.498	0.880

Table 2 Compromising

	Frequency	Percent	Cumulative percent
Compromised	585	56.6	56.6
Non-compromised	448	43.4	100.0

Table 3 Gender

	Frequency	Percent	Cumulative percent
Male	575	55.7	55.7
Female	458	44.3	100.0

Table 4 Education

	Frequency	Percent	Cumulative percent
High school	188	18.2	18.2
Diploma	195	18.9	37.1
Bachelor	568	55.0	92.1
Post graduate	82	7.9	100.0

Table 5 Age and seniority descriptive statistics

	Min.	Max.	Mean	Standard deviation	Skewness	Kurtosis
Age	18	66	37.27	13.046	0.552	−0.859
Seniority	1	9	4.88	2.470	0.071	1.142

5.2 Descriptive Statistics for Continuous Variables

Table 5 presents descriptive information for both scores of the continuous variables' age and seniority. The range of employees' ages from 18 to 66 years had 37.27 years for the mean and 13.05 years for the standard deviation.

In addition, the range of seniority from one to nine years had 4.88 years for the mean and 2.47 years for the standard deviation. The positive skewness values indicate positive skews with 0.55 for age and 0.07 for seniority and appear as a non-symmetrical distribution. The value of kurtosis for age is below zero, which means the distribution is relatively flat with many instances in the extremes. On the other hand, kurtosis value above zero for seniority indicates a distribution that is not entirely normal with short, thin tails.

The negative values of the kurtosis distribution indicate that it peaked and clustered in the center, as shown in the histogram of Fig. 7.

5.3 Assessing Normality and Identifying Outliers

This section is about assessing normality and identifying outliers of the variable for cybersecurity compromising, other variables not discussed because of the lack of space available in this paper.

Table 6 divided into two sections corresponding to the two groups, compromised and non-compromised, as shown in the histograms of Figs. 8 and 9 respectively. The descriptive statistics recognize the following: mean = 371.2, median = 372.5, standard deviation = 230.5, minimum = 12 and maximum = 812 for compromised employees. Furthermore, descriptive information for non-compromised persons in the sample shows that mean = 442.84, median = 449, deviation = 196.07, minimum = 11, and maximum = 815.

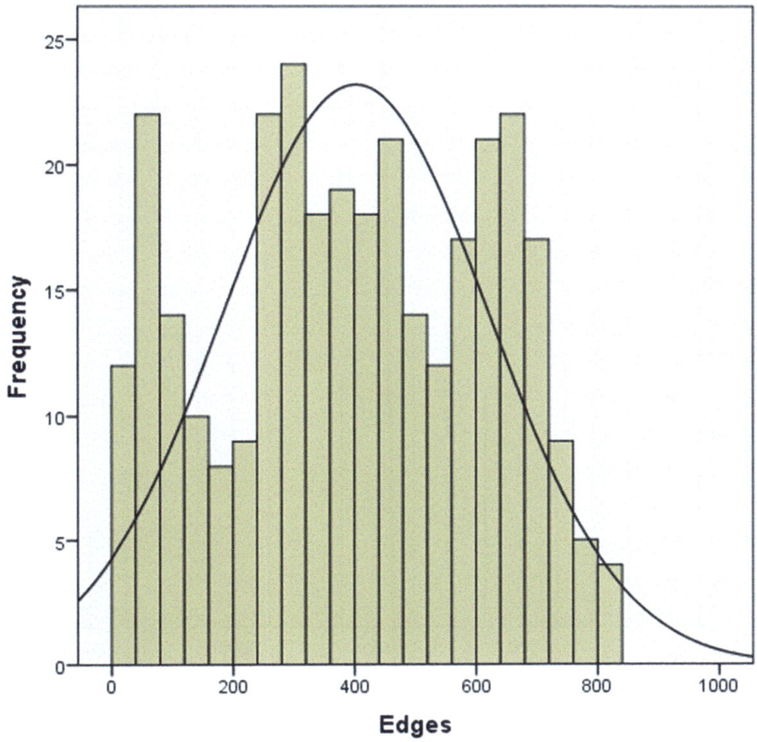

Fig. 7 Histogram of descriptive statistics for categorical variables

Table 6 Compromised and non-compromised description

	Compromising	Mean	Standard deviation	Median	Min.	Max.
Edges	Compromised	371.22	230.559	372.50	12	812
	Non-compromised	442.84	196.070	449.00	11	815

Table 7 shows the highest and lowest extreme values of the edges. The cybersecurity awareness of the nodes represent the active employees whether cybersecurity compromised or not in the sample, N6 is the highest compromised node with 812 edges value which is a little less than N4, the highest non-compromised node with 815 edges value. On the contrary, N114 is the lowest non-compromised node with 11 edges value, which is a little less than N112, the lowest compromised node with 12 edges value. In the following section, this paper shows topology structure for these nodes and edges.

Table 8 tests the normality of the distribution; given the scores of Shapiro-Wilk statistic, significant value is 0.012 for compromised and 0.023 for non-compromised. The results indicate normality of (p-value <0.05).

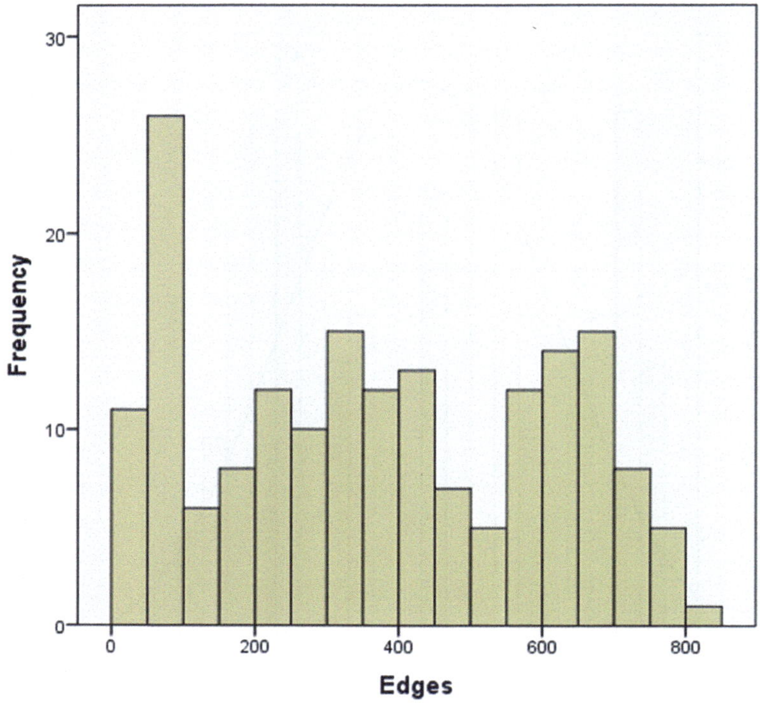

Fig. 8 Histogram of descriptive statistics for compromised

The value supported by probability examination of normal Q-Q plots shown in Figs. 10 and 11 respectively. In these plots, perceived scores for values plotted compared to expected scores within the normal distribution. A straight line indicated reasonably to a normal distribution.

In addition, the boxplot in Fig. 12 is providing the distribution of scores for both groups, which were a reasonably normal distribution. The line inside the rectangle is the median value.

The rectangle represents 50% of the edges with no values outside the range to be classified as outliers, although there are extreme points.

5.4 *Hypothesis (1) Testing*

$1H_0$ (Null Hypothesis)	Topology structure does not affect the ability to fulfill the objectives of cybersecurity awareness.
$1H_1$ (Alternative Hypothesis)	Topology structure affects the ability to fulfill the objectives of cybersecurity awareness.

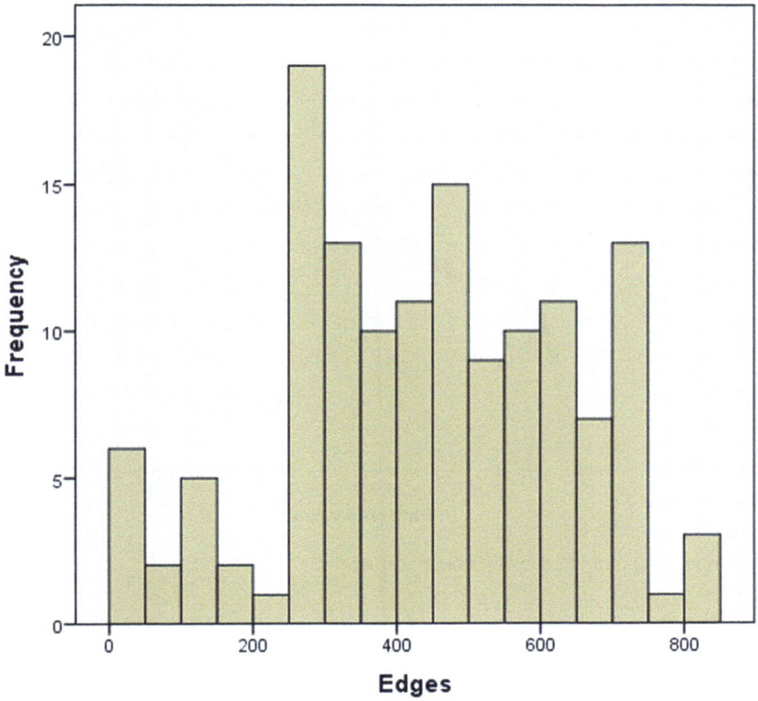

Fig. 9 Histogram of descriptive statistics for non-compromised

Table 7 Extreme values

	Compromising		Case number	Nodes	Value
Edges	Non-compromised	Highest	177	N4	815
		Lowest	65	N114	11
	Compromised	Highest	178	N6	812
		Lowest	64	N112	12

Table 8 Tests of normality

	Compromising	Kolmogorov-Smirnov			Shapiro-Wilk		
		Statistic	df	Sig.	Statistic	df	Sig.
Edges	Compromised	0.101	180	0.110	0.941	180	0.012
	Non-compromised	0.053	138	0.200	0.978	138	0.023

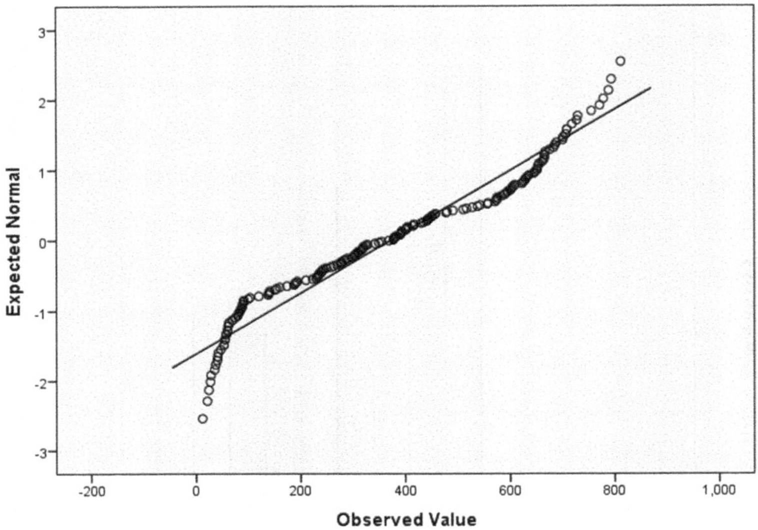

Fig. 10 Normal Q-Q plot of edges for compromised

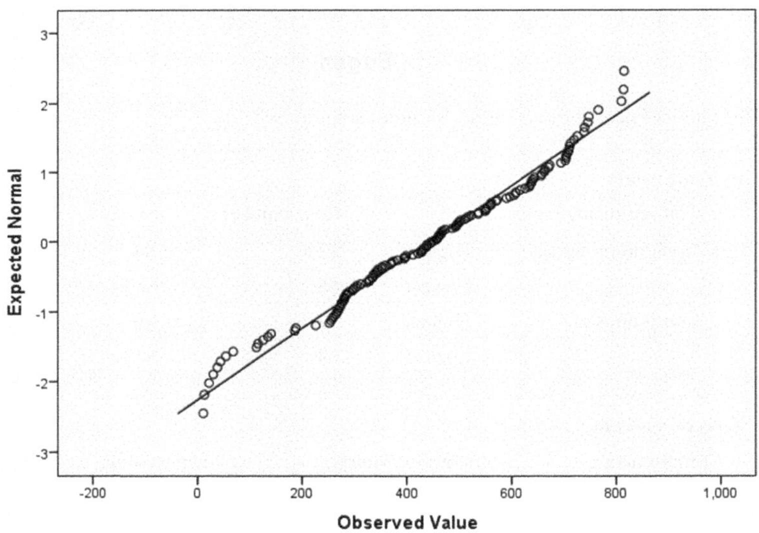

Fig. 11 Normal Q-Q plot of edges for non-compromised

For hypothesis (1) testing, Fig. 13 shows the total population of the online social network of the random nodes layout visualization for the Facebook employees group through opening the GDF file in Gephi.

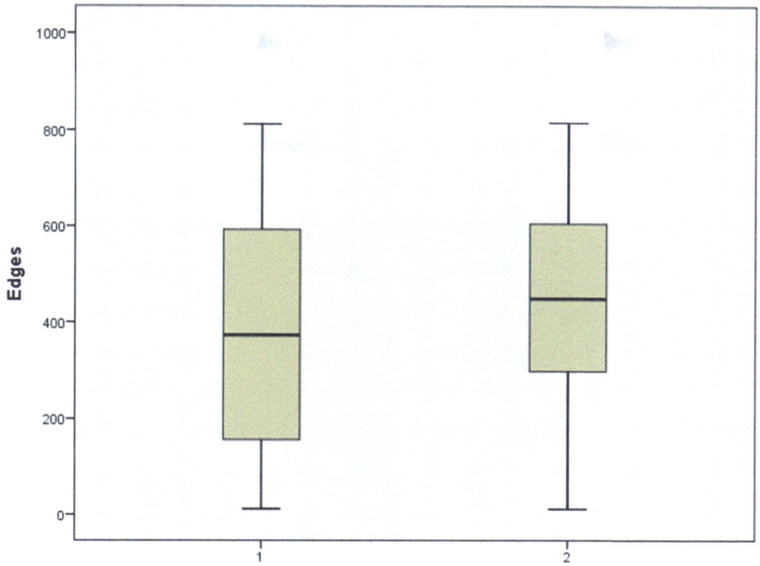

Fig. 12 Boxplot of edges for compromised and non-compromised

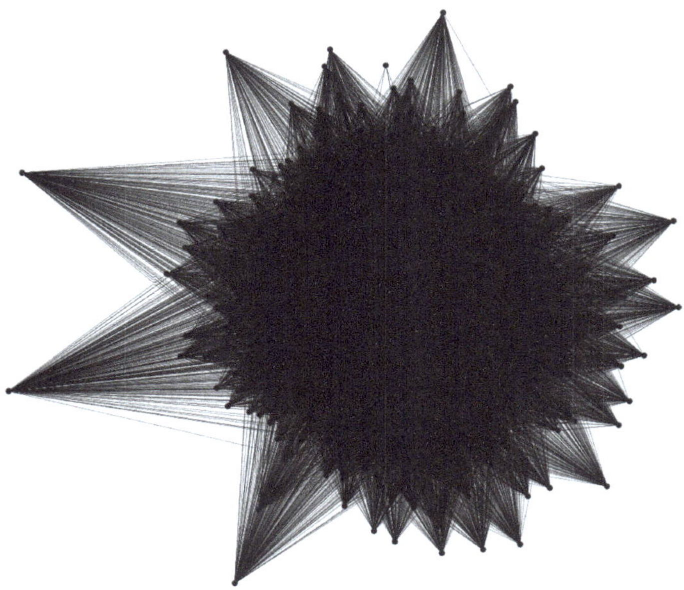

Fig. 13 Random nodes layout visualization

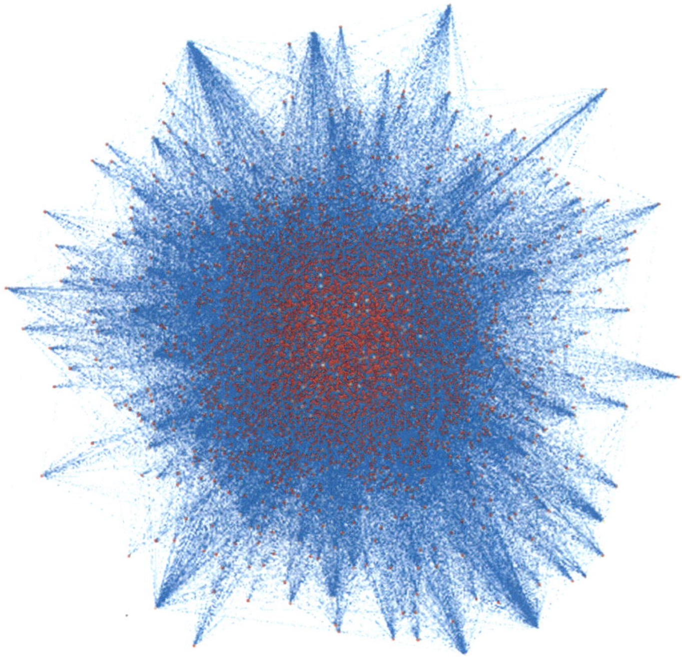

Fig. 14 Force atlas layout visualization

Figure 14 has a more meaningful visualization for the population of social network dataset through applying the Force Atlas layout with specific parameters to identify the employees, thus, the nodes, and their connections to the edges.

By applying the topology, filter in specific settings to hide all nodes and their edges with less than ten connections to obtain the sample through removing the edges in the social network that not connected to many other nodes, the topology shown in Fig. 15.

Figure 16 produces a much more unobstructed view of the community structure detection for the sample of Facebook employees group through pushing the most connected hub nodes apart from each other with colorizing the partitions and positioning. The other nodes that are connected in-group clusters around the hub nodes, creating the modularity class values for the clusters of the nodes that have the densest of the connections than the rest of the social network.

In Fig. 17, the final layout visualization shows the topology structure of the sample of the Facebook employee group and labels the most influential nodes that are the biggest ones with the extreme values of the connection edges in this online social network. The final layout in Fig. 17 is compatible with the results of Table 7 shown above for the extreme values. The topology structure of the Facebook employee group has high scores in all visualization graph metrics that had measured above by using Gephi. There is significant evidence to reject the null hypothesis whereas

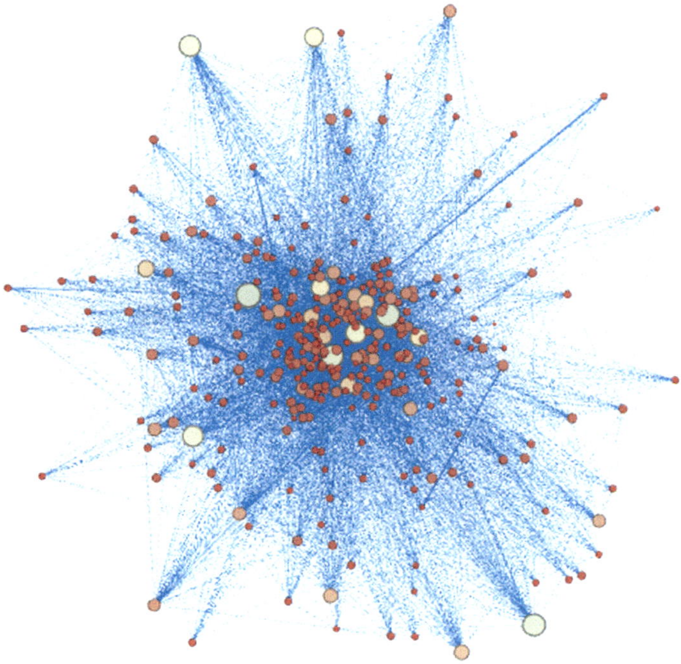

Fig. 15 Topology filter layout visualization of sample

the topology structure is responsible for connecting group clusters in the community together and allowing the more prominent nodes with higher influence to affect the overall nodes in the entire social network.

5.5 Hypothesis (2) Testing

$2H_0$ (Null Hypothesis)	Seniority of employees and active cybersecurity awareness of cybersecurity awareness are not related.
$2H_1$ (Alternative Hypothesis)	Seniority of employees and active cybersecurity awareness of cybersecurity awareness are related.

For hypothesis (2) testing, Table 9 shows that Pearson's r has calculated the correlation coefficient between seniority of employees and edges. It is 0.847 with p-level <0.001, as shown under the significant 2-tailed by 0.000. It means that the actual is p-value <0.001 but not precisely equal to zero. There is significant evidence to reject the null hypothesis, whereas the seniority of employees and edges were correlated and had a statistically significant positive correlation with $r = 0.85$, $p < 0.001$.

Fig. 16 Modularity layout visualization of community structure

5.6 *Hypothesis (3) Testing*

$3H_0$ (Null Hypothesis)	Higher-education employees cannot predict their influences.
$3H_1$ (Alternative Hypothesis)	Higher-education employees can predict their influences (Table 10).

For hypothesis (3) testing, the results show in Table 11: A positive correlation found between education and edges, $r = 0.92$.

The regression equation formula for predicting the employees' influence is as follows: edges $= 2.67 \times$ education $+ 22.8$, $R^2 = 0.846$ as shown in Table 12.

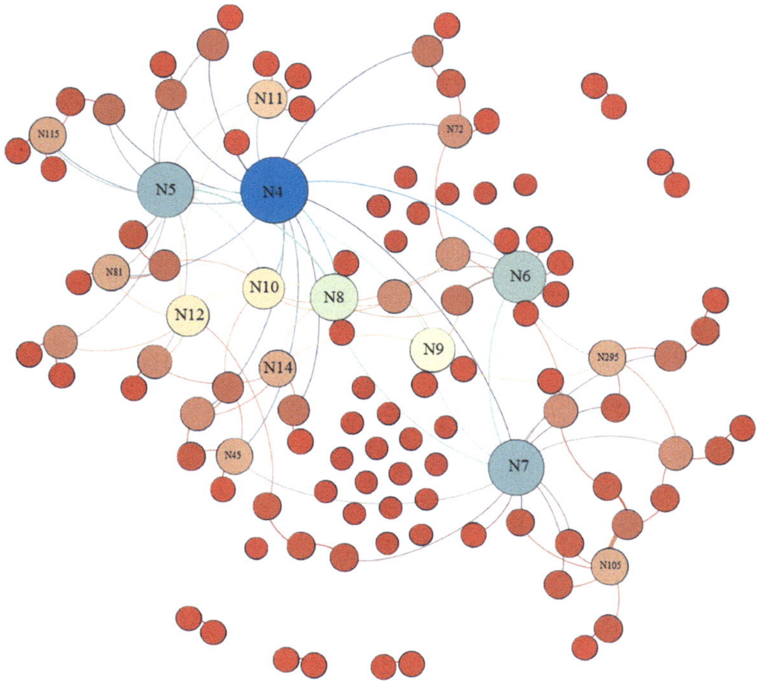

Fig. 17 Final layout visualization of most influential nodes

Table 9 Correlations

		Seniority	Edges
Seniority	Pearson r correlation	1	0.847
	Sig. (2-tailed)		0.000
	N	1033	1033

Table 10 Education descriptive statistics

	Mean	Standard deviation	N
Edges	402.30	218.849	1033
Education	2.53	0.880	1033

Table 11 Correlations

		Edges	Education
Pearson r correlation	Edges	1.000	0.920
	Education	0.920	1.000
Sig. (1-tailed)	Edges	.	.052
	Education	0.052	.

Table 12 Model summary

Model	R	R square	Adjusted R square	Standard error of the estimate
Edges	0.920	0.846	0.681	218.274
Education	0.000	0.000	0.000	218.849

Table 13 ANOVA

Model		df	Mean square	F	Sig.
Edges	Regression	1	127291.782	2.672	0.003
	Residual	1031	47643.560		
Education	Regression	0	0.000	0.000	0.000
	Residual	1031	47894.816		

Table 14 Coefficients

Model	Unstandardized coefficient	Standard coefficient	t	Sig.	95% coefficient interval for B		Correlation
	B	Beta			Lower	Upper	Order
Education	22.780	0.92	1.635	.003	50.200	4.640	0.92

Summary of the regression model in Table 12 predicted approximately 68% of the variance in edges the influence of employees accounted by a linear relationship with their education level and provided high accuracy in predicting the influence of employees.

The model was a good fit for the data F (1.317) = (2.67), ($p < 0.01$) as shown in Table 13.

The 95% confidence interval within Table 14 was 4.64–50.2. There was significant evidence to reject the null hypothesis and conclude that the education level of employees (M = 2.53, SD = 0.88) significantly predicted the edges the employees' influence (M = 402.3, SD = 218.9) as shown in Table 10.

5.7 *Hypothesis (4) Testing*

$4H_0$ (Null Hypothesis)	Compromised employees likely exert less influence on cybersecurity awareness than those non-compromised.
$4H_1$ (Alternative Hypothesis)	Compromised employees likely exert more influence on cybersecurity awareness than those non-compromised.

Table 15 Compromising * edges cross-tabulation

		Edges		Total
Compromised	Count	104	481	585
	Expected count	106.9	478.1	585.0
	% within compromising	18.8	81.2	100.0
	% within edges	37.6	46.0	56.6
	% of total	21.1	35.5	56.6
Non-compromised	Count	91	357	448
	Expected count	92.3	355.7	448.0
	% within compromising	20.2	79.8	100.0
	% within edges	62.2	54.0	43.4
	% of total	23.4	45.7	43.4

Table 16 Chi-square tests

	Value	df	Asymp. Sig. (2-sided)
Pearson chi-square	1016.716	28	0.277
Likelihood ratio	1128.097	28	0.000

For hypothesis (4) testing, the chi-squared sample data that analyzed passes the following two assumptions:

(#1) the variables measured at an ordinal level.
(#2) It consisted of two categorical, independent variables that meet criteria including compromising groups (compromised and non-compromised) as shown in Table 15.

The calculated Pearson Chi-Square value from the top row of Table 16 was 1016.71. It had 28 degrees of freedom ((3 − 1) × (15 − 1)). The Asymp. Significant (2-sided) value was at (0.277) probability level.

There was no significant relationship between compromising and edges of the influence ($\chi^2 = 1016.7$, $df = 28$, $p > 0.001$). Therefore, there is statistically significant evidence to accept the null hypothesis with 99.9% confidence and conclude that no association between compromising and edges.

5.8 *Hypothesis (5) Testing*

$5H_0$ (Null Hypothesis)	Cybersecurity awareness of males is not better than the cybersecurity awareness of females.
$5H_1$ (Alternative Hypothesis)	Cybersecurity awareness of males is better than the cybersecurity awareness of females.

Table 17 Group statistics

	Gender	N	Mean	Standard deviation	Standard error mean
Edges	Male	575	410.64	195.428	14.689
	Female	458	391.83	245.431	20.669

Table 18 Independent samples test

Edges	Levene's test		t-test for equality of means			
	F	Sig.	t	df	Sig. (2-tailed)	95% conf. int. of diff.
Equal variances assumed	17.496	0.062	6.761	316	0.017	−29.828
Equal variances not assumed			6.742	263.6	0.023	−31.120

For hypothesis (5) testing, the independent samples t-test performed by comparing scores of the edges for both male and female employees. There was statistically significant evidence to reject null hypothesis with scores for males (M = 410.6, SD = 195.43, n = 575) reported to be significantly higher in terms of cybersecurity awareness compared to female employees (M = 391.8, SD = 245.4, n = 458), (t (316) = 6.76), ($p = 0.017$) less than 0.05 as shown in Tables 17 and 18 respectively.

The differences degree in the means were very high (eta squared = 0.124) based on the formula of eta squared = $t^2/t^2 + (N1 + N2-2)$.

5.9 *Hypothesis (6) Testing*

$6H_0$ (Null Hypothesis)	Efficient communications do not differ between the employees based on their ages.
$6H_1$ (Alternative Hypothesis)	Efficient communications differ between the employees based on their ages.

For hypothesis (6) testing, Table 19 provides common descriptive statistics for six age groups on the edges dependent variable. The ANOVA sample data that analyzed passes the following three assumptions:

Assumption (#1) Testing: Table 20 shows the tests of normality distributions, given that $p = 0.325$ for 18–25 age group, $p = 0.305$ for 26–33 age group, ($p = 0.286$) for 34–41 age group, $p = 0.52$ for 42–49 age group, $p = 0.39$ for 50–57 age group, and $p = 0.407$ for 58–66 age group at $\alpha = 0.001$. It concluded that each of the K levels of the independent variable age groups normally distributed. Therefore, assumption (#1) of normality met for this sample (Table 21).

Table 19 Age description

Edges	Age	Mean	Standard deviation	Standard error
	18–25	462.72	194.950	23.817
	26–33	425.99	198.776	21.190
	34–41	397.69	198.259	28.616
	42–49	285.30	236.834	33.493
	50–57	457.86	226.310	42.769
	58–66	366.64	241.150	40.192

Table 20 Tests of normality

	Age	Kolmogorov-Smirnov			Shapiro-Wilk		
		Statistic	df	Sig.	Statistic	df	Sig.
Edges	18–25	0.099	67	0.172	0.959	67	0.325
	26–33	0.102	88	0.024	0.957	88	0.305
	34–41	0.094	48	0.200	0.958	48	0.286
	42–49	0.168	50	0.001	0.876	50	0.520
	50–57	0.162	28	0.058	0.937	28	0.390
	58–66	0.139	36	0.077	0.912	36	0.407

Table 21 Normality description

Residual for edges	Mean	Standard deviation	Standard error	95% confidence interval for mean	
				Lower	Upper
	113	210.183	11.805	−23.226	23.226

Assumption (#2) Testing: Table 22 shows the residual tests of normality distributions; there is evidence to suggest that age groups data normally distributed whereas the Shapiro-Wilk test significance level ($p = 0.216$) scored higher than 0.1. Therefore, assumption (#2) of normality met for this sample.

Assumption (#3) Testing: Table 23 assesses the assumption of the variances for the six age groups based on Levene's Test. It indicates that the variances were equal; Levene's test is not significant, (F (7.311) = 2.40), ($p = 0.237$) at 0.05 alpha level.

Table 22 Tests of normality

	Kolmogorov-Smirnov			Shapiro-Wilk		
	Statistic	df	Sig.	Statistic	df	Sig.
Residual for edges	0.086	1033	0.203	0.971	1033	0.216

Table 23 Test of homogeneity of variances

Levene statistic	df1	df2	Sig.
2.400	7	1026	0.237

Table 24 ANOVA

	df	Mean square	F	Sig.
Between groups	7	222253.797	4.951	0.006
Within groups	1026	44887.296		

Therefore, assumption (#3) of variance homogeneity met for this sample. All the above results have shown that all three assumptions can be accepted, so can use normal ANOVA analysis from Table 24 that demonstrate the efficient communications were significant, (F (7.311) = 4.95), ($p = 0.006$), (N = 1033).

Table 24 ANOVA gives an F statistic of 4.95 and a p-value of <0.01 that is statistically significant evidence to reject the null hypothesis with 99% confidence and conclude that there is substantial evidence that means efficient communications between the six groups of employees are unequal. Post hoc analysis using the Scheffé-Bonferroni post hoc comparisons in Table 25 for significance reported. The average number of errors was significantly lower in 18–25 age group (M = 462.72, SD = 194.95), 26–33 age group (M = 425.99, SD = 198.78), and 34–41 age group (M = 397.69, SD = 198.26) than in 42–49 age group (M = 285.3, SD = 236.83), 50–57 age group (M = 457.86, SD = 226.31), and 58–66 age group (M = 366.64, SD = 241.15).

However, the Tukey HSD post hoc test performed to establish statistically significant differences between age pairs of means and revealed. There are significant differences between 18–25 and 42–49 age groups, and 26–33 and 42–49 age groups. Finally, 50–57 and 42–49 age groups with substantial evidence ($p < 0.01$) compared to 18–25, 26–33, and 50–57 age groups that have efficient communications more than 42–49 age group. There was no statistically significant evidence of efficient communications differences between other age groups because all the other confidence intervals for Tukey HSD contain zero and statistically significant levels had higher levels of confidence interval bounds while being less than that for Bonferroni, as expected.

6 Findings and Discussion

The purpose of the discussion in this section is to interpret the significant findings in light of an understanding of the problem statement by answering questions and testing hypotheses.

Table 25 Post hoc tests multiple comparisons

Age	Age	Sig.	95% confidence interval	
			Lower bound	Upper bound
18–25	26–33	0.893	−61.78	135.23
	34–41	0.584	−49.86	179.92
	42–49	0.000	63.88	290.96
	50–57	1.000	−131.86	141.58
	58–66	0.243	−29.47	221.62
26–33	18–25	0.893	−135.23	61.78
	34–41	0.976	−80.71	137.32
	42–49	0.003	33.09	248.28
	50–57	0.983	−163.69	99.95
	58–66	0.717	−60.85	179.55
34–41	18–25	0.584	−179.92	49.86
	26–33	0.976	−137.32	80.71
	42–49	0.094	−10.38	235.16
	50–57	0.839	−204.64	84.30
	58–66	0.986	−102.90	165.00
42–49	18–25	0.000	−290.96	−63.88
	26–33	0.003	−248.28	−33.09
	34–41	0.094	−235.16	10.38
	50–57	0.008	−315.96	−29.15
	58–66	0.496	−214.14	51.46
50–57	18–25	1.000	−141.58	131.86
	26–33	0.983	−99.95	163.69
	34–41	0.839	−84.30	204.64
	42–49	0.008	29.15	315.96
	58–66	0.527	−61.87	244.30
58–66	18–25	0.243	−221.62	29.47
	26–33	0.717	−179.55	60.85
	34–41	0.986	−165.00	102.90
	42–49	0.496	−51.46	214.14
	50–57	0.527	−244.30	61.87

6.1 Findings and Discussion of Hypothesis (1)

The topology structure of Facebook employees group consists of six main dense clusters. It has accumulated and represented by their distinct color, 23% purple, 16% green, 12% blue, 9% magenta, 5% black, and 3% pink of nodes consequently. The most prominent nodes in the Facebook group indicated the most critical employees based on the visualization graph in the Gephi that have the highest scores of Betweenness centrality and have the shortest paths between dual randomly selected nodes within the social network that connected clusters compared to the rest of the online group. Each group cluster has multi-influential employees. It means that a single employee does not lead them and the interactions among clusters were high. The main hubs in the employee's social network give preferences to endorse the cybersecurity awareness that shared or liked by their Facebook friends that prompts immediate surrounding neighbors in their community to support the cybersecurity awareness. The overall observation indicates that the topology structure affects the employee group. One observation is the tendency to form more links between the nodes around a specific action, such as activities of cybersecurity awareness. It can have a better interconnection with efficient communication, proliferating information and bringing the social network into action, and helping keep the social network dynamic, sustainable, and open to new influences. Meanwhile, about 32% of employees in the social network who were not Facebook friends with any other members within the employee group, so the key employees had to involve these lonely employees to become more active through integrating them into existing clusters and encourage them to exert more efforts for cybersecurity awareness. Therefore, the topology structure of the online social network has a high effect on the ability of the Facebook employee group to fulfill the objectives of cybersecurity awareness.

6.2 *Findings and Discussion of Hypothesis (2)*

They are exploring the relationship between the seniority of employees and active cybersecurity awareness investigated by using the Pearson r correlation coefficient to provide a test of the null hypothesis. Preliminary analysis conducted to ensure that there was no violation of the assumptions of normality and linearity. There was a high and positive correlation coefficient between the two variables from the output given. The R-value, $r = 0.85$, the significance level reported, $p < 0.001$, and the number of cases, $N = 1033$, that scored in this analysis are an indication of a strong relationship between seniority of employees and active cybersecurity awareness. The statistical significance of r strongly influenced by the large size of this sample ($N = 1000+$). It means a small correlation will be significant enough to be considered as evidence of statistical significance. Therefore, senior employees have higher levels of influence on cybersecurity awareness than others on the social network. Therefore, seniority of the employees has a high effect on the topology structure of the online social network and enhances the ability of the Facebook employee group to fulfill the objectives of cybersecurity awareness.

6.3 Findings and Discussion of Hypothesis (3)

The simple linear regression has used to provide the correlation matrix along with probability values for the predictor and the independent variable of education levels of the employees, which are highly correlated with the dependent variable of their influence over the social network. Both variables are significantly correlated with strong and positive relationship, $r = 0.92$, $p < 0.01$, and $N = 1033$. The output of the correlation coefficient ($R = 0.92$) was shown in the summary of the linear regression model. The result of the coefficient of determination ($R^2 = 0.846$) also suggests there was a strong and positive relationship between education levels of employees and their influence within the Facebook group employees. The adjusted R^2, which is equal to 0.681, indicates that 68% of the variance in the online influence over the social network can explain the levels of education for employees. In addition, the coefficient adjustment affected by the sample size. In other words, the magnitude of the influence on this social network is strongly predicted significantly well by the levels of education that employees have. ANOVA statistical significance generated regression model with ($F (1.317) = 2.67$), ($p < 0.01$) that predicts the employees' influence variable and explains a statistically significant proportion of the variance. In the education scores row, the B column provides the gradient of the regression line, which is the regression coefficient ($B = 23$). It means that for every twenty-three score increase in education level score (standard deviation = 13), the model predicts an increase of standardized beta coefficients ($\beta = 0.92$) scores of employees' influence over the social network which is interpreted similarly to correlation coefficients or factor weight. The t value ($t = 1.6$) and the ($Sig. = 0.003$) counter the independent variable of education level, indicating that this variable is contributing significantly to the predicting equation of employees influence (employees influence $= 2.67 \times$ education level $+ 23$) and making a statistically significant contribution to the predictive power of the model. Therefore, the magnitude of employees' influence over the online social network can be predicted by the education levels of the employees in the Facebook group. Moreover, it could be attributed to the kind of people in the employee social network with higher education levels leads to much more influence, whereas these two measures are highly correlated to each other.

6.4 Findings and Discussion of Hypothesis (4)

A chi-squared test used for the independent normally distributed data of this sample that have the output of the Pearson chi-square value equal to 302 and continuity correction value of 413 which compensates for the overestimate of the chi-square value, with an associated Asymp. Significance (2-sided) level of 0.277, which is larger than the alpha value of 0.05. Therefore, the concluded result is that it is not statically significant. It means that the influence of compromised people is likely to exert is not significantly different from the influence of non-compromised people for cybersecurity awareness in the Facebook employee group. The summary information provided by compromising and influence cross-tabulation to convey that the percentage of compromised people are exerting within 18.8%, while 20.2% were non-compromised people. For influence, 46% compromised people, and 45.7% were non-compromised people. According to the test results, the percentage of the sample as a whole summarized across compromised or non-compromised was 18.9% of the sample with a high influence on cybersecurity awareness and 81.1% of profound influence for cybersecurity awareness. The contingency outputs show that more compromised people than expected did not exert the influence for cybersecurity awareness (37.6 vs. 46.0) and more non-compromised people than expected exerted the influence for cybersecurity awareness (62.2 vs. 54.0). Moreover, with the values of the chi-square test of $p = 0.277$ and df = 28, both variables are independent, and the compromised people in the employee social network do not likely exert influence for cybersecurity awareness more than those non-compromised people.

6.5 *Findings and Discussion of Hypothesis (5)*

The independent samples t-test were performed to evaluate the impact of the gender on the cybersecurity awareness through comparing the different average values of cybersecurity awareness of scores for both groups of males and females in the employee social network. The significance level of Levene's test is (F (17.5) = 0.06), which is larger than the cutoff value of $p > 0.05$. It means that the assumption of the equality of variances of scores for the two groups of males and females is the same and not violated. The t-value is 6.76, and the degrees of freedom, df, is 316. The sig. (2-tailed) value is 0.017, which is less than the required cutoff of (0.05). The conclusion is that there is a statistically significant difference in the mean values of the cybersecurity awareness scores for males and females in this sample. In addition, the magnitude of the eta, which is an indicator of the proportion of variance differences in the means of cybersecurity awareness, between both groups was very high within effect size statistic (eta squared = 0.124). It expressed as 12.4% of the variance in the cybersecurity awareness attributed to gender, based on the following guidelines proposed by (Cohen 1988) to interpret eta-squared values: small size effect = 0.010, moderate-size-effect = 0.060 and large-size-effect = 0.140. Therefore, there is a substantial significant difference in cybersecurity awareness between male and female employees. Also, the vital male employees were labeled (N4, N5, N6, N7) respectively have the most influential cybersecurity awareness and play a focal role in the six clusters within the employees' community more than the key female employees labeled (N8, N9) respectively. Consequently, the cybersecurity awareness of male employees of the online groups was more than that of female employees. Furthermore, the male employees can communicate messages and cybersecurity awareness, such as posts, likes, shares, and comments, to the rest of the Facebook employee group in an efficient manner more than the female employees of the same group.

6.6 Findings and Discussion of Hypothesis (6)

One way of analysis of variance among age groups conducted to explore the impact of age of employees on the efficient communications scores within the employees' social network. Employees in this sample divided into six groups according to their age: first age group 18–25, second age group 26–33, third age group 34–41, fourth age group 42–49, fifth age group 50–57, and sixth age group 58–66. There was a statistically significant difference at the $p < 0.01$ level. Inefficient communications scores for three age groups were determined with (F (7.311) = 4.95, $p < 0.01$). The multiple post hoc comparisons using the Tukey HSD test indicated that there were differences in mean scores between four age groups. The first age group (18–25) with M = 462.72, SD = 194.95, second age group (26–33) with M = 425.99, SD = 198.78, and the fifth age group (50–57) with M = 457.86, SD = 226.31, were all significantly different from the fourth age group (42–49) with M = 285.3, SD = 236.83. It means that the effect of employees on efficient communications within an online social network depends on the employees' ages. For young employees whose age groups of 18–25 and 26–33, it is shown that they have highly efficient communications and cybersecurity awareness because of their familiarity with new technologies and social networks more than the age group of 42–49. Similarly, the old age employees, whose age group of 50–57, are efficient communications than the age group of 42–49 since most compromising cases of cybersecurity discovered after the age of fifty. Therefore, the compromised employees in this stage have more communications with other employees to face the side effects of cybersecurity through cybersecurity awareness and social interactions. Consequently, the efficient communications differ between the employees based on their age whereas the young and the old compromised employees within the online social network have efficient communications than other employees in the rest of Facebook groups.

7 Conclusion

This paper analyzed the topology structure, and critical influences on Facebook for the employee group of cybersecurity awareness through hypotheses testing and conducted the appropriate statistical analysis for a set of dependent and independent variables such as gender, age, and cybersecurity awareness. The main conclusions derived from the implementation of statistical analysis and conducting test of hypotheses. It is clear that the topology structure of the employee group has high effects on the ability to fulfill the objectives of cybersecurity awareness, and there is a strong correlation between the seniority of employees and effectiveness of their cybersecurity awareness on the Facebook group. In addition, employees with higher education have more influences than others, and there is a significant ability to predict that. The tested data provides evidence that communications for young employees are more efficient than elderly employees in the same group. Moreover, cybersecurity awareness of male employees is better than the cybersecurity awareness of the female ones. On the other hand, the results indicate that there is no apparent difference between the influences for cybersecurity awareness exerted by the compromised employees and those non-compromised.

References

1. Liu, W., Sidhu, A., Beacom, A.M., Valente, T.W.: Social Network Theory. In: The International Encyclopedia of Media Effects, pp. 1–12 (2017)
2. Lonsdale, M.D.S., Lonsdale, D., Lim, H.W.: The impact of delivering online information neglecting user-centered information design principles. cyber security awareness websites as a case study. Inf. Des. J. **24**(2), 3–41 (2019)
3. Fire, M., Puzis, R.: Organization mining using online social networks. Netw. Spat. Econ. **16**(2), 545–578 (2016)
4. Knoke, D., Yang, S.: Social Network Analysis, vol. (154), pp. 201–209. SAGE Publications, Incorporated (2019)
5. Hassandoust, F., Techatassanasoontorn, A.: Understanding users' information security awareness and intentions: a full nomology of protection motivation theory. In: Cyber Influence and Cognitive Threats, pp. 129–143. Academic Press (2020)
6. Winner, J.B., Galbreath, N.: Online content delivery based on information from social networks. Patent No. 10,373,173. U.S. Patent and Trademark Office, Washington, DC (2019)
7. Fu, X., Passarella, A., Quercia, D., Sala, A., Strufe, T.: Online social networks. Comput. Commun. **73**(PB), 163–166 (2016)
8. Bada, M., Sasse, A.M., Nurse, J.R.: Cyber security awareness Campaigns: why do they fail to change behaviour? arXiv preprint arXiv:1901.02672 (2019)
9. Franchi, E., Poggi, A., Tomaiuolo, M.: Social media for online collaboration in firms and organizations. In: Information Diffusion Management and Knowledge Sharing: Breakthroughs in Research and Practice, pp. 473–489. IGI Global (2020)
10. Froehlich, D.E., Rehm, M., Rienties, B.C.: Mixed Methods Approaches To Social Network Analysis, 1st edn, pp. 167–176. The Open University of London, UK. Taylor & Francis, Book (2020)
11. Asim, Y., Malik, A.K.: A survey on access control techniques for social networks. In: Information Diffusion Management and Knowledge Sharing: Breakthroughs in Research and Practice, pp. 319–342. IGI Global (2020)
12. Alotaibi, F., Furnell, S., Stengel, I., Papadaki, M.: A review of using gaming technology for cyber-security awareness. Int. J. Inf. Secur. Res. (IJISR) **6**(2), 660–666 (2016)
13. Längler, M., Brouwer, J., Gruber, H.: Data collection for mixed method approaches in social network analysis. Mixed methods approaches to social network analysis, pp. 25–37 (2020)
14. Dey, P., Roy, S.: Social network analysis. In: Advanced Methods for Complex Network Analysis, pp. 237–265. IGI Global (2016)
15. Bastian, M., Heymann, S., Jacomy. M.: Gephi: an open source software for exploring and manipulating networks. In: Proceedings AAAI International Conference on Weblogs and Social Media (2014)
16. Öğütçü, G., Testik, Ö.M., Chouseinoglou, O.: Analysis of personal information security behavior and awareness. Comput. Secur. **56**, 83–93 (2016)
17. Ekta, D., Singh, K.: Feature extraction and content investigation of Facebook users' using Netviz and Gephi. Adv. Comput. Sci. Inf. Technol. **3**(4), 262–265 (2016)
18. IBM Corporation Released: IBM SPSS Statistics for Windows, Version 25.0. IBM Corporation, Armonk, NY (2017)

Mining Dubai Government Tweets to Analyze Citizens' Engagement

Zainab Alkashri, Omar Alqaryouti, Nur Siyam, and Khaled Shaalan

Abstract Social media plays a critical role in the public sector as it allows the government to interact with the citizens. With the United Arab Emirates being active on social media platforms, this study aims to identify the level of citizen engagement in Dubai government's Twitter through the use of data mining techniques. Post engagement is the total number of citizens' interactions with a tweet and can be measured using different tweet attributes including retweets, mentions, and likes. Moreover, this study investigates the impact of the twitter post characteristics on the citizens' engagements level. Thus, we collected, prepared and processed 74,037 tweets that represents all tweets for Dubai government twitter accounts during 2018. These tasks were followed by statistical analyses of the impact of post characteristics on the citizens' engagement level. Next, we implemented various machine learning models to evaluate the performance of using the post characteristics and post content to predict the engagement level of citizens. Results indicate that citizen engagement level in Dubai government's Twitter is significantly impacted by all post characteristics. It is also revealed in the study that citizen engagement is higher during weekdays compared to weekends. Furthermore, the machine learning models achieved promising results to predict the citizens' engagement with highest accuracy for Random Forest and Linear Support Vector Machine of 78.3% and 78.2% respectively.

Keywords Data mining · Government twitter · Dubai government · Machine learning · Twitter data analysis · Citizens' engagement

Z. Alkashri (✉) · O. Alqaryouti · N. Siyam · K. Shaalan
The British University in Dubai, Dubai, United Arab Emirates
e-mail: zainabalkashri@gmail.com

O. Alqaryouti
e-mail: omar.alqaryouti@gmail.com

K. Shaalan
e-mail: khaled.shaalan@buid.ac.ae

K. Shaalan
School of Informatics, University of Edinburgh, Edinburgh, UK

M. Al-Emran et al. (eds.), *Recent Advances in Intelligent Systems and Smart Applications*, Studies in Systems, Decision and Control 295,
https://doi.org/10.1007/978-3-030-47411-9_33

1 Introduction

With the rise of technology and the Internet, the use of social media has consequently become widespread. Social media has shaped societies and the way people go about their lives. According to Siddiqui and Singh [1], social media has transformed communication and has improved human connections. Furthermore, social media is used in several fields including education, business and in public service. This explains the reason why governments are also adopting social media to increase their efficiency in the service delivery to the public. As indicated by Mishaal and Abu-Shanab [2], the critical role of social media in the government lies on how it enables collaboration and interaction between the citizens and the public sector.

The significant developments of social media are aligned with the drastic improvement in technology. As pointed out by Chukwuere and Chukwuere [3], social media, as a widely used tool for communication, has become an integral part of today's society. It allows people to interact freely among each other—sharing information, provide feedback and participate in social discussions online. There are several social media platforms used nowadays such as Facebook, Twitter, Instagram and YouTube. Akram and Kumar [4] entailed that these popular social media platforms have shaped the way people communicate. For one, Twitter, as a microblogging site, is commonly used by businesses for advertisements and for interacting with potential clients. Even with restriction on the number of characters per tweet, there are still a lot of active Twitter users who make use of the platform to share information and engage in relevant social discussions.

The use of social media in the government is not at all a new concept. With the rise of social media comes the recognition that it can be an effective tool to be used in the public sector. According to Karakiza [5], government agencies consider social media as part of their strategy in serving their mission better. As a result, governments were able to take advantage of several opportunities that come along with using social such as increased government transparency and trust, increased engagement and participation of the citizens in public issues and improved collaboration within the public sector and between the government and the stakeholders. In that context, it is subsequently deemed that social media positively influences the relationship between the government and the public.

This paper aims to determine the level of citizen engagement in Dubai government's Twitter. Since the agency has been using Twitter as social media has long been promoted in UAE, the study aims to explore the effectiveness of the initiatives implemented by Dubai government to increase its presence on said social media platform. Data mining techniques are used in order to further delve into the study, denoting that it is the most suitable method to use to measure citizen engagement with Dubai government on Twitter. Based on the main purpose and the underlying objectives mentioned, the following are the research questions that it aims to answer:

- RQ1: Do post characteristics impact the users' engagement in Dubai Government Twitter accounts?
- RQ2: Can we predict the user's engagement based on post characteristics?

- RQ3: Can we identify the main keywords resulting in higher engagement?

2 Background and Related Work

The concept of eParticipation, as defined by Sanford and Rose [6], is relevant to "the extension and transformation of participation in societal democratic and consultative processes, mediated by information and communication technologies (ICTs)" (p. 406). It is significantly urged by governments as this will help enhance the overall legitimacy, acceptance and efficiency of political process. More so, it is a response to the public's demand for their involvement in democratic initiatives. The emergence of eParticipation is associated with the increased acceptance and utilization of the Internet, thereby prompting the government to extend its efforts in using social media to increase citizen engagement. In relation to this, Kassen underpinned that eParticipation, from the government perspective, consists of interactions between the government and other key players including businesses, citizens and the public sector itself [7]. In particular, the government-to-citizens interactions denote an approach that is citizen centered, with emphasis on civic engagement and collaboration. With that said, it can be purported that the primary aim of eParticipation, based on the government context, is to encourage and promote a special regulatory and technology avenue for cooperation between the government and its stakeholders.

In analyzing and identifying the level of user engagement on social media platforms such as Twitter, data mining is performed. Also called Knowledge Mining, Information Harvesting, and Data Dredging, among others, data mining is "a process of analyzing data from many different dimensions or angles and summarizing it into useful information that can be applied in different fields to take proper decision [8]. In other words, data mining is basically a computing process of varied patters obtained from databases that includes the use of several techniques such as AdaBoost, Artificial Neural Network (ANN), Apriori, Bayesian Networks (BN), Support Vector Machine (SVM), Decision Trees (DT), and Markov. Of all these data mining techniques, it is indicated by Injadat, Salo and Nassif that the most applied in social media are SVM, BN and DT [9].

Data mining in social media, according to Zatari [10], involves analysis and extraction of patterns or trends from raw social media data, thereby giving significant and valid information to governments and organizations in designing and implementing strategies and services. For instance, the government can use data mining techniques to analyze and determine the influence of social media posts like those on Twitter in increasing user engagement. As pertained in the study of Siyam et al. [11] posts can be classified into different content types and level of interaction of citizens can then be measured based on the number of retweets, comments and likes divided by the number of the followers. It is through this simple method that the government can maximize data mining process in determining the extent of its social media presence. However, the impact of the government on social media is yet to be determined, denoting that there is still a lack of enough metrics to use. The most important consideration in data mining in social media is the need to identify the indicators to be used. On Twitter, indicators including the number of likes, retweets and comments can be used. With the possible inclusion of hashtags (#) on the tweets, interaction between users and the government can be increased [11].

Because there are several things that can be done on Twitter, posts or tweets can be classified into several categories such as status update, link, photo, and video. As indicated by Siyam et al. [11], status updates are composed text only which could possibly include hashtags (#) and/or mentions (@). Those tweets identified as photo or video may also contain text along with the media content. Post vividness, which these types of posts are known as, is deemed to have a positive impact on user engagement. On the other hand, link tweets are posts consist of a link which when clicked, will direct users to another page. Unlike the photos and/or video posts, link tweets are found to have a contrasting impact on user engagement.

With the application of data mining in this study, it is therefore explored how the indicators mentioned for analyzing Twitter data can be translated to measure the engagement of citizens with the Dubai government. Since comments on the tweets are most likely interpreted using text mining, it can be consequently determined whether the sentiments of the public are positive, negative or neutral, thereby using this information to design strategies for improvement of the policies and services provided by the government to the public. According to Kavitha [12], use of data mining techniques in social media can be complex because of the large datasets included; however, the process is very much essential in fostering the effectiveness and productivity of the government and in enhancing government-citizen relationships as well.

This study focuses on eParticipation and the impact of Dubai Government's Twitter presence on citizen engagement. In that regard, the empirical results will show the level of citizen engagement on the Twitter account of Dubai Government, based on the characteristics of the posts which include time, day, the type of post, language, mentions and hashtags used.

3 Research Methodology

The ground objective of this work is to clarify and identify the relationship between the tweet features and the expected citizens' engagement level. This study focuses on the tweets that are posted through the government sector in Dubai. Accordingly, the aim to clarify what kind of topics and in which format the citizens are normally interested in.

In this research, the impact of the twitter post characteristics on the citizens' engagements level is studied to answer the research questions. To achieve this objective, the following hypothesis have been formulated:

- **H1a** Post Type impacts Citizens' Engagement (Photo, Status, Video).
- **H1b** Post Time impacts Citizens' Engagement.
- **H1c** Post Day impacts Citizens' Engagement.
- **H1d** Post Weekday impacts citizen's Engagement.
- **H1e** Links in Tweets impacts Citizens' Engagement.
- **H1f** Mentions in Tweets impacts Citizens' Engagement.
- **H1g** Hashtags in Tweets impacts Citizens' Engagement.
- **H1h** Tweet Language impacts the Citizen's Engagement.
- **H2** Post engagement level can be predicted using machine learning techniques according to the various post attributes.
- **H3** The main keywords that results from higher engagement can be identified.

The conceptual framework of this study is shown in Fig. 1. Figure 1 includes the research questions, hypothesis and the dependent and independent variables. This study adopts the proposed Knowledge Discovery in Databases (KDD) methodology by Azevedo and Santos [13]. The KDD methodology utilizes data mining techniques to discover hidden valuable knowledge from data. The KDD comprises of set of processes that are used to extract the desired knowledge such as selecting the target dataset, preparing the data, analyzing the data, interpreting the data and evaluating the results as shown in Fig. 2

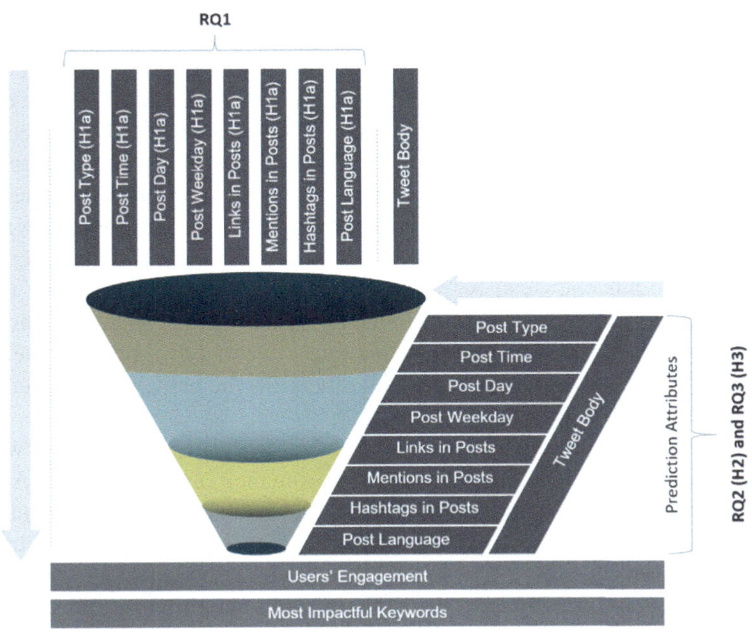

Fig. 1 Conceptual framework

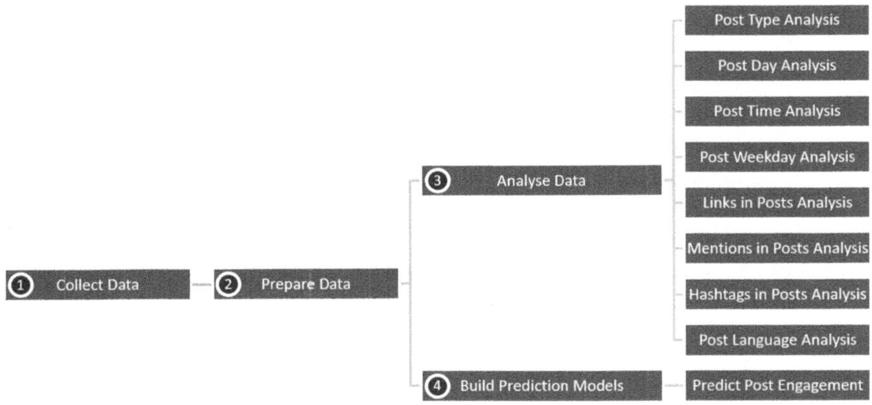

Fig. 2 Research methodology

3.1 Twitter Data Collection

The data used in this study are collected using the publicly available open source utility tool termed as OrgneatUI.[1] OrgneatUI tool is used to extract historical tweets from twitter accounts. To extract the desired tweets for each of the government accounts, the twitter account name and date range parameters were provided to the tool. The collected data comprises of 74,037 tweets that represents the entire tweets history originated from government accounts during that year of 2018. The dataset was collected in Microsoft Excel file format. The dataset includes both tweets that were posted by the government entities as well as reply tweets by twitter users. This is identified in the dataset through three attributes; is part of conversation, is reply and reply user id. Each tweet consists of various attributes. Among these attributes are the tweet ID, date which also includes time as part of it, tweet body, and the government authority. These attributes can be recognized as the main features of the tweets. For instance, tweet ID represents the unique identifier for the tweet originator. The datetime stamp shows original post date and time. Tweet body represents the actual text of the tweet. The government authority represents the name of the government body who posted the tweet.

The government authorities in Dubai were identified using the dubai.ae website and their official twitter accounts were obtained from their official websites. According to dubai.ae, the government departments in Dubai consist of 47 departments [14]. Those government department with official twitter accounts were selected for data collection.

These government departments play different roles and provide different kind of services according to their role and category. However, the government departments

[1] OrgneatUI (https://github.com/rahulha/OrgneatUI).

post similar kind of tweets in term of context. For instance, the tweets can represent promotions for new services, department achievements, polls, entities activities and achievements [11]. Table 1 shows some examples of tweets from different government entities along with the corresponding context.

Table 1 Samples for government entities tweets context

Entity	Tweet	Context
Smart Dubai	*"Tired of having to physically sign and process documents? Allow #UAEPASS; The National Digital Identity and Signature Solution; to save you time; effort and paper! Digitally sign and validate documents at the comfort of your own home.#SmartDubaipic.twitter.com/K53AQSjmki"*	Service promotion
Smart Dubai	*"#DubaiNow has made it easy and quick to top up ENOC VIP accounts.#SmartDubaipic.twitter.com/KNIPPku3sZ"*	Service promotion
Dubai Chamber	*"Dubai free zones witnessed 22% trade growth in 2018; Free zones accounted for 41% of Dubai's total trade during the first nine months of 2018 https://bit.ly/2LyS6en "*	Achievements
Dubai Courts	*"تطبيق نبراس بـ #محاكم_دبي https://youtu.be/WnNmP6fURQY "*	Service promotion
Dubai Municipality	*"Vote for our initiative "Smart Park" Al Mamzar that has been nominated for the Hamdan Bin Mohammed Smart Government Award. You can vote through the link below http://Vote.dtmc.gov.ae We care about your opinion! @DTMCentrepic.twitter.com/H9ABQsfjmU"*	*Poll*
Smart Dubai	*"We would like YOU to give your opinion about "What makes a city Smarter" by being part of a Smart Dubai video. DM us if you are interested! #SmartDubaipic.twitter.com/o6VLLodEMg"*	*Poll*
Dubai Customs	*"إدارة الابتكار في الخدمات بجمارك دبي تنظم ورشة عمل لمدراء المراكز الجمركية وفرق السبع نجوم الداخلية حول متطلبات برنامج النجوم العالمي لتصنيف الخدمات#جمارك_دبي #الابتكار #دبي pic.twitter.com/D38A2zQbFv"*	Activities
Dubai Customs	*"#DubaiCustoms wins Organization-wide Innovation Award at Global Organisational Excellence Congress 2018. #Dubaipic.twitter.com/kk8mTnbVuV"*	Achievements
Land Department	*"Free real estate panels were held during Dubai Property Show in London; featuring experts and consultants in real estate and law that enabled visitors to understand the real estate investment options which will help make Dubai their first choice.pic.twitter.com/nKR1GDuRQ1"*	*News*

3.2 Data Preparation

The collected data included 20 attributes. These attributes were used to produce the final dataset that is required to conduct our analysis and experiments. The data preparation stage is considered important due to its impact on the overall performance. The first stage in the data preparation included consolidating the collected data from the 36 government departments which resulted in 74,037 tweets. This study is focused to study the impact of post characteristics on the citizens' engagement level for Dubai Government departments. Thus, the reply tweets were eliminated and only tweets that are posted by the government departments are considered. Moreover, duplicate and blank tweets were removed. The finally resulted dataset comprises 66,832 tweets.

Form the final dataset, some of the attributes were used without performing any processing such as the tweet body, language, number of replies, number of retweets, number of favorites and mentions. Whereas, other attributes were created from other native existing attributes such as "Is Weekday", "Time of Day", "Day", "Has Link", "Has Hashtag", "Has Mention" and "Tweet Type" as shown in Table 2.

The engagement value was computed for each twitter post based on the number of interactions divided by the total number of followers of the corresponding twitter account. The number of interactions represents the summation of the total number of retweets, the total number of favorites and the total number of replies. Due to the total number of interactions variation from one government twitter account to another, wrong conclusions may take place because of the different number of followers for each account. Thus, the resulted engagement value is adjusted to reduce the impact of this variation. Formula 1 shows the computation of the engagement value [11]. For instance, 1000 engagement value after adjustment means that a twitter post is able to produce total interactions of 1000 with one million followers. The resulted engagement value is then classified into two different groups. The first group is the "Low" engagement when the engagement value is less than or equal to (57.888796). And the second group is the "High" engagement when the engagement value is greater than to (57.888796). This value was computed to divide the tweets into low and high engagement equally.

$$adjusted\,engagement = \frac{total\ post\ interactions}{page\ followers} * 1million \tag{1}$$

Table 2 shows the dataset existing and the newly created attributes after applying the preparation tasks.

3.3 Data Mining Experiments

The data mining experiments comprises two parts, namely the statistical analysis part and the machine learning prediction models. In the statistical analysis part, the aim is

Table 2 Dataset attributes after preparing the data

Attribute	Description
Weekday	The day on which the tweet was posted (Sunday, Monday, Tuesday, Wednesday, Thursday, Friday and Saturday)
Is Weekday	This attribute indicates if the tweet was posted in a week day (Yes) or during the weekend (No)
Time	The time on which the tweet was posted in 24 h format "hh" (00–23)
Time of Day	The value of the time of day on which the tweet was posted. The possible values are: Early morning, Morning, Afternoon, Evening, and Late night
Post Type	The media type of the post which includes status, photo and video
Has Link	This attribute indicates if the tweet includes at least one link or not
Has Hashtag	This attribute indicates if the tweet includes at least one hashtag or not
Has Mention	This attribute indicates if the tweet includes at least one mention or not
Tweet body	The textual content of the posted tweet
Number of followers	The total number of followers for the government entity's twitter account of the posted tweet
Number of replies	The total number of replies for a particular tweet
Number of favourites	The total number of favourites (likes) for a particular tweet
Number of retweets	The total number of retweets for a particular tweet
Engagement per 1 million	The adjusted value of Post engagement for the number of page followers
Engagement level	The engagement level value. There are two values: low and high engagement

to verify the eight hypotheses to test the extent to which the post characteristics (Post Type, Post Time, Post Day, Post Weekday, Links in Tweets, Mentions in Tweets, Hashtags in Tweets and Tweet Language) impact the engagement level for citizens on the Dubai government tweets. One-way ANOVA was used to analyse the impact of the Post Type (H1a), Post Day (H1c) and Post Language (H1h) on the citizens' engagement. To test the influence of the Post Time (H1b) on the citizens' engagement, the simple linear regression was used. Finally, a two-independent samples t test was used to test the impact of Post Weekday (H1d), Links in Tweets (H1e), Mentions in Tweets (H1f) and Hashtags in Tweets (H1g) on the citizens' engagement.

In the machine learning prediction models, the aim is to evaluate various machine learning models to test the second and third hypotheses (H2 and H3). In this vein, six machine learning models were implemented to test the combination of the tweet's standard features with the linguistic characteristics of the tweets' bodies. The six implemented machine learning models are: K-nearest Neighbour, Linear Support

Vector Machine (SVM), Naïve Bayes, Multilayer Perceptron (MLP) Classifier, Decision Tree and Random Forest. These models were selected to cover a various machine learning models and approaches. Prior to the implementation of each machine learning model, data pre-processing tasks such as tokenization, removing stop words and text vectorization were performed. The final processed dataset will be used to train the machine learning models and then test the performance for each model.

4 Experiments and Results

Study variables comprised both types, categorical and scale. Categorical variables included "Day", "Is Weekday?", "Time of Day", "Language", "Has Link", "Has Hashtag?", "Has Mention", "Tweet Type", and "Engagement Level". Scale variables included "Time", "Engagement", and "Engagement per 1 M". Findings of descriptive analysis can be briefly summarized in the following points:

- The majority of tweets (85.3%) are posted during weekdays. The peek day is Tuesday, with 18.3% of tweets posted that day.
- 50% of the tweets are posted in the morning, 27.2% in the afternoon, while very few (0.1%) are posted at late night.
- The majority of the tweets (64%) are in Arabic language. The second language used is English 28.1%. And 7.6% of the tweets are written in both Arabic and English.
- The majority of the tweets (87.1%) have links, and 85.3% have hashtags, while only 20% mention other users.
- The most used tweet type is Photo (77.4%), followed by Status (21.1%), and finally Video 1.5%.
- Engagement Level is almost equally represented with 51.8% of tweets receiving low engagement and 48.2% of tweets receiving high engagement.

In this study, we consider Time as a category (morning, afternoon, night, etc.) and as a scale from 0 to 23. As mentioned, it was found that most of the tweets were posted in the morning (6–11 am) and the number of posts decreases in the afternoon (12–16 pm) and then in the night (17–23 pm).

The dependent variable "Engagement" or "Engagement per 1 M" is not normal. That is, either variables have a highly skewed distribution to the right, which can be attributed to the existence of very high engagement records.

4.1 Data Analysis

The eight hypotheses of the current study are tested with the dependent variable "Engagement per 1 M" and the independent variables: "Post Type", "Post Time", "Post Day", "Post Weekday", "Links", "Mentions", "Hashtags", and "Tweet Language".

A one-way ANOVA was used to test H1, H3, and H8 since the independent variables "Post Type", "Post Day", and "Tweet Language" measure more than two categories. Simple linear regression was used to test H2. Simple linear regression is a technique that predicts a scale variable from a linear relation with another scale variable. The predicted (dependent) variable is the "Engagement per 1 M" (log transformed) and the independent variable is "Post Time". Two-independent samples t test was used to test H4, H5, H6, and H7. The two-independent samples t test is used to compare the means of two independent samples on a given variable [15]. For these hypotheses, the independent variables "Post Weekday", "Links", "Mentions", and "Hashtags" are divided into only two samples: "Yes" and "No".

In the current study, the distribution of "Engagement per 1 M" has been shown to have significantly different group variances across all dataset groups, p-value <0.05. Therefore, Welch and Brown-Forsythe tests were used instead of the F test. The Welch and Brown-Forsythe tests are Robust tests of equality of means.

H1a: Post Type impacts Citizens' Engagement (Photo, Status, Video)

A one-way ANOVA was conducted and revealed that there was a significant difference among three groups of Post Type, in terms of mean Log(Engagement per 1 M), Welch $= 537.859$ and p-value <0.001. Post hoc tests of multiple comparisons were performed using Games-Howell correction for unequal variances (based on results of test of homogeneity of variances), to find significant differences among the pairs of groups. The multiple comparisons revealed that Video and Photo posts have significantly higher mean Log(Engagement per 1 M) than Status posts, p-value <0.001.

H1b: Post Time impacts Citizens' Engagement

Simple linear regression was performed using the log transformed "Engagement per 1 M". The analysis results revealed a significant regression model that explained 0.3% of the variance in log(Engagement per 1 M), $F(1,63647) = 161.893$ and p-value <0.001. The regression coefficient is significant with a value equal to $B = -0.014$, $t = -12.724$ and p-value <0.001. This indicates that a one unit increase in Time decreases Engagement by 1.014 unit. The B value was transformed by taking its exponential to interpret the actual impact of Time on Engagement per 1 M.

H1c: Post Day impacts Citizens' Engagement

H3 is tested by performing a one-way ANOVA to find significant differences among week days in terms of Engagement per 1 M. The test revealed significant differences among week days in terms of log(Engagement per 1 M), Welch $= 42.94$ and p-value <0.001. That is, post day significantly impacts citizens' engagement.

The post hoc multiple comparisons tests revealed that Friday and Saturday had significantly lower mean log(Engagement per 1 M) than Sunday, Monday, Tuesday, Wednesday, and Thursday, p-value <0.001. Moreover, Sunday had significantly higher mean log(Engagement per 1 M) than Thursday, p-value $= 0.022$.

H1d: Post Weekday impacts citizen's Engagement

A two-independent samples t test was performed to test whether there is a significant impact of weekday on citizens' engagement or not. The t test revealed that there is a statistically significant difference between the two groups: Weekend and Not Weekend in terms of mean Engagement per 1 M (log transformed), $M = 4.138$ and $M = 3.93$, respectively; $t = -15.332$ and p-value <0.001. The Levene's test for equality of variances revealed that the two groups had significantly unequal variances, p-value <0.001, and hence t test statistics for the case "equal variances not assumed" was used. From these results, we can conclude that post weekday significantly impacts citizens' engagement, as mean engagement per 1 M was higher when the day is a weekday, while the mean engagement per 1 M was lower when the day is not a weekday.

H1e: Links in Tweets impacts Citizens' Engagement.

A two-independent samples t test was performed to test the hypothesis that Links significantly impacts citizens' Engagement. The test revealed that there is a statistically significant difference between the two cases: has link and doesn't have link, $t = -37.136$ and p-value <0.001. Posts with link had significantly higher mean log(Engagement per 1 M) than posts without link.

H1f: Mentions in Tweets impacts Citizens' Engagement.

A two-independent samples t test was performed to test whether or not mentions significantly impacts citizens' engagement. The test was statistically significant, $t = -15.68$ and p-value <0.001, indicating that there is a significant difference between posts with mentions and posts without. That is, posts with mentions had significantly higher mean log(Engagement per 1 M) than posts without mentions, $M = 4.255$ and $M = 4.071$, respectively.

H1g: Hashtags in Tweets impacts Citizens' Engagement.

A two-independent samples t test was conducted to test whether or not hashtags significantly impacts citizens' engagement. The test revealed that there was a statistically significant difference between posts with hashtags and posts without, $t = 23.283$ and p-value <0.001. That is, the mean log(Engagement per 1 M) for posts with hashtags ($M = 4.067$) was significantly lower than the mean log(Engagement per 1 M) for posts without hashtags ($M = 4.353$).

H1h: Tweet Language impacts the Citizen's Engagement.

A one-way ANOVA was conducted to test whether or not Tweet Language significantly impacts citizens' engagement. The test revealed that there was significant differences among the different groups of languages in terms of log(Engagement per 1 M), Welch = 475.773 and p-value <0.001. Post hoc multiple comparisons tests revealed that posts in "und" had significantly higher mean log(Engagement per 1 M) than posts in en, ar, and other languages, p-value <0.05. Moreover, posts in ar had significantly lower mean log(Engagement per 1 M) than posts in en and other languages, p-value <0.05.

4.2 Hypothesis Testing Summary

Table 3 summarizes the key findings and results of statistical tests used in hypothesis testing of the current study is reported. It is worthy to mention that all study hypotheses were significantly supported with p-value below 0.001, which is a very strong evidence of the significant effect reported of each independent variable.

4.3 Predicting Post Engagement

In this part, the dataset is used to predict the citizens' engagement using the various tweet features before the tweet is posted. This can help the government entities in taking the decision when to post a tweet to gain the maximum engagement level. Various machine learning models were explored. These machine learning models are: K-nearest Neighbour, Linear Support Vector Machine (SVM), Naïve Bayes, Multilayer Perceptron (MLP) Classifier, Decision Tree and Random Forest. In this study, the models are designed to predict the engagement value for both Arabic and English languages as it represents 92% of the overall tweets. The dataset includes 61,546 tweets with Arabic and English languages. Figure 3 shows the approach that is followed to implement the machine learning models. The dataset was split into two equally groups based on the engagement value. The threshold value that splits the data equally is 57.888796. The following subsections illustrate the carried-out tasks and experiments to predict and evaluate the citizens' engagement.

4.3.1 Data Pre-processing

Each tweet consists of several attributes, where the types of these attributes are mainly string and numerical numbers. To perform the experiments using a machine learning model, the user data must be only numerical values, since any other input type is not acceptable by any machine learning model Thus, in this step, we begin by processing the tweets to originate only numerical types attributes. The text attributes that are converted to numerical values is the day of the tweet, the type of the tweet, and the text of the tweet.

The post day of the tweet attribute (Sunday to Saturday) is converted to numerical value by mapping each day to its corresponding numeric value (1–7). For instance, Sunday is mapped to 1, Monday is mapped to 2, and Tuesday is mapped to 3 and so on. With respect to the post type of tweets, the values of this attribute are Photo, Status,

Table 3 Key findings of hypothesis testing

Hypothesis	Statistical test	Sig.	Result	Impact
H1a: Post Type impacts Citizens' Engagement (Photo, Status, Video)	One-way ANOVA	$P < .001$	Post Type (Photo, Status, Video) significantly impacts citizens' engagement	Photo and video posts have significantly higher engagement than status posts
H1b: Post Time impacts Citizens' Engagement	Simple linear regression	$P < .001$ $R^2 = 3\%$	Post Time significantly impacts citizens' engagement	A one unit increase in time decreases engagement by 1.014 units
H1c: Post Day impacts Citizens' Engagement	One-way ANOVA	$P < .001$	Post Day significantly impacts citizens' engagement	Friday and Saturday had significantly lower engagement than Sunday, Monday, Tuesday, Wednesday, and Thursday Sunday had significantly higher engagement than Thursday
H1d: Post Weekday impacts citizen's Engagement	Two-independent samples t test	$P < .001$	Post Weekday significantly impacts citizen's Engagement	Higher engagement was reported for week days, while lower engagement was reported for other days
H1e: Links in Tweets impacts Citizens' Engagement	Two-independent samples t test	$P < .001$	Links in tweets significantly impacts citizens' engagement	Posts with links have higher engagements than posts without links
H1f: Mentions in Tweets impacts Citizens' Engagement	Two-independent samples t test	$P < 0.001$	Mentions in tweets significantly impacts citizens' engagement	Posts with mentions have higher engagements than posts without mentions

(continued)

Table 3 (continued)

Hypothesis	Statistical test	Sig.	Result	Impact
H1g: Hashtags in Tweets impacts Citizens' Engagement	Two-independent samples t test	$P < 0.001$	Hashtags in tweets significantly impacts citizens' engagement	Posts with hashtags have lower engagements than posts without hashtags
H1h: Tweet Language impacts the Citizen's Engagement	One-way ANOVA	$P < 0.001$	Tweet Language significantly impacts the citizens' engagement	Posts in "mixed" had significantly higher Engagement than posts in "en", "ar", and "other" languages Posts in "ar" had significantly lower engagement than posts in "en" and "other" languages

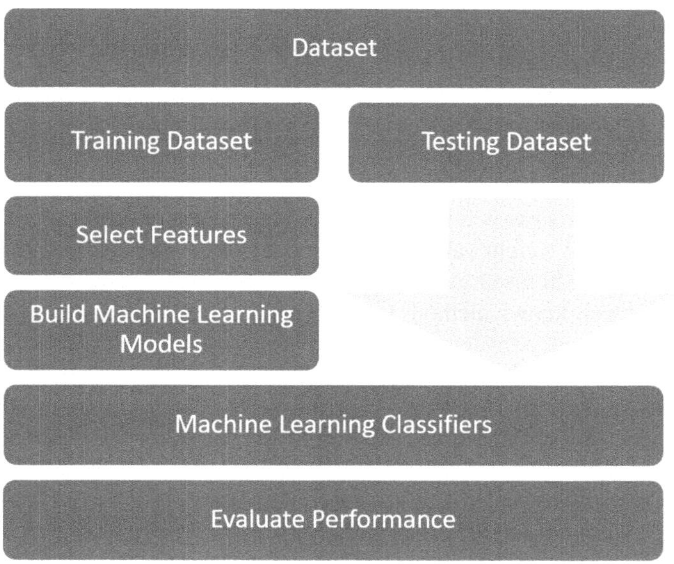

Fig. 3 Machine learning prediction approach for post engagement

and Video. Accordingly, the tweets types are mapped to one, two, and three; respectively. The text conversion of the tweets is conducted by performing the tokenization process followed by the text vectorization process.

4.3.2 Tokenization

In the tokenization process, we start by converting the text into lowercase. Then, we proceed by converting each word to its root. For instant "computing" and "computation" are converted to "compute". This aims to reduce the number of the words in the used vocabulary, and reduces the computational complexity, since each word is mapped to a unique value in the vectorization process. Next, we reduce the size of the input text by eliminating the punctuations and the "stopwords". Stop words are those words that are commonly used and do not have significant impact to change the meaning or change the context. The filtration of these stop words has significant impact on the performance of the NLP tasks. For instance, in the English language, the words "an", "the", "you", "he", "they", "however", "as" and "by" are recognized as common stop worlds, wherein the Arabic language, the words "الى"," "في", "هو", "هي", "بينما", "كان" and "خلال" are examples of common stop words.

4.3.3 Text Vectorization

Further to the text tokenization, the vectorization process aims to assign a numerical value for each word in the available text. Based on the employed vectorization process, a word numerical value represents the importance of the word in the document. The output of this process is represented as a matrix, where each row matrix represents the numerical weight values (importance) of a tweet. To obtain this matrix, several methods can be used to calculate the weights for the tweet values, where two of the most well-known methods to perform this calculation are the world count and the Term Frequency-Inverse Document Frequency (TF-IDF). The word count method (also known as bag of words) works by counting the number of times each word appeared in the text. This method is typically applied in situations where only the number times a word appears in the text (tweet) plays an important role in the classification process [16].

The TF-IDF method aims to determine the importance of each word in the tweets. Accordingly, words appear in small number of tweets are expected to have higher weight compared to words that appear in most of the tweets. This method has great advantage in situation where relatively unique words are expected to help in the classification process. Therefore, in this work, we adopt this method as part of the pre-processing step. The TF-IDF starts by calculating the frequency of each word appearance. The term frequency for word $w_i((w_i))$ is calculated by dividing the number of times a word appeared in the text over the total number of words in the text (TF). Then, this method proceeds by calculating the IDF for each word. This inverse is calculated as follows:

$$IDF(w_i) = \log\left(\frac{NT}{NTw_i}\right)$$

where NT refers to the total number of documents (tweets), and NTw_i refers to the number of documents (tweets) that has the word w_i. The IDF is typically used to scale down the weight of the words that appear in a high percentage of the tweets. Then, then the TD-IDF score for a word w_i is calculated as follows:

$$TD - IDF(w_i) = TF(w_i) \times IDF(w_i)$$

As we can see from the equation, the process of $TD - IDF$ aims to increase the weight of the words that appear in small percentage of tweets [17].

4.3.4 Experiments Settings

In this section, several sets of experiments were performed. These experiments aim to address the proposed research questions. In these experiments, the pre-processed input tweets have been divided into training and testing sets. The training set consists of 77% of the tweets, and the testing set has 33% of the tweets. The engagement level of the tweet (high:2, low:1) is used as the class label. Whereas the tweets features represent the text, day, time, category of the tweets.

To analyze the relationship between the tweets features and the expected engagement level, the following machine learning models were adopted and implanted: K-nearest Neighbour, Support Vector Machine (SVM), Multilayer Perceptron, Random Forest, Decision Tree, and Naïve Bayes. The performance metrics that are used to capture the performance of the used models are the following:

Precision: for each class label (low and high), precision refers to the percentage of correctly classified tweets from this class. For example, for the high engagement label class, precision refers to the percentage of correctly classified high engagement level tweets, and it is calculated as follows:

$$P = \frac{TP}{TP + FP}$$

where TP (True Positive) refers to the number of tweets that are classified as high engagement level, where FP (False Positive) refers to the number of low engagement level tweets that are misclassified as high-level engagement tweets.

For each classification engagement level, the **recall** score represents the percentage of correctly classified tweets of this level. For example, the recall score for the high engagement label class refers to the percentage of correctly classified high engagement level tweets, and it is calculated as follows:

$$R = \frac{TP}{TP + FN}$$

where in this situation, FN (false negative), refers the high-level engagement tweets that are misclassified as low-level engagement tweets.

The **F1-score** is used to analyze the quality of the classification, and it is calculated as follows:

$$F1 = 2 \times \frac{R \times P}{R + P}$$

where the higher the F1 score, the better the classifier quality. Next, we divide the performed experiments based on the used classifiers.

Accuracy: refers to the percentage of the correctly classified tweets. The accuracy can be calculated as per the formula below:

$$Accuracy = \frac{TP + TN}{(TP + TN + FP + FN)}$$

K-Nearest Neighbour

In this classifier, tweets features are represented as points in multidimensional Euclidean space. Additionally, each point (tweet) is labelled by the tweet engagement level (low or high). Once a new tweet (point) is submitted to the classifier, the k-nearest neighbours for this unique point will be identified. Then, a voting process will be followed to determine the label of the new point, where the label of the majority neighbours points will be assigned as the label of the new point. In the presented experiments, we have noticed that algorithm performance reaches stability once the value of k reaches ten, and therefore in this section, we have used $k = 10$. According to the results in Table 4, we can see that the accuracy of the k-nearest neighbours' classifier is relatively low (65%). This classifier uses distance to interpret the similarity between the tweets (points), and in the presented experiments, the number of used dimensions is relatively high, and this reduces the efficiency of using distance to interpret the relationship between the points (tweets).

Linear Support Vector Machine (SVM)

SVM is a discriminative classifier that aims to determine the optimal location for a hyperplane, which will be used to categorise the new point. In binary classifier, the hyperplane is a line that will be used to separate the two classes, where each class will be located at a different side of the line. In this classifier, the main mechanism aims to determine the location of the line such that the distance between the line and the nearest point on each is maximised (margin). From the results in Table 4, we can see that the SVM classifier achieves significantly higher accuracy 78% compared to the k-nearest neighbours' classifier. This is mainly due its objective of maximising the margin.

Table 4 Machine learning models performance results

Machine learning model	Class	Precision (%)	Recall (%)	F1-score (%)	Accuracy (%)
K-Nearest Neighbour	Low	67	70	68	65
	High	64	62	63	
Linear support vector machine	Low	78	81	80	78.2
	High	78	75	77	
Naïve Bayes	Low	72	81	76	73
	High	75	64	69	
Multilayer Perceptron	Low	74	69	72	71.9
	High	69	75	72	
Decision tree	Low	80	78	77	76
	High	73	74	76	
Random forest	Low	77	84	80	78.3
	High	80	72	76	

Naïve Bayes

The results for the Naïve Bayes classifier are shown in Table 4. Naïve Bayes classifier is a probabilistic classifier, which works by mainly calculating the probability of each factor, and assumes that the input features are independent. For any given new tweet, this classifier starts by calculating the conditional probabilities that this tweet belongs to a class label given each input feature. Then, the label for the new tweet is selected based on the highest probability. From the results we can see that the Naïve Bayes classifier achieves higher accuracy compared to the k-nearest neighbour classifier, and lower accuracy compared to the SVM classifier. This highlight that each of the class labels tweet sets has unique words that help in the classification process. Additionally, compared to the other two classifiers, in Naïve Bayes classifier, the precision and recall values are not identical for each class label, and this is due to the probabilistic nature of this classifier.

Multilayer Perceptron (MLP) Classifier

MLP classifier is recognized as feedforward neural network model, which eventually aims to map set of inputs to set of outputs. This classifier consists of at least three layers, where the backpropagation technique is used for training purposes. Training is established by modifying the weight of the connection between the nodes (neurons) in order to minimize the gap between the expected results and the actual results. Table 4 shows the results for the MLP classifier experiments, were to reduce the computational complexity, we sat the number of optimization iterations to 200. From the results we can see that with regard to the low-class label, the MLP classifier is able to achieve higher precision score compared to the high-class label. This suggest that used data from low-level engagement class have more distinct features, which simplify their classification.

Decision Tree

Decision tree classifier represents the historical training data as a tree. Whereas starting from the root, the nodes that will be located in each level are considered as the "best" available input feature. In this context, best refers to the impact of the feature on the classification process. For instance, assume that we have a binary classification problem (labels high and low) with three binary input features (A, B, and C). In this example, assume that feature B has the most impact on the classification process, where 80% of the training data has $B = 1$, and 20% of the training data has $B = 0$ (the highest percentage of the other labels is less than 80%). In this example, feature B will be selected as the root of the decision tree, since it has the most impact on the classification process. From the results in Table 4, we can see that the decision tree classifier is able to achieve 76% classification accuracy. The accuracy of this classifier is 1% less than the SVM classifier, and significantly higher compared to the rest of the classifiers. The results of this classifier suggested that the solution space has an organized structure, where there is a direct relationship between the values of the features and the classes points locality. Additionally, from the results we can that the percentage of the other performance metrics (precision, recall, and F1-score) is lower by less than 1% in the decision tree classifier compared to the SVM classifier. This is due to the fact that this classifier is expected to have difficulties correctly classifying the points located near the decision boundaries. Whereas in SVM, the representation of the points in higher dimensional space can help in addressing this problem.

Random Forest

Random forest classifier is an ensemble learning technique, in which a set of decision trees is created at random from the training data, and a voting mechanism is applied for these decision trees results in order to determine the classification process outcome. The number of used decision trees is sat to 20. According to the results in Table 4, it is clear that this classifier outperforms the previously discussed classifiers. This is mainly due to the mechanism of employing several decision trees since this has the advantage of determining efficient boundaries for the classes.

According to the achieved performance results of the implemented machine learning models, **Hypothesis 2** is supported as these adopted models were able to predict the citizens' engagement level with minimum accuracy of 65% for K-Nearest Neighbor and maximum accuracy of 78.3% for Random Forest.

4.3.5 Most Impactful Keywords

To further analyze the behavior of the classifier, we ran the experiments to extract the most common words in the high engagement level tweets. Figures 4 and 5 show the results of these experiments. The result shows that classifier relatively identified that same words as the most common ten words. This suggests that the used input tweets have distinct features that are being captured by all classifiers. In addition, the presented results suggested that the tweets that have the most impact are related to the public services. Additionally, the results show the tweets related the ruler of Dubai sheikh Mohammed Bin Rashid and his vision of improving the public and private sectors have achieved the highest engagements. From these figures, it is clear that citizens are more to be engaged with those tweets that talks about Dubai emirate, the influence of the leadership, the government entities' services, the digital transformation, the Pioneership of the emirate on the global level. Thus, **Hypotheses 3** is supported as the machine learning models were able to identify the most impactful keywords that are part of tweets with high engagement levels.

Fig. 4 Word cloud for keywords in Arabic tweets with highest engagement

Fig. 5 Word cloud for keywords in English tweets with highest engagement

5 Discussion

This study aims to determine the level of citizen engagement in Dubai government's Twitter. Through the use data mining techniques, results indicated that the post variables such as post type, post time, post day, links, mentions, hashtags and tweet language have a significant effect on citizen engagement, thereby supporting all the hypotheses. The main findings showed that: (1) photo and video posts have higher engagement than status posts; (2) engagement is higher during weekdays than weekends; (3) posts with links have higher engagements than posts without links; (4) posts with mentions have higher engagements than posts without mentions; (5) posts with hashtags have lower engagements than post without hashtags; and (6) posts in English and Arabic languages have higher engagements than posts in English and other languages.

In line with these primary findings, the study also emphasizes the increased interaction and engagement in Dubai government's Twitter which consequently entails the effectiveness of the said social media platform in enhancing citizen engagement. Furthermore, the study has revealed the content of the posts that have higher engagements, denoting that the citizens tweets talk more about the services, leadership and overall efficiency of the Dubai government. This means that social media platforms such as Twitter are primarily used by citizens to know and discuss more about the relevant activities of the government which are of direct importance to them. Needless to say, Twitter is an effective tool for enabling communication, interaction and engagement between the public and the government.

This study aims to determine the level of citizen engagement in Dubai government's Twitter. Through the use data mining technique, results indicated that the post variables such as post type, post time, post day, links, mentions, hashtags and tweet language have a significant effect on citizen engagement, thereby supporting all the hypotheses. In response to the study's research questions, the following are noted:

- **Do post characteristics impact the users' engagement in Dubai Government Twitter accounts?**

The results of the study indicate that post characteristics have a significant impact on the level of citizen engagement in Dubai Government's Twitter. For one, it was revealed that: (1) photo and video posts have higher engagement than status posts; (2) engagement is higher during weekdays; (3) posts with links have higher engagements than posts without links; (4) posts with mentions have higher engagements than posts without mentions; (5) posts with hashtags have lower engagements than post without hashtags; and (6) posts in English and Arabic languages have higher engagements than posts in English and other languages.

- **Can we predict the user's engagement based on post characteristics?**

Based on the results achieved from using the machine learning models, post characteristics can be used as indicators in predicting user engagement. As a matter of fact, post variables including post day, links, mentions and hashtags, among others having an effect on citizen engagement, Dubai Government can then make use of these attributes to increase eParticipation, underpinning the notion that Twitter is an effective platform for making conversations and for monitoring interactions [11].

- **Can we identify the main keywords resulting in higher engagement?**

According to [18], social media platforms such as Twitter enable the public sector to create discussions with the citizens regarding specific issues and at the same time, obtain feedback. This can be associated with the results of the study which point out that citizens show higher engagement in discourses relevant to government services. Keywords such as leadership, government services, digital transformation, and ruler of Dubai sheikh Mohammed Bin Rashid receive higher citizen engagements.

Similar to what has been said about posting updates and relevant information regarding government services, leadership, and the efficiency of the entire public sector, Dubai Government should also take note of post attributes that receive higher engagement. Understanding the three stages of eParticipation will also be of huge help. As stated in [11], these three stages, namely information stage, consultation stage, and decision-making stage, could help the Dubai Government have a more thorough and a more effective approach in motivating increased eParticipation among citizens.

6 Conclusion

The study, through the use of data mining, reveals the role of Twitter in increasing engagement between the citizens and the Dubai government. It is noted that post attributes have a significant effect on citizen engagement and that citizens are more actively engaged on topics including services, leadership and government transformation. The significance of this study is reflected on how the information provided can help governments enhance their citizen engagement by following Dubai government's approach. To accomplish this, we collected, prepared and processes 74,037 tweets, representing all tweets for Dubai Government's Twitter in 2018. Through the use of statistical analyses of the impact of post characteristics on the level of citizen engagement and of the various machine learning models, results indicate that citizen engagement level in Dubai Government's Twitter is significantly impacted by all post characteristics. It is also revealed in the study that citizen engagement is higher during weekdays compared to weekends. Furthermore, the machine learning models achieved promising results to predict the citizens' engagement with highest accuracy for Random Forest and Linear Support Vector Machine of 78.3% and 78.2% respectively. This study can be used for future research especially relating to the impact of citizen engagement in Twitter on the improvement of government services. It can also be used to explore strategies in improving citizen engagement on social media. This study can be expanded by analyzing the sentiments of tweets and replies [19].

References

1. Siddiqui, S., Singh, T.: Social media its impact with positive and negative aspects. Int. J. Comput. Appl. Technol. Res. **5**(2), 71–75 (2016)
2. Mishaal, D., Abu-Shanab, E.: The effect of using social media in governments: framework of communication success. In: ICIT 2015 The 7th International Conference on Information Technology. Amman, Jordan, 12–15 May (2015)
3. Chukwuere, J.E. Chukwuere, P.C.: The impact of social media on social lifestyle: a case study of university female students. Gend. Behav. 9928–9940 (2017)
4. Akram, W., Kumar, R.: A Study on Positive and Negative Effects of Social Media on Society. Int. J. Comput. Sci. Eng. **5**(10). 347–354 (2017)
5. Karakiza, M.: The impact of social media in the public sector. Procedia—Soc. Behav. Sci. **175**, 384–392 (2015)
6. Sanford, C., Rose, J.: Characterizing eParticipation. Int. J. Inf. Manage. **27**, 406–421 (2007)
7. Kassen, M.: E-participation actors: understanding roles, connections, partnerships. Knowl. Manag. Res. Pract. 1–22 (2018)
8. Pushpam, C.A., Jayanthi, J.G.: Overview on data mining in social media. Int. J. Comput. Sci. Eng. **5**(11), 147–157 (2017)
9. Injadat, M.N., Salo, F., Nassif, A.B.: Data mining techniques in social media: a survey. Neurocomputing 214, n.p (2016)
10. Zatari, T.: Data mining in social media. Int. J. Sci. Eng. Res. **6**(7), 152–154 (2015)
11. Siyam, N., Alqaryouti, O., Abdalla, S.: Mining government tweets to identify and predict citizens engagement. Technol. Soc. (2019). https://doi.org/10.1016/j.techsoc.2019.101211
12. Kavitha, D.: Survey of data mining techniques for social networking websites. Int. J. Comput. Sci. Mob. Comput. **6**(4), 418–426 (2017)
13. Azevedo, A.I.R.L., Santos, M.F.: KDD, SEMMA and CRISP-DM: a parallel overview. IADS—DM (2008). https://recipp.ipp.pt/handle/10400.22/136. Accessed 12 Oct 2019
14. Government of Dubai, Government Entities: (2019). http://www.dubai.ae/en/government/pages/default.aspx?category=Government. Accessed 26 April 2019
15. Urdan, T.: Statistics in Plain English, 2nd edn. Psychology Press, New Jersey (2005)
16. Alqaryouti, O., Siyam, N., Abdel Monemb, A., Shaalan, K.: Aspect-based sentiment analysis using smart government review data. Appl. Comput. Inf. (2019). https://doi.org/10.1016/j.aci.2019.11.003
17. Alqaryouti, O., Khwileh, H., Farouk, T., Nabhan, A., Shaalan, K.: Graph-based keyword extraction. In: Intelligent Natural Language Processing: Trends and Applications, pp. 159–172. Springer, Cham (2018)
18. Al-Badi, A.H.: The adoption of social media in government agencies: Gulf Cooperation Council case study. J. Technol. Res. **5**. 1–26 (2013)
19. Mhamdi, C., Al-Emran, M., Salloum, S.A.: Text mining and analytics: a case study from news channels posts on Facebook. In: Intelligent Natural Language Processing: Trends and Applications, pp. 399–415. Springer, Cham (2018)

The Impact of WhatsApp on Employees in Higher Education

Jasiya Jabbar, Sohail Iqbal Malik, Ghaliya AlFarsi, and Ragad M. Tawafak

Abstract The inexorable forward March of technology and the advent of smartphone applications offer effortless communication. WhatsApp is one of the most common and widely accepted mobile applications for communication and collaboration. We conduct this research to identify the influence of WhatsApp among the employees in an educational institution and to compare the means of communication used at workplace. A web-based survey was conducted among employees at AlBuraimi University College, Oman. The study revealed that respondents preferred WhatsApp for informal communication and email for formal communication. Email was preferred for communication with students and academic works. The results also showed that WhatsApp enables faster and easy communication, sharing ideas, chatting and suggested WhatsApp has a positive impact on them. The results also shed light on some of the consequences of WhatsApp usage like privacy intervention, concentration reduction in work which may lead to delayed work completion. Majority of respondents didn't suggest WhatsApp in learning and teaching process. The results affirmed the influence of WhatsApp in quick and effective communication. Regardless of the advantages and acceptance of WhatsApp, the study also revealed the reluctance of respondents in using WhatsApp for formal communication, academic works and in learning process.

Keywords Communication · Influence · Social media · WhatsApp · Workplace

---

J. Jabbar (✉) · S. I. Malik · G. AlFarsi · R. M. Tawafak
Department of Information Technology, Al Buraimi University College, Al Buraimi, Oman
e-mail: jasiya@buc.edu.om

S. I. Malik
e-mail: sohail@buc.edu.om

G. AlFarsi
e-mail: galfarsi@buc.edu.om

R. M. Tawafak
e-mail: raghad@buc.edu.om

M. Al-Emran et al. (eds.), *Recent Advances in Intelligent Systems and Smart Applications*, Studies in Systems, Decision and Control 295,
https://doi.org/10.1007/978-3-030-47411-9_34

1 Introduction

Social media sites utilize information and communication technology enhancements for effective and timely communication from anywhere and anytime. The efficacy of Social media is reflecting on almost all areas such as education, marketing, advertisement, communication and information sharing.

WhatsApp is a proprietary, cross-platform instant messaging subscription service for smartphones [1]. Various researches shed light on the fact that it is widely accepted and recognized by the people globally. Its growing popularity and practice led Facebook to acquire WhatsApp for around $19 billion. It is a fast, simple and reliable mode of communication across the world. As in whatsapp.com, more than 1.5 billion people in over 180 countries use it to stay in touch with friends and family.

WhatsApp is an application which is supported by almost all mobiles and is available even in low connectivity areas. It is a very user-friendly tool which provides multimedia messaging features and can even convey location details by using integrated mapping features. WhatsApp can be easily installed from play store in smart phones. Users need not pay additional charge for WhatsApp messaging, as it sends message through internet data connection. It also provides provision for audio and video calls. It can also be accessed from browsers in desktop by URL web.WhatsApp.com; by scanning QR code on the site [1].

WhatsApp is an active application that promote cultural collaboration among the participants with its personal, linguistic and social assistances [2]. Irrespective of age, qualification and profession, people wield this tool to support and simplify their tasks. Besides means of communication, students employ it in education to mediate, support active learning and collaboration among them. Employees exploit it in interacting each other, disseminating notifications, reminders, sharing ideas and successes, etc.

Around 1993, email changed the human's communication practice forever. Emails are the principal means of communication and collaboration since then. As the time passes the way we communicate also changes. Even though emails and hardcopies are approaches followed by professionals for their collaboration and communication, due to the wide acceptance and quick response to WhatsApp, it can be used efficiently and effectively in workplace.

In our workplace, the notifications and announcements are circulated through emails. The employees read and acknowledge it, when they receive it. But in most cases we observe that remainders invoke the employees to meet deadlines. In such cases, WhatsApp are more comfortable as informal remainders can be send. People are font of checking WhatsApp in-between and this updates and reminds them about the important facts. So WhatsApp can be a means of great support in today's busy schedule and mechanical life. This point shed light into this research. A study was conducted in Al Buraimi University College to analyze the impact of Google Apps in academic and administrative staffs, which revealed the significant impact of those on their works [3]. The aim of this research is to assess, compare and analyze disseminating applications and to evaluate the effectiveness of WhatsApp in our workplace (Al Buraimi University College).

This paper is organized into a number of segments. The introduction is followed by a literature study exploring various researches on WhatsApp application. The research questions are pinpointed, the methodology followed in this work, furthermore the findings of the survey are incorporated and ultimately the summary of this research.

2 Literature Study

2.1 WhatsApp as a Communication Tool

WhatsApp and social networks are fast growing domain in which various researches are conducted. Social media has a high impact as a means of communication [4]. Avi et al. in their research designed a model to understand the usage of WhatsApp and to predict the demographic details of the users. They analyzed the data, deployed Weka and pROC data mining packages and studied models created from decision trees, Bayesian networks and logistics regression algorithms. Without analyzing the message content, they successfully quantified and predicted user's gender and age demographic, the different types of activities they might perform, the frequency of activities, the quicker response times, length of messages. The results showed that both genders widely use WhatsApp and the purpose and usage habits differ for both groups [5]. The study shows that irrespective of gender and age, people use the WhatsApp tool for communication.

Shahid et al. analyzed WhatsApp as a messaging tool in a university of Pakistan. The results showed it is used for one to one messaging, broadcasting and group communication. They use it for sharing academic information, chatting, picture and video sharing, texting and calling family members irrespective of distance and country [6].

Muhammad and Ammaarah conducted a survey on 100 Abu Dhabi University students. The study focused on whether WhatsApp can replace face-to-face communication, the effectiveness of emojis, voice notes, text messaging in communication and whether there is any miscommunication by using this tool. The conclusion was WhatsApp can be employed as an addition to and not replacement to face-to-face communication [7].

WhatsApp consumption expands daily and its rate is alarming. In Northern India, Naveen and Sudhansh conducted an internet based survey using open-source survey software to collect responses. The results showed that people are shifting from social networking sites to WhatsApp. 66% of WhatsApp users believe that it has improved the relationship with friends, 63% thinks it is not harmful for them. Majority thinks that it enables communication instantly anytime and anywhere and gets instant response while using WhatsApp [8].

2.2 WhatsApp Among Students

Traditional classrooms and style of teaching and learning are shifting nowadays to virtual classrooms by the intervention of social networks [9, 10]. Youtube is the most popular site and facebook is in second position by the opinion of students in different universities, colleges and institutions in Oman [11]. Recently, WhatsApp is widespread, offering students an opportunity to build sense of connectedness and a basis for developing a learning community [12]. Moreover, Al-Qaysi et al. concluded that WhatsApp is one of the most predominant application used for educational purposes in Oman [13, 14].

Begona et al. in their research demonstrates that WhatsApp can be used as a communicational tool in educational domain. They performed an analysis on a sample of 200 university students. The students were divided into groups, they were accommodated in different places and they could communicate only using WhatsApp and was given complex decision making assignment. Findings confirm the WhatsApp usefulness in team effectiveness and coordination [15].

According to Sana Shahid, students used WhatsApp not just for daily communication but also for many other activities such as to share pictures, notes, discuss about attendance, assignment discussions, to arrange group studies, share past question papers. It is observed that they ask each other to mark attendance by responding to teacher with their sound. WhatsApp groups even guide and engage students [16].

WhatsApp has both positive and negative impact on students. Johnson and George conducted a study with students of Tertiary Institutions in Ghana to disclose the impact of WhatsApp on them. The study unveiled that in addition to making communication easier and faster and sharing ideas, WhatsApp spoils students study time, students spelling and grammatical construction of sentences are destroyed, distracts students in all sort of activities like their studies, managing timetable and concentrating in classes [17].

Levent, Cetinkaya discovered that WhatsApp increased the success rate in students. The study was conducted with 10th grade secondary education students. In this research, the traditional educational system was supported by WhatsApp usage. Important notes, images and descriptions were send to the student sample. Tests were conducted to analyze whether this enhance the learning and teaching environment. Some of the observations of the study are: WhatsApp enhanced the teaching process, multimedia messages increased the student's interest in topics and students strongly recommend using the same in all courses to support courses. Some of the recommendations given by students were related to the timing of messaging-untimely messages leads to distraction in students [18].

2.3 WhatsApp as a Tool for Maintaining Employee Relationship in Industries

Communication plays a vital role in all domains. Industries deal with projects which are done by teams. The team comprises team members, managers, stakeholders, business partners and suppliers. Effective communication among these members decides the project's success. A recent study in Malaysian construction field shows that WhatsApp tool has enhanced communication and data sharing. They used semi structured questions in interviewing the sample of 5 project managers [19].

Studies were conducted to explore WhatsApp and employee relationship correlation. Two departments from a telecommunication company at Cyberjaya, Selangor were selected for a survey. The members of the department were not available all time physically on desk but they need to communicate each other. The study revealed the positive and vital role of WhatsApp in both personal and professional intercommunication [20].

Businessmen and professionals share management decisions with subordinates, plan meetings, use location feature to inform their customers, advertise products and even to business using WhatsApp tool [16]. A study was conducted among the Malaysian employees understand how WhatsApp is utilized for the superiors and subordinate communication. It also identified the factors that motivate them to use the application. The results showed that internalization, compliance and identification were the main factors that influenced employees to use it [21, 22]. An understanding of WhatsApp group usage in a Malaysian university was studied. The impact of WhatsApp group was revealed in the study [23]. The social network sites especially WhatsApp plays a vital role in the employees in business environment to promote and strengthen business and enhance employee relationships. The employee can send and publish any topic specific to their business environment through WhatsApp and in a very short time, they get responses and the reaction of other employees [24, 25].

3 Research Questions

It is evident from the previous section that WhatsApp application can be used to enhance communication and collaboration. In this study, we purpose two research questions to guide the study and to determine the staff preferences among different applications in performing various tasks at workplace.

The two research questions being addressed in this research study are as follows:

RQ1 which applications the staffs' prefer in performing various tasks at workplace?
RQ2 what is the influence of WhatsApp application on staff at workplace?

4 Methodology

Both research questions were investigated by conducting a web-based survey with academic and administrative staff members. The population being surveyed consists of academic and administrative staff members at AlBuraimi University College, Oman. The population consisted of 56 staff members including 32 academics and 24 administrative members.

The data were collected after obtaining an ethical approval from Al Buraimi University College. Data collection was anonymous and the staff participation in the survey was optional and voluntary. The collected data was analyzed by using IBM SPSS (version 21) software.

The survey was prepared to determine the staff preferences in performing various tasks at workplace. The survey consists of 28 questions which comprising open and closed ended questions. The survey has three parts. The first part of the survey covers demographic questions relating to staff department, gender, age, qualification and experience. The second part included questions related to the comparison of WhatsApp with other applications in performing different tasks at workplace. Thirteen questions were asked in this section. For the first eleven questions, the audience might enter multiple answers. The main alternatives provided are WhatsApp and Email; moreover, respondents can enter their choice too. The questions focused on application preference on daily activities such as formal and informal communication, sharing ideas and achievements, chatting and broadcasting. The last two questions in this section were closed-ended question to inquire WhatsApp consumption and impact. The third part focuses on questions related to influence of WhatsApp in performing different tasks at workplace. Nine closed–ended which are dichotomous questions to elicit yes or no answers and one open–ended questions were included in this section. This section inquires the contribution of WhatsApp in personal and professional life.

5 Results

This section describes the results of this study and probes Research Questions 1 and 2. The first research question is as follows:

RQ1 which applications the staffs' prefer in performing various tasks at workplace?

A web-based survey was conducted with academic and administrative staff members to determine their preferences in choosing different applications for performing various tasks at workplace.

5.1 Demographic Details

The demographic details of the respondents are as follows:

- 57.1% respondents are academic staff, compared to 42.9% from administrative staff.
- 39.3% respondents have master degree, compared to 28.6% Ph.D., 26.8% bachelor, 3.6% secondary school and 1.8% diploma/advanced diploma.
- 41.1% respondents are between 26 and 35 years old, compared to 33.9% between 35 and 45 years, 23.2% above 46 years and 1.8% below 25 years.
- The majority of the respondents (60.7%) have more than 10 years' experience, compared to 26.8% between 5 and 10 years and 12.5% less than 5 years.
- 55.4% respondents are male compared to 44.6% female.

5.2 Comparison of WhatsApp with Other Applications

Table 1 shows the respondents' preferences of choosing applications in performing different tasks in our workplace. Respondents preferred WhatsApp application for 'Informal communication with colleagues (87.5%)', 'Sharing general information (83.9%)', 'Enhance collaboration at the workplace (60.7%)', 'Sharing ideas (83.9%)', 'Instant communication (83.9%)', 'Distribution of information to multiple people (64.3%)', and 'For Chatting (91.1%)'. On the other side, respondents preferred Email application for 'Formal communication with colleagues (89.3%)', 'Communication with students (75%)', 'For academic work (94.6%)' and 'Sharing achievements with others (71.4%)'.

5.3 Influence of WhatsApp Application

This section describes the result for Research Question 2. Table 2 shows the responses on influence of WhatsApp on staff at workplace. The second research question is as follows:

RQ2 what is the influence of WhatsApp application on staff at workplace?

This research question is investigated by asking questions in the survey from the respondents related to influence of WhatsApp in performing different tasks at workplace. Table 2 shows the respondents preferences of using WhatsApp application for performing various tasks at workplace.

Most of the respondents (78.6%) are member of WhatsApp groups at workplace. Respondents perceived WhatsApp application useful for faster communication (76.8%), relaxation from the work pressure (62.5%), and easy communication

Table 1 Comparison of WhatsApp with other applications

Which application you prefer for the following tasks (tick all applicable)	WhatsApp (%/N)	Email (%/N)	Other (%/N)	
Informal communication with colleagues	87.5%	28.6%	3.6%	
	49	16	2	In-person Snapchat
Formal communication with colleagues	23.2%	89.3%	0%	
	13	50	0	
Sharing general information	83.9%	33.9%	0%	
	47	19	0	
Enhance collaboration at the workplace	60.7%	57.1%	0%	
	34	32	0	
Sharing ideas	83.9%	39.3%	3.6%	
	47	22	2	In-person(2)
Instant communication	83.9%	16.1%	7.2%	
	47	9	4	In-person (2) Facebook Phone
Communication with students	26.8%	75%	21.6%	
	15	42	12	In-person (2) Google classroom (10)
Distribution of information to multiple people	64.3%	60.7%	3.6%	
	36	34	2	Text message Facebook
For chatting	91.1%	3.6%	7.2%	
	51	2	4	Phone Facebook Snapchat Viber
For academic work	25%	94.6%	0	
	14	53	0%	
Sharing achievements with others	55.4%	71.4%	0	
	31	40	0%	

(82.1%). On the other hand, respondents pointed out that WhatsApp application interfere in their privacy (67.9%), reduces the concentration in work (69.6%), distracts and affects the completion of work (58.9%), makes mobile-phone addiction (71.4%) for them. 60.7% respondents suggested not to use WhatsApp in the teaching and learning process.

Two more close-ended questions were asked in this section to determine the influence of WhatsApp on staff at workplace as shown in Table 3.

Table 2 Influence of WhatsApp application on staff at workplace

Questions	Yes (%/N)	No (%/N)
I am member of WhatsApp groups at workplace	78.6%	21.4%
	44	12
I suggest using WhatsApp on workplace for faster communication	76.8%	23.2%
	43	13
I suggest using WhatsApp on workplace for easy communication	82.1%	17.9%
	46	10
I suggest using WhatsApp in the learning and teaching process	39.3%	60.7%
	22	34
WhatsApp gives relaxation from the work pressure	62.5%	37.5%
	35	21
Does WhatsApp interfere in your privacy?	67.9%	32.1%
	38	18
The use of WhatsApp reduces the concentration in work	69.6%	30.4%
	39	17
The use of WhatsApp distracts and affects the completion of work	58.9%	41.1%
	33	23
The use of WhatsApp makes mobile-phone addiction in me	71.4%	28.6%
	40	16

Table 3 Impact of WhatsApp application

Questions	Preferences	(%/N)
How much time you spend on WhatsApp per day:	1–2 h	64.3%
		36
	3–5 h	21.4%
		12
	6–7 h	10.7%
		6
	Over 8 h	3.6%
		2
Overall, WhatsApp has positive or negative impact on my performance at workplace	Positive	62.5%
		35
	Negative	8.9%
		5
	No impact	28.6%
		16

Most of the respondents (64.3%) spend 1–2 h on WhatsApp per day, compared to 21.4% spend 3–5 h, 10.7% spend 6–7 h, and 3.6% spend over 8 h. 62.5% respondents perceived that WhatsApp has positive impact on their performance at workplace, compared to 28.6% has no impact, and 8.9% has negative impact.

In response to the open-ended question "Give your comment on 'A day without WhatsApp'", respondents mentioned it will be difficult day, can't achieve our tasks quickly, no faster communication, and feel like isolated from the world or out of reach. On the other hand, some respondents commented that it will be a great idea and best day, feel more freedom, relaxation, and relief.

Moreover, Respondents also commented that WhatsApp application should be used at workplace for faster, instant and quick communication among staff. Moreover, WhatsApp is considered as a social media tool so it is better to use it by media department to announce general information and announcements. Some respondents disagreed to use WhatsApp for educational purposes in a current shape.

6 Discussions

The above section depicts the descriptive and tabular representation of the survey results. The study found that respondents have a mix of opinions. Respondents prefer WhatsApp for informal communication and email for formal communication. It is evident from the results that respondents use WhatsApp for informal communication with colleagues, sharing general information, sharing ideas and for chatting which is consistent with the findings of [8]. On the other hand the respondents prefer email for formal communication which contradicts the findings of [16].

Respondents prefer email for communication with students which contradicts the findings of [18] which recommends using WhatsApp for in all courses to support course. Emails are preferred as it makes communication formal, documented and backup of all information is available any time conversely the messages for a limited period can only be viewed in WhatsApp.

Respondents prefer email for academic works which contradicts the findings in [16, 19]. This means WhatsApp has a positive impact in telecommunication, business and industrial areas whereas this research deals with educational sector. In academic work email is preferred due to the backup it provides, ease of searching, draft option and another important factor is it doesn't utilize phone memory.

Majority prefers WhatsApp for faster and easy communication, for sharing ideas, chatting which supports the findings of [8] and on the other hand majority suggests it interfere in privacy. Reason can be the time messages are send which intrudes in the family life, people will be busy with official works and discussions in WhatsApp even at home which negatively affects the family cohesion. Some of the other reasons are messages send in WhatsApp groups may be irrelevant to some, it also causes concentration reduction in work and leads to delayed work completion.

Majority of respondents didn't suggest WhatsApp in learning and teaching process which supports the findings of [17]. WhatsApp spoils students spelling and grammatical construction of students, distract the students in their studies furthermore decreases the face-to-face communication skills.

Most respondents perceive WhatsApp to have a positive impact on them because it enhances collaboration, relationship with friends and coordination. This finding is consistent with [15]. The pros of WhatsApp overwhelm the cons of it which distracts both personal and professional life.

Respondents mentioned it will be a difficult day if they are disconnected from WhatsApp, they can't achieve their tasks quickly, no faster communication, and feel like isolated from the world or out of reach. As now-a-days technology is indispensable in our day-to-day activities and WhatsApp makes communication easier, informal and faster, a day without it will be incredible. Some respondents has a different perspective on a day without WhatsApp, they commented that it will be a great idea and best day, feel more freedom, relaxation, and relief as technology engages and occupies all the time which disturbs the daily routine. Some respondents disagreed to use WhatsApp for educational purposes in a current shape and recommended for media department to announce general information and announcements.

While comparing WhatsApp with other means of communication, Gmail was specifically mentioned as it is main mode of communication used in our institution. The other options that the respondents claimed to be used for communication were Facebook, by direct calling, snapchat, viber and text messaging. But the respondents supporting those are negligible compared to WhatsApp and Gmail. For the communication with students, even though WhatsApp was not supported by respondents, they recommended using email and google classroom instead.

7 Limitations

Similar to most research, this paper also has various limitations too. The research was conducted among employees in one college in Oman; the opinions in this regard might be collected from a larger sample if the survey was disseminated to other educational institutions to infer the results with more precision. The focus was only on employees and the perspective of students in this regard was not considered. Thus there is a scope of research in this concern in future.

8 Conclusion

WhatsApp has already proved to be one of the best applications to ease communication and collaboration. The research was conducted among academic and administrative staff in AlBuraimi University College to understand the influence of WhatsApp at their workplace and to compare the applications used in workplace for collaboration and communication.

From the preceding section, it is evident that WhatsApp has a positive impact on employees in our workplace and they prefer it for enhancing communication and collaboration among employees. Previous studies suggest using WhatsApp in enhancing the learning and teaching process [18] but this study acclaims not to integrate it in education process due to the possible adverse impact of it on students. However, this research advocates the positive impact and influence of WhatsApp as a communicational tool.

The aim of this research was to compare WhatsApp with other means of communication and to study the influence of it in or workplace. The study found that WhatsApp was highly recommended for easy and fast communication, enhance collaboration and for informal communication. It also revealed that emails are preferred for formal communication and academic works. The study also suggested not to use WhatsApp for education process and in communication with students.

References

1. Mefolere, K.F.: WhatsApp and information sharing: proscpect and challenges. Int. J. Soc. Sci. Hum. Res. **4**(1) (2016)
2. Tulgar, A.T.: WHATSAPP as a tool for sustainable glocal linguistic. Turkish Online J. Distance Educ.-TOJDE **20**(3) (2019)
3. Al-Emran M., Malik, S.I.: The impact of Google Apps at work: higher educational perspective. iJIM **10**(4) (2016)
4. DowerahBaruah, T.: Effectiveness of social media as a tool of communication and its potential for technology enabled connection: a mirco-level-study. Int. J. Sci. Res. Publ. **2**(5) (2012)
5. Rosenfeld, A., Sina, S., Sarne, D., Avidov O., Kraus, S.: WhatsApp usage patterns and prediction of demographic characteristics without access to message content. Demogr. Res. **39**. Article 22 (2016)
6. Minhas, S., Ahmed M., Ullah, Q.F.: Usage of WhatApp: a study of University Of Peshawar, Pakistan. Int. J. Hum. Soc. Sci. Invent. **5**(7) (2016)
7. Ali, M.S.S., Kootbodien, A.: The effectiveness of WhatsApp as an interpersonal communication medium among Abu Dhabi University students. Int. J. Media, Journal. Mass Commun. **3**(1) (2017)
8. Kumar N., Sharma, S.: Survey analysis on the usage and Impact of WhatsApp Messenger. Glob. J. Enterp. Inf. Syst **8**(3) (2016)
9. Alabdulkareem, S.A.: Exploring the use and the impacts of social media on teaching and learning science in Saudi. Procedia—Soc. Behav. Sci. **182**, 213–22 (2012)
10. Abir A.H.-B., Al-Harrasi, S.: The impact of social networking: a study of the influence of smartphones on college students. Contemp. Issues Educ. Res. **7**(2) (2014)
11. Al-Mukhaini, E.M., Al-Qayoudhi, W.S., Al-Badi, A.H.: Adoption of social networking in education: a study of the use of social networks by higher education students in Oman. J. Int. Educ. Res. **10**(2) (2014)
12. Suzanne Stone, A.: Exploring students use of the social networking site Whatsapp to foster connectedness in the online learning experience. Irish J. Technol. Enhanc. Learn. **3**(1) (2018)
13. Al-Qaysi, N., Mohamad-Nordin N., Al-Emran, M.: What leads to social learning? Students' attitudes towards using social media applications in Omani higher education. Educ. Inf. Technol. 1–18 (2019)

14. Al-Qaysi, N., Mohamad-Nordin N., Al-Emran, M.: An empirical investigation of students' attitudes towards the use of social media in Omani Higher Education. In: International Conference on Advanced Intelligent Systems and Informatics, 2019
15. Urien, B., Erro-Garces A., Osca, A.: WhatsApp usefulness as a communicational tool in an educational context. Educ. Inf. Technol. **4** (2019)
16. Shahid, S.: Content analysis of WhatsApp conversations: an analytical study to evaluate the effectiveness of WhatsApp application in Karachi. Int. J. Media, Journal. Mass Commun. **4**(1) (2018)
17. Yeboah J., Ewur, G.D.: The impact of Whatsapp messenger usage on students performance in tertiary institutions in Ghana, J. Educ. Pract. **5**(6) (2014)
18. Cetinkaya, L.: The impact of whatsapp use on success in education process. Int. Rev. Res. Open Distrib. Learn. **18**(7) (2017)
19. Ahmad Pozin, M.A., Mohd Nawi, M.N.: Effective of communication using WhatsApp: industialised building system (IBS) construction. In: AIP Conference Proceedings, January 2016
20. Ariffin Z., Omar, S.Z.: Usage of WhatsApp in relation to employee engagement in a telecommunication company. Int. J. Acad. Res. Bus. Soc. Sci. **8**(1) (2018)
21. Mustaffa, C.S., Soh, C.H.C., Abu Bakar H., Mohamad, B.B.: Understanding the usage of Whatsapp among Malaysian employees through the lens of social influence theory. Int. J. Eng. Technol. **7**, 1039–1043, 2018
22. Idris, A.H.M.: Deploying mobile trend in Higher Education Institutions in Malysiai. Int. J. Eng. Technol. **8**(1), 59–62 (2019)
23. Nuredayu, O., Chee, M., Zuraidah, M.M.A.T.: Reciving and responding to WhatsApp official group message among employees: an early interpertation analysis. Int. J. Eng. Technol. **7**, 1030–1033 (2018)
24. Datt P., Fernandes, T.: Entrepreneur & entrepreneurship. In: A Guide for Entrepreneurship Education The Omani Case Studies. Al Roya Press and Publishing House, Oman (2019)
25. Chotrani, A.: Fashion entrepreneurship. In: A Guide for Entrepreneurship Education The Omani Case Studies, Al Roya Press and Publishing House, Oman (2019)

Effects of Facebook Personal News Sharing on Building Social Capital in Jordanian Universities

Mohammed Habes, Mahmoud Alghizzawi, Said A. Salloum, and Chaker Mhamdi

Abstract This study examines the relationship between types of social capital and the use of Facebook to share personal news. This relationship is significant because, as the research shows, there is a link between personal news sharing and personal self-esteem and life satisfaction. The quantitative study comprised a survey of 415 students at Yarmouk University and Amman University, Jordan, and was based on a theoretical model that employed structural equation modeling for measuring relations and correlations between variables and the effects of news sharing in promoting student social capital. The results showed a positive relationship between the intensity of news sharing using Facebook and social capital, bridging social capital, and self-esteem. The strongest relationship was with self-esteem; however, the results showed no significant relationship between social capital and the degree of personal life satisfaction.

Keywords Jordanian universities · Personal news sharing · Facebook · Social capital · Life satisfaction · Self-esteem · College students

1 Introduction

Social media sites, such as Facebook, YouTube, Twitter, and Instagram have attracted millions of users across all age groups due to their exponential growth in recent years [1–5]. Many users have integrated these sites into their daily social practices

M. Habes · M. Alghizzawi
University Sultan Zainal Abidin, Kuala Nerus, Terengganu, Malaysia

S. A. Salloum (✉)
Faculty of Engineering & IT, The British University in Dubai, Dubai, UAE
e-mail: ssalloum@sharjah.ac.ae

Research Institute of Sciences & Engineering, University of Sharjah, Sharjah, UAE

C. Mhamdi
University of Manouba, Manouba, Tunisia

Al Buraimi University College, Al Buraimi, Oman

M. Al-Emran et al. (eds.), *Recent Advances in Intelligent Systems and Smart Applications*, Studies in Systems, Decision and Control 295,
https://doi.org/10.1007/978-3-030-47411-9_35

[6–11]. Many studies have examined the relations between individual and social needs in terms of media use for the promotion of social capital and the influence of news sharing through social media [12, 13]. Facebook has become a popular social network site in most countries, including Jordan, especially among young people [14]. However, Facebook is popular among all age groups with access to the internet [15]. There are many functions, features, and tools in Facebook that allow users to connect, establish and maintain relations by communicating with friends, colleagues, and family members irrespective of their geographical location [7, 16–19]. Social media users can share information by using "share buttons" provided on news sites or by "retweeting" or "reposting" them.

Further, they can share links to news found on a Facebook fan's page or a friend's Twitter feed. Online news sites increasingly rely on the share feature in social media to increase the breadth and depth of target audiences [20, 21]. As broadband services in Jordan have increased, more people are accessing the internet and eventually become Facebook users [22]. One of the motivations for high-use frequency by Facebook users is to build social capital. This enables individuals to build new personal networks and maintain and strengthen them [23, 24]. News sharing by social media is important, and information relevant to social capital is available to individuals or groups who share the same interests. Knowledge sharing through social media has become very popular [25]. Facebook is a popular platform for building social capital [26]. The purpose of this study is to examine the relationship between Bonding Social Capital and the promotion of social capital and Facebook use to share news [27]. This study also examines the relationship between the degrees of a person's self-esteem and the use of Facebook to share the news. As well as it examines the relationship between a people's life satisfaction and the use of Facebook to share news.

2 Literature Review

Social media is defined as the amalgamation of services, websites, and practices on the internet that nurture sharing, participation, collaboration, and community building [28]. Facebook is the largest social media company, and of its 1.49 billion global users in 2015 [29], a large proportion is students. The acceptance of this form of social interaction is significant because Ref. [30] argued that motivations to participate in social and collective behaviors on social media platforms can be used to develop strategies that enhance and promote environments for the betterment of society in general. These include building social capital and improving self-esteem, life satisfaction, exchange of information, and, in a commercial context, the provision of enhanced customer services [14, 31–33].

Facebook is a personal news site that includes stories and events about people known to the user, where information concerning an individual is shared by multimedia, such as news items, posts, photos, and videos [9, 10]. This helps to build online relationships by communication and, in turn, creates social capital. There are two types of social capital: promotional social capital and Bridging Social Capital [34], and the literature has established these are motivating factors for Facebook usage [35, 36].

A literature review was conducted on the relationship between news sharing in social media and social capital by the use of personal Facebook pages as news content [37]. The literature concerning independent variables, which include Bonding Social Capital, self-esteem and life satisfaction was reviewed, and this was followed by the dependent variable, which is the use of Facebook to share news [20, 34, 35, 38–40]. It was found that the factors include Bridging Social Capital, knowledge, emotional convictions, continual feedback, keeping in contact, and personal behavior influences, life satisfaction. Reference [32] studied the impact of social capital on social media usage and found a strong association between the use of Facebook to share information and the three types of social capital. The strongest relationship was between news sharing and Bridging Social Capital. There are many ways to measure the positive or negative influence of Facebook news sharing for the promotion of social capital, such as self-esteem, personal motives, Bridging Social Capital [35] consumption news, behavioral news, Bonding Social Capital [41], Social factors, and frequency of usage [37]. A number of theoretical models were found to examine the acceptance of social medial [42].

Many theories can be used to explain the relationship between personal news sharing and social capital, such as Social Behavior Theory, Mass Communication Theory, and Social Penetration Theory, etc. [26, 32, 39, 40, 43]. Many studies deal with various forms of social capital in using social media for news sharing.

3 Social Penetration Theory

Social Penetration Theory was developed by Taylor to provide a conceptual framework to describe the evolution of personal relationships. Miller (2002) explains that the theory refers to reciprocity in behaviors between two people who are in the process of developing a relationship. These behaviors include verbal and non-verbal exchanges, personal perceptions, and the environment in which they occur. Behaviors vary depending on the levels of intimacy. Reference [44] pointed out that Almas and Taylor's original statement of Social Penetration Theory was laid out in their book *Social Penetration: The Development of Interpersonal Relationships*. The theory is significant because it explains factors contributing to the development, maintenance, and deterioration of social relationships. Reference [31] pointed out that the main variables in this theory were the breadth and depth of communication. Social Penetration Theory is used in this study because it indicates that there is a positive relationship between the use of Facebook to share news and an individual's perceived Bonding and Bridging Social Capital, personal self-esteem, and personal satisfaction with life.

4 Research Model and Hypothesis

A number of studies were conducted to examine the students' attitudes towards the use of social media in education [45–47]. To investigate how social capital motivated personal news sharing among students at Yarmouk and Amman Arab Universities, the study employed a research framework with hypotheses [44] based on Social Penetration Theory. See Fig. 1, which shows how the variables of social capital, including Bonding Social Capital (BSC), Bridging Social Capital (BRSC), Self-esteem (SE), and Life Satisfaction (LS) were hypothesized to motivate personal news sharing on Facebook.

4.1 News Sharing and Social Capital

Facebook is used to share news and interact with other social media users and, thereby, facilitate social interaction and cooperation, contribute to the development of personal relationships between the students, and create an opportunity to promote and improve social capital. According to [43], the use of Facebook is a means of communication and exchange of ideas in learning activities, such as discussions. It enables students to explore new topics and interact with the content and, thereby, build self-esteem, which leads to enhanced life satisfaction in general [29]. Shared learning activities involving externally sourced educational material, therefore, increases social capital beyond that derived from personal news or social interaction. A related study confirms that Facebook usage promotes the exchange of information and Bridging Social Capital by providing students with experience in the virtual community; consequently, students can improve their understanding of educational content and network across different sources [48]. However, the effectiveness of news sharing

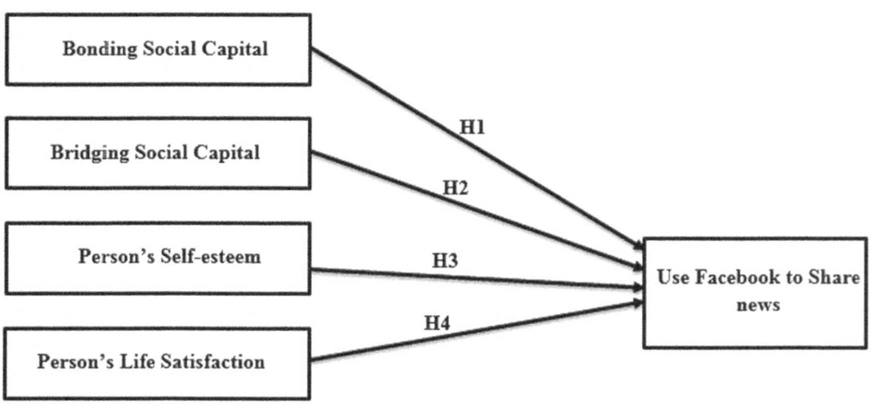

Fig. 1 Conceptual framework

on social capital depends on the motivation of the student, and the type of news exchanged [49].

4.2 *Bonding and Bridging Social Capital and News Sharing by Facebook*

Bonding Social Capital "refers to the links between like-minded people or their enforcement of homogeneity; it builds strong ties, but can also result in higher walls, excluding those who do not qualify" [50]. Social media networks allow users to examine contents, create personal files, share events and news, and conduct conversations with other users [43]. According to [51], news sharing and engagement via Facebook, in the context of Bonding Social Capital, results in an interactive environment which improves the social capital of all users [15]. Further, previous studies show that news sharing by social media develops a sense of student community via collaboration and bonding among peers on social topics [52]. Moreover, news sharing through Facebook, social capital, and psychological well-being offers an additional link to the interconnection of students through the mechanism of participation and the increased interaction with social ties and the Internet [53]. Other studies used the term Web 2.0 when dealing with various social media networks to organize and mobilize effectively for collective action [14, 44]. Reference [24] described "Bridging Social Capital" that Facebook propagates through shared information as "loose connections between individuals which may provide useful information or new perspectives for one another". In their study on social relations in networked societies, Ref. [26] stated that societies with higher levels of social capital enjoy a higher quality of life than those with low social capital. This is because they have high levels of social trust and well-established norms of mutuality, dense social media, and shared information [53]. The following hypotheses were, therefore, proposed:

H1 There is a significant relationship between Bonding Social Capital and news sharing by Facebook.

H2 There is a significant relationship between Bridging Social Capital and news sharing by Facebook.

4.3 *Life Satisfaction and Self-esteem and News Sharing by Facebook*

Various studies have shown that the intensity of personal news sharing activity on Facebook in the first year strongly predicted positive effects on Boding Social Capital, increased self-esteem and life satisfaction over the next two years [14, 37]. Further,

Table 1 Cronbach's alpha values for the pilot study (Cronbach's alpha ≥0.70)

Construct	Cronbach's alpha
Bonding social capital	0.726
Bridging social capital	0.800
Person's self-esteem	0.835
Person's life satisfaction	0.875
Use Facebook to share news	0.798

various studies have shown that exchange of information on Facebook and the resulting bridging of social capital resulted in improved life satisfaction and self-esteem [41, 43, 44, 49]. Hence, the following hypotheses were proposed:

H3 There is a significant relationship between a Person's Life Satisfaction and news sharing by Facebook.

H4 There is a significant relationship between a Person's Self-esteem and news sharing by Facebook.

5 Research Methodology

5.1 Pre-test of the Questionnaire

Before conducting the final survey, a pilot study was carried out to measure the reliability of the survey items. Fifty students were randomly selected from the target population [54]. Cronbach's alpha (CA) was used to measure the internal reliability of the construct items. According to [55], a reliability coefficient of 0.7 or above is deemed acceptable. As shown in Table 1, CA values for all constructs were above 0.7, and therefore, all constructs were reliable for use in the final study.

5.2 Data Collection

A quantitative study [56] was developed using an online survey instrument developed from the literature review and focus groups. The sample has been collected from two well-known Jordanian universities; Yarmouk and Amman Universities. Four-hundred-and-fifteen students completed the survey, of which 110 were male, and 305 were female [57, 58]. Random sampling was applied using the formula of a 95% confidence level to determine the correct sample size for a reliable representation of total students [14, 59].

6 Findings and Discussion

6.1 Data Analysis

Partial Least Squares-Structural Equation Modeling (PLS-SEM) was applied to the structural and measurement models using industry-standard Smart PLS software [60–62]. Smart PLS software, which was developed by [63], is used in this study for Partial Least Squares Structural Equation Modeling (PLS-SEM). It is a renowned software and is extensively used in academia. The software is easily accessible to academics and researchers. In addition to this, the software involves an extremely user-friendly interface and common reporting characteristics. Due to these characteristics, the software has been gaining a lot of popularity since its introduction in 2005 [64]. This study follows the general guidelines for employing the PLS-SEM in IS research [65]. The link among latent constructs is indicative of structural model validity (inner model), while the relationship among the indicators themselves indicates measurement model validity (outer model). To establish convergent validity and reliability, several measurements were performed comprising Factor Loadings (FL), Composite Reliability (CR) and Average Variance Extracted (AVE). FRs were used to determine the correlation value and weight of every questionnaire variable as a perceived indicator, whereas factor dimensionality was represented by load value.

Further, CR was used to measure reliability since a precise value was offered through the use of factor loadings in the constructed formula. The average quantity of variance in a given variable, which explains the latent construct, is referred to as the AVE. The convergence of every factor can be analyzed through its AVE and can be calculated when the discriminate validity is greater than one factor. Table 1 above showed the condition for convergent validity and reliability were exceeded by the validity tests. Moreover, Table 2 below contains a summary of the validity and reliability of the questionnaire; it shows the findings for every construct by presenting the variable obtained from the questionnaire analyzed by FL, CA, CR, and AVE.

6.1.1 Convergent Validity

After PLS-SEM [66], AVE, FL, and reliability (comprising Cronbach's Alpha (CA), composite reliability (CR), and Dijkstra-Henseler's (PA)) were used as values to determine relative convergent validity, and these exceeded 0.7 for reliability coefficient and CR, which points to internal consistency among multiple measurements of the constructs [55]. Table 2 shows that the acceptable value of 0.7 was surpassed by CA scores [67, 68]. Further, the CR values of constructs ranged between 0.744 and 0.887. Instead, researchers should evaluate and report construct reliability by means of the Dijkstra-Henseler's rho (pA) reliability coefficient [69]. Similar to CA and CR, the reliability coefficient ρA should indicate values of 0.70 or higher in exploratory research and values above 0.80 or 0.90 for more advanced stages of research [70–72]. As shown in Table 2, the reliability coefficient ρA of each measurement construct

Table 2 Convergent validity results validating acceptable values (FL, CA, CR, and PA $\geq$ 0.70 and AVE > 0.5)

Constructs	Items	Factor loading	Cronbach's alpha	CR	PA	AVE
Bonding social capital	BSC1	0.788	0.856	0.748	0.798	0.598
	BSC2	0.826				
	BSC3	0.710				
	BSC4	0.736				
Bridging social capital	BRI1	0.810	0.702	0.790	0.747	0.622
	BRI2	0.769				
	BRI3	0.777				
	BRI4	0.896				
Person's self-esteem	PERSEL1	0.889	0.863	0.744	0.793	0.581
	PERSEL2	0.804				
	PERSEL3	0.750				
	PERSEL4	0.914				
Person's life satisfaction	PERLIF1	0.764	0.777	0.789	0.795	0.661
	PERLIF2	0.720				
	PERLIF3	0.770				
	PERLIF4	0.799				
Use Facebook to share news	USEFACE1	0.748	0.706	0.887	0.889	0.522
	USEFACE2	0.739				

is above 0.70. The condition of explaining 50% of variance extracted among a set of items was fulfilled by all AVE values, which ranged from 0.522 to 0.661 [73], and each can be assumed as the cause triggering the latent construct. Consequently, convergent validity is achieved according to scales for assessing the constructs.

6.1.2 Discriminant Validity

It is recommended that three criteria be used for measuring discriminate validity: the Fornell-Larker scale and the Heterotrait-Monotrait ratio (HTMT) [74]. Table 3 shows that the Fornell-Larker scale fulfilled the first criteria because all AVEs had a square root that exceeded its correlation with the rest of the constructs [75]. The HTMT method is evidently better than the traditional approaches to discriminate validity assessment, like the Fornell-Larcker criterion [75] and (partial) cross-loadings, which are mostly incapable of determining an absence of discriminate validity [76]. When HTMT values are less than 1, the true correlation between the two constructs should be shown to be distinct. When the HTMT value is greater than this threshold, then discriminate validity is not present. A threshold of 0.85 has been implied by a few researchers [77], while others have suggested a value of 0.90 [78]. The HTMT ratio

Table 3 Fornell-Larcker scale

	BSC	BRI	PERSEL	PERLIF	USEFACE
BSC	**0.845**				
BRI	0.200	**0.798**			
PERSEL	0.406	0.446	**0.911**		
PERLIF	0.236	0.548	0.369	**0.857**	
USEFACE	0.494	0.336	0.598	0.498	**0.885**

Note BSC bonding social capital, *BRI* bridging social capital, *PERSEL* person's self-esteem, *PERLIF* person's life satisfaction, *USEFACE* use Facebook to share news
Bold indicates Measurement model demonstrated that each value of AVE was found to exceed the value of squared correlation among constructs

Table 4 Heterotrait-Monotrait ratio (HTMT)

	BSC	BRI	PERSEL	PERLIF	USEFACE
BSC					
BRI	0.156				
PERSEL	0.356	0.499			
PERLIF	0.638	0.123	0.298		
USEFACE	0.287	0.266	0.692	0.338	

Note BSC bonding social capital, *BRI* bridging social capital, *PERSEL* person's self-esteem, *PERLIF* person's life satisfaction, *USEFACE* use Facebook to share news

findings are presented in Table 4, which shows that each construct value was less than 0.85—the threshold value [79], and this confirmed the HTMT ratio. Hence, the measurement model is valid and reliable.

6.1.3 Model Fit

SmartPLS offers the following fit measures: The standard root mean square residual (SRMR), exact fit criteria, d_ULS, d_G, Chi-Square, NFI, and RMS_theta indicate the model fit in PLS-SEM [80]. The SRMR implies the difference between observed correlations and model implied correlation matrix [60] and values less than 0.08 are considered as good model fit measures [81]. The NFI values higher than 0.90 indicate a good model fit [82]. The NFI is a ratio of Chi-square value of the proposed model to the null model or benchmark model [83]. The larger the parameters, the higher the NFI, and therefore NPI is not recommended as a model fit indicator [60]. The squared Eucledian distance, d_ULS, and the geodesic distance d_G are two metrics that provide discrepancy between empirical covariance matrix and covariance matrix implied by composite factor model [60, 69]. RMS theta is only applicable for the reflective models, and evaluates the degree of outer model residuals correlation [83]. The closer the RMS theta value to the zero, the better the PLS-SEM model, and their

Table 5 Model fit indicators

	Complete model	
	Saturated model	Estimated mod
SRMR	0.051	0.072
d_ULS	0.988	2.307
d_G	0.664	0.735
Chi-square	489.936	491.568
NFI	0.826	0.827
RMS_theta	0.089	

values less than 0.12 are considered a good fit, and anything else suggest lack of fit [84]. Reference [60] suggested that saturated model evaluates correlation between all constructs, whereas, the estimated model takes total effects and model structure into account.

Table 5 shows that RMS_theta value was found to be 0.089 which implies that the concerned goodness-of-fit for PLS-SEM model was large enough to demonstrate global PLS model validity.

6.1.4 Coefficient of Determination

The validity of the structural model was tested by the coefficient of determination (R^2 value) measure. This coefficient is treated as the squared correlation between a specific endogenous construct's actual and predicted values. Moreover, the predictive accuracy of the model can also be gauged through the R^2 value, which denotes the combined effect between exogenous and endogenous latent variables. The squared correlation between the actual and predicted values of the variables is referred to as the coefficient, and it involves the degree of variance in the endogenous constructs as protected by every discovered exogenous construct. Reference [85] showed that a value above 0.67 was high, weak values range from 0.19 to 0.33, and direct values were 0.33–0.67. Further, values below 0.19 were inadmissible. Figure 2 shows that overall the model had high predictive power at 73.4%. Also, according to Table 6, the R^2 values for the variable Use Facebook to Share news was above 0.67; and hence, the predictive power of the constructs was high.

6.1.5 Hypotheses Testing—Path Coefficient

To validate the proposed hypotheses, a structural equation model and PLS-SEM with maximum possibility estimates were used so that relations among the theoretical constructs could be evaluated. The model was made to run through a bootstrap re-sampling routine to obtain the path significances as per the previous relevant studies [86–89]. A total of 5000 re-samples were used. A one-tailed t-test was employed

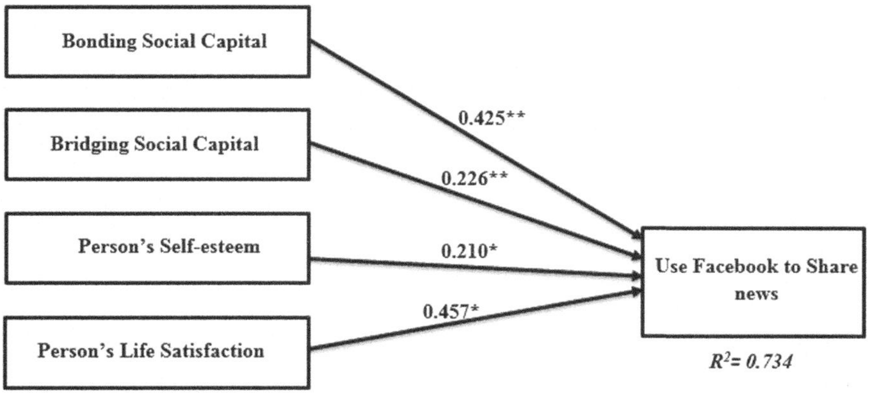

Fig. 2 Path coefficient of the model. Significant at $p^{**} \leq 0.01$, $p^{*} < 0.05$

Table 6 R^2 of the endogenous latent variables

Constructs	R^2	Results
Use Facebook to share news	0.734	High

in this study as all the hypotheses are directional. The one-tailed t-test (df = 415) entails that the 0.05 significance level (i.e., $p < 0.05$) requires a t-value >1.657, the 0.01 significance level (i.e., $p < 0.01$) requires a t-value >2.354, and the 0.001 significance level (i.e., $p < 0.001$) requires a t-value >3.152 [90]. Table 7 shows that all values were in the acceptable range. In addition, few direct hypotheses showed support to the previous literature [27, 36, 89–93]. The resulting path coefficients of the suggested research model are shown in Fig. 2. Based on the data, the analysis shows that hypotheses H1, H2, H3, and H4 were supported. Specifically, the results showed that Use Facebook to Share news variable significantly influenced BSC ($\beta = 0.425$, $P < 0.001$), BRI ($\beta = 0.226$, $P < 0.001$), PERSEL ($\beta = 0.210$, $P < 0.05$), and PERLIF ($\beta = 0.457$, $P < 0.05$). A summary of the hypotheses testing results is shown in Table 7.

Table 7 Hypotheses-testing of the research model

H	Relationship	Path	*t*-value	*p*-value	Direction	Decision
H1	BSC → USEFACE	0.425	15.658	0.000	Positive	Supported**
H2	BRI → USEFACE	0.226	11.524	0.000	Positive	Supported**
H3	PERSEL → USEFACE	0.210	3.115	0.011	Positive	Supported*
H4	PERLIF → USEFACE	0.457	2.105	0.040	Positive	Supported*

Significant at $p^{**} \leq 0.01$, $p^{*} < 0.05$

Note BSC bonding social capital, *BRI* bridging social capital, *PERSEL* person's self-esteem, *PERLIF* person's life satisfaction, *USEFACE* use Facebook to share news

7 Conclusion

The objective of this study was to analyze the effects of Facebook news sharing in developing student social capital in Jordanian universities. Web 2.0 usage for communication, learning, education, and entertainment is highly significant. The findings of this research showed that, in the case of students at Yarmouk and Amman Universities in Jordan, there was a significant relationship between the intensity of Facebook users to share personal news and social capital, with Bridging Social Capital scoring slightly higher than Bonding Social Capital. These relations suggest that Facebook is used to build social capital among college students. Further, Facebook may help Jordanian students maintain contact by sharing news after they have graduated. Therefore, Jordanian universities may want to explore ways to encourage the use of Facebook among graduates. Additionally, news sharing by social media opens the door to develop a greater sense of student community via collaboration among peers and mutual bonding. The findings of this research showed that all the hypotheses were supported. The findings indicated that $R^2 = 73.4\%$. These results support previous studies [18, 27, 37, 94–96]. The findings suggest that Facebook serves as a "virtual place" for students to establish and maintain relations with close friends; however, the findings of this research cannot be generalized to other communities. Yarmouk and Amman Arab universities are exclusively residential, and the findings may differ with non-residential campuses. Further, ideas of social capital may differ between student communities. Since news sharing on Facebook can have a positive effect on building student social capital, further studies could explore other universities and educational contexts, including high schools.

7.1 Theoretical Implications

This study has a number of theoretical implications on the existing literature which could be taken into consideration in future research on the topic under examination. To begin with, the current investigation sought to scrutinize the main factors that may have an impact on the constant intent to utilize Facebook as a medium for sharing private news. The study is supposed to serve as an illustration of proposing a model to examine continuous intention in the context of higher educational levels. Second, the suggested model envisioned to ensure a strong predictive power to unveil reasons behind the constant intent to utilize Facebook and the relationship between people's satisfaction and the use of Facebook to share the news. Therefore, the proposed model of the current study can be a useful theoretical basis for future research seeking to explore people's continuous intention to use any type of other prevalent platforms. Third, the findings of this study elucidate the fact that, within the context of long term usage, users' satisfaction may raise the likelihood of using technology to accomplish their educational targets.

7.2 *Managerial and Practical Implications*

This study has several managerial and practical implications. Users' satisfaction is thought to be a vital factor in gauging the constant use of Facebook. Hence, studies, such as the current investigation, offer a significant identification of the most important factor based on the users' preferences. Additionally, technology providers can benefit from the findings of such studies by focusing on the users' needs and preferences when using Facebook to share personal news. To be precise, divulging the essential rationale behind the continuous use of Facebook can bring about higher satisfaction levels to the user. Taking into account factors, including concentration, usefulness, ease of use, and high information quality, can undoubtedly add a more comprehensive understanding of the rationale lying behind the continuous use of Facebook to share personal news.

7.3 *Limitations and Further Research*

Though the findings of the current study are fairly interesting and provide an essential basis for the exploration of the effects of Facebook news sharing in developing student social capital in Jordanian universities, it undeniably poses some limitations. First, the sole target of the investigation was students. Therefore, if instructors' responses had also been examined, it would have become possible to obtain a more comprehensive perspective and comparisons between the responses of instructors and students could have been conducted. That is why further research should consider this lacuna. Second, the proposed model is cross-sectional and defines users' perceptions and intentions for a single point in time. Consequently, it is expected that future inquiries on the topic under investigation be carried out using the longitudinal survey as it is probable that the people's perceptions and preferences change over time. Third, as the primary focus of the current has been two Jordanian universities, the conclusions cannot be generalized to the private and public sectors of higher education in Jordan. Fourth, in the same vein, the sample population of this study has been attained from a restricted number of universities. Hence, considering larger and more representative populations, having diverse aspects including income, demographical, and psychological attitudes, will allow for valid generalizability of the research findings.

References

1. Salloum, S.A., Mhamdi, C., Al-Emran, M., Shaalan, K.: Analysis and classification of Arabic Newspapers' Facebook pages using text mining techniques. Int. J. Inf. Technol. Lang. Stud. **1**(2), 8–17 (2017)
2. Mhamdi, C.: Transgressing media boundaries: news creation and dissemination in a globalized world. Mediterr. J. Soc. Sci. **7**(5), 272 (2016)

3. Mhamdi, C., Al-Emran, M., Salloum, S.A.: Text mining and analytics: a case study from news channels posts on Facebook **740** (2018)
4. Salloum, S.A., Maqableh, W., Mhamdi, C., Al Kurdi, B., Shaalan, K.: Studying the social media adoption by university students in the United Arab Emirates. Int. J. Inf. Technol. Lang. Stud. 2(3), 83–95 (2018)
5. Salloum, S.A., Mhamdi, C., Al Kurdi, B., Shaalan, K.: Factors affecting the adoption and meaningful use of social media: a structural equation modeling approach. Int. J. Inf. Technol. Lang. Stud. **2**(3), 96–109 (2018)
6. Edelman, L.F., Bresnen, M., Newell, S., Scarbrough, H., Swan, J.: The benefits and pitfalls of social capital: empirical evidence from two organizations in the United Kingdom. Br. J. Manag. **15**(S1), S59–S69 (2004)
7. Mhandi, C.: Translating news texts during wars and conflicts: challenges and strategies. ANGLICA-An Int. J. English Stud. **28**(2), 141–151 (2019)
8. Mhamdi, C.: Interpreting games: meaning creation in the context of temporality and interactivity. Mediterr. J. Soc. Sci. 8(4) (2017)
9. Mhamdi, C.: What can video add to the learning experience? Challenges and opportunities. Int. J. Inf. Technol. Lang. Stud. **1**(1), 17–24 (2017)
10. Mhamdi, C.: Framing 'the other' in times of conflicts: CNN's coverage of the 2003 Iraq War. Mediterr. J. Soc. Sci. **8**(2), 147–153 (2017)
11. Mhamdi, C.: The use of political cartoons during popular protests: the case of the 2011 Tunisia uprising. J. English Stud. **15**, 193–220 (2017)
12. Alghizzawi, M., Salloum, S.A., Habes, M.: The role of social media in tourism marketing in Jordan. Int. J. Inf. Technol. Lang. Stud., **2**(3) (2018)
13. Alghizzawi, M., Habes, M., Salloum, S.A.: The relationship between digital media and marketing medical tourism destinations in Jordan: Facebook Perspective. In: International Conference on Advanced Intelligent Systems and Informatics, pp. 438–448, 2019
14. Habes, M., Salloum, S.A., Alghizzawi, M., Alshibly, M.S.: The role of modern media technology in improving collaborative learning of students in Jordanian universities. Int. J. Inf. Technol. Lang. Stud. **2**(3), 71–82 (2018)
15. Yaakop, A., Mohamed Anuar, M., Omar, K., Liaw, A.: Consumers' perceptions and attitudes towards advertising on Facebook in Malaysia. In: World Business and Economics Research Conference, Auckland New Zealand, 2012
16. Derbel, E.: The African Novel: the ongoing battle against literary and national neo-colonialism. Int. J. Inf. Technol. Lang. Stud. **1**(1) (2017)
17. Derbel, E.: Teaching literature through comics: an innovative pedagogical tool. Int. J. Appl. Linguist. English Lit. **8**(1), 54–61 (2019)
18. Habes, M., Salloum, S.A., Alghizzawi, M., Mhamdi, C.: The relation between social media and students' academic performance in Jordan: YouTube perspective. In: International Conference on Advanced Intelligent Systems and Informatics, pp. 382–392, 2019
19. Derbel, E.: Feminist graphic narratives: the ongoing game of eluding censorship. Mediterr. J. Soc. Sci. **10**(1), 49 (2019)
20. Kümpel, A.S., Karnowski, V., Keyling, T.: News sharing in social media: A review of current research on news sharing users, content, and networks. Soc. Media + Soc. **1**(2), 2056305115610141 (2015)
21. Habes, M.: The influence of personal motivation on using social TV: a uses and gratifications approach. Int. J. Inf. Technol. Lang. Stud. **3**(1), (2019)
22. Alghizzawi, M., et al.: The impact of Smartphone adoption on marketing therapeutic tourist sites in Jordan. Int. J. Eng. Technol. 7(4.34), 91–96 (2018)
23. Al-Shibly, M.S., Alghizzawi, M., Habes, M., Salloum, S.A.: The impact of de-marketing in reducing Jordanian youth consumption of energy drinks. In: International Conference on Advanced Intelligent Systems and Informatics, pp. 427–437, 2019
24. Su, C.C., Chan, N.K.: Predicting social capital on Facebook: the implications of use intensity, perceived content desirability, and Facebook-enabled communication practices. Comput. Human Behav. **72**, 259–268 (2017)

25. Salloum, S.A., Al-Emran, M., Shaalan, K.: The impact of knowledge sharing on information systems: a review. In: International Conference on Knowledge Management in Organizations, pp. 94–106, 2018
26. Miyata, K., Kobayashi, T.: Causal relationship between Internet use and social capital in Japan. Asian J. Soc. Psychol. **11**(1), 42–52 (2008)
27. Salloum, S.A., Al-Emran, M., Habes, M., Alghizzawi, M., Ghani, M.A., Shaalan, K.: Understanding the impact of social media practices on E-learning systems acceptance. In: International Conference on Advanced Intelligent Systems and Informatics, pp. 360–369, 2019
28. Junco, R., Heiberger, G., Loken, E.: The effect of Twitter on college student engagement and grades. J. Comput. Assist. Learn. **27**(2), 119–132 (2011)
29. Habes, M., Alghizzawi, M., Khalaf, R., Salloum, S.A., Abd. Ghani, M.: The relationship between social media and academic performance: Facebook perspective. Int. J. Inf. Technol. Lang. Stud., **2**(1), 12–18 (2018)
30. Chadwick, A., Vaccari, C., O'Loughlin, B.: Do tabloids poison the well of social media? Explaining democratically dysfunctional news sharing. New Media Soc. **20**(11), 4255–4274 (2018)
31. McNair, B.: An Introduction to Political Communication. Routledge (2017)
32. Burke, M., Marlow, C., Lento, T.: Social network activity and social well-being. In: Proceedings of the SIGCHI Conference on Human Factors in Computing Systems, pp. 1909–1912, 2010
33. Salloum, S.A., Al-Emran, M., Monem, A.A., Shaalan, K.: Using text mining techniques for extracting information from research articles. In: Studies in Computational Intelligence, vol. 740. Springer (2018)
34. Ellison, N.B., Steinfield, C., Lampe, C.: The benefits of Facebook 'Friends:' social capital and college students' use of online social network sites. J. Comput. Commun. **12**(4), 1143–1168 (2007)
35. Burke, M., Kraut, R., Marlow, C.: Social capital on Facebook: differentiating uses and users. In: Proceedings of the SIGCHI Conference on Human Factors in Computing Systems, pp. 571–580, 2011
36. Salloum, S.A., Al-Emran, M., Khalaf, R., Habes, M., Shaalan, K.: An innovative study of E-payment systems adoption in higher education: theoretical constructs and empirical analysis. Int. J. Interact. Mob. Technol. **13**(6) (2019)
37. Phua, J., Ahn, S.J.: Explicating the 'like' on Facebook brand pages: the effect of intensity of Facebook use, number of overall "likes", and number of friends" "likes" on consumers' brand outcomes'. J. Mark. Commun. **22**(5), 544–559 (2016)
38. Salloum, S.A., Al-Emran, M., Shaalan, K.: Mining social media text: extracting knowledge from Facebook. Int. J. Comput. Digit. Syst. **6**(2), 73–81 (2017)
39. Hurlbert, J.S., Beggs, J.J., Haines, V.A.: Social networks and social capital in extreme environments. In: Social Capital, pp. 209–231. Routledge (2017)
40. Ellison, N.B., Vitak, J.: Social network site affordances and their relationship to social capital processes. Handb. Psychol. Commun. Technol. **32**, 205–228 (2015)
41. Mehdizadeh, S.: Self-presentation 2.0: Narcissism and self-esteem on Facebook. Cyberpsychology, Behav. Soc. Netw. **134**, 357–364 (2010)
42. Al-Qaysi, N., Mohamad-Nordin, N., Al-Emran, M.: A systematic review of social media acceptance from the perspective of educational and information systems theories and models. J. Educ. Comput. Res. **57**(8), 2085–2109 (2020)
43. Phulari, S.S., Khamitkar, S.D., Deshmukh, N.K., Bhalchandra, P.U., Lokhande, S.N., Shinde, A.R.: Understanding formulation of social capital in online social network sites (SNS). arXiv Prepr. arXiv1002.1201 (2010)
44. Kays, K.M., Miles, R.E., Koch, C.J.: Is social media like an onion? Exploring the social penetration theory as an explanation for viral responses to intimate self-disclosures (2016)
45. Al-Qaysi, N., Mohamad-Nordin, N., Al-Emran, M., Al-Sharafi, M.A.: Understanding the differences in students' attitudes towards social media use: a case study from Oman. In: 2019 IEEE Student Conference on Research and Development (SCOReD), pp. 176–179 (2019)

46. Al-Qaysi, N., Mohamad-Nordin, N., Al-Emran, M.: What leads to social learning? Students' attitudes towards using social media applications in Omani higher education. Educ. Inf. Technol. (2019)
47. Al-Qaysi, N., Mohamad-Nordin, N., Al-Emran, M.: An empirical investigation of students' attitudes towards the use of social media in Omani Higher Education. In: International Conference on Advanced Intelligent Systems and Informatics, pp. 350–359, 2019
48. Kim, B., Kim, Y.: College students' social media use and communication network heterogeneity: implications for social capital and subjective well-being. Comput. Human Behav. **73**, 620–628 (2017)
49. Keppler, S., Leonardi, P.M.: Strengthening ties on social networking sites: the effects of social learning and commitment building (2017)
50. Fredriksson M., Ihlen, Ø.: Introduction: public relations and social theory. In: Public Relations and Social Theory, pp. 13–28. Routledge (2018)
51. Kavanaugh, M.B., Reese, A.L., Carroll, D.D., Rosson, J.M.: Weak ties in networked communities. Inf. Soc. **21**(2), 119–131 (2005)
52. Phua, J., Jin, S.V., Kim, J.J.: Uses and gratifications of social networking sites for bridging and Bonding Social Capital: a comparison of Facebook, Twitter, Instagram, and Snapchat. Comput. Human Behav. **72**, 115–122 (2017)
53. Fu, P.-W., Wu, C.-C., Cho, Y.-J.: What makes users share content on facebook? Compatibility among psychological incentive, social capital focus, and content type. Comput. Human Behav. **67**, 23–32 (2017)
54. Krejcie, R.V., Morgan, D.W.: Determining sample size for research activities. Educ. Psychol. Meas. **30**(3), 607–610 (1970)
55. Hair, J.F., Black, W.C., Babin, B.J., Anderson, R.E.: Multivariate Data Analysis. New Jersey (2010)
56. Creswell J.W., Creswell, J.D.: Research Design: Qualitative, Quantitative, and Mixed Methods Approaches. Sage Publications (2017)
57. Hew, J.-J., Leong, L.-Y., Tan, G.W.-H., Ooi, K.-B., Lee, V.-H.: The age of mobile social commerce: an artificial neural network analysis on its resistances. Technol. Forecast. Soc. Change **144**, 311–324 (2019)
58. Hew T.-S., Syed, S.L., Kadir, A.: Applying channel expansion and self-determination theory in predicting use behaviour of cloud-based VLE. Behav. Inf. Technol. **36**(9), 875–896 (2017)
59. Alnawafleh, H., Alghizzawi, M., Habes, M.: The impact of introducing international brands on the development of Jordanian tourism. Int. J. Inf. Technol. Lang. Stud. **3**(2) (2019)
60. Hair, Jr. J.F., Hult, G.T.M., Ringle, C., Sarstedt, M.: A Primer on Partial Least Squares Structural Equation Modeling (PLS-SEM). Sage Publications (2016)
61. Leong, L.-Y., Hew, T.-S., Ooi, K.-B., Wei, J.: Predicting mobile wallet resistance: a two-staged structural equation modeling-artificial neural network approach. Int. J. Inf. Manage. p. 102047 (2019)
62. Leong, L.-Y., Jaafar, N.I., Sulaiman, A.: What triggers impulse purchase in Facebook commerce? Int. J. Mob. Commun. **16**(4), 459–492 (2018)
63. Ringle, C.M., Wende, S., Will, S.: SmartPLS 2.0 (M3) Beta, Hamburg 2005. (2005)
64. Wong, K.K.-K.: Partial least squares structural equation modeling (PLS-SEM) techniques using SmartPLS. Mark. Bull. **24**(1), 1–32 (2013)
65. Al-Emran, M., Mezhuyev, V., Kamaludin, A.: PLS-SEM in information systems research: a comprehensive methodological reference. In: 4th International Conference on Advanced Intelligent Systems and Informatics (AISI 2018), pp. 644–653, 2018
66. Hair, J.F., Black, W.C., Babin, B.J., Anderson, R.E., Tatham, R.L.: Multivariate Data Analysis, vol. 5. Prentice hall Upper Saddle River, NJ (1998)
67. Gefen, D., Straub, D., Boudreau, M.-C.: Structural equation modeling and regression: guidelines for research practice. Commun. Assoc. Inf. Syst. **4**(1), 1–77 (2000)
68. Nunnally J.C., Bernstein, I.H.: Psychometric Theory (1978)
69. Dijkstra, T.K., Henseler, J.: Consistent and asymptotically normal PLS estimators for linear structural equations. Comput. Stat. Data Anal. **81**, 10–23 (2015)

70. Hair, J.F., Ringle, C.M., Sarstedt, M.: PLS-SEM: indeed a silver bullet. J. Mark. theory Pract. **19**(2), 139–152 (2011)
71. Henseler, J., Ringle, C.M., Sinkovics, R.R.: The use of partial least squares path modeling in international marketing. In: New Challenges to International Marketing, pp. 277–319. Emerald Group Publishing Limited (2009)
72. Nunnally J.C., Bernstein, I.H.: Psychometric Theory (1994)
73. Falk R.F., Miller, N.B.: A Primer for Soft Modeling. University of Akron Press (1992)
74. Hair, J., Hollingsworth, C.L., Randolph, A.B., Chong, A.Y.L.: An updated and expanded assessment of PLS-SEM in information systems research. Ind. Manag. Data Syst. **117**(3), 442–458 (2017)
75. Fornell, C., Larcker, D.F.: Evaluating structural equation models with unobservable variables and measurement error. J. Mark. Res. **18**(1), 39–50 (1981)
76. Campbell D.T., Fiske, D.W.: Convergent and discriminant validation by the multitrait-multimethod matrix. Psychol. Bull. **56**(2), 81 (1959)
77. Kline, R.B.: Convergence of Structural Equation Modeling and Multilevel Modeling. na (2011)
78. Teo, T.S.H., Srivastava, S.C., Jiang, L.: Trust and electronic government success: an empirical study. J. Manag. Inf. Syst. **25**(3), 99–132 (2008)
79. Henseler, J., Ringle, C.M., Sarstedt, M.: A new criterion for assessing discriminant validity in variance-based structural equation modeling. J. Acad. Mark. Sci. **43**(1), 115–135 (2015)
80. Hair, J. M., et al.: A primer on partial least squares structural equation modeling (PLSSEM). Sage Publications, (2016)
81. Hu L., Bentler, P.M.: Fit indices in covariance structure modeling: Sensitivity to underparameterized model misspecification. Psychol. Methods **3**(4), 424 (1998)
82. Bentler P.M., Bonett, D.G.: Significance tests and goodness of fit in the analysis of covariance structures. Psychol. Bull. 88(3), 588 (1980)
83. Lohmöller, J.B.: Latent Variable Path Modeling with Partial Least Squares. Physica-Verlag, Heidelberg, Germany (1989)
84. Henseler, J., et al.: Common beliefs and reality about PLS: comments on Rönkkö and Evermann (2013). Organ. Res. methods **17**(2), 182–209 (2014)
85. Chin, W.W.: The partial least squares approach to structural equation modeling. Mod. methods Bus. Res. **295**(2), 295–336 (1998)
86. Efron B., Tibshirani, R.J.: The jackknife. In: An Introduction to the Bootstrap, pp. 141–152. Springer (1993)
87. Efron B., Tibshirani, R.J.: An Introduction to the Bootstrap. CRC Press (1994)
88. Alghizzawi, M., Habes, M., Salloum, S.A., Ghani, M.A., Mhamdi, C., Shaalan, K.: The effect of social media usage on students' e-learning acceptance in higher education: a case study from the United Arab Emirates. Int. J. Inf. Technol. Lang. Stud. **3**(3) (2019)
89. Al-Emran, M., Arpaci, I., Salloum, S.A.: An empirical examination of continuous intention to use m-learning: an integrated model. Educ. Inf. Technol. (2020)
90. Salloum, S.A., Alhamad, A.Q.M., Al-Emran, M., Monem, A.A., Shaalan, K.: Exploring students' acceptance of E-learning through the development of a comprehensive technology acceptance model. IEEE Access **7**, 128445–128462 (2019)
91. Salloum S.A.S., Shaalan, K.: Investigating Students' Acceptance of E-Learning System in Higher Educational Environments in the UAE: Applying the Extended Technology Acceptance Model (TAM). The British University in Dubai (20180
92. Alshurideh, M., Salloum, S.A., Al Kurdi, B., Al-Emran, M.: Factors affecting the social networks acceptance: an empirical study using PLS-SEM approach. In: 8th International Conference on Software and Computer Applications, pp. 414–418, 2019
93. Salloum, S.A., Al-Emran, M.: Factors affecting the adoption of E-payment systems by university students: extending the TAM with trust. Int. J. Electron. Bus. **14**(4), 371–390 (2018)
94. Stromberg, K., Westerlund, G.: Level of self esteem, social network and experience of school education among girls. Diss. Public Heal. Educ. (2005)

95. Sousa, S., Lyubomirsky, L.: Encylopedia of women and gender: sex similarities and differences and the impact of society on gender. Life Satisf. J. Worell, **2**, 667–676 (2001)
96. Raacke, J., Bonds-Raacke, J.: MySpace and Facebook: applying the uses and gratifications theory to exploring friend-networking sites. Cyberpsychology Behav. **11**(2), 169–174 (2008)

The manufacturer's authorised representative in the EU is Springer Nature Customer Service Centre GmbH, Europaplatz 3, 69115 Heidelberg, Germany. If you have any concerns regarding our products, please contact ProductSafety@springernature.com

Printed and bound by CPI Group (UK) Ltd, Croydon, CR0 4YY
30/09/2026
02225503-0018